DAVID RAYNER

5th edition

HIGHER

GCSE
Mathematics
Revision and Practice

OXFORD
UNIVERSITY PRESS

OXFORD
UNIVERSITY PRESS

Great Clarendon Street, Oxford OX2 6DP

Oxford University Press is a department of the University of Oxford.
It furthers the University's objective of excellence in research, scholarship, and education by publishing worldwide in

Oxford   New York

Auckland   Cape Town   Dar es Salaam   Hong Kong   Karachi
Kuala Lumpur   Madrid   Melbourne   Mexico City   Nairobi
New Delhi   Shanghai   Taipei   Toronto

With offices in

Argentina   Austria   Brazil   Chile   Czech Republic   France   Greece
Guatemala   Hungary   Italy   Japan   Poland   Portugal   Singapore
South Korea   Switzerland   Thailand   Turkey   Ukraine   Vietnam

British Library Cataloguing in Publication Data

Data available

ISBN 9780199139262

10 9 8 7 6 5 4 3 2 1

Printed in Great Britain by Bell and Bain Ltd; Glasgow

Paper used in the production of this book is a natural, recyclable product made from wood grown in sustainable forests. The manufacturing process conforms to the environmental regulations to the country of origin.

Acknowledgements
The Publisher would like to thank the following for permission to reproduce photographs:

Photo Disc/OUP Picture Bank p 46, 154, 155; Ingram/OUP Picture Bank p 46; Val Bakhtin/Dreamstime p 95; Image Source/OUP Picture Bank p 155; Moodboard/OUP Picture Bank p 155; Jakezc/Dreamstime p 284; Maurizio Camerin/Dreamstime p 285.

Additional artwork by Stephen Hill, Nick Bawden, Oxford Illustrators and Julian Page.

Cover image: Yurok Aleksandrovich/Dreamstime

Preface

This Higher book is for candidates working through Key Stage 4 towards a GCSE in Mathematics, and has been adapted for the new specifications for first examination in 2010. It can be used both in the classroom and by students working on their own. There are explanations, worked examples and numerous exercises which, it is hoped, will help students to build up confidence. The questions are graded in difficulty throughout the exercises.

The book can be used either as a course book over the last one or two years before the GCSE examinations or as a revision text in the final year.

Elements of functional mathematics are highlighted at the beginning of each chapter, in exercises throughout the book (F) and in four functional tasks that focus on problem-solving in context. New problem-solving material has been added.

Exercises gave been graded according to their level of demand, with Medium (M), Medium/High (M/H) and High (H) material corresponding to grade ranges D/C, B and A/A* respectively. Questions marked * are intended as a challenge or extension.

At the end of the book there are several revision exercises which provide mixed questions across the curriculum.

The author is indebted to the many students and colleagues who have assisted him in this work. He is particularly grateful to Christine Godfrey for her help and many suggestions.

Thanks are also due to the following examination boards for kindly allowing the use of questions from their past mathematics papers and sample questions:

Edexcel Foundation	(Edexcel)
Assessment and Qualification Alliance	(AQA)
Oxford Cambridge and RSA Examinations	(OCR)
Northern Ireland Council for the Curriculum Examinations and Assessment	(CCEA)
Welsh Joint Education Committee	(WJEC)

D. Rayner   2010

Contents

1 Number 1

<table>
<tr><td>

In this unit you will:
- revise place value
- revise factors, multiples and prime numbers
- learn about highest common factor and least common multiple
- revise square and cube numbers, square and cube roots and reciprocals
- revise fractions
- practise estimating
- revise ratio and proportion
- revise negative numbers.

Functional skills coverage and range:
- Understand and use positive and negative numbers of any size in practical contexts
- Carry out calculations with numbers of any size in practical contexts, to a given number of decimal places
- Understand, use, and calculate ratio and proportion, including problems involving scale
- Understand and use equivalences between fractions, decimals and percentages.

</td><td>

Links
Knowing how to use numbers efficiently is useful for estimating bills, setting up and using bank accounts, everyday shopping and technical calculations at work.

</td></tr>
</table>

1.1   Place value

1.1.1   Whole numbers
- In the number 3264

> the digit 3 means 3 thousands
> the digit 2 means 2 hundreds
> the digit 6 means 6 tens
> the digit 4 means 4 units (ones).

- In words you write 'three thousand, two hundred and sixty-four'.

Exercise 1 Ⓜ

In questions **1** to **8** write the value of the figure that is underlined.

1 2<u>7</u>	**2** 4<u>1</u>6	**3** 23<u>8</u>2	**4** 51<u>6</u>
5 <u>6</u>008	**6** <u>2</u>6 104	**7** <u>5</u> 250 000	**8** <u>8</u>26 111

In questions **9** to **12** write the number that goes in each box.

9 $293 = \square + 90 + 3$ **10** $574 = 500 + \square + 4$

11 $816 = 800 + \square + 6$ **12** $899 = \square + 90 + 9$

13 Write these numbers in figures
   a seven hundred and twenty
   b five thousand, two hundred and six
   c sixteen thousand, four hundred and thirty
   d half a million
   e three hundred thousand and ninety
   f eight and a half thousand.

14 Here are four number cards.

   a Use all the cards to make the largest possible number.
   b Use all the cards to make the smallest possible number.

15 Here are five number cards.

   a Use all the cards to make the largest possible **odd** number.
   b Use all the cards to make the smallest possible **even** number.

16 Write the number that is ten more than
   a 247   **b** 3211   **c** 694

17 Write the number that is one thousand more than
   a 392   **b** 25 611   **c** 256 900

18 a Prini puts a 2-digit whole number into her calculator. She multiplies the number by 10. Copy and fill in **one** other digit which you know must now be on the calculator.

   b Prini starts again with the same 2-digit number and this time she multiplies it by 1000. Copy and fill in all five digits on the calculator this time.

1.1.2 Decimals

These are decimal numbers.

Notice that $53 \cdot 62 = 50 + 3 + \dfrac{6}{10} + \dfrac{2}{100}$

$0 \cdot 873 = \dfrac{8}{10} + \dfrac{7}{100} + \dfrac{3}{1000}$

tens	units	.	$\frac{1}{10}$	$\frac{1}{100}$	$\frac{1}{1000}$
5	3	.	6	2	
	0	.	8	7	3

Exercise 2 (M)

In questions **1** to **8** write each statement and decide whether it is true or false.

1 0·3 is less than 0·31   **2** 0·82 is more than 0·825

3 0·7 is equal to 0·70   **4** 0·17 is less than 0·71

5 0·02 is more than 0·002   **6** 0·6 is less than 0·06

7 0·1 is equal to $\dfrac{1}{10}$ **8** 5 is equal to 5·00

9 The number 43·6 can be written $40 + 3 + \dfrac{6}{10}$.

Write the number 57·2 in this way.

10 Write these additions as decimal numbers.

a $200 + 30 + 5 + \dfrac{1}{10}$ **b** $60 + 7 + \dfrac{2}{10} + \dfrac{3}{100}$

c $90 + 8 + \dfrac{3}{10} + \dfrac{2}{100}$ **d** $3 + \dfrac{1}{10} + \dfrac{6}{100} + \dfrac{7}{1000}$

In questions **11** to **18** arrange the numbers in order of size, smallest first.

11 0·41, 0·31, 0·2   **12** 0·75, 0·58, 0·702   **13** 0·43, 0·432, 0·41

14 0·609, 0·61, 0·6   **15** 0·04, 0·15, 0·2, 0·35   **16** 1·8, 0·18, 0·81, 1·18

17 0·7, 0·061, 0·07, 0·1   **18** 0·2, 0·025, 0·03, 0·009

19 Here are numbers with letters.

Put the numbers in order and write the letters to make a word.

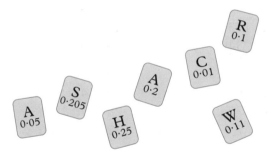

20 Increase these numbers by $\dfrac{1}{10}$.

a 32·41   **b** 0·753   **c** 1·06

21 Increase these numbers by $\dfrac{1}{100}$.

a 5·68   **b** 0·542   **c** 1·29

22 Write these amounts in pounds.

a 350 pence   **b** 15 pence   **c** 3 pence

d 10 pence   **e** 1260 pence   **f** 8 pence

23 Write each statement and say whether it is true or false.

a £5·4 = £5 + 40p   **b** £0·6 = 6p

c 5p = £0·05   **d** 50p is more than £0·42

1.2 Arithmetic without a calculator

1.2.1 Whole numbers

Here are examples to remind you of non-calculator methods.

EXAMPLE

a
```
    4 2 7
+ 5 1 8 6
---------
  5 6 1 3
---------
    1  1
```

b
```
  2 7⁷8¹4̶
-   6 3 5
---------
  2 1 4 9
```

c $57 \times 100 = 5700$

d
```
    3 7 4
  ×     6
---------
  2 2 4 4
---------
    4 2
```

e
```
        5 4 2
7)3 7²9¹4
```

f
```
        6 3 8   r 4   or   138⁴⁄₅
5)6¹9⁴4
```

Exercise 3 Ⓜ

Work out, without using a calculator.

1 653 + 2844	**2** 2106 + 329	**3** 64 + 214 + 507	**4** 65 941 + 25 804
5 387 − 175	**6** 527 − 486	**7** 927 − 68	**8** 1024 − 816
9 27 × 10	**10** 5 × 1000	**11** 73 × 5	**12** 214 × 4
13 316 × 8	**14** 9224 × 7	**15** 340 ÷ 4	**16** 1944 ÷ 6
17 3195 ÷ 5	**18** 2600 ÷ 8	**19** 365 ÷ 7	**20** 520 ÷ 10
21 289 + 15 + 1714	**22** 9704 − 5135	**23** 6001 − 5994	**24** 54 × 20
25 2906 − 1414	**26** 4716 ÷ 9	**27** 725 × 8	**28** 1504 ÷ 8
29 7 + 1609 + 25	**30** 289 + 154 − 78	**31** 7 + 295 − 48	**32** 53 × 400

Exercise 4 Ⓜ

Copy these and fill in the missing digits.

1 a
```
    2 8 5
+ ☐ 1 4
---------
  7 ☐ ☐
```

b
```
    6 3 ☐
+ ☐ 5 2
---------
  8 ☐ 9
```

c
```
  ☐ 3 5
+ 3 4 ☐
---------
  9 ☐ 9
```

2 a
```
    3 5 6
+ 5 ☐ 6
---------
  ☐ 8 ☐
```

b
```
    2 ☐ 4
+ 5 3 7
---------
  ☐ 6 1
```

c
```
    3 8 8
+ ☐ 2 ☐
---------
  8 ☐ 3
```

3 a
```
      4 □
  ×     3
  ─────────
  1 4 4
```
b
```
      3 □
  ×     7
  ─────────
  2 3 1
```
c
```
    □ □ 1
  ×     5
  ─────────
  1 6 0 5
```

4 a □□□ ÷ 3 = 50   **b** □□ × 4 = 60

   c 9 × □ = 81   **d** □□□□ ÷ 6 = 192

5 a
```
    4 □ 5
  + 2 8 □
  ─────────
  □ 3 0
```
b
```
    4 □ 7
  + □ 7 □
  ─────────
  6 0 4
```
c
```
    □ 3 □
  + 2 □ 4
  ─────────
  7 9 9
```

6 a □□ × 7 = 245   **b** □□ × 10 = 580

   c 32 ÷ □ = 8   **d** □□□ ÷ 5 = 190

1.2.2 Adding and subtracting decimals

a 4·2 + 1·76   **b** 26 − 1·7   **c** 0·24 + 5 + 12·7

....

a
```
   4·20 ←put a zero
 + 1·76
 ──────
   5·96
```
b
```
  26·0
 −  1·7
 ──────
  24·3
```
c
```
    0·24
    5·00
 + 12·70 ←extra zeros
 ───────
   17·94
```

Remember to line up the decimal points.

Exercise 5 Ⓜ ⭐F

Work out

1 2·84 + 7·3   **2** 18·6 + 2·34   **3** 25·96 + 0·75   **4** 212·7 + 4·25

5 3·6 + 6   **6** 7 + 16·1   **7** 8 + 0·34 + 0·8   **8** 12 + 5·32

9 0·004 + 0·058   **10** 4·81 − 3·7   **11** 6·92 − 2·56   **12** 8·27 − 5·86

13 3·6 − 2·24   **14** 8·4 − 2·17   **15** 8·24 − 5·78   **16** 15·4 − 7

17 8 − 5·2   **18** 13 − 2·7   **19** 0·5 − 0·32   **20** 5 − 0·99

21 6 + 0·06 + 0·6   **22** 12·4 + 28·71   **23** 11 − 7·4   **24** 8·2 + 9·54 − 11·3

25 Four rods are joined end to end. Their individual lengths are
18·3 cm, 75·2 cm, 11 cm and 0·7 cm.
What is their combined length?

 26 Sue buys three computer games costing £11·45, £23·99 and £5·60.
How much change does she get from £50?

1.2.3 Multiplying decimals

● When you multiply two decimal numbers, the answer has the same number of decimal places as the total number of decimal places in the question.

EXAMPLE

a 0.2×0.8 **b** 0.4×0.07 **c** 5×0.06

· ·

a $2 \times 8 = 16$ **b** $4 \times 7 = 28$ **c** $5 \times 6 = 30$
   So $0.2 \times 0.8 = 0.16$ So $0.4 \times 0.07 = 0.028$ So $5 \times 0.06 = 0.3$

Exercise 6 Ⓜ

Do these multiplications.

1 0.2×0.3	**2** 0.5×0.3	**3** 0.4×0.3	**4** 0.2×0.03
5 0.6×3	**6** 0.7×5	**7** 0.9×2	**8** 8×0.1
9 0.4×0.9	**10** 0.02×0.7	**11** 2.1×0.6	**12** 4.7×0.5
13 21.3×0.4	**14** 5.2×0.6	**15** 4.2×0.03	**16** 212×0.6
17 0.85×0.2	**18** 3.27×0.1	**19** 12.6×0.01	**20** 0.02×17
21 0.05×1.1	**22** 52×0.01	**23** 65×0.02	**24** 0.5×0.002

1.2.4 Dividing by a decimal

EXAMPLE

a $9.36 \div 0.4$ **b** $0.0378 \div 0.07$

· ·

a $9.36 \div 0.4$
   Multiply both numbers by 10 so that you can divide by a **whole number**. (Move the decimal points to the right.)
   So work out $93.6 \div 4$

$$\begin{array}{r} 2\ 3 \cdot\ 4 \\ 4\overline{)9^13 \cdot\ {}^16} \end{array}$$

b $0.0378 \div 0.07$
   Multiply both numbers by 100 so that you can divide by a whole number. (Move the decimal points to the right.)
   So work out $3.78 \div 7$

$$\begin{array}{r} 0 \cdot 5\ 4 \\ 7\overline{)3 \cdot 7^28} \end{array}$$

Exercise 7 Ⓜ

Do these divisions.

1 $0.84 \div 0.4$	**2** $0.93 \div 0.3$	**3** $0.872 \div 0.2$	**4** $0.8 \div 0.2$
5 $2.8 \div 0.7$	**6** $1.25 \div 0.5$	**7** $8 \div 0.5$	**8** $40 \div 0.2$

9 $7 \div 0·1$	**10** $0·368 \div 0·4$	**11** $0·915 \div 0·03$	**12** $0·248 \div 0·04$
13 $0·625 \div 0·05$	**14** $8·54 \div 0·07$	**15** $1·272 \div 0·006$	**16** $4·48 \div 0·08$

17 $0·12 \div 0·002$	**18** $7·5 \div 0·005$	**19** $0·09 \div 0·3$	**20** $0·77 \div 1·1$
21 $0·055 \div 0·11$	**22** $21·28 \div 7$	**23** $22·48 \div 4$	**24** $3·12 \div 4$
25 $0·7 \div 5$	**26** $3 \div 0·8$	**27** $0·3 \div 4$	**28** $1·2 \div 8$
29 $0·732 \div 0·6$	**30** $0·1638 \div 0·001$	**31** $1·05 \div 0·6$	**32** $7·52 \div 0·4$

Exercise 8 Ⓜ (Mixed questions)

Do these calculations.

1 $7·6 + 0·31$	**2** $15 + 7·22$	**3** $7·004 + 0·368$
4 $0·06 + 0·006$	**5** $4·2 + 42 + 420$	**6** $3·84 - 2·62$
7 $11·4 - 9·73$	**8** $4·61 - 3$	**9** $17 - 0·37$
10 $8·7 + 19·2 - 3·8$	**11** $25 - 7·8 + 9·5$	**12** $3·6 - 8·74 + 9$
13 $20·4 - 20·399$	**14** $2·6 \times 0·6$	**15** $0·72 \times 0·04$
16 $27·2 \times 0·08$	**17** $0·1 \times 0·2$	**18** $(0·01)^2$

19 Sharon gets a basic weekly wage of £115 and then a further 30p for each item she completes. How many items did she complete in a week when she earned a total of £221·50?

20 James has the same number of 10p and 50p coins. He has £12 altogether. How many of each coin does he have?

21 A pile of 10p coins is 25·6 cm high. Calculate the total **value** of the coins if each coin is 0·2 cm thick.

25·6 cm

Work out

22 $0·34 \times 100\,000$	**23** $3·6 \div 0·2$	**24** $56·9 \div 100$
25 $0·1401 \div 0·06$	**26** $3·24 \div 0·002$	**27** $0·968 \div 0·11$
28 $600 \div 0·5$	**29** $0·007 \div 4$	**30** $2640 \div 200$
31 $1100 \div 5·5$	**32** $(11 + 2·4) \times 0·06$	**33** $(0·4)^2 \div 0·2$
34 $77 \div 1000$	**35** $(0·3)^2 \div 100$	**36** $(0·1)^4 \div 0·01$

$(0·4)^2$ means
$0·4 \times 0·4$

> **EXAMPLE**
>
> The calculation $\dfrac{480 \times 55}{22}$ is easier to perform after cancelling.
>
> $$\dfrac{480 \times \overset{5}{\cancel{55}}}{\underset{2}{\cancel{22}}} = \dfrac{\overset{240}{480} \times 5}{\underset{1}{\cancel{2}}} = 1200$$
>
>   cancel by 11   cancel by 2

37 Use cancelling to help you work out these.

a $\dfrac{180 \times 4}{36}$ b $\dfrac{92 \times 4 \cdot 6}{2 \cdot 3}$ c $\dfrac{35 \times 0 \cdot 81}{45}$ d $\dfrac{72 \times 3000 \times 24}{150 \times 64}$ e $\dfrac{63 \times 600 \times 0 \cdot 2}{360 \times 7}$

38 Copy the diagram and fill in the missing numbers so that the answer is always 0·7.

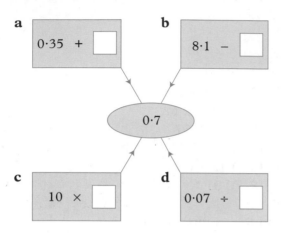

a $0 \cdot 35 \;+\; \square$

b $8 \cdot 1 \;-\; \square$

$0 \cdot 7$

c $10 \;\times\; \square$

d $0 \cdot 07 \;\div\; \square$

● When you multiply by a number greater than 1 you make the answer bigger.

so $5 \cdot 3 \times 1 \cdot 03 > 5 \cdot 3$ but $6 \cdot 75 \times 0 \cdot 89 < 6 \cdot 75$

● When you divide by a number greater than 1 you make the answer smaller.

so $8 \cdot 92 \div 1 \cdot 13 < 8 \cdot 92$ but $11 \cdot 2 \div 0 \cdot 73 > 11 \cdot 2$

---

39 State whether these are true or false.

a $3 \cdot 72 \times 1 \cdot 3 > 3 \cdot 72$ b $253 \times 0 \cdot 91 < 253$

c $0 \cdot 92 \times 1 \cdot 04 > 0 \cdot 92$ d $8 \cdot 5 \div 1 \cdot 4 > 8 \cdot 5$

e $113 \div 0 \cdot 73 < 113$ f $17 \cdot 4 \div 2 \cdot 2 < 17 \cdot 4$

g $0 \cdot 73 \times 0 \cdot 73 < 0 \cdot 73$ h $2511 \div 0 \cdot 042 < 2511$

i $614 \times 0 \cdot 993 < 614$ j $3201 \div 0 \cdot 35 > 3201$

> $<$ means is less than
> $>$ means is more than

40 Write each statement with either $>$ or $<$ in place of the box.

a $17 \cdot 83 \div 0 \cdot 07 \;\square\; 17 \cdot 83$ b $9 \cdot 04 \times 1 \cdot 13 \;\square\; 9 \cdot 04$

c $207 \times 0 \cdot 96 \;\square\; 207$ d $308 \div 2 \cdot 8 \;\square\; 308$

e $1 \cdot 19 \div 0 \cdot 06 \;\square\; 1 \cdot 19$ f $0 \cdot 74 \div 0 \cdot 95 \;\square\; 0 \cdot 74$

1.2.5 Long multiplication and division

EXAMPLE

Work out 327×52.

....

a Use the fact that
   $3\ 2\ 7 \times 53 = (327 \times 50) + (327 \times 3)$

```
      3 2 7
  ×   5 3
  ─────────
  1 6 3 5 0  → This is 327 × 50
      9 8 1  → This is 327 × 3
  ─────────
  1 7 3 3 1  → This is 327 × 53
  ─────────
```

b Here is another method using grids.
   Many people prefer this method.
   $327 \times 53 = 17\,331$

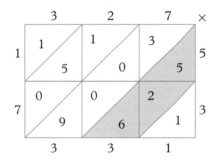

Exercise 9 Ⓜ

Work out these multiplications.

1 35×23	**2** 27×17	**3** 26×25
4 31×43	**5** 45×61	**6** 52×24
7 323×14	**8** 416×73	**9** 504×56
10 306×28	**11** 624×75	**12** 839×79
13 694×83	**14** 973×92	**15** 415×235

With ordinary 'short' division, you divide and find remainders. The method for 'long' division is really the same but you set it out so that the remainders are easier to find.

EXAMPLE

Work out $736 \div 32$.

....

```
         23
    32)736
       64↓
       ───
        96
        96
       ───
         0
```

i   32 into 73 goes 2 times
ii $2 \times 32 = 64$
iii $73 - 64 = 9$
iv 'bring down' 6
v   32 into 96 goes 3 times

Exercise 10

Work out these divisions.

1 672 ÷ 21	**2** 425 ÷ 17	**3** 576 ÷ 32
4 247 ÷ 19	**5** 875 ÷ 25	**6** 574 ÷ 26
7 806 ÷ 34	**8** 748 ÷ 41	**9** 666 ÷ 24
10 707 ÷ 52	**11** 951 ÷ 27	**12** 803 ÷ 31
13 2917 ÷ 45	**14** 2735 ÷ 18	**15** 56 274 ÷ 19

Exercise 11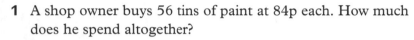

Solve each problem.

1 A shop owner buys 56 tins of paint at 84p each. How much does he spend altogether?

2 Eggs are packed eighteen to a box.

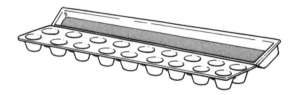

How many boxes are needed for 828 eggs?

3 Mr Eichoff eats 146 mints a week. How many does he eat in a year?

4 Sally wants to buy as many 32p stamps as possible. She has £5 to buy them. How many can she buy and how much change is left?

5 How many 49-seater coaches will be needed for a school trip for a party of 366?

 6 A syndicate of 17 people won a lottery prize of £238 million. They shared the prize equally between them. How much did each person receive?

7 It costs £7905 to hire a plane for a day. A trip is organised for 93 people. How much does each person pay?

8 An office building has 24 windows on each of eight floors. A window cleaner charges 42p for each window. How much is he paid for the whole building?

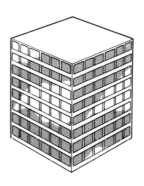

9 The headmaster of a school discovers an oil well in the school playground. As is the custom in such cases, he receives all the money from the oil. The oil comes out of the well at a rate of £15 for every minute of the day and night.
How much does the headmaster receive in a 24-hour day?

10 Tins of peaches are packed 24 to a box.
How many boxes are needed for 1285 tins?

11 Work out
   a $2·1 \times 3·2$ **b** $0·65 \times 4·3$ **c** $0·71 \times 5·6$

12 Work out these, giving your answers correct to 1 decimal place.
   a $8·32 \div 1·7$ **b** $0·7974 \div 0·23$ **c** $1·4 \div 0·71$

13 Susie breeds gerbils for pet shows. The number of gerbils trebles each year. Susie has 40 gerbils at the end of year 1.
   a Copy and complete the table below.

End of year	1	2	3	4	5
Number of gerbils	40				

   b A gerbil cage can hold 16 gerbils. Work out the minimum number of cages Susie will need by the end of year 6.

Exercise 12 Ⓜ Crossnumbers

Make four copies of this crossnumber grid and work out the answers using the clues. You can check your working by doing **all** the across and **all** the down clues.

A

Across

1 327 + 198
3 245 ÷ 7
5 3146 − 729
6 248 − 76
7 2^6
8 850 ÷ 5
10 $10^2 + 1^2$
11 3843 ÷ 7
12 1000 − 913
13 37 × 5 × 3
16 152 300 ÷ 50
19 3^6
20 $100 - \left(\dfrac{17 \times 10}{5}\right)$

Down

1 3280 + 1938
2 65 720 − 13 510
3 3·1 × 1000
4 1284 ÷ 6
7 811 − 127
9 65 × 11
10 $(12^2 - 8) \div 8$
11 $(7^2 + 1^2) \times 11$
12 7 + 29 + 234 + 607
14 800 − 265
15 1 + 2 + 3 + 4 + 5 + 6 + 7 + 8 + 13
17 (69 × 6) ÷ 9
18 $3^2 + 4^2 + 5^2 + 2^4$

B

Draw decimal points on the lines between squares where necessary.

Across

1 4·2 + 1·64
3 7 × 0·5
5 20·562 ÷ 6
6 $(2^3 \times 5) \times 10 - 1$
7 0·034 × 1000
8 61 × 0·3
10 8 − 0·36
11 19 × 50
12 95·7 ÷ 11
13 8·1 × 0·7
16 (11 × 5) ÷ 8
19 (44 − 2·8) ÷ 5
20 Number of inches in a yard

Down

1 62·6 − 4·24
2 48·73 − 4·814
3 25 + 7·2 + 0·63
4 2548 ÷ 7
7 0·315 × 100
9 169 × 0·05
10 770 ÷ 100
11 14·2 + 0·7 − 5·12
12 11·4 − 2·64 − 0·18
14 $0·0667 \times 10^3$
15 0·6 + 0·7 + 0·8 + 7·3
17 0·73 m written in cm
18 0·028 × 200

C

Across

1 Eleven squared take away six
3 Next in the sequence 21, 24, 28, 33
5 Number of minutes in a day
6 $2 \times 13 \times 5 \times 5$
7 Next in the sequence 92, 83, 74
8 5% of 11 400
10 $98 + 11^2$
11 $(120 - 9) \times 6$
12 $1\frac{2}{5}$ as a decimal
13 $2387 \div 7$
16 $9 \cdot 05 \times 1000$
19 8 m − 95 cm (in cm)
20 3^4

Down

1 Write 18·6 m in cm
2 Fifty-one thousand and fifty-one
3 Write 3·47 km in m
4 $1\frac{1}{4}$ as a decimal
7 7 m − 54 cm (in cm)
9 $0 \cdot 0793 \times 1000$
10 2% of 1200
11 $\frac{1}{5}$ of 3050
12 $127 \div 100$
14 Number of minutes between 12:00 and 20:10
15 4% of 1125
17 $7^2 + 3^2$
18 Last two digits of (67×3)

D

Across

1 $1\frac{3}{4}$ as a decimal
3 Two dozen
5 Forty less than ten thousand
6 Emergency
7 5% of 740
8 Nine pounds and five pence
10 1·6 m written in cm
11 $5649 \div 7$
12 One-third of 108
13 $6 - 0 \cdot 28$
16 A quarter to midnight on the 24h clock
19 $5^3 \times 2^2 + 1^5$
20 $3300 \div 150$

Down

1 Twelve pounds 95 pence
2 Four less than sixty thousand
3 245×11
4 James Bond
7 Number of minutes between 09:10 and 15:30
9 $\frac{1}{20}$ as a decimal
10 Ounces in a pound
11 8·227 to two decimal places
12 4 m − 95 cm (in cm)
14 Three to the power 6
15 20·64 to the nearest whole number
17 $\left(6\frac{1}{2}\right)^2$ to the nearest whole number
18 Number of minutes between 14:22 and 15:14

1.3 Properties of numbers

1.3.1 Factors and multiples

- **Factors**   Any number which divides exactly into 8 is a **factor** of 8.
  The factors of 8 are 1, 2, 4, 8.
- **Multiples**   Any number in the 8-times table is a **multiple** of 8.
  The first five multiples of 8 are 8, 16, 24, 32, 40.

- **Prime**   A **prime** number has just two different factors: 1 and itself.
  The number 1 is **not** prime. It does not have two different factors.
  The first five prime numbers are 2, 3, 5, 7, 11.
- **Prime factor**   The factors of 8 are 1, 2, 4, 8. The only **prime factor** of 8 is 2. It is the only prime number which is a factor of 8.

- **Square numbers**   A square number can be written in the form $n \times n$, where n is an integer.
  Examples: 4, 9, 100
- **Cube numbers**   A cube number can be written in the form $n \times n \times n$, where n is an integer.
  Examples: 1, 8, 27, 64

> An integer is another name for a whole number.

EXAMPLE

a Express 6930 as a product of primes.

b Express in prime factors
   i 90   **ii** 42

....

a
$2)\overline{69^13^10}$ (divide by 2)
$3)\overline{34^16^15}$ (divide by 3)
$3)\overline{11^25^15}$ (divide by 3)
$5)\underline{3\ 8^35}$ (divide by 5)
$7)\underline{7\ 7}$ (divide by 7)
       $1\ 1$ (stop because 11 is prime)

So $6930 = 2 \times 3 \times 3 \times 5 \times 7 \times 11$

b   **i** $90 = 2 \times 3 \times 3 \times 5$
    ii $42 = 2 \times 3 \times 7$

> This is called prime factor decomposition.

Exercise 13 (M)

1 Write all the factors of these numbers.
   a 6   **b** 15   **c** 18   **d** 21   **e** 40

2 Write all the prime numbers less than 20.

3 Write two prime numbers which add up to another prime number. Do this in **two** ways.

4 Which of these numbers are prime?
3, 11, 15, 19, 21, 23, 27, 29, 31, 37, 39, 47, 51, 59, 61, 67, 72, 73, 87, 99

5 Express each of these numbers as a product of primes.
   a 600   **b** 693   **c** 2464
   d 3510   **e** 4000   **f** 22 540

6 Copy the grid and then write the numbers 1 to 9, one in each box, so that all the numbers satisfy the conditions for both the row and the column.

	Prime number	Multiple of 3	Factor of 16
Number greater than 5			
Odd number			
Even number			

7 Find each of these mystery numbers.
   a I am an odd number and a prime number. I am a factor of 14.
   b I am a two-digit multiple of 50.
   c I am one less than a prime number which is even.
   d I am odd, greater than one and a factor of both 20 and 30.

8 Write the first 15 square numbers. **Learn** these numbers.

9 Write the first 5 cube numbers.

10 Shirin has five number cards   1   2   4   5   7

   a Using two of the cards, show how Shirin can make a square number.

   b Using two of the cards, show how Shirin can make a cube number.

   c Using three of the cards, show how Shirin can make a larger cube number.

11 Find the smallest value of n for which
$1^2 + 2^2 + 3^2 + 4^2 + 5^2 + \ldots + n^2 > 800$.

12 Put three different numbers in the circles so that when you add the numbers at the end of each line you always get a square number.

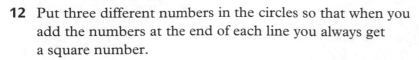 **13** The odd numbers can be added in groups to give an interesting sequence.

1	= 1	= 1^3
3 + 5	= 8	= 2^3
7 + 9 + 11	= 27	= 3^3

Write the next three rows of the sequence to see if the sum of each row always gives a cube number.

14 The letters p and q represent prime numbers.
Is $p + q$ always an even number? Explain your answer.

15 Work out which of these numbers are prime.
    a 91   **b** 101   **c** 143   **d** 151   **e** 293

> Divide by the prime numbers 2, 3, 5, 7, 11 etc.

16 The first of three consecutive numbers is n. Find the smallest value of n if the three numbers are all not prime.

> Consecutive means next to, for example, 3, 4, 5.

17 Make six prime numbers using the digits 1, 2, 3, 4, 5, 6, 7, 8, 9 once each.

*18 Apart from 1, 3 and 5, all odd numbers less than 100 can be written in the form $p + 2^n$ where p is a prime number and n is greater than or equal to 2.

For example $43 = 11 + 2^5$
                $27 = 23 + 2^2$

For the odd numbers 7, 9, 11, ... 27 write as many as you can in the form $p + 2^n$.

*19 Look at the expression $5^n + 3$.
Substitute six different values for n and find out if the result is a multiple of 4 each time.

*20 Substitute six pairs of values of a and n in the expression $a^n - a$.
Is the result always a multiple of n?

1.3.2 L.C.M. and H.C.F.

The first few multiples of 4 are   4, 8, 12, 16, ⑳, 24, 28 ...
The first few multiples of 5 are   5, 10, 15, ⑳, 25, 30, 35 ...

- The **Least Common Multiple** (L.C.M.) of 4 and 5 is 20.
  It is the lowest number which is in both lists.

The factors of 12 are   1, 2, 3, ④, 6, 12
The factors of 20 are   1, 2, ④, 5, 10, 20

- The **Highest Common Factor** (H.C.F.) of 12 and 20 is 4.
  It is the highest number which is in both lists.

Exercise 14 Ⓜ

1 a Write the first six multiples of 2.
  b Write the first six multiples of 5.
  c Write the L.C.M. of 2 and 5.

2 a Write the first four multiples of 4.
  b Write the first four multiples of 12.
  c Write the L.C.M. of 4 and 12.

3 Find the L.C.M. of
  a 6 and 9   **b** 8 and 12   **c** 14 and 35
  d 4 and 6   **c** 5 and 10   **f** 7 and 9

4 The table shows the factors and common factors of 24 and 36.

Number	Factors	Common factors
24	1, 2, 3, 4, 6, 8, 12, 24	} 1, 2, 3, 4, 6, 12
36	1, 2, 3, 4, 6, 9, 12, 18, 36	

Write the H.C.F. of 24 and 36.

5 The table shows the factors and common factors of 18 and 24.

Write the H.C.F. of 18 and 24.

Number	Factors	Common factors
18	1, 2, 3, 6, 9, 18	} 1, 2, 3, 6
24	1, 2, 3, 4, 6, 8, 12, 24	

6 Find the H.C.F. of
  a 12 and 18   **b** 22 and 55   **c** 45 and 72
  d 18 and 30   **e** 60 and 72   **f** 40 and 50

7 **a** Find the H.C.F. of 12 and 30.
   b Find the L.C.M. of 8 and 20.
   c Write two numbers whose H.C.F. is 11.
   d Write two numbers whose L.C.M. is 10.

> Don't confuse your L.C.M.s with your H.C.F.s!

8 Given that $30 = 2 \times 3 \times 5$ and $165 = 3 \times 5 \times 11$, find the highest common factor of 30 and 165. (That is, the highest number that goes into 30 and 165.)

9 If $315 = 3 \times 3 \times 5 \times 7$ and $273 = 3 \times 7 \times 13$, find the highest common factor of 315 and 273.

10 **a** Express 1008 and 840 as products of their prime factors.
    b Find the H.C.F. of 1008 and 840.
    c Find the smallest number which can be multipied by 1008 to give a square number.

11 **a** Express 19 800 and 12 870 as products of their prime factors.
    b Find the H.C.F. of 19 800 and 12 870.
    c Find the smallest number which can be multiplied by 19 800 to give a square number.

1.3.3 Operations and inverses

Operation	Inverse	Example
Add n	Subtract n	$9 + 7 = 16, \quad 16 - 7 = 9$
Subtract n	Add n	$12 - 5 = 7, \quad 7 + 5 = 12$
Multiply by n	Divide by n	$20 \times 4 = 80, \quad 80 \div 4 = 20$
Divide by n	Multiply by n	$12 \div 3 = 4, \quad 4 \times 3 = 12$
Square	Square root	$7^2 = 49, \quad \sqrt{49} = 7$
Raise to power n	Find nth root $\left(\text{raise to the power } \frac{1}{n}\right)$	$10^4 = 10\,000,$ $\sqrt[4]{10\,000} = 10$

> See the section on indices, page 350.

- The **reciprocal** of n is $\frac{1}{n}$ ($n \neq 0$). The reciprocal of $\frac{1}{n}$ is n.
  The reciprocal is sometimes called the **multiplicative inverse**.

> Note that zero has no reciprocal because you cannot divide by zero (division by zero is not defined).

Exercise 15 Ⓜ

In the flow charts in each question, the boxes A, B, C and D each contain a single mathematical operation (like $+5$, $\times 4$, -15, $\div 2$).

Look at flow charts **i** and **ii** together and work out what operation happens in box A. Copy and complete the flow chart by replacing B, C and D.

Now copy and complete each flow chart on the right for each question, using the same operations.

1 i 2 → [A] 17 → [B] 34 → [C] 12 → [D] 3 →

ii 4 → [A] 19 → [B] 38 → [C] 16 → [D] 4 →

a 7 → [A] ? → [B] ? → [C] ? → [D] ? →

b 10 → [A] ? → [B] ? → [C] ? → [D] ? →

c ? → [A] ? → [B] 62 → [C] ? → [D] ? →

d ? → [A] $15\frac{1}{2}$ → [B] ? → [C] ? → [D] ? →

2 i 2 → [A] 4 → [B] 12 → [C] 2 → [D] 1 →

ii 3 → [A] 9 → [B] 27 → [C] 17 → [D] $8\frac{1}{2}$ →

a 4 → [A] 16 → [B] ? → [C] ? → [D] ? →

b 5 → [A] ? → [B] ? → [C] ? → [D] ? →

c ? → [A] ? → [B] 108 → [C] ? → [D] ? →

d ? → [A] ? → [B] ? → [C] 182 → [D] ? →

In questions **3**, **4** and **5** find the operations A, B, C, D.

3 i 4 → [A] 16 → [B] 4 → [C] -6 → [D] 12 →

ii 9 → [A] 36 → [B] 6 → [C] -4 → [D] 8 →

5 i -1 → [A] 2 → [B] 8 → [C] -4 → [D] 96 →

ii 7 → [A] 10 → [B] 1000 → [C] -500 → [D] -400 →

4 i 5 → [A] $\frac{1}{5}$ → [B] 1·2 → [C] 1·44 → [D] 0·48 →

ii $\frac{1}{2}$ → [A] 2 → [B] 3 → [C] 9 → [D] 3 →

1.4 Fractions

1.4.1 Equivalent fractions

The fractions $\frac{2}{3}$, $\frac{4}{6}$ and $\frac{10}{15}$ are **equivalent** fractions.

- To find a fraction equivalent to $\frac{2}{3}$ multiply the numerator (top) and the denominator (bottom) by the same number.

$$\overset{\times 2}{\frac{2}{3}} = \underset{\times 2}{\frac{4}{6}} \qquad \overset{\times 3}{\frac{3}{5}} = \underset{\times 3}{\frac{9}{15}} \qquad \overset{\times 4}{\frac{3}{7}} = \underset{\times 4}{\frac{12}{28}}$$

- Some fractions can be simplified by cancelling.

$$\overset{\div 5}{\frac{20}{25}} = \underset{\div 5}{\frac{4}{5}} \qquad \overset{\div 8}{\frac{16}{24}} = \underset{\div 8}{\frac{2}{3}} \qquad \overset{\div 7}{\frac{21}{28}} = \underset{\div 7}{\frac{3}{4}}$$

Exercise 16

1 Copy and complete by finding the missing numbers.

a $\frac{2}{5} = \frac{\square}{10}$ b $\frac{3}{4} = \frac{\square}{8}$ c $\frac{2}{3} = \frac{\square}{9}$ d $\frac{5}{6} = \frac{\square}{18}$

e $\frac{2}{7} = \frac{6}{\square}$ f $\frac{3}{5} = \frac{12}{\square}$ g $\frac{1}{6} = \frac{3}{\square}$ h $\frac{7}{10} = \frac{21}{\square}$

2 Copy and simplify these fractions.

a $\frac{9}{12}$ b $\frac{6}{24}$ c $\frac{8}{20}$ d $\frac{9}{36}$ e $\frac{9}{15}$

f $\frac{32}{36}$ g $\frac{24}{30}$ h $\frac{14}{42}$ i $\frac{27}{45}$ j $\frac{56}{64}$

3 Here are some number cards.   5   4   9   2   7

a Use two cards to make a fraction which is equal (equivalent) to $\frac{1}{2}$.

b Use two cards to make a fraction which is equal to $\frac{20}{36}$.

c Use three of the cards to make the smallest possible fraction.

4 Write these fractions in pairs of equivalent fractions.

$\frac{3}{12}$ $\frac{6}{15}$ $\frac{10}{45}$ $\frac{5}{6}$ $\frac{3}{7}$ $\frac{1}{4}$ $\frac{12}{28}$ $\frac{30}{36}$ $\frac{2}{5}$ $\frac{2}{9}$

5 a Copy and complete.

$\frac{2}{3} = \frac{\square}{12}$, $\frac{1}{2} = \frac{\square}{12}$, $\frac{1}{4} = \frac{\square}{12}$

b Hence write $\frac{2}{3}, \frac{1}{2}, \frac{1}{4}$ in order of size, smallest first.

6 a Copy and complete.

$\frac{5}{6} = \frac{\square}{12}$, $\frac{2}{3} = \frac{\square}{12}$, $\frac{1}{4} = \frac{\square}{12}$

b Hence write $\frac{5}{6}, \frac{2}{3}, \frac{1}{4}$ in order of size, smallest first.

In questions **7** to **10** write the fractions in order of size.

7 $\frac{3}{5}, \frac{7}{10}, \frac{1}{2}$ **8** $\frac{7}{12}, \frac{5}{6}, \frac{3}{4}$

9 $\frac{1}{2}, \frac{3}{8}, \frac{3}{4}$ **10** $\frac{7}{15}, \frac{1}{3}, \frac{2}{15}$

11 Write the larger fraction in each pair.

a $\frac{2}{3}, \frac{4}{7}$ **b** $\frac{1}{3}, \frac{2}{5}$ **c** $\frac{5}{8}, \frac{3}{5}$ **d** $\frac{3}{4}, \frac{7}{10}$

12 a Copy and complete $\frac{7}{4} = 1\frac{\square}{4}$.

b Write these as mixed fractions.

i $\frac{8}{3}$ **ii** $\frac{12}{5}$ **iii** $\frac{9}{4}$ **iv** $\frac{7}{6}$ **v** $\frac{15}{2}$

$2\frac{1}{2}$ is a mixed fraction.

13 Write these as improper ('top heavy') fractions.

a $1\frac{2}{5}$ **b** $2\frac{1}{2}$ **c** $3\frac{3}{4}$ **d** $4\frac{1}{3}$ **e** $2\frac{1}{7}$

$\frac{5}{3}$ is an improper fraction.

1.4.2   Arithmetic with fractions

● You can only add and subtract fractions if they have the same denominator.

a $\dfrac{3}{4} + \dfrac{2}{5}$ **b** $\dfrac{3}{7} - \dfrac{1}{3}$ **c** $2\dfrac{3}{8} - 1\dfrac{5}{12}$

....

a $\dfrac{3}{4} + \dfrac{2}{5} = \dfrac{15}{20} + \dfrac{8}{20}$ **b** $\dfrac{3}{7} - \dfrac{1}{3} = \dfrac{9}{21} - \dfrac{7}{21}$ **c** $2\dfrac{3}{8} - 1\dfrac{5}{12} = \dfrac{19}{8} - \dfrac{17}{12}$

$\qquad\quad = \dfrac{23}{20}$ $\qquad\quad = \dfrac{2}{21}$ $\qquad\qquad\quad = \dfrac{57}{24} - \dfrac{34}{24}$

Find equivalent fractions with the same denominator.

$\qquad\quad = 1\dfrac{3}{20}$ $\qquad\qquad\quad = \dfrac{23}{24}$

It is easier to multiply fractions if you 'cancel down'.

a $\dfrac{2}{5} \times \dfrac{6}{7}$ **b** $\dfrac{12}{13} \times \dfrac{5}{8}$

....

a $\dfrac{2}{5} \times \dfrac{6}{7} = \dfrac{12}{35}$ **b** $\dfrac{\cancel{12}^{3}}{13} \times \dfrac{5}{\cancel{8}_{2}} = \dfrac{15}{26}$

(no cancelling here) (cancel down)

Exercise 17 Ⓜ

1 Copy and complete

a $\dfrac{1}{4} + \dfrac{1}{8}$ **b** $\dfrac{1}{2} + \dfrac{2}{5}$ **c** $\dfrac{2}{5} + \dfrac{1}{3}$

$= \dfrac{\square}{8} + \dfrac{1}{8} =$ $= \dfrac{\square}{10} + \dfrac{\square}{10} =$ $= \dfrac{\square}{15} + \dfrac{\square}{15} =$

2 Work out

a $\dfrac{1}{2} + \dfrac{1}{3}$ **b** $\dfrac{1}{4} + \dfrac{1}{6}$ **c** $\dfrac{1}{3} + \dfrac{1}{4}$ **d** $\dfrac{1}{4} + \dfrac{1}{5}$

e $\dfrac{3}{4} + \dfrac{4}{5}$ **f** $\dfrac{5}{6} + \dfrac{2}{3}$ **g** $1\dfrac{1}{4} + \dfrac{1}{3}$ **h** $2\dfrac{1}{3} + 1\dfrac{1}{4}$

3 Work out

a $\dfrac{2}{3} - \dfrac{1}{2}$ **b** $\dfrac{3}{4} - \dfrac{1}{3}$ **c** $\dfrac{4}{5} - \dfrac{1}{3}$

d $\dfrac{1}{2} - \dfrac{2}{5}$ **e** $1\dfrac{3}{4} - \dfrac{2}{3}$ **f** $1\dfrac{2}{3} - 1\dfrac{1}{2}$

4 Copy and complete the addition square on the right.

+			$\frac{1}{3}$
$\frac{1}{2}$	$\frac{3}{4}$		$\frac{5}{6}$
$\frac{1}{8}$	$\frac{3}{8}$	$\frac{1}{2}$	
$\frac{1}{5}$			

5 Work out

  a $\frac{1}{8} \times \frac{8}{9}$ 　　　　**b** $\frac{2}{5} \times \frac{1}{2}$ 　　　　**c** $\frac{3}{4} \times \frac{5}{6}$

  d $\frac{3}{10} \times \frac{3}{10}$ 　　　**e** $\frac{1}{5}$ of 35 　　　**f** $\frac{2}{5}$ of 35

  g $\frac{2}{3}$ of £33 　　　**h** $\frac{5}{7}$ of £28

6 Work out these multiplications.

  a $\frac{2}{3} \times \frac{4}{5}$ 　　**b** $\frac{1}{7} \times \frac{5}{6}$ 　　**c** $\frac{5}{8} \times \frac{12}{13}$ 　　**d** $1\frac{3}{4} \times \frac{2}{3}$

  e $2\frac{1}{2} \times \frac{1}{4}$ 　　**f** $3\frac{2}{3} \times 2\frac{1}{2}$ 　　**g** $\frac{2}{3}$ of $\frac{2}{5}$ 　　**h** $\frac{3}{7}$ of $\frac{1}{2}$

1.4.3 Multiplying and dividing with fractions

$2\frac{1}{2} \div \frac{1}{4}$ means

'How many $\frac{1}{4}$s are there in $2\frac{1}{2}$?'

The diagram shows that the answer is 10.

You can divide $2\frac{1}{2}$ by $\frac{1}{4}$ by multiplying $2\frac{1}{2}$ by the **reciprocal** of $\frac{1}{4}$.

So $2\frac{1}{2} \div \frac{1}{4}$ $\left(\text{write } 2\frac{1}{2} \text{ as } \frac{5}{2}\right)$

$= \frac{5}{2} \div \frac{1}{4}$ $\left(\text{multiply by the reciprocal of } \frac{1}{4}\right)$

$= \frac{5}{2} \times \frac{4}{1} = 10$

> Most people remember this as: 'Invert the second fraction and multiply.'

EXAMPLE

  a $\frac{2}{3} \div \frac{3}{5}$ 　　　**b** $\frac{4}{5} \div \frac{9}{10}$ 　　　**c** $2\frac{2}{5} \div 6$

....

  a $\frac{2}{3} \div \frac{3}{5}$ 　　　**b** $\frac{4}{5} \div \frac{9}{10}$ 　　　**c** $2\frac{2}{5} \div 6$

  $= \frac{2}{3} \times \frac{5}{3}$ 　　　$= \frac{4}{\underset{1}{\cancel{5}}} \times \frac{\cancel{10}^{2}}{9}$ 　　　$= \frac{12}{5} \div 6$

  $= \frac{10}{9}$ 　　　　　$= \frac{8}{9}$ 　　　　　$= \frac{\cancel{12}^{2}}{5} \times \frac{1}{\underset{1}{\cancel{6}}} = \frac{2}{5}$

> Write $2\frac{2}{5}$ as $\frac{12}{5}$.

> 'Dividing by 6' is the same as 'multiplying by $\frac{1}{6}$'.

Exercise 18 Ⓜ

1 Copy and complete

a $4 \div \frac{2}{3} = \frac{4}{1} \times \frac{\square}{\square} =$ **b** $\frac{2}{3} \div \frac{3}{5} = \frac{2}{3} \times \frac{\square}{\square} =$

2 Work out

a $\frac{2}{3} \div \frac{3}{4}$ **b** $\frac{3}{5} \div \frac{1}{10}$ **c** $\frac{2}{3} \div \frac{5}{6}$ **d** $\frac{3}{8} \div 2$

e $4 \div \frac{3}{4}$ **f** $8 \div \frac{1}{2}$ **g** $\frac{3}{7} \div \frac{3}{7}$ **h** $\frac{7}{10} \div \frac{1}{100}$

3 Work out these divisions.

a $\frac{1}{3} \div \frac{4}{5}$ **b** $\frac{3}{4} \div \frac{1}{6}$ **c** $\frac{5}{6} \div \frac{1}{2}$ **d** $\frac{5}{8} \div 2$

e $\frac{3}{8} \div \frac{1}{5}$ **f** $1\frac{3}{4} \div \frac{2}{3}$ **g** $3\frac{1}{2} \div 2\frac{3}{5}$ **h** $6 \div \frac{3}{4}$

4 Copy the diagram and fill in the missing numbers so that the answer is always $\frac{3}{8}$.

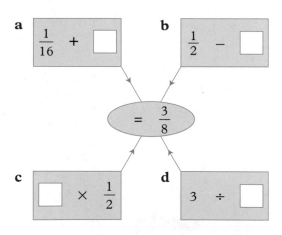

? **5** A rubber ball is dropped from a height of 300 cm. After each bounce, the ball rises to $\frac{4}{5}$ of its previous height.

How high, to the nearest cm, will it rise after the fourth bounce?

6 Work out

a $\frac{5}{6} + \frac{1}{2}$ **b** $\frac{5}{6} - \frac{1}{2}$ **c** $\frac{5}{6} \times \frac{1}{2}$ **d** $\frac{5}{6} \div \frac{1}{2}$

e $\frac{2}{5} + \frac{3}{10}$ **f** $\frac{2}{5} - \frac{3}{10}$ **g** $\frac{2}{5} \times \frac{3}{10}$ **h** $1\frac{2}{5} \div \frac{3}{10}$

7 Steve Braindead spends his income as follows

   a $\frac{2}{5}$ of his income goes in tax

   b $\frac{2}{3}$ of what is left goes on food, rent and transport

   c he spends the rest on cigarettes, beer and betting.

   What fraction of his income is spent on cigarettes, beer and betting?

8 If $a = \frac{3}{4}$, $b = \frac{2}{5}$, $c = \frac{1}{3}$, work out

   a a^2 **b** $a - b$ **c** $\frac{b}{c}$ **d** $\frac{1}{b} - \frac{1}{a}$

   e abc **f** $2b - c$ **g** $\frac{1}{a + b}$ **h** $\frac{b}{ac}$

> For quick revision on algebra, see page 97.

9 In this equation all the asterisks stand for the same number. What is the number?

$$\frac{\star}{\star} - \frac{\star}{6} = \frac{\star}{30}$$

*10 Find the value of n if $\left(1\frac{1}{3}\right)^n - \left(1\frac{1}{3}\right) = \frac{28}{27}$

? ***11** A formula used by opticians is $\frac{1}{f} = \frac{1}{u} + \frac{1}{v}$

     Given that $u = 3$ and $v = 5\frac{1}{2}$ find the exact value of f.

*12 When it hatches from its egg, the shell of a certain crab is 1 cm across. When fully grown the shell is approximately 10 cm across. Each new shell is one-third bigger than the previous one. How many shells does a fully grown crab have during its life?

*13 In each equation find the missing numbers a and b.

    a $\frac{a}{4} + \frac{b}{3} = \frac{11}{12}$ **b** $\frac{a}{6} + \frac{b}{21} = \frac{17}{42}$ **c** $\frac{a}{12} + \frac{b}{18} = \frac{29}{36}$

14 Copy and complete the multiplication square.

×	$\frac{2}{5}$		
$\frac{1}{3}$		$\frac{1}{4}$	
		$\frac{3}{8}$	
$\frac{1}{4}$			$\frac{1}{6}$

? ***15** A cylinder is $\frac{1}{4}$ full of water. After 60 ml of water is added the cylinder is $\frac{2}{3}$ full. Calculate the total volume of the cylinder.

*16 **a** Here is a sequence.

> line 1 $\quad \frac{1}{2} + \frac{1}{4} = \frac{3}{4}$
>
> line 2 $\quad \frac{1}{2} + \frac{1}{4} + \frac{1}{8} = \frac{7}{8}$
>
> line 3 $\quad \frac{1}{2} + \frac{1}{4} + \frac{1}{8} + \frac{1}{16} = \square$
>
> line 4 $\quad \frac{1}{2} + \frac{1}{4} + \frac{1}{8} + \frac{1}{16} + \frac{1}{32} = \square$

Copy the sequence and fill in the missing answers.

b Write the answer to the nth line of the sequence.

1.5 Estimating

1.5.1 Estimations

It is always sensible to check that the answer to a calculation is 'about the right size'.

EXAMPLE

Estimate the value of $\dfrac{57 \cdot 2 \times 110}{2 \cdot 146 \times 46 \cdot 9}$, correct to one significant figure.

$$\frac{57 \cdot 2 \times 110}{2 \cdot 146 \times 46 \cdot 9} \approx \frac{60 \times 100}{2 \times 50} = \frac{60 \times \overset{1}{\cancel{100}}}{\underset{1}{\cancel{2}} \times \underset{1}{\cancel{50}}} = 60$$

On a calculator the answer is $62 \cdot 52$ (to 4 significant figures).

This agrees with the estimate so it is probably correct!

Exercise 19 Ⓜ/Ⓗ

Write each question and decide (by estimating) which answer is closest to the exact answer. Do not do the calculation exactly.

	Question	Answer A	Answer B	Answer C
1	$7 \cdot 2 \times 9 \cdot 8$	50	100	70
2	$2 \cdot 03 \times 58 \cdot 6$	120	90	140
3	$23 \cdot 4 \times 19 \cdot 3$	210	300	450
4	$313 \times 107 \cdot 6$	3600	4300	34 000
5	$6 \cdot 3 \times 0 \cdot 098$	0·6	0·06	6
6	$1200 \times 0 \cdot 89$	700	1000	100
7	$0 \cdot 21 \times 93$	40	9	20
8	$88 \cdot 8 \times 213$	19 000	1700	2000
9	$0 \cdot 04 \times 968$	40	20	90
10	$0 \cdot 11 \times 0 \cdot 089$	0·1	0·9	0·01
11	$13 \cdot 92 \div 5 \cdot 8$	0·5	4	2·4
12	$105 \cdot 6 \div 9 \cdot 6$	9	11	15
13	$8405 \div 205$	4	400	40
14	$881 \cdot 1 \div 99$	4·5	9	88
15	$4 \cdot 183 \div 0 \cdot 89$	4·7	48	51
16	$6 \cdot 72 \div 0 \cdot 12$	6	20	50
17	$20 \cdot 301 \div 1010$	0·02	0·2	0·002
18	$0 \cdot 28896 \div 0 \cdot 0096$	300	100	30
19	$0 \cdot 143 \div 0 \cdot 11$	2·3	1·3	11
20	$159 \cdot 65 \div 515$	0·1	3·6	0·3
21	$(5 \cdot 6 - 0 \cdot 21) \times 39$	400	200	20
22	$\dfrac{17 \cdot 5 \times 42}{2 \cdot 5}$	300	500	90
23	$(906 + 4 \cdot 1) \times 0 \cdot 31$	500	280	30
24	$\dfrac{543 + 472}{18 \cdot 1 + 10 \cdot 9}$	70	35	85
25	$\dfrac{112 \cdot 2 \times 75 \cdot 9}{6 \cdot 9 \times 5 \cdot 1}$	240	20	25

Exercise 20 Ⓜ/Ⓗ ⭐ **F**

1 For a wedding the caterers provided food at £39·75 per head. There were 207 guests. Estimate the total cost of the food.

2 985 people share the cost of hiring an ice rink. About how much does each person pay if the total cost is £6017?

3 On a charity walk, Susie walked 31 miles in 11 hours 7 minutes. Estimate the number of minutes it took her to walk one mile.

In questions **4** to **7**, estimate which answer is closest to the actual answer.

4 The top speed of a Grand Prix racing car

A	B	C
600 km/h	80 km/h	300 km/h

5 The number of times your heart beats in one day (24 h)

A	B	C
10 000	100 000	1 000 000

6 The thickness of one page in this book

A	B	C
0·01 cm	0·001 cm	0·0001 cm

7 The number of cars in a traffic jam 10 km long on a 3-lane motorway

A	B	C
3000	30 000	300 000

(Assume each car takes up 10 m of road.)

In questions **8** and **9** there are six calculations and six answers. Write each calculation and insert the correct answer from the list given. Use estimation.

8 a $8·9 \times 10·1$ **b** $7·98 \div 1·9$ **c** $112 \times 3·2$
    d $11·6 + 47·2$ **e** $2·82 \div 9·4$ **f** $262 \div 100$
    Answers: 2·62, 58·8, 0·3, 89·89, 358·4, 4·2

9 a $49·5 \div 11$ **b** 21×22 **c** $9·1 \times 104$
    d $86 - 8·2$ **e** $2·4 \div 12$ **f** $651 \div 31$
    Answers: 21, 946·4, 0·2, 4·5, 462, 77·8

10 Mr Gibson, the famous maths teacher, has won the lottery. He decides to give a rather unusual prize for the person who comes top in his next maths test.

The prize winner receives his or her own weight in coins and they can choose to have either 1p, 2p, 5p, 10p, 20p, 50p or £1 coins. All the coins must be the same.
Shabeza is the winner and she weighs 47 kg. **Estimate** the highest value of her prize.

Approximate masses	
1p	3·6 g
2p	7·2 g
5p	3·2 g
10p	6·5 g
20p	5·0 g
50p	7·5 g
£1	9·0 g

11 The largest tree in the world has a diameter of 11 m. Estimate the number of 'average' 15-year-olds required to circle the tree so that they form an unbroken chain.

12 There are about 7000 cinemas in the UK and every day about 300 people visit each one. The population of the UK is about 60 million.

A film magazine report said:
Is the magazine report fair?
Show the working you did to decide.

'Every day, over 3% of British people go to the cinema.'

13 The petrol consumption of a large car is 4 miles per litre and petrol costs 79p per litre.
Jasper estimates that the petrol costs of a round trip of about 1200 miles will be £240. Is this a reasonable estimate? Show your working.

14 The 44 teachers in a rather sporty school decide to buy 190 footballs at £2·42 each. They share the cost equally between them. The headmaster used a calculator to work out the cost per teacher and got an answer of £1·05 to the nearest penny.
Without using a calculator, work out an estimate for the answer to check whether or not he got it right. Show your working.

*15 The surface area, A, of an object is given by the formula

$$A = \pi r(10r + \sqrt{(r^2 + d^2)}).$$

Estimate the value of A correct to one significant figure if $r = 1·08$ and $d = 5·87$.

1.5.2 Rounding numbers

● You can round numbers in several ways:

   1 you can round to **the nearest whole number**, ten, hundred, thousand
   2 you can round to one or more **significant figures**
   3 you can round to one or more **decimal places**.

a To round a number to the nearest whole number, look at the first digit after the decimal point to see if it is '**five or more**'. If that number is '**five or more**' you round **up**. Otherwise you round **down**.

$13\cdot82 = 14$ to the nearest whole number

$211\cdot7 = 200$ to the nearest hundred

b Reminders:

$35\cdot2\,|\,6 = 35\cdot3$ to 3 sf $0\cdot041\,|\,2 = 0\cdot041$ to 2 sf

'5 or more'   Do not count zeros at the beginning.

$15\cdot26\,|\,66 = 15\cdot27$ to 2 dp $0\cdot349\,|\,7 = 0\cdot350$ to 3 dp

> sf means significant figures.

> dp means decimal places.

c Suppose you timed a race with a stopwatch and got $13\cdot2$ seconds while your friend with an electronic watch got $13\cdot20$ seconds. Is there any difference? Yes!
When you write $13\cdot20$, the figure is accurate to 2 decimal places even though the last figure is a zero.
The figure $13\cdot2$ is accurate to only one decimal place.

d Similarly a weight, given as $12\cdot00$ kg, is accurate to 4 significant figures while '12 kg' is accurate to only 2 significant figures.

e Suppose you measure the length of a line in cm and you want to show that it is accurate to one decimal place. The measured length using a ruler might be 7 cm. To show that it is accurate to one decimal place, you must write 'length = $7\cdot0$ cm'.

Exercise 21 Ⓜ

1 Round to the nearest hundred.

   a 5632   **b**   644   **c** 28 245
   d   86   **e** 11 352   **f**   174

2 Round to the nearest whole number.

   a $18\cdot32$ **b** $224\cdot9$ **c** $3\cdot511$

3 Round to the number of decimal places indicated.

   a $0\cdot672$ (1 dp)   **b** $8\cdot814$ (2 dp)   **c** $0\cdot7255$ (3 dp)
   d $1\cdot1793$ (2 dp)   **e** $0\cdot863$ (1 dp)   **f** $8\cdot2222$ (2 dp)
   g $0\cdot07518$ (3 dp)   **h** $11\cdot7258$ (3 dp)   **i** $20\cdot154$ (1 dp)
   j $6\cdot6666$ (2 dp)   **k** $0\cdot342$ (1 dp)   **l** $0\cdot07248$ (4 dp)

4 Round to the number of significant figures indicated.

    a 2·658 (2 sf)   **b** 188·79 (3 sf)   **c** 2·87 (1 sf)   **d** 0·3569 (2 sf)

    e 1·7231 (2 sf)   **f** 0·041 551 (3 sf)   **g** 0·0371 (1 sf)   **h** 811·1 (1 sf)

    i 9320 (2 sf)   **j** 8·051 (2 sf)   **k** 6·0955 (3 sf)   **l** 8·205 (1 sf)

Common sense

- When you find the answer to a problem, you should choose the degree of accuracy appropriate for that situation.

a Suppose you were calculating how much tax someone should pay in a year and your actual answer was £2153·6752. It would be sensible to give the answer as £2154 to the nearest pound.

b In calculating the average speed of a car journey from Bristol to Cardiff, it would not be realistic to give an answer of 38·241 km/h. A more sensible answer would be 38 km/h.

1.6 Ratio and proportion

1.6.1 Ratio

- The word 'ratio' is used to describe a fraction. If the **ratio** of a boy's height to his father's height is 4 : 5, then he is $\frac{4}{5}$ as tall as his father.

EXAMPLE

a Express each ratio in its simplest form.

    i 10 : 4 = 5 : 2 (÷2)   **ii** 25 : 40 = 5 : 8 (÷5)

b Change the ratio 2 : 5 into the form

    i 1 : n **ii** m : 1

    i $2 : 5 = 1 : \dfrac{5}{2}$ **ii** $2 : 5 = \dfrac{2}{5} : 1$

          $= 1 : 2 \cdot 5$ $= 0 \cdot 4 : 1$

c Divide £60 between two people, A and B, in the ratio 5 : 7.

    Consider £60 as 12 equal parts (that is, 5 + 7). Then A receives 5 parts and B receives 7 parts.

    So   A receives $\dfrac{5}{12}$ of £60 = £25

              B receives $\dfrac{7}{12}$ of £60 = £35

Check: £25 + £35 = £60

Exercise 22 Ⓜ ⭐

1 Express each ratio in its simplest form.
   a 9:6   **b** 15:25   **c** 10:40   **d** 30:12
   e 18 to 24   **f** 40 to 25   **g** 9:6:12   **h** 18:12:30

2 Express these ratios in the form $1:n$.
   a 5:30   **b** 2:100   **c** 5:8   **d** 4:3

3 Express these ratios in the form $n:1$.
   a 12:5   **b** 5:2   **c** 4:5

In questions **4** to **9**, divide the quantity in the ratio given.
   4 £40; (3:5)   **5** £120; (3:7)   **6** £98; (5:2)
   7 £8·40; (1:3)   **8** 180 kg; (1:5:6)   **9** 184 minutes; (2:3:3)

10 When £143 is divided in the ratio 2 : 4 : 5, what is the difference
    between the largest share and the smallest share?

11 If $\frac{5}{8}$ of the students in a school are boys, what is the ratio of boys to girls?

12 Find the ratio (shaded area) : (unshaded area) for each diagram.

   a   **b**    **c**

13 A man and a woman share a bingo prize of £1000 between them
    in the ratio 1 : 4. The woman shares her part between herself, her
    mother and her daughter in the ratio 2 : 1 : 1.
    How much does her daughter receive?

14 A man and his wife share a sum of money in the ratio 3 : 2. If the
    sum of money is doubled, in what ratio should they divide it so
    that the man still receives the same amount?

15 In a herd of x cattle, the ratio of the number of bulls to cows is 1 : 6.
    Find the number of bulls in the herd in terms of x.

16 If $x : 3 = 12 : x$, calculate the positive value of x.

❓ 17 £400 is divided between Ann, Brian and Carol so that Ann has
    twice as much as Brian, and Brian has three times as much as
    Carol. How much does Brian receive?

18 A cake weighing 550 g has three ingredients: flour, sugar and raisins. There is twice as much flour as sugar and one and a half times as much sugar as raisins. How much flour is here?

*19 A brother and sister share out their collection of 5000 stamps in the ratio 5 : 3. The brother then shares his stamps with two friends in the ratio 3 : 1 : 1, keeping most for himself. How many stamps do each of his friends receive?

1.6.2   Map scales

> EXAMPLE
>
> A map is drawn to a scale of 1 to 50 000.
> Calculate the length of a road which appears as 3 cm long on the map.
> ....
> 1 cm on the map is equivalent to 50 000 cm on the earth.
> So 3 cm ≡ 3 × 50 000 = 150 000 cm
> $$= 1500 \, m$$
> $$= 1 \cdot 5 \, km$$
> The road is 1·5 km long.

Exercise 23 Ⓜ ⭐

1 On a map of scale 1 : 100 000, the distance between Tower Bridge and Hammersmith Bridge is 12·3 cm.
What is the actual distance in km?

2 On a map of scale 1 : 15 000, the distance between Buckingham Palace and Brixton Underground Station is 31·4 cm.
What is the actual distance in km?

3 If the scale of a map is 1 : 10 000, what will be the length on this map of a road which is 5 km long?

4 The distance from Hertford to St Albans is 32 km.
How far apart will they be on a map of scale 1 : 50 000?

5 The 17th hole at the famous St Andrews golf course is 420 m in length.
How long will it appear on a plan of the course of scale 1 : 8000?

6 The scale of a map is 1 : 1000. What are the actual dimensions of a rectangle which appears as 4 cm by 3 cm on the map? What is the area on the map in cm²? What is the actual area in m²?

7 The scale of a map is $1:100$. What area does $1\,cm^2$ on the map represent? What area does $6\,cm^2$ represent?

8 The scale of a map is $1:20\,000$. What area does $8\,cm^2$ represent?

9 This is a plan of Mr. Cooper's house and gardens. It has been drawn to a scale of 1 cm for every 2 m.

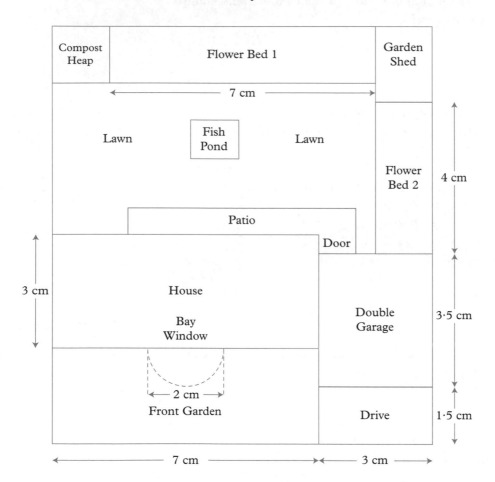

a How wide is:
  i the front garden
  ii the drive
  iii the bay window?
b How long is flower bed 2?
c How wide is flower bed 1?
d If the fish pond is 4 m wide, what size should it be on the plan?
e Measure carefully the width of the patio on the plan. How many *metres* wide is the real patio?
f What is the real *area* of the double garage?

1.6.3 Proportion

a If 9 litres of petrol costs £5·76, find the cost of 20 litres.

b If five men can paint a bridge in 12 days, how long would it take three men?

....

a The cost of petrol is **directly** proportional to the quantity bought.

9 litres costs £5·76
So 1 litre costs
  £5·76 ÷ 9 = £0·64
So 20 litres costs
  £0·64 × 20 = £12·80

b The length of time required to paint the bridge is **inversely** proportional to the number of men who paint.

5 men take 12 days.
  (5 × 12 = 60)
So 1 man takes 60 days.
  (1 × 60 = 60)
So 3 men take 20 days.
  (3 × 20 = 60)

> In part **a** find the cost of **one** litre of petrol.
> In part **b** find the time needed for **one** man.

Exercise 24 Ⓜ

Decide if each question uses **direct** or **inverse** proportion before you start.

1 If 7 CDs cost £1·54, find the cost of 5 CDs.

2 Twelve people are needed to harvest a crop in 2 hours. How long would it take 4 people?

3 Three men build a wall in 10 days. How long would it take five men?

4 Nine milk bottles contain $4\frac{1}{2}$ litres of milk between them. How much do five bottles hold?

5 A car uses 10 litres of petrol in 75 km. How far will it go on 8 litres?

6 A wire 11 cm long has a mass of 187 g. What is the mass of 7 cm of this wire?

7 A train travels 30 km in 120 minutes. How long will it take to travel 65 km at the same speed?

8 A ship has sufficient food to supply 600 passengers for 3 weeks. How long would the food last for 800 people?

9 Usually it takes 12 hours for 5 men to tarmac a road. How many men are needed to tarmac the same road in 10 hours?

10 80 machines can produce 4800 identical pens in 5 hours.
At this rate
  a how many pens would one machine produce in one hour?
  b how many pens would 25 machines produce in 7 hours?

1.7 Negative numbers

1.7.1 Adding and subtracting negative numbers

For adding and subtracting use a number line.

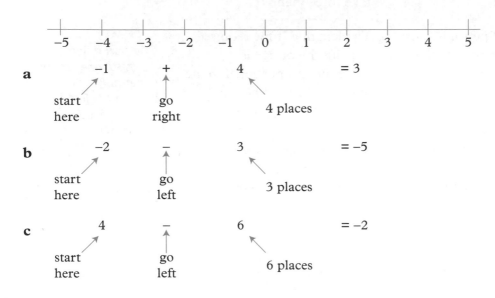

a start here -1 go right $+$ 4 places 4 $= 3$

b start here -2 go left $-$ 3 places 3 $= -5$

c start here 4 go left $-$ 6 places 6 $= -2$

Exercise 25 (M)/(L)

Work out

1 $-6 + 2$	**2** $-7 - 5$	**3** $-3 - 8$	**4** $-5 + 2$
5 $-6 + 1$	**6** $8 - 4$	**7** $4 - 9$	**8** $11 - 19$
9 $4 + 15$	**10** $-7 - 10$	**11** $16 - 20$	**12** $-7 + 2$
13 $-6 - 5$	**14** $10 - 4$	**15** $-4 + 0$	**16** $-6 + 12$
17 $-7 + 7$	**18** $2 - 20$	**19** $8 - 11$	**20** $-6 - 5$
21 $-8 - 4$	**22** $-3 + 7$	**23** $-6 + 10$	**24** $-5 + 5$
25 $-11 + 3$	**26** $7 - 10$	**27** $-5 + 8$	**28** $-12 + 0$
29 $-1 + 19$	**30** $20 - 25$	**31** $-6 - 60$	**32** $-2 + 100$

● When you have two (+) or (−) signs together use these rules.

$+ \ + = +$ $+ - = -$
$- \ - = +$ $- + = -$

$3 - (-6) = 3 + 6 = 9$
$-4 + (-5) = -4 - 5 = -9$
$-5 - (+7) = -5 - 7 = -12$

Exercise 26 Ⓜ/Ⓛ

Work out
1 $-3 + (-5)$ 2 $-5 - (+2)$ 3 $4 - (+3)$ 4 $-3 - (-4)$
5 $6 - (-3)$ 6 $16 + (-5)$ 7 $-4 + (-4)$ 8 $20 - (-22)$
9 $-6 - (-10)$ 10 $95 + (-80)$ 11 $-3 - (+4)$ 12 $-5 - (+4)$
13 $6 + (-7)$ 14 $-4 + (-3)$ 15 $-7 - (-7)$ 16 $3 - (-8)$
17 $-8 + (-6)$ 18 $7 - (+7)$ 19 $12 - (-5)$ 20 $9 - (+6)$
21 $-3 - (-2)$ 22 $8 + (-11)$ 23 $10 - (-2)$ 24 $-7 + (-2)$
25 $9 - (+6)$ 26 $7 + (-7)$ 27 $0 - (-8)$ 28 $-6 - (-8)$

29 Copy and complete these addition squares.

a

+	−5	1	6	−2
3	−2	4		
−2				
6				
−10				

b

+			−4	
5			1	
		0		5
	7		6	17
		−4		

1.7.2 Multiplying and dividing negative numbers

A directed number is a positive or a negative number.

- When two directed numbers with the same sign are multiplied together, the answer is positive.

$+7 \times (+3) = +21$
$-6 \times (-4) = +24$

- When two directed numbers with different signs are multiplied together, the answer is negative.

$-8 \times (+4) = -32$
$+7 \times (-5) = -35$

When dividing directed numbers, the rules are the same as in multiplication.
$-70 \div (-2) = +35$
$+12 \div (-3) = -4$
$-20 \div (+4) = -5$

Exercise 27 Ⓜ/Ⓛ

1 $-3 \times (+2)$	**2** $-4 \times (+1)$	**3** $+5 \times (-3)$	**4** $-3 \times (-3)$
5 $-4 \times (+2)$	**6** $-5 \times (+3)$	**7** $6 \times (-4)$	**8** $3 \times (+2)$
9 $-3 \times (-4)$	**10** $6 \times (-3)$	**11** $-7 \times (+3)$	**12** $-5 \times (-5)$
13 $6 \times (-10)$	**14** $-3 \times (-7)$	**15** $8 \times (+6)$	**16** $-8 \times (+2)$
17 $-7 \times (+6)$	**18** $-5 \times (-4)$	**19** $-6 \times (+7)$	**20** $11 \times (-6)$
21 $8 \div (-2)$	**22** $-9 \div (+3)$	**23** $-6 \div (-2)$	**24** $10 \div (-2)$

25 $-12 \div (-3)$	**26** $-16 \div (+4)$	**27** $4 \div (-1)$	**28** $8 \div (-8)$
29 $16 \div (-8)$	**30** $-20 \div (-5)$	**31** $-16 \div (+1)$	**32** $18 \div (-9)$
33 $36 \div (-9)$	**34** $-45 \div (-9)$	**35** $-70 \div (+7)$	**36** $-11 \div (-1)$

37 $-16 \div (-1)$	**38** $1 \div \left(-\frac{1}{2}\right)$	**39** $-2 \div \left(+\frac{1}{2}\right)$	**40** $50 \div (-10)$
41 $-8 \times (-8)$	**42** $-9 \times (+3)$	**43** $10 \times (-60)$	**44** $-8 \times (-5)$
45 $-12 \div (-6)$	**46** $-18 \times (-2)$	**47** $-8 \div (+4)$	**48** $-80 \div (+10)$

49 Copy and complete these multiplication squares.

a

$\times$	4	-3	0	-2
-5				
2				
10				
-1				

b

$\times$			-1	
3			-3	
		-15		18
	-14		-7	-42
		10		

Questions on negative numbers are more difficult when the different sorts are mixed together. The remaining questions are four short tests.

Test 1

1 $-8 - 8$	**2** $-8 \times (-8)$	**3** -5×3	**4** $-5 + 3$
5 $8 - (-7)$	**6** $20 - 2$	**7** $-18 \div (-6)$	**8** $4 + (-10)$
9 $-2 + 13$	**10** $+8 \times (-6)$	**11** $-9 + (+2)$	**12** $-2 - (-11)$
13 $-6 \times (-1)$	**14** $2 - 20$	**15** $-14 - (-4)$	**16** $-40 \div (-5)$
17 $5 - 11$	**18** -3×10	**19** $9 + (-5)$	**20** $7 \div (-7)$

Test 2

1 $-2 \times (+8)$	**2** $-2 + 8$	**3** $-7-6$	**4** $-7 \times (-6)$
5 $+36 \div (-9)$	**6** $-8 - (-4)$	**7** $-14 + 2$	**8** $5 \times (-4)$
9 $11 + (-5)$	**10** $11 - 11$	**11** $-9 \times (-4)$	**12** $-6 + (-4)$
13 $3 - 10$	**14** $-20 \div (-2)$	**15** $16 + (-10)$	**16** $-4 - (+14)$
17 $-45 \div 5$	**18** $18 - 3$	**19** $-1 \times (-1)$	**20** $-3 - (-3)$

Test 3

1 $-10 \times (-10)$	**2** $-10 - 10$	**3** $-8 \times (+1)$	**4** $-8 + 1$
5 $5 + (-9)$	**6** $15 - 5$	**7** $-72 \div (-8)$	**8** $-12 - (-2)$
9 $-1 + 8$	**10** $-5 \times (-7)$	**11** $-10 + (-10)$	**12** $-6 \times (-3 +4)$
13 $6 - 16$	**14** $-42 \div (+6)$	**15** $-13 + (-6)$	**16** $-8 - (-7)$
17 $5 \times (-1)$	**18** $2 - 15$	**19** $21 + (-21)$	**20** $-16 \div (-2)$

Test 4

Write down each statement and find the missing number.

1 $(-6) \times \square = 30$　　**2** $\square + (-2) = 0$　　**3** $\square \div (-2) = 10$

4 $\square - (-3) = 7$　　**5** $(-1) \times \square = \frac{1}{2}$　　**6** $(-2) - \square = 3$

7 $(-2) \div \square = -2$　　**8** $\square - 8 = -6$　　**9** $(-12) \times \left(-\frac{1}{2}\right) = \square$

10 $\square + (-10) = 2$　　**11** $6 \times \square = 0$　　**12** $\square - (-8) = 0$

13 $(-1)^4 = \square$　　**14** $0 \cdot 2 \times \square = -200$　　**15** $(-1)^{13} = \square$

1.8 Mixed numerical problems 1

Exercise 28 Ⓜ

1 I have lots of 1p, 2p, 3p and 4p stamps. How many different combinations of stamps can I make which total 5p?

2 Find n if
$$8 + 9 + 10 + \ldots + n = 5^3$$

3 Copy and complete.
$$3^2 + 4^2 + 12^2 = 13^2$$
$$5^2 + 6^2 + 30^2 = 31^2$$
$$6^2 + 7^2 + \quad = \quad$$
$$10^2 + \quad + \quad = \quad$$

4 The pattern 23456 23456 23456 ... is continued to form a number with one hundred digits. What is the sum of all one hundred digits?

5 The total mass of a jar one-quarter full of jam is 250 g. The total mass of the same jar three-quarters full of jam is 350 g. What is the mass of the empty jar?

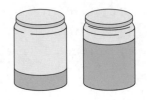

 6 A piece of wire 48 cm long is bent to form a rectangle in which the length is twice the width. Find the area of the rectangle.

7 Evaluate
$$\frac{1}{3} \times \frac{2}{4} \times \frac{3}{5} \times \ldots \times \frac{9}{11} \times \frac{10}{12}.$$

8 Find the smallest positive integer n for which $(829\,642 + n)$ is exactly divisible by 7.

9 Copy and find all the missing digits in these multiplications.

a
$$\begin{array}{r} 5\,\square \\ 9 \times \\ \hline \square\,\square\,6 \end{array}$$

b
$$\begin{array}{r} \square\,7 \\ \square \times \\ \hline 4\,\square\,6 \end{array}$$

c
$$\begin{array}{r} 5\,\square \\ \square \times \\ \hline 1\,\square\,4 \end{array}$$

10 a Work out $\frac{1}{4} + \frac{1}{12}$ as a single fraction in its lowest terms.

    b Find integers a and b such that $\frac{1}{a} + \frac{1}{b} = \frac{5}{8}$.

Exercise 29 Ⓜ

1 Copy and complete this multiplication square.

2 Look at this number pattern.

$$7^2 = 49$$
$$67^2 = 4489$$
$$667^2 = 444\,889$$
$$6667^2 = 44\,448\,889$$

    a Write the next line of the pattern.
    b Use the pattern to work out $6\,666\,667^2$.
    c What is the square root of $4\,444\,444\,488\,888\,889$?

3 Find a pair of positive integers a and b for which
    a $18a + 65b = 1865$
    b $23a + 7b = 2314$

$\times$	$\frac{2}{3}$		
$\frac{1}{2}$		$\frac{3}{8}$	
		$\frac{3}{16}$	
$\frac{2}{5}$			$\frac{2}{25}$

4 Copy these numbers. If the number in the space is prime, write
PRIME next to it. If it is not prime, write it as the product of its
prime factors.
The first two have been done for you.

47 ..`PRIME`..   40 ....   63 ....

26 ..`2 × 13`..   25 ....   71 ....

5 Express 419 965 in terms of its prime factors. You may use a
calculator.

6 How many prime numbers are there between 120 and 130?

7 Find three consecutive square numbers whose sum is 149.

8 The diagrams show magic squares in which the sum of the
numbers in any row, column or diagonal is the same. Copy the
squares and find the value of x in each square.

a

	x	6
3		7
		2

b

4		5	16
x		10	
	7	11	2
1			13

9 Work out $100 - 99 + 98 - 97 + 96 - \ldots + 4 - 3 + 2 - 1$.

10 Choose the correct answer. The smallest three-digit product
of a one-digit prime and a two-digit prime is
  a 102   **b** 103   **c** 104   **d** 105   **e** 106

Exercise 30 Ⓜ

1 A group of 17 people share some money equally and each person
gets £415. What is the total amount of money they have shared?

2 The 'reciprocal' of 2 is $\frac{1}{2}$. The reciprocal of 7 is $\frac{1}{7}$. The reciprocal
of x is $\frac{1}{x}$.

Find the square root of the reciprocal of the square root of the
reciprocal of ten thousand.

3 Use a calculator to work out 11^2, 111^2 and 1111^2. Use your
answers to predict the values of $11\,111^2$ and $111\,111^2$.

4 Put four different numbers in the circles so that when you add the numbers at the end of each line you always get a square number.

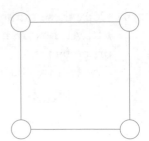

5 You are given that $41 \times 271 = 11\,111$. Work out these **in your head**.

   a 246×271

   b $22\,222 \div 271$

   c This time you can write down a (little!) working. Work out $41^2 \times 271$.

6 Copy and complete the crossnumber puzzle.

1	2		3
	4	5	
6			
		7	

Across

1 Square number

4 (1 across) × (3 down) − 7000

6 Next in the sequence
1, 3, 7, 15, 31, 63, 127, __

7 Sum of the prime numbers between 30 and 40

Down

2 South-east as a bearing

3 Number of days in 13 weeks

5 One-eighth of 6048

6 Cube number

7 S_1 is the sum of all the even numbers from 2 to 1000 inclusive. S_2 is the sum of all the odd numbers from 1 to 999 inclusive. Work out $S_1 - S_2$.

8 Pages 6 and 27 are on the same (double) sheet of a newspaper. What are the page numbers on the opposite side of the sheet? How many pages are there in the newspaper altogether?

9 Use the numbers 1, 2, 3, 4, 5, 6, 7, 8, 9 once each and in their natural order to obtain an answer of 100. You may use only the operations +, −, ×, ÷.

Exercise 31 Ⓜ

1 **a** The sum of the factors of n (including 1 and n) is 7. Find n.

   b The sum of the factors of p (including 1 and p) is 6. Find p.

2 A student in Mr Gibson's class bet him that he couldn't complete this magic square in less than 10 seconds. Mr Gibson eventually completed it in 47.4 seconds.

		1
−2		2
		−3

   a Solve the magic square and time yourself.

   b Did you beat Mr Gibson?

3 Two of the angles of an isosceles triangle are $x°$ and $(x + 9)°$.
Form equations to find two possible values of x.

4 The sum of edges of a cube is 72 cm.
Find the volume of the cube.

5 The digits 1, 9, 9, 4 are used to make fractions less than 1. Each of
the four digits must be used. For example $\dfrac{9}{419}$ or $\dfrac{4}{919}$

  a Write, smallest first, the three smallest fractions that can be made.
  b Write the largest fraction less than one that can be made.

6 a 411 fans went to a Radio One pop concert. The tickets cost £16 each.
    Calculate the total amount paid by the fans.
  b i All of these fans travelled on 52-seater coaches.
     What is the least number of coaches that could be used?
   ii On the return journey two coaches failed to turn up.
     How many fans had to be left behind?

7 Find the smallest positive whole number n for which
582 416 035 + n is exactly divisible by 11.

Exercise 32 Ⓜ Operator squares

Each empty square contains either a number or a mathematical
symbol (+ , − , × , ÷). Copy each square and fill in the details.

1

0·5	−	0·01	→	
		×		
	×		→	35
↓		↓		
4	÷	0·1	→	

2

	−	1·8	→	3·4
−		÷		
	×		→	
↓		↓		
	+	0·36	→	1·36

3

	×	30	→	21
×			−	
	−		→	29
↓		↓		
	−	49	→	

4

	×	−6	→	72
÷		+		
	+		→	
↓		↓		
	+	1	→	−2

5

	÷	4	→	
×		÷		
−2	+		→	6
↓		↓		
16	−		→	

6

	+	−7	→	20
			÷	
$\frac{1}{3}$			→	1
↓		↓		
9	+		→	−12

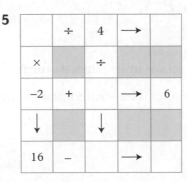

Test yourself

1 Linda works in a sandwich factory.

  a She makes 30 sandwiches every hour.
     How long does it take her to make 220 sandwiches?
     Give your answer in hours and minutes.

  b Linda's wage is £360 a week. She receives a 5% wage rise.
     Work out Linda's new weekly wage.

  c Linda makes cheese sandwiches and chicken sandwiches in the
     ratio 2:3. She makes 200 sandwiches altogether.
     How many of these are cheese sandwiches?

(OCR, 2009)

2 **a** You are given that $3x^3 = 648$
     Without using a calculator, find the value of x.

  b Write 240 as the product of its prime factors.

3 Paul and Kelly each buy a can of drink.

Volume	500 ml	Volume	330 ml
Sugar per can	35 g	Sugar per can	28 g

Paul drinks 100 ml of the Blackcurrant juice.
Kelly drinks 100 ml of the Fizzy orange.

Who drinks more sugar?
You **must** show your working.

(AQA, 2007)

4 **a** Work out $2\frac{3}{4} + 3\frac{2}{3}$
     Give your answer as a fraction in its simplest form.

  b **i** Which of these fractions can be written as a recurring decimal?

        $\frac{1}{2}$ $\frac{1}{3}$ $\frac{1}{4}$ $\frac{1}{5}$

    ii Explain your answer.

(Edexcel, 2007)

5 The answers to the following calculations have been rounded to the nearest whole number.

In each case the first digit of the answer is missing.

Use estimation to work out the missing first digit in each calculation.

Make clear the estimates you use.

a $8 \cdot 98^2 \times 2 \cdot 43 = \boxed{} 96$ **b** $\dfrac{402 \times 87}{47} = \boxed{} 44$

(OCR, 2003)

6 a A group of 17 people win £59 372 in a lottery.

They share the money equally between them.

Estimate how much money they will each receive.

Show how you worked out your estimate.

b Work out an **estimate** for the value of

$$\frac{51 \times 38}{0 \cdot 47}$$

Show how you worked out your estimate.

(OCR, 2004)

7 Mrs Jones inherits £12 000.

She divides the £12 000 between her three children Laura, Mark and Nancy in the ratio $7 : 8 : 9$, respectively.

How much does Laura receive?

(AQA, 2003)

8 Work out, without a calculator

a $-7 + 4$ **b** $(-2)^2 \times -3$ **c** $8 + (-11)$

d $(2 \times (-2))^2$ **e** $-4 - (-2)$ **f** $-12 \div 3$

9 The sizes of the interior angles of a quadrilateral are in the ratio $3 : 4 : 6 : 7$

Calculate the size of the largest angle.

(AQA, 2003)

10 Find the value of

a the cube of 4

b $0 \cdot 3 \times 0 \cdot 2$

c $3^2 \times 2^4$

d $8 \cdot 7 - 3 \cdot 24$

(WJEC, 2003)

Functional task 1

Starting a business

Ayesha is trying to start a new business to publish science books for schools and colleges. After some research, she has found out about the main costs involved in the business. In the first year she plans to publish six books.

Costs

- Designing the books at £13·20 per page
- Printing the books at £2·65 per book
- Payments to the authors of the book at 10% of the selling price of the book.
- Payments on a bank loan to start the business: £2160 per month
- Storage of books at a warehouse: £1400 per month
- Wages for office staff: £3200 per month.

Income

The income is from the sale of the books.

In the first year she plans to print 10 000 copies of each of the six books. Two of the books have 380 pages and the other four books have 406 pages.

She hopes to sell all the books at £7·95 each.

Task

Would she make a profit or loss if she sold all the books in her first year?

2 Number 2

In this unit you will:
- revise percentages, fractions and decimals
- calculate percentage increase and decrease
- calculate percentage profit and loss
- work out reverse percentages
- calculate compound interest
- learn how to use your calculator effectively
- learn how to write very large and very small numbers in standard form
- practise substituting numerical values into expressions and formulae
- do calculations with measures
- learn about lower and upper bounds.

Functional skills coverage and range:
- Understand and use simple equations and simple formulae involving one- or two-step operations
- Carry out calculations with numbers of any size in practical contexts, to a given number of decimal places
- Understand and use equivalences between fractions, decimals and percentages
- Use, convert and calculate using metric and, where appropriate, imperial measures.

Links
Percentages are used for many business and government calculations, including interest rates on bank accounts and tax calculations. Scientists use algebra and formulae to describe the world around us using mathematics, for example, how a population of bacteria increases with time, or how fast a stone will move when it is dropped from a roof.

2.1 Percentages

2.1.1 Percentages, fractions and decimals

Percentages are a convenient way of expressing fractions or decimals. You can change between fractions, decimals and percentages easily.

$$25\% = \frac{25}{100} = \frac{1}{4} = 0.25$$

To change a fraction or a decimal to a percentage you multiply by 100%.

EXAMPLE

Change

a $\dfrac{3}{8}$ to a percentage   **b** $\dfrac{7}{8}$ to a decimal   **c** $\dfrac{3}{11}$ to a decimal

d 0·35 to a fraction   **e** 6% to a decimal.

....

a $\dfrac{3}{8} = \left(\dfrac{3}{8} \times \dfrac{100}{1}\right)\%$

    $= 37\dfrac{1}{2}\%$

b $\dfrac{7}{8}$ means 'divide 8 into 7'

$\begin{array}{r} 0.875 \\ 8)\overline{7.000} \end{array}$, $\dfrac{7}{8} = 0.875$

c $\begin{array}{r} 0.\ 2\ 7\ 2\ 7\ 2\ 7 \dots \\ 11)\overline{3.^{3}0^{8}0^{3}0^{8}0^{3}0^{8}0} \dots \end{array}$

The calculation is never going to end.

You write $\dfrac{3}{11} = 0.\dot{2}\dot{7}$, which is a **recurring** decimal.

d $0.35 = \dfrac{\overset{7}{\cancel{35}}}{\underset{20}{\cancel{100}}} = \dfrac{7}{20}$ **e** $6\% = \dfrac{6}{100} = 0.06$

Exercise 1 Ⓜ

1 Change these fractions to decimals.

   a $\dfrac{1}{4}$ **b** $\dfrac{2}{5}$ **c** $\dfrac{3}{8}$ **d** $\dfrac{5}{12}$ **e** $\dfrac{1}{6}$ **f** $\dfrac{2}{7}$

2 Change these decimals to fractions and simplify.

   a 0·2   **b** 0·45   **c** 0·36   **d** 0·125   **e** 1·05   **f** 0·007

3 Change to percentages.

   a $\dfrac{1}{4}$ **b** $\dfrac{1}{10}$ **c** 0·72   **d** 0·075   **e** 0·02   **f** $\dfrac{1}{3}$

4 What percentage of this shape is shaded?

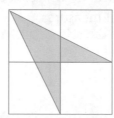

In questions **5** to **10** change the fractions to decimals and then do the calculations, giving your answer to 2 decimal places.

5 $\dfrac{1}{4} + \dfrac{1}{3}$ **6** $\dfrac{2}{3} + 0.75$ **7** $\dfrac{8}{9} - 0.24$

8 $\dfrac{7}{8} - \dfrac{5}{9}$ **9** $\dfrac{1}{3} \times 0.2$ **10** $\dfrac{5}{8} \times 1.1$

11 What percentage of the integers from 1 to 10 inclusive are prime numbers?

1 is **not** prime.

12 Copy and complete the table.

	a	b	c	d	e	f
Fraction	$\frac{3}{4}$			$\frac{1}{1000}$		$\frac{1}{3}$
Decimal		0·2				
Percentage			64%		2%	

In questions **13** to **16** arrange the numbers in order of size (smallest first).

13 $\frac{1}{2}$, 45%, 0·6

14 0·38, $\frac{6}{16}$, 4%

15 0·111, 11%, $\frac{1}{9}$

16 32%, 0·3, $\frac{1}{3}$

17 The pass mark in a physics test was 60%.
Sam got 23 out of 40. Did she pass?

18 At the end of 2008 the prison population was 48 700 men and 1600 women.
What percentage of the total prison population were men?

19 The table shows the results of a test for pulse rate conducted
on 274 children.

	Boys	Girls	Total
High pulse	58	71	129
Low pulse	83	62	145
Total	141	133	274

a What percentage of the boys had a low pulse rate?
b What percentage of the children with a high pulse rate were girls?

20 Pete reads a newspaper and then says,
'Great! Now prices will fall for a change'.
Explain whether or not you agree with
Pete's statement.

> **Inflation falls
> from 5% to 3·5%**

Recurring decimals Ⓗ

EXAMPLE

> This is how you write the recurring decimal 0·631 631 631 ... in
> the form $\frac{a}{b}$, where a and b are integers (whole numbers).
>
> Let $\quad\quad\quad\quad\quad\quad\quad r = 0·631\ 631\ 631 ...$
> Multiply by 1000 $\quad 1000r = 631·631\ 631\ 631 ...$
> Subtract $\quad\quad\quad\quad 999r = 631$
> $\quad\quad\quad\quad\quad$ So $r = \frac{631}{999}$

*21 Copy and complete this calculation to write $0.\dot{4}$ in the form $\frac{a}{b}$.

Let $\qquad\qquad r = 0.44444\ldots$
Multiply by 10 $\qquad 10r = 4.44444\ldots$
Subtract $\qquad\qquad 9r = \boxed{}$

So $\quad r = \dfrac{\boxed{}}{\boxed{}}$

> Fractions with denominators that have only prime factors of 2 and 5 do not change into recurring decimals, e.g. $\frac{3}{8}, \frac{2}{5}, \frac{7}{40}$

*22 Use the method of question **21** to write these numbers in the form $\frac{a}{b}$.

a $0.\dot{2}$ 　　b $0.\dot{7}$ 　　c $0.2\dot{9}$ 　　d $0.\dot{5}4\dot{1}$

2.1.2 Working out percentages

> **EXAMPLE**
>
> Work out 7% of £3200.
>
> Either $\quad$ 7% of £3200 $\qquad\qquad$ or $\quad$ 7% of £3200
>
> $= \dfrac{7}{100} \times \dfrac{3200}{1} = £224 \qquad\qquad = 0.07 \times 3200 = £224$

Exercise 2 Ⓜ

Work out

1 20% of £60	**2** 10% of £80	**3** 5% of £200
4 6% of £50	**5** 4% of £60	**6** 30% of £80
7 9% of £500	**8** 18% of £400	**9** 61% of £400
10 12% of £80	**11** 6% of $700	**12** 11% of $800
13 5% of 160 kg	**14** 20% of 60 kg	**15** 68% of 400 g
16 15% of 300 m	**17** 2% of 2000 km	**18** 71% of $1000
19 26% of 19 kg	**20** 1% of 6000 g	**21** 8·5% of £2400

> **EXAMPLE**
>
> Work out 6·5% of £17·50 correct to the nearest penny.
>
> $\dfrac{6\cdot5}{100} \times \dfrac{17\cdot5}{1} \quad = \dfrac{113\cdot75}{100} \quad$ or $\ldots \quad 6\cdot5\% = \dfrac{6\cdot5}{100} = 0.065$
>
> 6·5% of £17·50 $\quad = £1\cdot1375 \qquad\qquad 0.065 \times 17\cdot5 = 1\cdot1375$
> $\qquad\qquad\qquad = £1\cdot14$ (to nearest penny)

Exercise 3 (M)

Give answers to the nearest penny where necessary.

1 4·5% of £6·22   **2** 17% of £6·84   **3** 15% of £8·11

4 17% of £17·07   **5** 37% of £9·64   **6** 3·5% of £12·90

7 8% of £11·64   **8** 68% of £54·45   **9** 73% of £23·24

10 2·5% of £15·20   **11** 6·3% of £12·50   **12** 8·2% of £19·50

13 87% of £15·40   **14** 80% of £62·50   **15** 12% of £24·50

16 $12\frac{1}{2}$% of £88·50   **17** $7\frac{1}{2}$% of £16·40   **18** $5\frac{1}{2}$% of £80

19 $12\frac{1}{2}$% of £90   **20** 19% of £119·50   **21** 8·35% of £110

> **EXAMPLE**
>
> The price of a car costing £8400 is increased by 5%.
> Find the new price.
>
> ··
>
> Either 5% of £8400 = $\dfrac{5}{100} \times 8400$ or   New price = 105% of
> $$= £420$$
> $$\text{New price} = £8400 + £420$$
> $$= £8820$$
>   £8400
> $$= 1·05 \times 8400$$
> $$= £8820$$

> 1·05 is a **percentage multiplier**.

Exercise 4 (M)

1 Increase a price of £60 by 5%   **2** Reduce a price of £800 by 8%

3 Reduce a price of £82·50 by 6%   **4** Increase a price of £65 by 60%

5 Reduce a price of £2000 by 2%   **6** Increase a price of £440 by 80%

7 Increase a price of £66 by 100%   **8** Reduce a price of £91·50 by 50%

9 Increase a price of £88·24 by 25%   **10** Reduce a price of £63 by $33\frac{1}{3}$%

In these questions give the answers to the nearest penny.

11 Increase a price of £8·24 by 46%   **12** Increase a price of £7·65 by 24%

13 Increase a price of £5·61 by 31%   **14** Reduce a price of £8·99 by 22%

15 Increase a price of £11·12 by 11%   **16** Reduce a price of £17·62 by 4%

17 Increase a price of £28·20 by 13%   **18** Increase a price of £8·55 by $5\frac{1}{2}$%

19 Reduce a price of £9·60 by $7\frac{1}{2}$%   **20** Increase a price of £12·80 by $10\frac{1}{2}$%

Exercise 5 Ⓜ ★

1 Calculate
    a 30% of £50   **b** 45% of 2000 kg
    c 4% of $70   **d** 2·5% of 5000 people

2 In a sale, a jacket costing £40 is reduced by 20%. What is the sale price?

3 In July 2009, 360 000 people visited Bali for their holiday.
    a One-eighth of the people were American.
       Find the number of American vistors.
    b 11% of the people were French. How many people was that?
    c There were 12 000 people from Japan. What fraction of the total were from Japan?

4 In peeling potatoes 4% of the mass of the potatoes is lost as 'peel'. How much is **left** for use from a bag containing 55 kg?

5 Copy these and write the missing number, as a **decimal**.
    a 53% of 650 = ☐ × 650   **b** 3% of 2600 = ☐ × 2600
    c 8·5% of 700 = ☐ × 700   **d** 122% of 285 = ☐ × 285

6 Work out these, to the nearest penny.
    a 6·4% of £15·95   **b** 11·2% of £192·66
    c 8·6% of £25·84   **d** 2·9% of £18·18

7 Find the total bill.
    5 golf clubs at £18·65 each
    60 golf balls at £16·50 per dozen
    1 bag at £35·80
  V.A.T. at $17\frac{1}{2}$% is added to the total cost.

8 In 2008 a club had 250 members who each paid a £95 annual subscription. In 2009 the membership increased by 4% and the annual subscription was increased by 6%. What was the total income from subscriptions in 2009?

9 The cash price for a car was £7640. Alan Elder bought the car on the following hire purchase terms: 'A deposit of 20% of the cash price and 36 monthly payments of £191·60.' Calculate the total amount Alan paid.

10 A quarterly telephone bill consists of £19·15 rental plus 4·7p for each dialled unit. V.A.T. is added at $17\frac{1}{2}$%. What is the total bill for Mrs Jones who used 915 dialled units?

11 A motor bike costs £820. After a 12% increase the new price is
112% of £820. The 'quick' way to work this out is as follows:

New price = 112% of £820
        = 1·12 × 820
        = £918·40

A lorry costs £6500. The price is increased by 5%.
Use this quick method to find the new price.

12 Find the new prices of these items.

Item	Old price	Price change
Video	£190	6% increase
House	£150 000	11% increase
Boat	£2500	8% increase
Tree	£210	5% decrease
Phone	£65	15% decrease

13 Over a period of 6 months, a colony of rabbits increased in
number by 25% and then by a further 30%. If there were
originally 200 rabbits in the colony how many were there at the end?

14 In the last two weeks of a sale, prices are reduced first by 30% and
then by a **further** 40% of the new price. What is the final sale
price of a shirt which originally cost £15?

15 Work out

    a   8% of 3·2 kg (in grams)   **b**   16% of £4·50 (in pence)

    c   28% of 5 cm (in mm)   **d**   5% of 10 hours (in minutes)

    e   12% of 2 minutes (in seconds)

*16 A 12% increase in the value of N, followed by a 12% decrease in the
new value can be calculated as $1·12 \times N \times 0·88$. Copy and complete.

    a   An 8% increase in the value of P, followed by a 10% decrease

        in the new value can be calculated as ☐ .

    b   A ☐ increase in the value of X, followed by a further ☐
        increase in value can be calculated as $1·15 \times X \times 1·06$.

    c   A 25% increase in the value of Q, followed by a 5% decrease

        in the new value can be calculated as ☐ .

2.1.3 Percentage profit/loss

- Percentage profit = $\dfrac{\text{actual profit}}{\text{original price}} \times \dfrac{100}{1}\%$

- Percentage loss = $\dfrac{\text{actual loss}}{\text{original price}} \times \dfrac{100}{1}\%$

EXAMPLE

A radio is bought for £16 and sold for £20.
What is the percentage profit?

Actual profit $= £4$

So percentage profit $= \dfrac{\text{actual profit}}{\text{original price}} \times 100\%$

$= \dfrac{4}{16} \times \dfrac{100}{1}\% = 25\%$

The radio is sold at a 25% profit.

- Percentage increase = $\dfrac{\text{actual increase}}{\text{original price}}$

- Percentage decrease = $\dfrac{\text{actual decrease}}{\text{original price}}$

Exercise 6 Ⓜ

In questions **1** to **10** calculate the percentage increase.

	Original price	Final price
1	£50	£54
2	£80	£88
3	£180	£225
4	£100	£102
5	£75	£78
6	£400	£410
7	£5000	£6000
8	£210	£315
9	£600	£690
10	$4000	$7200

In questions **11** to **20** calculate the percentage decrease.

	Original price	Final price
11	£800	£600
12	£50	£40
13	£120	£105
14	£420	£280
15	£6000	£1200
16	$880	$836
17	$15 000	$14 100
18	$7·50	$6·00
19	£8·20	£7·79
20	£16 000	£15 600

Exercise 7 Ⓜ ⭐

1 The first figure is the cost price and the second figure is the selling price. Calculate the percentage profit or loss in each case.

a £20, £25   **b** £400, £500
c £60, £54   **d** £9000, £10 800
e £460, £598   **f** £512, £550·40

2 A car dealer buys a car for £500, gives it a clean, and then sells it for £640. What is the percentage profit?

3 A damaged carpet which cost £180 when new, is sold for £100. What is the percentage loss?

4 During the first four weeks of her life, a baby girl increases her weight from 3·2 kg to 4·7 kg. What percentage increase does this represent? (Give your answer to 3 sf.)

5 When V.A.T. is added to the cost of a car tyre, its price increases from £16·50 to £18·48. What is the rate at which V.A.T. is charged?

6 In order to increase the chance of a sale, the price of a country estate is reduced from £30 000 000 to £28 400 000. What percentage reduction is this?

7 A picture has the dimensions shown.

10 cm

20 cm

Calculate the percentage increase in the area of the picture after both the length and width are increased by 20%.

8 A box has a square base of side 30 cm and a height of 20 cm. Calculate the percentage increase in the volume of the box after the length and width of the base are both increased by 10% and the height is increased by 5%.

9 The rental for a television set changed from £80 per year to £8 per month. What is the percentage increase in the yearly rental?

❓ *10 If an employer reduces the working week from 40 hours to 35 hours, with no loss of weekly pay, calculate the percentage increase in the hourly rate of pay.

***11** Given that $G = ab$, find the percentage increase in G when both a and b increase by 10%.

*12 The formula connecting P, a and n is $\boxed{P = an^3}$

    a Calculate the value of P when $a = 200$ and $n = 5$.

    b Calculate the percentage increase in P when a is increased by 2% and n is increased by 20%.

2.1.4 Reverse percentages

> EXAMPLE
>
> After an increase of 6%, the price of a motor bike is £9010. What was the price before the increase?
>
> ....
>
> 106% of old price = £9010
>
> 1% of old price $= \dfrac{9010}{106}$
>
> 100% of old price $= \dfrac{9010}{106} \times 100\%$
>
> Original price $= £8500$

A common mistake here is to work out 6% of £9010. This is wrong because the increase is 6% of the **old** price, not 6% of the new price.

> EXAMPLE
>
> To increase sales, the price of a magazine is reduced by 5%. Find the original price if the new price is 190p.
>
> ....
>
> 95% of old price $= 190p$
>
> 1% of old price $= \dfrac{190}{95}$
>
> 100% of old price $= \dfrac{190}{95} \times 100\%$
>
> Original price $= 200p$

This is a reduction, so the new price is 95% of the old price.

Exercise 8 Ⓜ/Ⓗ

1 After an increase of 8%, the price of a car is £6696. Find the price of the car before the increase.

2 After a 12% pay rise, Jo Brown's salary was £28 560. What was her salary before the increase?

3 Find the missing prices.

	Item	Old price	New price	Percentage change
a	Jacket	?	£55	10% increase
b	Dress	?	£212	6% increase
c	CD player	?	£56·16	4% increase
d	TV	?	£195	30% increase
e	Car	?	£3960	65% increase

4 Between 2000 and 2007 the population of an island fell by 4%. The population in 2007 was 201 600. Find the population in 2000.

5 After being ill for 3 months, Steve's weight went down by 12%. Find Steve's original weight if he weighed 74·8 kg after the illness.

6 The diagram shows two rectangles. The width and height of rectangle B are both 20% greater than the width and height of rectangle A.

Use the figures given to find the width and height of rectangle A.

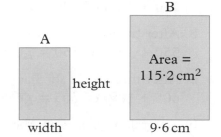

7 A heron has discovered Sergio's goldfish pond. It has eaten 70% of his collection of goldfish. If he has 60 survivors, how many did he have originally?

8 A television costs £376 including $17\frac{1}{2}$% V.A.T. How much of the cost is tax?

9 During a Grand Prix car race, the tyres on a car are reduced in weight by 3%. If they weigh 388 kg at the end of the race, how much did they weigh at the start?

10 The label on a carton of yoghurt is shown. The figure on the right is smudged out. Work out what the figure should be.

11 The average attendance at Alnwick City football club fell by 7% in 2009. If 2030 fewer people went to matches in 2009, how many went in 2004?

12 When an iron bar is heated it expands by 0·2%. If the increase in length is 1 cm, what was the original length of the bar?

13 An oven is sold for £600, thereby making a profit of 20% on the cost price. What was the cost price?

14 In 2009 Corpus plc had sales of £516 000 000 which was $7\frac{1}{2}$% higher than the figure for 2008. Calculate the value of sales in 2008.

2.1.5 Compound interest

A bank pays a fixed interest of 10% on money in deposit accounts.
Ann puts £500 in the bank. How much will she have after
a one year   **b** two years   **c** three years?

a After one year she has
500 + 10% of 500 = £550

b After two years she has
550 + 10% of 550 = £605   Check that this is $1 \cdot 10^2 \times 500$

c After three years she has
605 + 10% of 605 = £665·50   Check that this is $1 \cdot 10^3 \times 500$

In general, after n years the money in the bank will be £$(1 \cdot 10^n \times 500)$.

Exercise 9 Ⓜ/Ⓗ

1 A bank pays interest of 10% on money in deposit accounts.
Mrs Wells puts £2000 in the bank. How much does she
have after
    a one year   **b** two years   **c** three years?

2 A bank pays interest of 12%. Mr Olsen puts £5000 in the bank.
How much does he have after
    a one year   **b** three years?

3 A computer operator is paid £10 000 a year. Assuming her pay is
increased by 7% each year, what will her salary be in four years time?

4 Mrs Bergkamp's salary in 2007 is £30 000 per year. Every year her
salary is increased by 5%.
In 2008 her salary will be 30 000 × 1·05   = £31 500
In 2009 her salary will be 30 000 × 1·05 × 1·05   = £33 075
In 2010 her salary will be 30 000 × 1·05 × 1·05 × 1·05 = £34 728·75
And so on.
    a What will her salary be in 2011?
    b What will her salary be in 2012?

5 The price of a flat was £90 000. At the end of each year the price
has decreased by 5%.
    a Find the price of the flat after 1 year.
    b Find the price of the flat after 2 years.

6 Assuming an average inflation rate of 8%, work out the probable cost of the following items in 10 years' time.
  a car £8500   **b** TV £340   **c** house £500 000

7 A new bike is valued at £8000. At the end of each year its value is reduced by 15% of its value at the start of the year. What will it be worth after 3 years?

8 Twenty years ago a bus driver was paid £50 a week. He is now paid £185 a week. Assuming an average rate of inflation of 7%, has his pay kept up with inflation?

9 The population of an island increases by 10% each year. After how many years will the original population be doubled?

10 A bank pays interest of 11% on £6000 in a deposit account. After how many years will the money have trebled?

11 A tree grows in height by 21% per year. It is 2 m tall after one year. After how many more years will the tree be 20 m tall?

12 Which is the better investment over ten years
       £20 000 at 12% compound interest
   or £30 000 at 8% compound interest?

*13 **a** Draw the graph of $y = 1 \cdot 08^x$ for values of x from 0 to 10.
   b Solve approximately the equation $1 \cdot 08^x = 2$.
   c Money is invested at 8% interest. After how many years will the money have doubled?

x	1	2	3	$\cdots$	10
y	1·08			$\cdots$	

2.2 Using a calculator

2.2.1 Money and time on a calculator

● To work out £27·30 ÷ 7, key in ⬚2 ⬚7 ⬚. ⬚3 ⬚0 ⬚÷ ⬚7 ⬚=
  The answer is 3·9. **Remember** this means £3·90.

EXAMPLE

A machine takes 15 minutes to make one toy. How long will it take to make 1627 toys?

....

15 minutes is one quarter of an hour and $\frac{1}{4}$ = 0·25 as a decimal.

Key in 0·25   ×   1627   =

The answer is 406·75.
It will take 406·75 hours or 406 hours 45 minutes.

EXAMPLE

6 minutes is $\frac{6}{60}$ of an hour. $\frac{6}{60} = \frac{1}{10}$ = 0·1 hours.

Similarly 27 minutes = $\frac{27}{60}$ of an hour. $\frac{27}{60}$ = 0·45 hours.

And 11 minutes = $\frac{11}{60}$ of an hour = 0·183 hours (to 3 dp)

Exercise 10 Ⓜ

1 Work out these and give your answers in **pounds**.

a £1·22 × 5   b £153·60 ÷ 24   c £12·35 − £7·65
d 20p × 580   e 6p × 2155   f £10 ÷ 250

2 Write these time intervals in **hours** as decimals.

a 2 h 30 min   b 4 h 15 min   c 3 h 45 min
d 6 min   e 12 min   f 15 min
g 17 min   h 1 h 8 min   i 2 h 34 min

$6\,min = \frac{6}{60}$ hour
$= 0\cdot1$ hour
$15\,min = \frac{15}{60}$ hour
$= 0\cdot25$ hour

3 Work out these and give your answers in **hours**.

a 2 h 45 min × 9   b 3 h 15 min × 7   c 14 h 30 min ÷ 5
d 15 min × 11   e 30 min × 5   f 5 h 15 min ÷ 3

2.2.2 Order of operations

Sometimes you have to do a mixture of operations. To avoid uncertainty you must follow these rules.

a work out brackets first   **B**
b work out indices next   **I**
c work out ÷, × before +, −   **DMAS**

Remember
BIDMAS

Work out these using 'B I D M A S'.

a $5 + 8 \times 3 - 7$ **b** $99 \div (11 - 8)$ **c** $17 + 3^2$ **d** $\dfrac{5 \times 7 - 9}{13}$

• •

a $5 + 8 \times 3 - 7$ **b** $99 \div (11 - 8)$ **c** $17 + 3^2$ **d** $\dfrac{5 \times 7 - 9}{13}$

$= 5 + 24 - 7$ $= 99 \div 3$ $17 + 9$ $= \dfrac{26}{13} = 2$

$= 22$ $= 33$ $= 26$

Exercise 11 Ⓜ

Work out these without a calculator.

1 $8 + 7 \times 2$ **2** $8 - 1 \times 5$ **3** $17 - 4 \times 3$

4 $4 \times 5 - 8$ **5** $28 \div 7 - 4$ **6** $6 - 12 \div 4$

7 $25 - 7 \times 3 + 5$ **8** $8 + 15 \div 3 - 1$ **9** $2 \times 7 + 8 \div 2$

10 $50 + 6 \times 5 - 10$ **11** $12 \div (8 - 5)$ **12** $7 \times (2 + 3 + 5)$

13 $(8 + 3^2) - 11$ **14** $(7 - 2)^2 \times 10$ **15** $100 - (75 \div 25)$

16 $\dfrac{4 + 3 \times 2}{5}$ **17** $\dfrac{18 - 4^2}{4}$ **18** $7 + \dfrac{16}{4}$

19 $\dfrac{19 + 13}{7 - 3}$ **20** $11 - \dfrac{10}{20}$

Exercise 12 Ⓜ

Use a calculator and give the answers correct to one decimal place.

1 $2 \cdot 5 \times 1 \cdot 67$ **2** $19 \cdot 6 - 3 \cdot 7311$ **3** $0 \cdot 792^2$

4 $0 \cdot 13 + 8 \cdot 9 - 3 \cdot 714$ **5** $2 \cdot 4^2 - 1 \cdot 712$ **6** $5 \cdot 3 \times 1 \cdot 7 + 3 \cdot 7$

7 $0 \cdot 71 \times 0 \cdot 92 - 0 \cdot 15$ **8** $9 \cdot 6 \div 1 \cdot 72$ **9** $8 \cdot 17 - 1 \cdot 56 + 7 \cdot 4$

10 $\sqrt{4 \cdot 52}$ **11** $\sqrt{198}$ **12** $\sqrt{\dfrac{2 \cdot 63}{1 \cdot 9}}$

In questions **13** to **30** remember 'B I D M A S'.

13 $2 \cdot 5 + 3 \cdot 1 \times 2 \cdot 4$ **14** $7 \cdot 81 + 0 \cdot 7 \times 1 \cdot 82$ **15** $8 \cdot 73 + 9 \div 11$

16 $11 \cdot 7 \div 9 - 0 \cdot 74$ **17** $7 \div 0 \cdot 32 + 1 \cdot 15$ **18** $2 \cdot 6 + 5 \cdot 2 \times 1 \cdot 7$

19 $2 \cdot 9 + \dfrac{8 \cdot 3}{1 \cdot 83}$ **20** $1 \cdot 7^2 + 2 \cdot 62$ **21** $5 \cdot 2 + \dfrac{11 \cdot 7}{1 \cdot 85}$

22 $9 \cdot 64 + 26 \div 12 \cdot 7$ **23** $1 \cdot 27 + 3 \cdot 1^2$ **24** $4 \cdot 2^2 \div 9 \cdot 4$

25 $0 \cdot 151 + 1 \cdot 4 \times 9 \cdot 2$ **26** $1 \cdot 7^3$ **27** $8 \cdot 2 + 3 \cdot 2 \times 3 \cdot 3$

28 $3 \cdot 2 + \dfrac{1 \cdot 41}{6 \cdot 72}$ **29** $\dfrac{1 \cdot 9 + 3 \cdot 71}{2 \cdot 3}$ **30** $\dfrac{8 \cdot 7 - 5 \cdot 371}{1 \cdot 14}$

2.2.3 Using brackets

Most calculators have brackets buttons like these $\boxed{(}$ $\boxed{)}$.

When you press the right-hand bracket button you will see that the calculation inside the brackets has been performed. Try it. Don't forget to press the $\boxed{=}$ button at the end to give the final answer.

EXAMPLE

a $8\cdot72 - (1\cdot4 \times 1\cdot7)$ 　　　　　**b** $\dfrac{8\cdot51}{(1\cdot94 - 0\cdot711)}$

....

a $\boxed{8\cdot72}$ $\boxed{-}$ $\boxed{(}$ $\boxed{1\cdot4}$ $\boxed{\times}$

$\boxed{1\cdot7}$ $\boxed{)}$ $\boxed{=}$

$8\cdot72 - (1\cdot4 \times 1\cdot7) = 6\cdot3$ to 1 dp

b $\boxed{8\cdot51}$ $\boxed{\div}$ $\boxed{(}$ $\boxed{1\cdot94}$ $\boxed{-}$

$\boxed{0\cdot711}$ $\boxed{)}$ $\boxed{=}$

$\dfrac{8\cdot51}{(1\cdot94 - 0\cdot711)} = 6\cdot9$ to 1 dp

You can use the $\boxed{\text{ANS}}$ button as a 'short-term memory'. It holds the answer from the previous calculation.

EXAMPLE

Work out $\dfrac{5}{1\cdot2 - 0\cdot761}$, correct to 4 sf, using the $\boxed{\text{ANS}}$ button.

....

Find the bottom line first:

$\boxed{1\cdot2}$ $\boxed{-}$ $\boxed{0\cdot761}$ $\boxed{=}$ $\boxed{5}$ $\boxed{\div}$ $\boxed{\text{ANS}}$ $\boxed{=}$

The calculator reads $11\cdot38952164$

$\dfrac{5}{1\cdot2 - 0\cdot761} = 11\cdot39$ (to 4 sf)

Exercise 13 Ⓜ

Use a calculator to evaluate these, giving the answers to 4 significant figures. Use the brackets buttons or the $\boxed{\text{ANS}}$ button.

1 $\dfrac{7\cdot351 \times 0\cdot764}{1\cdot847}$ 　　**2** $\dfrac{0\cdot0741 \times 14\,700}{0\cdot746}$ 　　**3** $\dfrac{0\cdot0741 \times 9\cdot61}{23\cdot1}$

4 $\dfrac{417\cdot8 \times 0\cdot00841}{0\cdot07324}$ 　　**5** $\dfrac{8\cdot41}{7\cdot601 \times 0\cdot00847}$ 　　**6** $\dfrac{4\cdot22}{1\cdot701 \times 5\cdot2}$

7 $\dfrac{9\cdot61}{17\cdot4 \times 1\cdot51}$ 　　**8** $\dfrac{8\cdot71 \times 3\cdot62}{0\cdot84}$ 　　**9** $\dfrac{0\cdot76}{(0\cdot412 - 0\cdot317)}$

10 $\dfrac{81\cdot4}{(72\cdot6 + 51\cdot92)}$ 　　**11** $\dfrac{111}{27\cdot4 + 2960}$ 　　**12** $\dfrac{27\cdot4 + 11\cdot61}{5\cdot9 - 4\cdot763}$

13 $\dfrac{6 \cdot 51 - 0 \cdot 1114}{7 \cdot 24 + 1 \cdot 653}$ **14** $\dfrac{5 \cdot 71 + 6 \cdot 093}{9 \cdot 05 - 5 \cdot 77}$ **15** $\dfrac{0 \cdot 943 - 0 \cdot 788}{1 \cdot 4 - 0 \cdot 766}$

16 $\dfrac{2 \cdot 6}{1 \cdot 7} + \dfrac{1 \cdot 9}{3 \cdot 7}$ **17** $\dfrac{8 \cdot 06}{5 \cdot 91} - \dfrac{1 \cdot 594}{1 \cdot 62}$ **18** $\dfrac{4 \cdot 7}{11 \cdot 4 - 3 \cdot 61} + \dfrac{1 \cdot 6}{9 \cdot 7}$

2.2.4   Other useful buttons

To use a calculator efficiently you sometimes have to think ahead
and make use of the memory, inverse $1/x$ and $+/-$ buttons.
In the example below the buttons are:

$\sqrt{\ }$	square root	y^x	raises number y to the power x
x^2	square		
$1/x$ or x^{-1}	reciprocal	$\wedge$	raises to a power

> Your calculator may have different buttons to these. Refer to the manual or ask your teacher if you are unsure.

●   It is good practice to check your answers by estimation to see if the answer is 'about the right size'.

EXAMPLE

Evaluate these to 4 significant figures.

a $3 \cdot 5^2 + \dfrac{1}{0 \cdot 62}$ **b** $\left(\dfrac{1}{0 \cdot 084}\right)^4$ **c** $\sqrt[3]{[3 \cdot 2 \times (1 \cdot 7 - 1 \cdot 64)]}$

....

a | 3·5 | | x^2 | | + | | 0·62 | | $1/x$ | | = |

$3 \cdot 5^2 + \dfrac{1}{0 \cdot 62} = 13 \cdot 86$ (to 4 sf)

b | 0·084 | | $1/x$ | | ^ | | 4 | | = |

$\left(\dfrac{1}{0 \cdot 084}\right)^4 = 20\,090$ (to 4 sf)

c | 1·7 | | − | | 1·64 | | = | | × | | 3·2 | | = | | y^x | | 0·333333 | | = |

$\sqrt[3]{[3 \cdot 2 \times (1 \cdot 7 - 1 \cdot 64)]} = 0 \cdot 5769$ (to 4 sf)

> To find a cube root, raise to the power $\frac{1}{3}$ or as a decimal 0·333... Of course, if your calculator has a $\sqrt[3]{\ }$ button, use that.

Exercise 14 Ⓜ

Evaluate, correct to 4 significant figures

1 $2 \cdot 7^2 + 1 \cdot 7^2$

2 $\dfrac{1}{7 \cdot 2} - \dfrac{1}{5}$

3 $\dfrac{5 \cdot 89}{7 - 3 \cdot 83}$

4 $\dfrac{102}{58 \cdot 1 + 65 \cdot 32}$

5 $\dfrac{18 \cdot 8}{3 \cdot 72 \times 1 \cdot 86}$

6 $\dfrac{904}{65 \cdot 3 \times 2 \cdot 86}$

7 $12 \cdot 2 - \left(\dfrac{2 \cdot 6}{1 \cdot 95}\right)$

8 $8 \cdot 047 - \left(\dfrac{6 \cdot 34}{10 \cdot 2}\right)$

9 $14 \cdot 2 - \left(\dfrac{1 \cdot 7}{2 \cdot 4}\right)$

10 $\dfrac{9 \cdot 75 - 8 \cdot 792}{4 \cdot 31 - 3 \cdot 014}$

11 $\dfrac{19 \cdot 6 \times 3 \cdot 01}{2 \cdot 01 - 1 \cdot 958}$

12 $3 \cdot 7^2 - \left(\dfrac{8 \cdot 59}{24}\right)$

13 $8 \cdot 27 - 1 \cdot 56^2$

14 $111 \cdot 79 - 5 \cdot 04^2$

15 $18 \cdot 3 - 2 \cdot 841^2$

16 $(2 \cdot 93 + 71 \cdot 5)^2$

17 $(8 \cdot 3 - 6 \cdot 34)^4$

18 $54 \cdot 2 - 2 \cdot 6^4$

19 $(8 \cdot 7 - 5 \cdot 95)^4$

20 $\sqrt{68 \cdot 4} + 11 \cdot 63$

21 $9 \cdot 45 - \sqrt{8 \cdot 248}$

22 $3 \cdot 24^2 - \sqrt{1 \cdot 962}$

23 $\dfrac{3 \cdot 54 + 2 \cdot 4}{8 \cdot 47^2}$

24 $2065 - \sqrt{44\,000}$

25 $\sqrt{(5 \cdot 69 - 0 \cdot 0852)}$

26 $\sqrt{(0 \cdot 976 + 1 \cdot 03)}$

27 $\sqrt{\left(\dfrac{17 \cdot 4}{2 \cdot 16 - 1 \cdot 83}\right)}$

28 $\sqrt{\left(\dfrac{28 \cdot 9}{\sqrt{8 \cdot 47}}\right)}$

29 $257 - \dfrac{6 \cdot 32}{0 \cdot 059}$

30 $75\,000 - 5 \cdot 6^4$

31 $\dfrac{11 \cdot 29 \times 2 \cdot 09}{2 \cdot 7 + 0 \cdot 082}$

32 $85 \cdot 5 - \sqrt{105 \cdot 8}$

33 $\dfrac{4 \cdot 45^2}{8 \cdot 2^2 - 51 \cdot 09}$

34 $\left(\dfrac{8 \cdot 53 + 7 \cdot 07}{6 \cdot 04 - 4 \cdot 32}\right)^4$

35 $2 \cdot 75 + \dfrac{5}{8 \cdot 2} + \dfrac{11 \cdot 2}{4 \cdot 3}$

36 $8 \cdot 2 + \dfrac{6 \cdot 3}{0 \cdot 91} + \dfrac{2 \cdot 74}{8 \cdot 4}$

37 $\dfrac{18 \cdot 5}{1 \cdot 6} + \dfrac{7 \cdot 1}{0 \cdot 53} + \dfrac{11 \cdot 9}{25 \cdot 6}$

38 $\dfrac{83 \cdot 6}{105} + \dfrac{2 \cdot 95}{2 \cdot 7} + \dfrac{81}{97}$

39 $\left(\dfrac{98 \cdot 76}{103} + \dfrac{4 \cdot 07}{3 \cdot 6}\right)^2$

40 $\dfrac{(5 \cdot 843 - \sqrt{2 \cdot 07})^2}{88 \cdot 4}$

41 $\left(\dfrac{1}{7 \cdot 6} - \dfrac{1}{18 \cdot 5}\right)^3$

42 $\dfrac{\sqrt{4 \cdot 79} + 1 \cdot 6}{9 \cdot 63}$

43 $\dfrac{(0 \cdot 761)^2 - \sqrt{4 \cdot 22}}{1 \cdot 96}$

44 $\sqrt[3]{\left(\dfrac{1 \cdot 74 \times 0 \cdot 761}{0 \cdot 0896}\right)}$

45 $\left(\dfrac{8 \cdot 6 \times 1 \cdot 71}{0 \cdot 43}\right)^3$

46 $\dfrac{\sqrt[3]{86 \cdot 6}}{\sqrt[4]{4 \cdot 71}}$

47 $\dfrac{1}{8 \cdot 2^2} - \dfrac{3}{19^2}$

48 $\dfrac{100}{11^3} + \dfrac{100}{12^3}$

2.2.5 **Fractions**

You use the $\boxed{a^b/_c}$ key for fractions.

To enter $\dfrac{3}{4}$, press $\boxed{3}$ $\boxed{a^b/_c}$ $\boxed{4}$ You see $\boxed{3 \lrcorner 4}$

To enter $5\dfrac{1}{3}$, press $\boxed{5}$ $\boxed{a^b/_c}$ $\boxed{1}$ $\boxed{a^b/_c}$ $\boxed{3}$ You see $\boxed{5 \lrcorner 1 \lrcorner 3}$

> Your calculator may have a different fraction key.

Exercise 15 Ⓜ

Work out

1 $\frac{2}{5} + \frac{1}{3}$ **2** $\frac{5}{6} + \frac{1}{3}$ **3** $\frac{3}{7} + \frac{1}{3}$ **4** $\frac{4}{15} + \frac{1}{2}$

5 $\frac{3}{4} + \frac{1}{12}$ **6** $\frac{7}{8} - \frac{1}{16}$ **7** $\frac{3}{7} - \frac{1}{4}$ **8** $\frac{5}{6} - \frac{1}{5}$

9 $\frac{9}{10} + \frac{1}{20}$ **10** $\frac{11}{12} - \frac{3}{4}$ **11** $\frac{4}{9} \times \frac{1}{2}$ **12** $\frac{3}{11} \times \frac{1}{4}$

13 $2\frac{1}{2} + \frac{1}{3}$ **14** $3\frac{2}{3} - 1\frac{1}{2}$ **15** $4\frac{1}{2} + \frac{5}{8}$ **16** $\frac{1}{6} + 3\frac{3}{4}$

17 $3\frac{1}{4} \times 1\frac{1}{2}$ **18** $4\frac{1}{2} \div \frac{3}{4}$ **19** $3\frac{1}{2} \div \frac{2}{5}$ **20** $21 \div 5\frac{1}{4}$

21 Copy and complete

a $1\frac{1}{4} + 2\frac{1}{5} = \square$ **b** $\square + 1\frac{1}{3} = 4\frac{1}{2}$ **c** $\square + \frac{5}{7} = 1\frac{1}{2}$

d $\square - \frac{1}{5} = \frac{3}{8}$ **e** $\square \div \frac{3}{7} = 2$ **f** $\square \times 1\frac{2}{7} = \frac{1}{2}$

2.2.6 Rounding errors – early approximation

● A common error occurs when numbers are rounded in the intermediate steps of a calculation.

EXAMPLE

Find the area of a circle of circumference 25 cm, correct to 3 significant figures.

•••

Circumference = 25 cm

so $2\pi r = 25$

Radius = $\frac{25}{2\pi} = 3.9788736 \ldots$ ①

Suppose you round this number to 3 significant figures, that is, take the radius = 3.98 cm

Then area = $\pi \times 3.98^2$

= $49.764084 = 49.8$ cm^2 to 3 sf

The **correct** method is to use the radius shown in line ①.
Use the ANS button.

Then area = 49.7 cm^2 to 3 sf (check this on your calculator.)

See p194 for more on the area of a circle.

Avoid **early approximation** in calculations with several steps.

Exercise 16 Ⓜ

1 a Find the area of a circle of circumference 80 cm, giving your answer correct to 3 significant figures.

  b Repeat the calculation but this time round off the radius to 3 significant figures before you find the area. Compare your two answers.

Exercise 17 Ⓜ Calculator words

On a calculator work out $7770^2 + 250^2 + 25^2 + 9$.
If you turn the calculator upside down and use a little
imagination, you can see the word 'HEDGEHOG'.
Find the words given by the clues below.

1 $19 \times 20 \times 14 - 2.66$ (Not an upstanding man)

2 $(84 + 17) \times 5$ (Dotty message)

3 $904^2 + 89\,621\,818$ (Prickly customer)

4 $(559 \times 6) + (21 \times 55)$ (What a surprise!)

5 $566 \times 711 - 23\,617$ (Bolt it down)

6 $\dfrac{9999 + 319}{8.47 + 2.53}$ (Sit up and plead)

7 $\dfrac{2601 \times 6}{4^2 + 1^2}$; $(401 - 78) \times 5^2$ (Two words) (Not a great man)

8 $0.4^2 - 0.1^2$ (Little Sidney)

9 $\dfrac{(27 \times 2000 - 2)}{(0.63 \div 0.09)}$ (Not quite a mountain)

10 $(5^2 - 1^2)^4 - 14\,239$ (Just a name)

11 $48^4 + 102^2 - 4^2$ (Pursuits)

12 $615^2 + (7 \times 242)$ (Almost a goggle)

13 $(130 \times 135) + (23 \times 3 \times 11 \times 23)$ (Wobbly)

14 $164 \times 166^2 + 734$ (Almost big)

15 $8794^2 + 25 \times 342.28 + 120 \times 25$ (Thin skin)

16 $0.08 - (3^2 \div 10^4)$ (Ice house)

17 $235^2 - (4 \times 36.5)$ (Shiny surface)

18 $(80^2 + 60^2) \times 3 + 81^2 + 12^2 + 3013$ (Ship gunge)

19 $3 \times 17 \times (329^2 + 2 \times 173)$ (Unlimited)

20 $230 \times 230\frac{1}{2} + 30$ (Fit feet)

21 $33 \times 34 \times 35 + 15 \times 3$ (Beleaguer)

22 $0.32^2 + \dfrac{1}{1000}$ (Did he or didn't he?)

23 $(23 \times 24 \times 25 \times 26) + (3 \times 11 \times 10^3) - 20$ (Help)

24 $(16^2 + 16)^2 - (13^2 - 2)$ (Slander)

25 $(3 \times 661)^2 - (3^6 + 22)$ (Pester)

26 $(22^2 + 29 \cdot 4) \times 10$; $(3 \cdot 03^2 - 0 \cdot 02^2) \times 100^2$ (Four words) (Goliath)

27 $1 \cdot 25 \times 0 \cdot 2^6 + 0 \cdot 2^2$ (Tissue time)

28 $(710 + (1823 \times 4)) \times 4$ (Liquor)

29 $(3^3)^2 + 2^2$ (Wriggler)

30 $14 + (5 \times (83^2 + 110))$ (Bigger than a duck)

31 $2 \times 3 \times 53 \times 10^4 + 9$ (Opposite to hello, almost!)

32 $(177 \times 179 \times 182) + (85 \times 86) - 82$ (Good salesman)

33 $14^4 - 627 + 29$ (Good book, by God!)

34 $6 \cdot 2 \times 0 \cdot 987 \times 1\,000\,000 - 860^2 + 118$ (Flying ace)

35 $(426 \times 474) + (318 \times 487) + 22\,018$ (Close to a bubble)

36 $\dfrac{36^3}{4} - 1530$ (A German girl's name)

37 $(7^2 \times 100) + (7 \times 2)$ (Lofty)

38 $240^2 + 134$; $241^2 - 7^3$ (Two words) (Devil of a chime)

39 $(2 \times 2 \times 2 \times 2 \times 3)^4 + 1929$ (Unhappy ending)

40 $141\,918 + 83^3$ (Hot stuff in France)

2.3 Standard form

2.3.1 Large and small numbers

When dealing with either very large or very small numbers, it is not convenient to write them out in full in the normal way. It is better to use **standard form**. Most calculators represent large and small numbers in this way.

- The number $a \times 10^n$ is in **standard form** when $1 \leqslant a < 10$ and n is a positive or negative integer.

For example, this calculator shows $2 \cdot 3 \times 10^8$.

EXAMPLE

a $2000 = 2 \times 1000 = 2 \times 10^3$

b $150 = 1 \cdot 5 \times 100 = 1 \cdot 5 \times 10^2$

c $0 \cdot 0004 = 4 \times \dfrac{1}{10\,000} = 4 \times 10^{-4}$

d $450 \times 10^6 = (4 \cdot 5 \times 10^2) \times 10^6$
$= 4 \cdot 5 \times 10^8$

e $0 \cdot 07 \times 10^{-4} = (7 \times 10^{-2}) \times 10^{-4}$
$= 7 \times 10^{-6}$

Exercise 18 Ⓜ/Ⓗ

Write these numbers in standard form.

1 4000	**2** 500	**3** 70 000
4 60	**5** 2400	**6** 380
7 46 000	**8** 46	**9** 900 000
10 2560	**11** 0·007	**12** 0·0004
13 0·0035	**14** 0·421	**15** 0·000 055
16 0·01	**17** 564 000	**18** 19 million

19 A hydrogen atom has a mass of 0·000 000 000 000 000 000 000 001 67 grams. Write this mass in standard form.

20 There are 217 000 000 bricks in one of the pyramids in Egypt. Write this number in standard form.

21 The area of the surface of the Earth is about 510 000 000 km^2. Express this in standard form.

22 A certain virus is 0·000 000 000 25 cm in diameter. Write this in standard form.

23 Avogadro's number is 602 300 000 000 000 000 000 000. Express this in standard form.

24 The speed of light is 300 000 km/s. Express this speed in cm/s in standard form.

25 A very rich oil sheikh leaves his fortune of £$3·6 \times 10^8$ to be divided between 100 relatives. How much does each relative receive? Give the answer in standard form.

26 If $a = 512 \times 10^2$
$b = 0·478 \times 10^6$
$c = 0·0049 \times 10^7$

arrange a, b and c in order of size (smallest first).

27 If the number $2·74 \times 10^{15}$ is written out in full, how many zeros follow the 4?

28 If the number $7·31 \times 10^{-17}$ is written out in full, how many zeros would there be between the decimal point and the first significant figure?

EXAMPLE

To work out $(3 \times 10^2) \times (2 \times 10^5)$, **without** a calculator

a Multiply: $3 \times 2 = 6$
b Add the powers of 10: $10^2 \times 10^5 = 10^7$
So $(3 \times 10^2) \times (2 \times 10^5) = 6 \times 10^7$

29 Without a calculator, work out

   a $(2 \times 10^8) \times (4 \times 10^3)$ **b** $(2 \cdot 5 \times 10^3) \times (2 \times 10^{10})$

   c $(1 \cdot 5 \times 10^4) \times (3 \times 10^3)$ **d** $(3 \cdot 2 \times 10^4) \times (2 \times 10^{-2})$

   e $(1 \cdot 1 \times 10^{-6}) \times (5 \times 10^2)$ **f** $(8 \times 10^{12}) \div (2 \times 10^3)$

   g $(9 \times 10^6) \div (3 \times 10^2)$ **h** $(2 \times 10^6)^2$

30 Write out these numbers in full.

   a $2 \cdot 3 \times 10^4$ **b** 3×10^{-2} **c** $5 \cdot 6 \times 10^2$

   d 8×10^5 **e** $2 \cdot 2 \times 10^{-3}$ **f** 9×10^8

   g 6×10^{-1} **h** 7×10^3 **i** $3 \cdot 14 \times 10^6$

31 Without a calculator, work out these and give your answers in standard form.

   a $(6 \times 10^5) + (5 \times 10^4)$ **b** $(3 \times 10^3) - (3 \times 10^2)$

   c $(3 \cdot 5 \times 10^6) + (4 \times 10^4)$ **d** $(8 \times 10^{-3}) + (2 \times 10^{-2})$

   e $(6 \cdot 2 \times 10^{-3}) + (8 \times 10^{-4})$ **f** $(7 \times 10^{22}) + (7 \times 10^{20})$

> In part **a** write
> $6 \times 10^5 = 60 \times 10^4$.
> Now $(60 \times 10^4) +$
> (5×10^4)
> $= 65 \times 10^4$
> $= 6 \cdot 5 \times 10^5$

32 Rewrite these numbers in standard form.

   a $0 \cdot 6 \times 10^4$ **b** 11×10^{-2}

   c 450×10^6 **d** $0 \cdot 085 \times 10^{-4}$

   e 6 billion   **f** $\dfrac{1}{2}$

2.3.2 Using a calculator for standard form

EXAMPLE

a Work out $(5 \times 10^7) \times (3 \times 10^{12})$

Use the $\boxed{\text{EXP}}$ button

$\boxed{5}$ $\boxed{\text{EXP}}$ $\boxed{7}$ $\boxed{\times}$ $\boxed{3}$ $\boxed{\text{EXP}}$ $\boxed{12}$ $\boxed{=}$

Notice that you do **not** press the $\boxed{\times}$ button after the $\boxed{\text{EXP}}$ button!

$(5 \times 10^7) \times (3 \times 10^{12}) = 1 \cdot 5 \times 10^{20}$

b Work out $(3 \cdot 2 \times 10^3) \div (8 \times 10^{-7})$

$\boxed{3 \cdot 2}$ $\boxed{\text{EXP}}$ $\boxed{3}$ $\boxed{\div}$ $\boxed{8}$ $\boxed{\text{EXP}}$ $\boxed{(-)}$ $\boxed{7}$ $\boxed{=}$

$(3 \cdot 2 \times 10^3) \div (8 \times 10^{-7})$

$\quad = 4 \times 10^9$ ↑ Press this key for negative.

c Work out $(5 \times 10^{-8})^2$

$\boxed{5}$ $\boxed{\text{EXP}}$ $\boxed{(-)}$ $\boxed{8}$ $\boxed{x^2}$ $\boxed{=}$

$(5 \times 10^{-8})^2 = 2 \cdot 5 \times 10^{-15}$

The display on a Casio calculator looks like this:

Exercise 19 Ⓜ/Ⓗ

In questions **1** to **24** use a calculator and give the answer in standard form.

1 $5000 \times 300\,000$ **2** $60\,000 \times 5000$

3 $0.000\,07 \times 400$ **4** $0.0007 \times 0.000\,01$

5 $8000 \div 0.004$ **6** $(0.002)^2$

7 150×0.0006 **8** $0.000\,033 \div 500$

9 $0.007 \div 20\,000$ **10** $(0.0001)^4$

11 $(2000)^3$ **12** $0.005\,92 \div 8000$

13 $(1.4 \times 10^7) \times (3.5 \times 10^4)$ **14** $(8.8 \times 10^{10}) \div (2 \times 10^{-2})$

15 $(1.2 \times 10^{11}) \div (8 \times 10^7)$ **16** $(4 \times 10^5) \times (5 \times 10^{11})$

17 $(2.1 \times 10^{-3}) \times (8 \times 10^{15})$ **18** $(8.5 \times 10^{14}) \div 2000$

19 $(3.3 \times 10^{12}) \times (3 \times 10^{-5})$ **20** $(2.5 \times 10^{-8})^2$

21 $(1.2 \times 10^5)^2 \times (5 \times 10^{-3})$ **22** $(6.2 \times 10^{-4}) \times (1.1 \times 10^{-3})$

23 $(2 \times 10^{-4})^2 \times (3 \times 10^2)$ **24** $10^5 \div (2 \times 10^8)$

25 A certain dinosaur laid its eggs 30 million years ago. How many days ago was that? Round off your answer to 2 significant figures.

26 A pile of ten thousand sheets of paper is 1·3 m high. How thick is each sheet of paper in metres?

27 a Find the population densities of Europe and Asia.

Continent	Population	Area (m²)
Europe	6.82×10^8	1.05×10^{10}
Asia	2.96×10^9	4.35×10^{10}

> Population density
> $= \dfrac{\text{Population}}{\text{Area}}.$

b Which continent has the higher population density?

28 A patient in hospital is very ill. Between noon and midnight one day the number of viruses in her body increased from 3×10^7 to $5 \cdot 6 \times 10^9$. Work out the increase in the number of viruses, giving your answer in standard form.

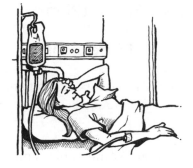

29 Here is a statement about roads.

> *At a constant 80 kilometres per hour it would take 35 years to travel over the entire road network of the world.*

Taking a year as 365 days, calculate the length of the road network of the world, giving your answer in standard form.

30 Oil flows through a pipe at a rate of $40 \, \text{m}^3/\text{s}$. How long will it take to fill a tank of volume $1 \cdot 2 \times 10^5 \, \text{m}^3$?

31 Given that $L = 2\sqrt{\dfrac{a}{k}}$, find the value of L in standard form when $a = 4 \cdot 5 \times 10^{12}$ and $k = 5 \times 10^7$.

32 The mean weight of all the men, women, children and babies in the UK is $42 \cdot 1$ kg. The population of the UK is 56 million. Work out the total weight of the entire population giving your answer in kg in standard form. Give your answer to a sensible degree of accuracy.

33 A light year is the distance travelled by a beam of light in a year. Light travels at a speed of approximately $3 \times 10^5 \, \text{km}/\text{s}$.

  a Work out the length of a light year in km.

  b Light takes about 8 minutes to reach the Earth from the Sun. How far is the Earth from the Sun in km?

*34 a The number 10 to the power 100 (10 000 sexdecillion) is called a 'googol'! Write 60 googols in standard form.

  b If it takes $\dfrac{1}{5}$ second to write a zero and $\dfrac{1}{10}$ second to write a 'one', how long would it take to write the number 100 'googols' in full?

  c The number 10 to the power of a 'googol' is called a 'googolplex'. Using the same speed of writing, how long in years would it take to write 1 'googolplex' in full? You may assume that your pen has enough ink.

2.4 Substitution

2.4.1 Expressions

- An **expression** contains algebraic terms and numbers but there is **no equals sign.**

Here are some expressions: $1 + 5x$, $2x(x - 1)$, $\dfrac{x + 1}{x}$.

EXAMPLE

Find the value of each expression when $x = 3$.

a $1 + 5x$ **b** $2x(x - 1)$ **c** $\dfrac{x + 1}{x}$

....

a $1 + 5x$ **b** $2x(x - 1)$ **c** $\dfrac{x + 1}{x}$

    $= 1 + 15$ $= 6(3 - 1)$ $= \dfrac{4}{3}$

    $= 16$ $= 12$

Exercise 20 Ⓜ

1 Find the value of each expression when $a = 4$.

   a $3a$ **b** $a - 3$ **c** $5a - 1$ **d** $a + a + a$

   e $8 - a$ **f** $18 - 3a$ **g** a^2 **h** $a^2 + 2a$

2 Evaluate each expression when $n = 5$.

   a $3n - 7$ **b** $2(n + 1)$ **c** $n + 2n + 3n$ **d** $n - 6$

   e $\dfrac{n}{5}$ **f** $\dfrac{3n + 1}{4}$ **g** n^2 **h** $15 - 2n$

3 Evaluate each expression when $t = 3$.

   a $t(t - 2)$ **b** $3t + 1 - t$ **c** $1 - t$ **d** t^3

   e $20 - 6t$ **f** $2(2 - t)$ **g** $\dfrac{t}{6}$ **h** $\dfrac{8}{t + 1}$

4 Copy and complete the puzzle using the clues.

Across

1 $4n$ when $n = 6$

2 $p^2 + 3$ when $p = 4$

4 $12(m + 2)$ when $m = 8$

5 $15 + n^2$ when $n = 20$

6 $a(a + 1)$ when $a = 4$

8 $7t - 11$ when $t = 8$

Down

1 $3a - 4$ when $a = 9$

2 x^3 when $x = 5$

4 $8 + 3t$ when $t = 1$

5 $4(n^2 + 1)$ when $n = 10$

6 $\dfrac{c + 39}{2}$ when $c = 5$

7 $\dfrac{x}{4} + 22$ when $x = 12$

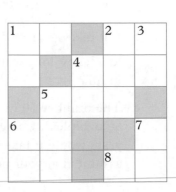

When $a = 3$, $b = -2$, $c = 5$, find the value of

a $3a + b$ **b** $ac + b^2$ **c** $a(c - b)$.

a $3a + b = (3 \times 3) + (-2)$ **b** $ac + b^2 = (3 \times 5) + (-2)^2$ **c** $a(c - b)$
           $= 9 - 2$ $= 15 + 4$ $= 3[5 - (-2)]$
           $= 7$ $= 19$ $= 3 \times 7$
                                                                  $= 21$

> Notice that working **down** the page is often easier to follow.

Exercise 21 Ⓜ

Evaluate these expressions.
For questions **1** to **12**

$$a = 3$$
$$c = 2$$
$$e = 5$$

1 $3a - 2$ **2** $4c + e$ **3** $2c + 3a$ **4** $5e - a$

5 $e - 2c$ **6** $e - 2a$ **7** $4c + 2e$ **8** $7a - 5e$

9 $c - e$ **10** $10a + c + e$ **11** $a + c - e$ **12** $a - c - e$

For questions **13** to **24**

$$h = 3$$
$$m = -2$$
$$t = -3$$

13 $2m - 3$ **14** $4t + 10$ **15** $3h - 12$ **16** $6m + 4$

17 $9t - 3$ **18** $4h + 4$ **19** $2m - 6$ **20** $m + 2$

21 $3h + m$ **22** $t - h$ **23** $4m + 2h$ **24** $3t - m$

For questions **25** to **36**

$$x = -2$$
$$y = -1$$
$$k = 0$$

25 $3x + 1$ **26** $2y + 5$ **27** $6k + 4$ **28** $3x + 2y$

29 $2k + x$ **30** xy **31** xk **32** $2xy$

33 $2(x + k)$ **34** $3(k + y)$ **35** $5x - y$ **36** $3k - 2x$

When $x = -2$, find the value of   **a** $2x^2 - 5x$ **b** $(3x)^2 - x^2$

$$2x^2 - 5x = 2(-2)^2 - 5(-2)$$
$$= 2(4) + 10$$
$$= 18$$

$$(3x)^2 - x^2 = (3 \times -2)^2 - 1(-2)^2$$
$$= (-6)^2 - 1(4)$$
$$= 36 - 4$$
$$= 32$$

Reminder:
$2x^2$ means $2(x^2)$
$(2x)^2$ means 'work out $2x$ and **then** square it'
$-7x$ means $-7(x)$
$-x^2$ means $-1(x^2)$

Exercise 22 Ⓜ

If $x = -3$ and $y = 2$, evaluate these expressions.

1 x^2
2 $3x^2$
3 y^2
4 $4y^2$
5 $(2x)^2$

6 $2x^2$
7 $10 - x^2$
8 $10 - y^2$
9 $20 - 2x^2$
10 $20 - 3y^2$

11 $5 + 4x$
12 $x^2 - 2x$
13 $y^2 - 3x^2$
14 $x^2 - 3y$
15 $(2x)^2 - y^2$

16 $4x^2$
17 $(4x)^2$
18 $1 - x^2$
19 $y - x^2$
20 $x^2 + y^2$

21 $x^2 - y^2$
22 $2 - 2x^2$
23 $(3x)^2 + 3$
24 $11 - xy$
25 $12 + xy$

26 $(2x)^2 - (3y)^2$
27 $2 - 3x^2$
28 $y^2 - x^2$

When $a = -2$, $b = 3$, $c = -3$, evaluate

a $\dfrac{2a(b^2 - a)}{c}$

b $\sqrt{(a^2 + b^2)}$

a $(b^2 - a) = 9 - (-2)$
$$= 11$$
$$\therefore \frac{2a(b^2 - a)}{c} = \frac{2 \times (-2) \times (11)}{-3}$$
$$= 14\frac{2}{3}$$

b $a^2 + b^2 = (-2)^2 + (3)^2$
$$= 4 + 9$$
$$= 13$$
$$\therefore \sqrt{(a^2 + b^2)} = \sqrt{13}$$

Notice that $\sqrt{13}$ means 'the positive square root of 13' **not** $\pm\sqrt{13}$.

Exercise 23 Ⓜ

Evaluate these.

In questions **1** to **16**, $a = 4$
$b = -2$
$c = -3$

1 $a(b + c)$
2 $a^2(b - c)$
3 $2c(a - c)$
4 $b^2(2a + 3c)$

5 $c^2(b - 2a)$
6 $2a^2(b + c)$
7 $2(a + b + c)$
8 $3c(a - b - c)$

9 $b^2 + 2b + a$
10 $c^2 - 3c + a$
11 $2b^2 - 3b$
12 $\sqrt{(a^2 + c^2)}$

13 $\sqrt{(ab + c^2)}$
14 $\sqrt{(c^2 - b^2)}$
15 $\dfrac{b^2}{a} + \dfrac{2c}{b}$
16 $\dfrac{c^2}{b} + \dfrac{4b}{a}$

In questions **17** to **32**, $k = -3$
$m = 1$
$n = -4$

17 $k^2(2m - n)$ **18** $5m\sqrt{(k^2 + n^2)}$ **19** $\sqrt{(kn + 4m)}$ **20** $kmn(k^2 + m^2 + n^2)$

21 $k^2m^2(m - n)$ **22** $k^2 - 3k + 4$ **23** $m^3 + m^2 + n^2 + n$ **24** $k^3 + 3k$

25 $m(k^2 - n^2)$ **26** $m\sqrt{(k - n)}$ **27** $100k^2 + m$ **28** $m^2(2k^2 - 3n^2)$

29 $\dfrac{2k + m}{k - n}$ **30** $\dfrac{kn - k}{2m}$ **31** $\dfrac{3k + 2m}{2n - 3k}$ **32** $\dfrac{k + m + n}{k^2 + m^2 + n^2}$

*33 Find $K = \sqrt{\left(\dfrac{a^2 + b^2 + c^2 - 2c}{a^2 + b^2 + 4c}\right)}$
    if $a = 3$, $b = -2$, $c = -1$.

*34 Find $W = \dfrac{kmn(k + m + n)}{(k + m)(k + n)}$
    if $k = \dfrac{1}{2}$, $m = -\dfrac{1}{3}$, $n = \dfrac{1}{4}$.

2.4.2 Formulae

When a calculation is repeated many times it is often helpful to use a formula.

When a building society offers a mortgage it may use a formula like '$2\frac{1}{2}$ times the main salary plus the second salary'.

For a polygon with n sides the sum of the angles, S, is given by the formula $S = (n - 2) \times 180°$.

In a formula letters stand for defined quantities or variables.

$n = 5$

Exercise 24 Ⓜ ⭐Ⓕ

1 The final speed v of a car is given by the formula $v = u + at$. [u = initial speed, a = acceleration, t = time taken]. Find v when $u = 15$, $a = 0.2$, $t = 30$.

2 The area, A, of a trapezium is given by the formula $A = \frac{1}{2}(a + b)h$. Find the area of each trapezium.

a
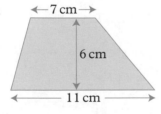
←7 cm→
6 cm
←— 11 cm —→

b
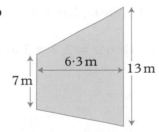
6·3 m
7 m
13 m

3 The time period T of a simple pendulum is given by the formula $T = 2\pi\sqrt{\left(\dfrac{l}{g}\right)}$, where l is the length of the pendulum and g is the gravitational acceleration.
Find T when $l = 0.65$, $g = 9.81$ and $\pi = 3.142$.

4 The total surface area A of a cone is related to the radius r and the
slant height l by the formula $A = \pi r (r + l)$.
Find A when $r = 7$ and $l = 11$.

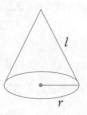

5 The sum S of the squares of the integers from 1 to n is given by
$S = \dfrac{1}{6}n(n + 1)(2n + 1)$. Find S when $n = 12$.

6 The acceleration a of a train is found using the formula
$a = \dfrac{v^2 - u^2}{2s}$. Find a when $v = 20$, $u = 9$ and $s = 2.5$.

7 Einstein's famous formula relating energy, mass and the
speed of light is $E = mc^2$. Find E when $m = 0.0001$ and
$c = 3 \times 10^8$.

*8 The area A of a parallelogram with sides a and b is given by
$A = ab\sin\theta$, where θ is the angle between the sides.
Find A when $a = 7$, $b = 3$ and $\theta = 30°$.

9 The distance s travelled by an accelerating rocket is given by
$s = ut + \dfrac{1}{2}at^2$. Find s when $u = 3$, $t = 100$ and $a = 0.1$.

10 The formula for the velocity of sound in air is $V = 72\sqrt{T + 273}$
where V is the velocity in km/h and T is the temperature of the
air in $°C$.
 a Find the velocity of sound where the temperature is $26\,°C$.
 b Find the temperature if the velocity of sound is $1200\,km/h$.
 c Find the velocity of sound where the temperature is $-77\,°C$.

?11 a Find a formula for the area, A, of this shape, in terms of
a, b and c.
 b Find the value of A when $a = 2$, $b = 7$ and $c = 10$.

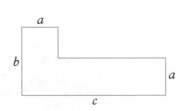

12 a Find a formula for the shaded part, S,
in the diagram, in terms of p, q and r.

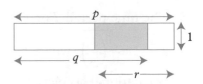

 b Find the value of S when $r = 8$, $p = 12.1$ and
$q = 8.2$.

2.5 Measures

2.5.1 Metric and imperial units
Metric units

Length: 10 mm = 1 cm   Mass: 1000 g = 1 kg   Volume: 1000 ml = 1 litre   Area:
100 cm = 1 m   1000 kg = 1 t   1000 l = 1 m^3 10 000 cm^2 = 1 m^2
1000 m = 1 km   (t for tonne)   Also 1 ml = 1 cm^3

Exercise 25 (M)

Copy and complete these conversions.
1 85 cm =   m   2 2·4 km =   m   3 0·63 m =   cm   4 25 cm =   m
5 7 mm =   cm   6 2 cm =   mm   7 1·2 km =   m   8 20 000 cm^2 =   m^2
9 0·58 km =   m   10 815 mm =   m   11 650 m =   km   12 12·5 m^2 =   cm^2
13 5 kg =   g   14 4·2 kg =   g   15 6·4 kg =   g   16 3 kg =   g
17 0·8 kg =   g   18 400 g =   kg   19 2 t =   kg   20 250 g =   kg
21 0·5 t =   kg   22 0·62 t =   kg   23 7 kg =   t   24 1500 g =   kg
25 800 ml = l 26 2 l =   ml   27 4·5 l =   ml   28 6 l =   ml
29 3 l =   cm^3 30 2 m^3 = l 31 5·5 m^3 = l 32 0·5 m^3 =   cm^3
33 1000 000 mm^3 =   m^3

34 Write down the most appropriate metric unit for measuring:
  a the distance between Glasgow and Leeds   b the capacity of a wine bottle
  c the mass of raisins needed for a cake   d the diameter of a small drill
  e the mass of a car   f the area of a football pitch.

Converting between imperial and metric measures

Although the metric system is generally replacing the Imperial system
you still need to be able to convert from one set of units to the other.

30 cm ≈ 1 foot   4·5 litres ≈ 1 gallon
1 kg ≈ 2.2 pounds   1 litre ≈ $1\frac{3}{4}$ pints
8 km ≈ 5 miles

EXAMPLE

  a Change 16 km into miles   b Change 1·5 m into feet

  1 km ≈ $\frac{5}{8}$ mile   1·5 m = 150 cm
  so 16 km ≈ $\frac{5}{8}$ × 16   30 cm ≈ 1 foot
  16 km ≈ 10 miles   ∴ 150 cm ≈ 5 feet

Exercise 26 Ⓜ ⭐

Copy each statement and fill in the missing numbers.

1 6 feet =   cm
2 10 gallons =   litres
3 22 pounds =   kg

4 4 kg =   pounds
5 8 km =   miles
6 1·1 pounds =   kg

7 32 km =   miles
8 4·5 feet =   cm
9 66 pounds =   kg

10 10 litres =   pints
11 40 km =   miles
12 100 litres =   pints

13 3 kg =   pounds
14 $\frac{1}{2}$ km =   mile
15 900 litres =   gallons

16 A car handbook calls for the oil to be changed every 8000 km.
How many miles is that?

17 On an Italian road the speed limit is 80 km/h.
Convert this into a speed in mph.

18 Tomatoes are sold in Tesco at 85p per kilo and at the market for
30p per pound. Which has the lower price?

In questions **19** to **23** copy each sentence and choose the number
which is the best estimate.

19 A one pound coin has a mass of about [1 g, 10 g, 1 kg].

20 The width of the classroom is about [100 inches, 7 m, 50 m].

There are 12
inches in a foot.

21 A can of Pepsi contains about [500 ml, $\frac{1}{2}$ gallon].

22 The distance from London to Birmingham is about [20 miles,
20 km, 100 miles].

23 The thickness of a one pound coin is about [3 mm, 6 mm, $\frac{1}{4}$ inch].

24 Here are scales for changing:
A kilograms and pounds,
B litres and gallons.
In this question give your answers to the
nearest whole number.

 a About how many kilograms are there
in 6 pounds?

 b About how many litres are there in
3·3 gallons?

 c About how many pounds are there
in 1·4 kilograms?

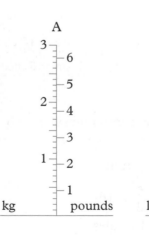
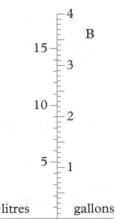

2.5.2 Speed, distance and time

Calculations involving these three quantities are simpler when the speed is **constant**. The formulae connecting distance, speed and time are:

- distance = speed × time

- speed = $\dfrac{\text{distance}}{\text{time}}$

- time = $\dfrac{\text{distance}}{\text{speed}}$

A helpful way of remembering these formulae is to write the letters D, S and T in a triangle, thus:

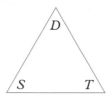

to find D, cover D and you get ST

to find S, cover S and you get $\dfrac{D}{T}$

to find T, cover T and you get $\dfrac{D}{S}$

Great care must be taken with the units in these questions.

EXAMPLE

a A man is running at a speed of 8 km/h for a distance of 5200 metres. Find the time taken in minutes.

b A boat sails at an average speed of 12 km/h for 2 days. How for does it travel?

....

a 5200 metres = 5·2 km

time taken in hours = $\left(\dfrac{D}{S}\right) = \dfrac{5\cdot2}{8}$

$= 0\cdot65$ hours

time taken in minutes = 0·65 × 60

$= 39$ minutes

b 2 days = 48 hours

$[D = S \times T]$

Distance travelled = 12 × 48

$= 576$ km

Exercise 27 Ⓜ ⭐

1 Find the time taken for these journeys. [Take care with units in parts **c** and **d**]

  a 100 km at a speed of 40 km/h

  b 250 miles at a speed of 80 miles per hour

  c 15 metres at a speed of 20 cm/s (answer in seconds)

  d 10^4 metres at a speed of 2·5 km/h

2 A Grand Prix driver completed a lap of 5400 m in exactly 2 minutes. What was his average speed in m/s?

3 Find the speeds of the celestial bodies which move
   a a distance of 600 km in 8 hours
   b a distance of 98 miles in 7 hours
   c a distance of 25 m in $\frac{1}{2}$ second.

   State clearly the units in which your answers are given.

4 Find the distance travelled
   a at a speed of 40 km/h for 3 hours
   b at a speed of 60 mph for 30 minutes
   c at a speed of 15 m/s for 5 minutes
   d at a speed of 14 m/s for 1 hour.

5 A car travels 60 km at 30 km/h and then a further 180 km at 160 km/h. Find
   a the total time taken
   b the average speed for the whole journey.

> average speed =
> $\frac{\text{(total distance travelled)}}{\text{(total time taken)}}$.

6 A cyclist travels 25 kilometres at 20 km/h and then a further 80 kilometres at 25 km/h.

   Find
   a the total time taken
   b the average speed for the whole journey.

7 Kelly Holmes ran two laps around a 400 m track. She completed the first lap in 60 seconds and then decreased her speed by 5% for the second lap. Find
   a her speed on the first lap
   b her speed on the second lap
   c her total time for the two laps.

8 An airliner flies 2000 km at a speed of 1600 km/h and then returns, due to bad weather, at a speed of 1000 km/h. Find the average speed for the whole trip.

9 A train travels from A to B, a distance of 100 km, at a speed of 20 km/h.
If it had gone two and a half times as fast, how much earlier would it have arrived at B?

10 Two men running towards each other at 4 m/s and 6 m/s respectively are one kilometre apart. How long will it take before they meet?

11 A car travelling at 90 km/h is 500 m behind another car travelling at 70 km/h in the same direction. How long will it take the first car to catch the second?

12 How long is a train which passes a signal in twenty seconds at a speed of 108 km/h?

> Convert the speed into m/s.

13 A train of length 180 m approaches a tunnel of length 620 m. How long will it take the train to pass completely through the tunnel at a speed of 54 km/h?

14 An earthworm of length 15 cm is crawling along at 2 cm/s. An ant overtakes the worm in 5 seconds. How fast is the ant walking?

15 A train of length 100 m is moving at a speed of 50 km/h. A horse is running alongside the train at a speed of 56 km/h. How long will it take the horse to overtake the train?

16 A car completes a journey at an average speed of 40 mph. At what speed must it travel on the return journey if the average speed for the complete journey (out and back) is 60 mph?

2.5.3 Density and other compound measures

● The **density** of a material is its mass per unit volume.

It is helpful to use an 'MDV' triangle similar to the one used for speed.

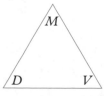

$$M = D \times V \qquad D = \frac{M}{V} \qquad V = \frac{M}{D}$$

Exercise 28 Ⓜ ⭐

1 The volume of a solid object is $5\,\text{cm}^3$ and its mass is $50\,\text{g}$.
Calculate the density of the material.

2 A statue is made of metal of density $8\,\text{g/cm}^3$.
Find the mass of the statue if its volume is $30\,\text{cm}^3$.

3 Water of density $1000\,\text{kg/m}^3$ fills a pool. The mass of the
water is 35 tonnes. Calculate the volume of the pool.
[1 tonne = $1000\,\text{kg}$]

4 This solid cuboid is made of metal with
density $10\,\text{g/cm}^3$. Calculate the mass
of the cuboid, giving your answer in kg.

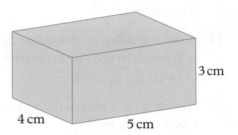

3 cm

4 cm   5 cm

5 The population of the United Kingdom is about
60 million and the land area is about $240\,000\,\text{km}^2$.
Work out the population density of the UK [number of
people/km^2].

6 Hong Kong has a population of about 6 million and
land area $1000\,\text{km}^2$. The USA has a land area of about
$9\,300\,000\,\text{km}^2$.
a Work out the population density of Hong Kong.
b Work out what the population of the United States would be if
it was as densely populated as Hong Kong.

7 One $1\cdot2\,\text{kg}$ box of grass seed covers $24\,\text{m}^2$ of lawn.
How much seed is needed for a field of area 2 hectares?
[1 hectare = $10\,000\,\text{m}^2$]

8 A certain plastic has a density of $3\,\text{g/cm}^3$. Convert the density
into kg/m^3.

9 Oil has a density of $800\,\text{kg/m}^3$. Convert this density into g/cm^3.

10 Change the units of the following speeds as indicated.
a $72\,\text{km/h}$ into m/s
b $30\,\text{m/s}$ into km/h
c $0\cdot012\,\text{m/s}$ into cm/s
d $9000\,\text{cm/s}$ into m/s

2.6 Measurements and bounds

2.6.1 Lower and upper bounds

● Measurement is approximate.

a If you measure the length of some fabric for curtains you might say the length is 145 cm to the nearest cm. The actual length could be anything from 144·5 cm to 145·49999... cm if you use the normal convention which is to round up a figure of 5 or more. Clearly 145·4999... is effectively 145·5 and you could use this figure.

● The upper bound often causes confusion. You use 145·5 as the upper bound simply because it is **inconvenient** to work with 145·49999...

b Suppose you measure a line as 14 mm, correct to the nearest mm. The actual length could be from 13·5 mm to 14·5 mm.

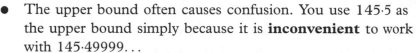

13·5 mm   14 mm   14·5 mm

● In both cases **a** and **b** the measurement expressed to a given unit is in **possible error** of **half a unit**.

c i Similarly if you say your weight is 57 kg to the nearest kg, you could actually weigh anything from 56·5 kg to 57·5 kg.

> The 'unit' is 1 so 'half a unit' is 0·5.

   ii If your brother was weighed on more sensitive scales and the result was 57·2 kg, his actual weight could be from 57·15 kg to 57·25 kg.

> The 'unit' is 0·1 so 'half a unit' is 0·05.

57·1   57·15   57·2   57·25   57·3

kg

Here are some more examples.

Measurement	Half unit	Lower bound	Upper bound
The diameter of a CD is 12 cm to the nearest cm.	0·5 cm	11·5 cm	12·5 cm
The mass of a coin is 6·2 g to the nearest 0·1 g.	0·05 g	6·15 g	6·25 g
The length of a fence is 330 m to the nearest 10 m.	5 m	325 m	335 m

Exercise 29 (H)

1 In a DIY store the height of a door is given as 195 cm to the nearest cm. Write down the upper bound for the height of the door.

2 A vet weighs a sick goat at 37 kg to the nearest kg. What is the least possible weight of the goat?

3 A cook's weighing scales weigh to the nearest 0·1 kg. What is the upper bound for the weight of a chicken which she weighs at 3·2 kg?

4 A surveyor using a laser beam device can measure distances to the nearest 0·1 m. What is the least possible length of a warehouse which he measures at 95·6 m?

5 In the county sports championships, Jill was timed at 28·6 s for the 200 m run. What is the upper bound for the time she could have taken?

6 Copy and complete the table.

	Measurement	Half unit	Lower bound	Upper bound
a	Temperature in a fridge = 2 °C to the nearest degree			
b	Mass of an acorn = 2·3 g to 1 dp			
c	Length of telephone cable = 64 m to nearest m			
d	Time taken to run 100 m = 13·6 s to nearest 0·1 s			

7 The length of a telephone is measured as 193 mm, to the nearest mm. The length lies between:

A	B	C
192 and 194 mm	192·5 and 193·5 mm	188 and 198 mm

8 Celine's cooking scales weigh to the **nearest 10 grams**. She weighs some ingredients and the scales show 180 grams. What are the minimum and maximum possible weights of the ingredients?

9 The length of a farmer's field is 2700 m, correct to the **nearest 100 m**. What is the minimum possible value of the length?

10 Liz and Julie each measure a different worm and they both say that their worm is 11 cm long to the nearest cm.
   a Does this mean that both worms are the same length?
   b If not, what is the maximum possible difference in the length of the two worms?

EXAMPLE

To the nearest cm, the length, l, of a stapler is 12 cm.

Write down the possible values for l.

....

As an inequality you can write $11 \cdot 5 \leqslant l < 12 \cdot 5$.

11 In parts **a** to **j** write the possible values of the measurements using an inequality as in the example.

   a mass = 17 kg   **b** $d = 256$ km
   c length = $2 \cdot 4$ m   **d** $m = 0 \cdot 34$ grams
   e $v = 2 \cdot 04$ m/s   **f** $x = 12 \cdot 0$ cm
   g $T = 81 \cdot 4 \,°C$ **h** $M = 0 \cdot 3$ kg
   i mass = $0 \cdot 7$ tonnes   **j** $n = 52\,000$ (nearest thousand)

12 A card measuring $11 \cdot 5$ cm long (to the nearest $0 \cdot 1$ cm) is to be posted in an envelope which is 12 cm long (to the nearest cm). Can you guarantee that the card will fit inside the envelope? Explain your answer.

11·5 cm   12 cm

2.6.2 Calculations with lower and upper bounds

This section is concerned with the inaccuracy of numbers used in a calculation.

EXAMPLE

Here is a rectangle with sides measuring 37 cm by 19 cm to the nearest cm.

19 cm

37 cm

What are the largest and smallest possible areas of the rectangle consistent with this data?

....

You know the length is between $36 \cdot 5$ cm and $37 \cdot 5$ cm and the width is between $18 \cdot 5$ cm and $19 \cdot 5$ cm.

largest possible area = $37 \cdot 5 \times 19 \cdot 5$ smallest possible area = $36 \cdot 5 \times 18 \cdot 5$
                   = $731 \cdot 25 \text{ cm}^2$ = $675 \cdot 25 \text{ cm}^2$

A common problem occurs when a measured quantity is multiplied by a large number.

> EXAMPLE
>
> A marble is weighed and found to be 12·3 grams. Find the weight of 100 identical marbles.
>
> ··
>
> A simple answer is $100 \times 12·3 = 1230\,g$. But each marble could weigh from 12·25 g to 12·35 g.
>
> $100 \times 12·25 = 1225\,g$
> $100 \times 12·35 = 1235\,g$
>
> So there is a possible error of up to 5 grams above or below the initial answer of 1230 g.

> EXAMPLE
>
> Given that the numbers 8·6, 3·2 and 11·5 are accurate to 1 decimal place, calculate the upper and lower bounds for this calculation.
>
> $$\left(\frac{8·6 - 3·2}{11·5}\right)$$
>
> ··
>
> For the upper bound, make the top line of the fraction as large as possible and make the bottom line as small as possible.
>
> upper bound $= \left(\dfrac{8·65 - 3·15}{11·45}\right)$ lower bound $= \left(\dfrac{8·55 - 3·25}{11·55}\right)$
>
> $\qquad = 0·480\,3493\ldots$ $\qquad\qquad = 0·458\,8744\ldots$

Exercise 30 (H)

1 The sides of this triangle are measured correct to the nearest cm.
   a Write the upper bounds for the lengths of the three sides.
   b Work out the maximum possible perimeter of the triangle.

7 cm   8 cm

10 cm

2 The dimensions of the photo are measured correct to the nearest cm. Work out the minimum possible area of the photo.

9 cm

6 cm

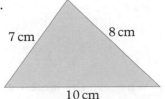

3 In this question the value of a is **either exactly** 4 or 5, and the value of b is **either exactly** 1 or 2.
Work out

   a the maximum value of $a + b$ **b** the minimum value of $a + b$

   c the maximum value of ab **d** the maximum value of $a - b$

   e the minimum value of $a - b$ **f** the maximum value of $\dfrac{a}{b}$

   g the minimum value of $\dfrac{a}{b}$ **h** the maximum value of $a^2 - b^2$.

4 A builder uses bricks which are 12 cm high to the nearest cm.
What is the maximum possible height of a stack of 20 bricks?

5 A house has three rooms of area $12\,\text{m}^2$, $17\,\text{m}^2$ and $20\,\text{m}^2$, each to the nearest m^2. What is the minimum possible value for the total area of the three rooms?

6 A rectangular room measures 3·2 m by 5·5 m, each to the nearest 0·1 m. Work out the maximum possible area of the room.

7 If $p = 7$ cm and $q = 5$ cm, both **to the nearest cm**, find:

   a the largest possible value of $p + q$

   b the smallest possible value of $p + q$

   c the largest possible value of $p - q$

   d the largest possible value of $\dfrac{p^2}{q}$.

8 If $a = 3·1$ and $b = 7·3$, both correct to 1 decimal place, find the largest possible value of

   a $a + b$ **b** $b - a$

9 If $x = 5$ and $y = 7$ to one significant figure, find the smallest possible values of

   a $x + y$ **b** $y - x$ **c** $\dfrac{x}{y}$

10 In the diagram, ABCD and EFGH are rectangles with AB = 10 cm, BC = 7 cm, EF = 7 cm and FG = 4 cm, all figures accurate to the nearest cm.
Find the largest possible value of the shaded area.

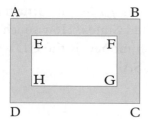

11 When a voltage V is applied to a resistance R the power consumed P is given by $P = \dfrac{V^2}{R}$.

If you measure V as 12·2 and R as 2·6, both correct to 1 dp, calculate the smallest possible value of P.

12 A cyclist was timed along a straight piece of road.
The time taken was 24·5 seconds, to the nearest 0·1 second,
and the distance was 420 metres, to the nearest metre.
Calculate the maximum possible value for the speed of the cyclist
consistent with this data.

13 The velocity v of a body is calculated from the formula $v = \dfrac{2s}{t} - u$
where u, s and t are measured correct to 1 decimal place.
Find the largest possible value for v when $u = 2·1$, $s = 5·7$ and
$t = 2·2$. Find also the smallest possible value for v consistent with
these figures.

2.7   Mixed numerical problems 2

Exercise 31 Ⓜ

Questions **1** to **5** are about foreign currency.
Use the rates of exchange given in the table.

1 Change these amounts of British money
into the foreign currency stated.

Country/Region	Rate of exchange
Europe (euro)	€1·54 = £1
Japan (yen)	¥210 = £1
South Africa (rand)	11 rand = £1
United States (dollar)	$1·64 = £1

    a £20 [euros]   **b** £70 [dollars]
    c £200 [rand]   **d** £1·50 [yen]
    e £2·50 [dollars]   **f** 90p [euros]

2 Change these amounts of foreign currency into
British currency.
    a € 500   **b** $2500
    c 2640 rand   **d** € 900
    e ¥ 8820   **f** ¥84

3 A CD costs £12·99 in Britain and $15 in
the United States. How much cheaper,
in British money, is the CD when bought
in the USA?

4 A Renault car is sold in several countries at the prices given below.

Britain   £15 000
France   €20 790
USA $24 882

Write, in order, a list of the prices converted into pounds.

5 A traveller in Switzerland exchanges 1300 Swiss francs for £400. What is the exchange rate?

6 A maths teacher bought 40 calculators at £8·20 each and a number of other calculators costing £2·95 each. In all she spent £387.
How many of the cheaper calculators did she buy?

7 At a temperature of 20 °C the common amoeba reproduces by splitting in half every 24 hours. If you start with a single amoeba, how many will there be after

   a 8 days   **b** 16 days?

8 A humming bird moves its wings 130 times per second. How many times will it move its wings in one hour? Give your answer in standard form.

9 Here are four fractions: $\dfrac{3}{19}, \dfrac{37}{232}, \dfrac{1}{6}, \dfrac{83}{511}$

   a Which of the fractions is the smallest?
   b Which of the fractions is the largest?

10 A wicked witch stole a newborn baby from its parents. On the baby's first birthday the witch sent the grief-stricken parents 1 penny. On the second birthday she sent 2 pence. On the third birthday she sent 4 pence and so on, doubling the amount each time. How much did the witch send the parents on the twenty-first birthday?

Exercise 32 Ⓜ

1 A videotape is 225 metres long. Its total playing time is three hours. Find the speed of the tape

   a in metres per minute   **b** in centimetres per second.

2 In 2009 there were 21 280 000 licensed vehicles on the road. Of these, 16 486 000 were private cars. What percentage of the licensed vehicles were private cars?

3 A map is 278 mm wide and 445 mm long. When reduced on a photocopier, the copy is 360 mm long. What is the width of the copy, to the nearest millimetre?

4 Taking 1 litre to be $1\frac{3}{4}$ pints, find

    a the number of pints in 20 litres

    b the number of litres in 9 pints, giving your answer as a decimal.

5 Measured to the nearest millimetre, the sides of a triangle are 11 mm, 14 mm and 17 mm. Work out the upper limit for the perimeter of the triangle.

6 a Fill in the empty boxes in the square so that each row, each column and each diagonal adds up to 0.

    b Multiply together the three numbers in the bottom row and write down your answer.

3	–4	1
	0	

7 In 2010 Anna picked 252 kg of plums. This was 20% more than she picked in 2009. Calculate how many kilograms of plums she picked in 2009.

8 Tiger arrived at the bus station at 11:15. His bus had left half an hour before.

    a At what time had his bus left?

    b Tiger's next bus is at 12:06. How long must he wait?

9 What is the smallest number greater than 1000 that is exactly divisible by 13 and 17?

10 Mr Giggs drove 8360 miles in his car. His car does 37 miles per gallon. Petrol costs 86 pence per litre.

    a By taking one gallon to be 4·55 litres, calculate, in pounds, how much Mr Giggs spent on petrol.

    b Show how you can use approximation to check that your answer is of the right order of magnitude. You **must** show all your working.

Exercise 33 Ⓜ/Ⓗ

1 A stick insect moves 3 centimetres in one minute.

    a Find the speed of the stick insect in kilometres per hour, giving your answer as a decimal.

    b Write your answer to part **a** in standard form.

2 Find the value of n for which
$$1^2 + 2^2 + 3^2 + 4^2 + 5^2 + \ldots + n^2 = 650$$

3 Pythagoras, the Greek mathematician, was also
a shrewd businessman. Suppose he deposited the
equivalent of £1 in the Bank of Athens
in the year 500 BC at 1% compound interest.
What would the investment be worth
to his descendants in the year 2000 AD?

4 Find the missing prices.

	Object	Old price	New price	Percentage change
a	Football	£9·50	?	3% increase
b	Radio	?	£34·88	9% increase
c	Roller blades	£52	?	15% decrease
d	Golf club	?	£41·40	8% decrease

5 Booklets have a mass of 19 g each and they are posted in an
envelope of mass 38 g. Postage charges are shown in the table.

Mass (in grams) not more than	60	100	150	200	250	300	350	600
Postage (in pence)	30	46	64	79	94	107	121	215

   a A package consists of 15 booklets in an envelope.
      What is the total mass of the package?
   b The mass of a second package is 475 g.
      How many booklets does it contain?
   c What is the postage charge on a package of mass 320 g?
   d The postage on a third package was £2·15.
      What is the largest number of booklets it could contain?

6 A rabbit runs at 7 m/s and a hedgehog at $\frac{1}{2}$ m/s.

They are 90 m apart and start to run towards each other.
How far does the hedgehog run before they meet?

7 Mark's recipe for a cake calls for 6 fluid ounces of milk. Mark
knows that one pint is 20 fluid ounces, that one gallon is 8 pints
and that 5 litres is roughly one gallon. He only has a measuring jug
marked in millilitres.
How many millilitres of milk does he need for the cake? Give your
answer to a sensible degree of accuracy.

8 8% of 2500 + 37% of P = 348. Find the value of P.

9 A train leaves Paris at 15:24 and arrives in Milan at 19:44.
A new train will cut 20% off the journey time. At what time will
the 15:24 train now arrive in Milan?

10 **a** Given that 1 mile is about 1600 m, convert a speed of 20 mph
into metres per second.
b The world's fastest car can travel at up to 342 metres per
second. Convert this into miles per hour.

11 A fitter is doing a job that needs a 3 mm drill. He has no metric
drills but he does have these Imperial sizes (in inches): $\frac{1}{16}, \frac{1}{8}, \frac{3}{16}$.
Which of these drills is the nearest in size to 3mm?

> 12 in $\approx$ 1 ft
> 30 cm $\approx$ 1 ft

? *12 A group of friends share a bill for £13·69 equally between them.
How many were in the group?

? *13 Which is larger: 7^{77} or 77^7? Have a guess before checking on a calculator.

Test yourself

1 *ABCD* is a rectangle with
length 25 cm and width 10 cm.

The length of the rectangle
is increased by 10%.
The width of the rectangle
is increased by 20%.
Find the percentage increase
in the area of the rectangle.

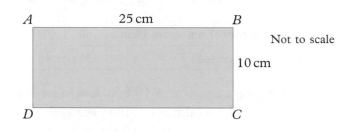

Not to scale

(AQA, 2003)

2 The table gives information about
Year 10 and Year 11 at Mathstown
School. Mathstown School had an
end of term party. 40% of the
students in Year 10 and 70% of the
students in Year 11 went to the party.

	Number of girls	Number of boys
Year 10	108	132
Year 11	90	110

Work out the percentage of all students in Years 10 and 11 who
went to the party. Give your answer correct to 3 significant figures.

(Edexcel, 2004)

3 The annual fees at an independent school were increased by 6%.
The fees after this increase are £8957.
Calculate the fees before the increase.

(OCR, 2004)

4 Arnie saw a camera priced at £250 in London.
He saw the same camera priced at $297·50 in New York.
This is a 30% saving on the London price.
How many dollars are there to the pound?

(AQA, 2007)

5 The value of a car depreciates by 35% each year.
At the end of 2007 the value of the car was £5460.
Work out the value of the car at the end of 2006.

(Edexcel, 2008)

6 Express $0·3\dot{4}$ as a fraction in its simplest form.

7 a Write the number 40 000 000 in standard form.
 b Write $1·4 \times 10^{-5}$ as an ordinary number.
 c Work out
  $(5 \times 10^4) \times (6 \times 10^9)$
  Give your answer in standard form.

(Edexcel, 2005)

8 Use your calculator to work out $\sqrt{\dfrac{8·21 + 3·9^2}{5·2}}$

 a Write down the full calculator display.
 b Give your answer to 3 significant figures.

9 Use a calculator to work out

$$\sqrt{\frac{21·6 \times 15·8}{3·8}}$$

 a Write down all the figures on your calculator display.
 b Give your answer to part (a) correct to 3 significant figures.

(Edexcel, 2008)

10 Calculate.
 a $\dfrac{16·5}{8·25 + 5·15}$

  Give your answer correct to 1 decimal place.

 b $\dfrac{45}{(0·3)^2}$

(OCR, 2009)

11 The following information is from a table on the side of a packet of biscuits.

NUTRITION INFORMATION		
AVERAGE VALUES	PER BISCUIT	PER 100g
PROTEIN	0·7g	5·5g

Use this information to work out the weight of one biscuit.
Give your answer to an appropriate degree of accuracy.

(AQA, 2007)

12 A toy car travels 180 cm, correct to the nearest 10 cm. It takes 7 seconds, correct to the nearest second, to travel this distance.

Work out the greatest possible value of the average speed of the toy car. You must show all your working.

(OCR, 2009)

13 Each side of a regular pentagon has a length of 101 mm, correct to the nearest millimetre.
  i Write down the **least** possible length of each side.
  ii Write down the **greatest** possible length of each side.

(Edexcel, 2004)

14 Peter transports metal bars in his van.
The van has a safety notice "Maximum Load 1200 kg".
Each metal bar has a label "Weight 60 kg".

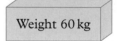

For safety reasons Peter assumes that 1200 is rounded correct to 2 significant figures and 60 is rounded correct to 1 significant figure. Calculate the greatest number of bars that Peter can **safely** put into the van if his assumptions are correct.

(Edexcel, 2005)

15 A cyclist travels 60 metres at a speed of 8 m/s.
Both values are measured correct to the nearest whole number.
What is the greatest possible time taken?

16 A sack of rice weighs 25 kg measured to the nearest kg. It is used to fill bags that contain 500 g of rice measured to the nearest 10 g. Find the maximum number of bags that could be filled.

(OCR, 2005)

Functional task 2

Building a house in Spain

Mrs Cameron is building a new house in Spain and then plans to sell it at a profit.

Her costs are:
- Buying a plot of land in Spain
- Construction costs (building the house)
- Estate agent's fee for selling the house.

Cost of land
The plot of land has an area of 2500 m².
The cost of land is €16·80 per m².

Estate agent's fee
5% of the selling price of the house.

Plan of house

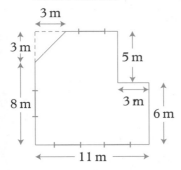

Construction costs
Her builder works in feet rather than metres, and tells her that the cost of the house will be £125 per square foot of the floor area of the house.

1 m² = 10·8 square feet
£1 = €1·20

Task 1
What is the total cost of buying the land and then building the house (in Euros)?

Task 2
Mrs Cameron sold the house for Euro 285000.

Work out whether Mrs Cameron made a profit or a loss on this project. How much was the profit/loss?

Task 3
Mrs Cameron liked the house so much that she decided to keep it and go there on holiday two or three times a year.
She decides to rent the house to other people when she is not there.
She could rent the house for £260 per week for 34 weeks every year.

How long will it take for the rental payments to cover the cost of building the house?

3 Algebra 1

In this unit you will:
- write, simplify, expand and factorise expressions
- set up and solve linear equations
- learn how to use trial and improvement to solve an equation
- learn how to find rules for linear sequences
- draw straight-line graphs
- learn how to find the midpoint of a line segment
- find the gradient and intercept of a straight-line graph
- draw and interpret travel graphs and other real-life graphs
- learn how to solve simultaneous equations.

Functional skills coverage and range:
- Understand and use simple equations and simple formulae involving one- or two-step operations.

Links
Mathematicians can model real-life situations using equations. Graphs are also useful for comparing information such as the amount of fuel consumed by different types of car over the same distance or exchange rates for different currencies.

3.1 Basic algebra

3.1.1 Using letters for numbers

Exercise 1 Ⓜ

In questions **1** and **2** find what expression I am left with if

1 **a** I start with x, double it and then subtract 6

$x \to \boxed{\times 2} \xrightarrow{2x} \boxed{-6} \to 2x - 6$

   b I start with x, add 4 and then square the result
   c I start with x, take away 5, double the result and then divide by 3
   d I start with x, multiply by 7 and then add n
   e I start with h, multiply by 4 and then subtract t
   f I start with x, add y and then double the result.

2 **a** I start with a, double it and then add b
   b I start with n, square it and then subtract n
   c I start with x, add 2 and then square the result
   d I start with w, subtract x and then square the result
   e I start with n, add p, cube the result and then divide by a
   f I start with t, subtract 1, treble the result and then square the new result.

3 A brick weighs w kg.
   a How much do 8 bricks weigh?
   b How much do n bricks weigh?

4 Susie shares n pounds equally between 3 people. How much does each person receive?

5 A man shares a sum of x pence equally between n children. How much does each child receive?

6 A cake weighing y kg is cut into n equal pieces. How much does each piece weigh?

7 A sum of £p is shared equally between you and four others. How much does each person receive?

8 Gary is n years older than Mark who is y years old. How old will Gary be in six years time?

9 Sangita used to earn £n per week. She then had a rise of £r per week. How much will she now earn in w weeks?

10 The height of a balloon increases at a steady rate of x metres in t hours.
   a How far will the balloon rise in 1 hour?
   b How far will the balloon rise in n hours?

11 How many drinks costing x pence each can be bought for £n?

12 A prize of £n is shared equally between you and x other people. How much does each person receive in **pence**?

3.1.2 Collecting like terms

'$3x + 4$' is an **expression**. It is **not** an equation because there is no equals sign. $3x$ and 4 are two **terms** in the expression.

- The expression $7x + 2x$ consists of two like terms, $7x$ and $2x$.
  Like terms can be added or subtracted. So $7x + 2x = 9x$.

- The expression $5x - 3y$ consists of two unlike terms, $5x$ and $3y$.
  Unlike terms cannot be added or subtracted to make one term.

Here are some examples:
- $8a + 7a - 3a = 12a$

- $3x + y + 2x = 5x + y$

- $x^2 + 3x^2 + 2x = 4x^2 + 2x$

- $5 + 3y + 6 + y^2 = 11 + 3y + y^2$

Exercise 2 Ⓜ

In questions **1** to **20** collect like terms together.

1 $2x + 3 + 3x + 5$

2 $4x + 8 + 5x - 3$

3 $5x - 3 + 2x + 7$

4 $6x + 1 + x + 3$

5 $4x - 3 + 2x + 10 + x$

6 $5x + 8 + x + 4 + 2x$

7 $7x - 9 + 2x + 3 + 3x$

8 $5x + 7 - 3x - 2$

9 $4x - 6 - 2x + 1$

10 $10x + 5 - 9x - 10 + x$

11 $4a + 6b + 3 + 9a - 3b - 4$

12 $8m - 3n + 1 + 6n + 2m + 7$

13 $6p - 4 + 5q - 3p - 4 - 7q$

14 $12s - 3t + 2 - 10s - 4t + 12$

15 $a - 2b - 7 + a + 2b + 8$

16 $3x + 2y + 5z - 2x - y + 2z$

17 $6x - 5y + 3z - x + y + z$

18 $2k - 3m + n + 3k - m - n$

19 $12a - 3 + 2b - 6 - 8a + 3b$

20 $3a + x + e - 2a - 5x - 6e$

21 Simplify where possible.

  a $x^2 - 3x + 1 + 2x^2 + 3x$

  b $5a + ab - 3a + 4ab$

  c $x^3 - 7x + 4x^2$

  d $ab + 3a^2 - 7a - ab + a^2$

22 Which of these expressions is equivalent to $x - 2$?

A $x^2 - 7x - 1 - x^2 + 8x + 3$

B $x^2 + 7x + 2x - 8x - x^2 - 2$

C $5 + 7x - x - 4 - 6x + x^2$

D $5x - 7 + 4 - x + 1 - 3x$

Collect like terms together.

23 $x^2 + 5x + 2 - 2x + 1$

24 $x^2 + 2x + 2x^2 + 4x + 5$

25 $x^2 + 5x + x^2 + x - 7$

26 $2x^2 - 3x + 8 + x^2 + 4x + 4$

27 $3x^2 + 4x + 6 - x^2 - 3x - 3$

28 $5x^2 - 3x + 2 - 3x^2 + 2x - 2$

29 $2x^2 - 2x + 3 - x^2 - 2x - 5$

30 $6x^2 - 7x + 8 - 3x^2 + 5x - 10$

31 $3y^2 - 6x + y^2 + x^2 + 7x + 4x^2$

32 $8 - 5x - 2x^2 + 4 + 6x + 2x^2$

33 $5 + 2y + 3y^2 - 8y - 6 + 2y^2 + 3$

34 $ab + a^2 - 3b + 2ab - a^2$

35 $3c^2 - d^2 + 2cd - 3c^2 - d^2$

36 $ab + 2a^2 + 3ab - 4a^2 + 2a$

37 $x^3 + 2x^2 - x + 3x^2 + x^3 + x$

38 $5 - x^2 - 2x^3 + 6 + 2x^2 + 3x^3$

39 $xy + ab - cd + 2xy - ab + dc$

40 $pq - 3qp + p^2 + 2qp - q^2$

3.1.3 Simplifying terms and brackets

Here are some examples of multiplying terms:

$3 \times 4x = 12x$ $5x \times 4x = 20x^2$ $-3 \times 2x = -6x$

$5(2a \times 3a) = 30a^2$ $3m \times 2n = 6mn$ $5b \times 2bc = 10b^2c$

$3(2x - 1) = 6x - 3$ $x(3x + 4) = 3x^2 + 4x$ $2(x + 3) + 5(x - 1)$

$a(a + 2b) = a^2 + 2ab$ $t^2(3t - 2) = 3t^3 - 2t^2$ $= 2x + 6 + 5x - 5$

                                                                   $= 7x + 1$

Exercise 3 Ⓜ

In questions **1** to **18** answer 'true' or 'false'.

1 $6 \times a = 6a$ **2** $3 \times n = 3 + n$ **3** $n \times n = n^2$

4 $a + b = b + a$ **5** $n \times n \times n = 3n$ **6** $a \times 7 = 7a$

7 $n - m = m - n$ **8** $a + a = 2a$ **9** $n + n + n = n^3$

10 $m + m^2 = m^3$ **11** $3c - c = 3$ **12** $a + 2b = 2b + a$

13 $a(m + n) = am + an$ **14** $n \div 2 = 2 \div n$ **15** $n \times n \times m = n^2 m$

16 $2n^2 = 4n^2$ **17** $(2n)^2 = 4n^2$ **18** $2n + 2n = 4n^2$

19 Here are some cards.

$n + n$		$n \times n^2$		$3n \div 3$		$n \div 4$

$n \times n \times n$		$n^2 \div n$		$4 \div n$		$5n - n$		$4n - 2n$

  a Which cards will always be the same as $\boxed{2n}$?

  b Which cards will always be the same as $\boxed{n^3}$?

  c Which cards will always be the same as $\boxed{n}$?

  d Which card will always be the same as $\dfrac{4}{n}$?

  e Draw a new card which will always be the same as $\boxed{n^2 + n^2}$.

20 In the expression $3n + 7$, two operations are performed in this order

$n \rightarrow \boxed{\times 3} \rightarrow \boxed{+ 7} \rightarrow 3n + 7$

Draw similar diagrams to show the correct order of operations for these expressions.

  a $6n - 1$ **b** $8n + 10$ **c** $\dfrac{n}{2} + 3$

  d $3(2n + 5)$ **e** $\dfrac{(2n - 4)}{5}$ **f** $\dfrac{(n^2 + 4)}{7}$

In questions **21** to **38** simplify the expressions.

21 $\dfrac{3n}{n}$ **22** $\dfrac{a}{a}$ **23** $\dfrac{n^2}{n}$

24 $6n - 5n$ **25** $a + b + c + a$ **26** $3n^2 - n^2$

27 $mn + mn$ **28** $\dfrac{n \times n \times n}{n}$ **29** $\dfrac{n + n + n}{n}$

30 $a \times a^2$ **31** $6n \div 6$ **32** $3t + 4 - 3p - 1$

33 $\dfrac{2a}{2a}$ **34** $2n + 2(n + 1)$ **35** $n + 4 + 4 + n$

36 $\dfrac{n}{2} + \dfrac{n}{2}$ **37** $\dfrac{a^2}{a^3}$ **38** $ab + ba + (a \times b)$

Exercise 4 Ⓜ ⭐ Ⓕ

1 Write in a simpler form.

a $2 \times 3x$	**b** $3a \times 5$	**c** $5t \times (-2)$	**d** $100 \times 10a$
e $x \times 2x$	**f** $x \times 4x$	**g** $x \times 2x^2$	**h** $4t \times 5t$

2 Remove the brackets.

a $2(2x + 1)$	**b** $5(2x + 3)$	**c** $a(a - 3)$	**d** $2n(n + 1)$
e $-2(2x + 3)$	**f** $2x(x + y)$	**g** $-3(a - 2)$	**h** $p(q - p)$
i $a(b + a)$	**j** $2b(b - 1)$	**k** $2n(n + 2)$	**l** $3x(2x + 3)$

3 Find four pairs of equivalent expressions or terms.

a $2a \times 3a^2$	**b** $6 \times a \times a$	**c** $4a^2$	**d** $6a^2$
e $(2a)^2$	**f** $3a \times 3a$	**g** $6a^3$	**h** $9a^2$

4 Write these terms more simply.

a $2a \times 3b$	**b** $x \times 3x^3$	**c** $y \times 2x$	**d** $2p \times 5q$
e $3x \times 5y$	**f** $6x \times 3x^2$	**g** $3a \times 8a^3$	**h** $ab \times 2a$
i $xy \times 3y$	**j** $cd \times 5c$	**k** $ab \times ab$	**l** $2xy \times xy$
m $3d \times d \times d$	**n** $5(2x \times xy)$	**o** $3a \times (-2a)$	**p** $-a \times (-2ab)$

5 A solid rectangular block measures x cm by x cm by $(x + 3)$ cm.
Find a simplified expression for its surface area in cm^2.

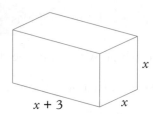

6 Remove the brackets and simplify.

a $3(x + 2) + 4(x + 1)$ **b** $5(x - 2) + 3(x + 4)$

c $2(a - 3) + 3(a + 1)$ **d** $5(a + 1) + 6(a + 2)$

e $7(a - 2) + (a + 4)$ **f** $3(t - 2) + 5(2 + t)$

g $3(x + 2) - 2(x + 1)$ **h** $4(x + 3) - 3(x + 2)$

7 Three rods A, B and C have lengths of x, $(x + 1)$ and $(x - 2)$ cm respectively, as shown.

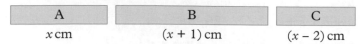

In the diagrams below express the length l in terms of x.
Give your answers in their simplest form.

a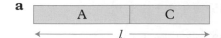

b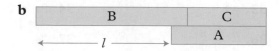

c  **d**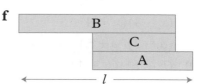

e 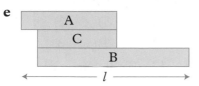 **f**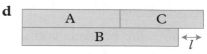

8 Simplify these expressions.

   a $x(2x + 1) + 3(x + 2)$ **b** $x(2x - 3) + 5(x + 1)$

   c $a(3a + 2) + 2(2a - 2)$ **d** $y(5y + 1) + 3(y - 1)$

   e $x(2x + 1) + x(3x + 1)$ **f** $a(2a + 3) - a(a + 1)$

3.1.4 Expanding brackets

When you work out the expression $(x + 3)(x + 2)$, the area of a rectangle can be a help.

From the diagram, the total area of the rectangle

$= (x + 3)(x + 2)$
$= x^2 + 2x + 3x + 6$
$= x^2 + 5x + 6$

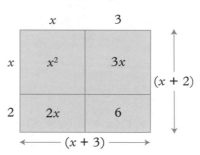

After a little practice, it is possible to do without the diagram.

Remove the brackets and simplify

a $(3x - 2)(2x - 1)$ **b** $(x - 3)^2$.

· ·

a $(3x - 2)(2x - 1) = 3x(2x - 1) - 2(2x - 1)$ Multiply $(2x - 1)$ by $3x$ and then by -2
$= 6x^2 - 3x - 4x + 2$
$= 6x^2 - 7x + 2$

b Be careful with this expression. It is not $x^2 - 9$, nor even $x^2 + 9$.

$(x - 3)^2 = (x - 3)(x - 3)$
$= x(x - 3) - 3(x - 3)$ Multiply $(x - 3)$ by x and then by -3
$= x^2 - 3x - 3x + 9$
$= x^2 - 6x + 9$

Exercise 5 Ⓜ/Ⓗ

Remove the brackets and simplify.

1 $(x + 1)(x + 3)$ **2** $(x + 3)(x + 2)$ **3** $(y + 4)(y + 5)$

4 $(x - 3)(x + 4)$ **5** $(x + 5)(x - 2)$ **6** $(x - 3)(x - 2)$

7 $(a - 7)(a + 5)$ **8** $(z + 9)(z - 2)$ **9** $(x - 3)(x + 3)$

10 $(k - 11)(k + 11)$ **11** $(2x + 1)(x - 3)$ **12** $(3x + 4)(x - 2)$

13 $(2y - 3)(y + 1)$ **14** $(7y - 1)(7y + 1)$ **15** $(x + 4)^2$

16 $(x + 2)^2$ **17** $(x - 2)^2$ **18** $(2x + 1)^2$

19 $(x + 1)^2 + (x + 2)^2$ **20** $(x - 2)^2 + (x + 3)^2$ **21** $(x + 2)^2 + (2x + 1)^2$

22 $(y - 3)^2 + (y - 4)^2$ **23** $(x + 2)^2 - (x - 3)^2$ **24** $(x - 3)^2 - (x + 1)^2$

 25 Four identical tiles surround the blue square.
  a Find an expression for the sum of the areas
     of the four tiles.
  b Find the area of square ABCD.
  c Hence find the area of the blue square.

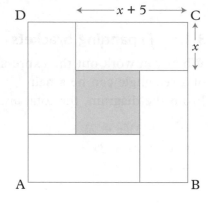

*26 You can expand the expression $(x + 1)(x + 2)(x + 3)$ in stages.
Copy and complete the working.

$$(x + 1)\,[(x + 2)\,(x + 3)]$$
$$= (x + 1)\,(x^2 + 5x + 6)$$
$$= \ldots$$

> Multiply the
> second bracket by
> x and then by 1.

*27 Remove the brackets and simplify.
  a $(x + 1)\,(x + 3)\,(x + 4)$ **b** $(x + 1)^3$ **c** $(x - 1)\,(x + 2)^2$

*28 **a** If n is an integer, explain why $(2n - 1)$ is an odd number.
  b Write the next odd number after $(2n - 1)$ and show that
     the sum of these two numbers is a multiple of 4.
  c Write the next odd number after these two and show that
     the sum of the three odd numbers is a multiple of 3.
  d Show that the sum of the squares of these three odd numbers
     is $12n^2 + 12n + 11$.
     Is it possible for this expression to be a multiple of 11? If so
     give two possible values of n.

3.1.5 Factorising expressions

Earlier you expanded expressions such as $x(3x - 1)$ to give
$3x^2 - x$.
The reverse of this process is called **factorising**.

> **EXAMPLE**
>
> Factorise   **a** $x^2 + 7x$ **b** $3y^2 - 12y$ **c** $6a^2b - 10ab^2$
>
> **a** x is common to x^2 and $7x$, so $x^2 + 7x = x(x + 7)$.
>   The factors are x and $(x + 7)$.
>
> **b** $3y$ is common, so $3y^2 - 12y = 3y(y - 4)$
>
> **c** $2ab$ is common, so $6a^2b - 10ab^2 = 2ab(3a - 5b)$

Exercise 6 Ⓜ/Ⓗ

Factorise these expressions completely.

1 $x^2 + 5x$	**2** $x^2 - 6x$	**3** $7x - x^2$
4 $y^2 + 8y$	**5** $2y^2 + 3y$	**6** $6y^2 - 4y$
7 $3x^2 - 21x$	**8** $16a - 2a^2$	**9** $6c^2 - 21c$
10 $15x - 9x^2$	**11** $56y - 21y^2$	**12** $ax + bx + 2cx$
13 $x^2 + xy + 3xz$	**14** $x^2y + y^3 + z^2y$	**15** $3a^2b + 2ab^2$
16 $x^2y + xy^2$	**17** $6a^2 + 4ab + 2ac$	**18** $ma + 2bm + m^2$
19 $2kx + 6ky + 4kz$	**20** $ax^2 + ay + 2ab$	**21** $7x^2 + x$
22 $4y^2 - 4y$	**23** $p^2 - 2p$	**24** $6a^2 + 2a$
25 $4 - 8x^2$	**26** $5x - 10x^3$	**27** $4\pi r + \pi h$
28 $\pi r^2 + 2\pi r$	**29** $3\pi r^2 + \pi rh$	**30** $3xy + 2x$
31 $x^2k + xk^2$	**32** $a^3b + 2ab^2$	**33** $abc - 3b^2c$
34 $2a^2e - 5ae^2$	**35** $a^3b + ab^3$	**36** $x^3y + x^2y^2$
37 $6xy^2 - 4x^2y$	**38** $3ab^3 - 3a^3b$	**39** $2a^3b + 5a^2b^2$
40 $ax^2y - 2ax^2z$	**41** $2abx + 2ab^2 + 2a^2b$	**42** $ayx + yx^3 - 2y^2x^2$

43   In these two rectangles you are given one side and the area.
     Find the perimeter of each rectangle.

a
area =
$2xy + 4y^2$ $2y$

b
area =
$6a^2 - 2ax$ $2a$

3.1.6 Mathematical language

In algebra, you use letter symbols to represent unknowns in a variety of situations.

- In an **equation** the letters stand for particular numbers (the solutions of the equation), for example, $3x - 1 = 2(x + 1)$.
- In an **expression** there is no equals sign, for example, $5x^2 - 3x + 1$.
- In **formulae** letters stand for defined quantities or variables, for example, $F = ma$. (F is force, m is mass and a is acceleration.)
- In an **identity** there is an equals sign but the equality holds for **all values** of the unknown, for example, $(x^2 - 1) = (x - 1)(x + 1)$
- In the definition of **functions**, for example, $y = x^2 + 7x$, $f(x) = 3x^2$.

> Identities are usually written with an identity sign, $\equiv$.

Exercise 7 Ⓗ

Copy and complete the table by putting the contents of each box in the correct column.

$$x(x + 1) = x^2 + x$$

$$7y + 10$$

$$V = IR$$

$$x^2 - 3x + 10$$

$$7x + 11 = x - 9$$

$$A = \pi r^2$$

$$(x + 1)^2 = x^2 + 2x + 1$$

$$x^2 - 7x = 0$$

Equation	Expression	Identity	Formula

3.2 Linear equations

Many questions in mathematics are easier to answer if algebra is used. It is often best to choose x for the unknown number and to try to translate the question into the form of an equation.

3.2.1 Solving equations

Basic rules

● **Rule 1** Do the same thing to both sides.

a You can add the same number to both sides.

$3x - 1 = 4$ add 1

$3x - 1 + 1 = 4 + 1$

$3x = 5$ $x = \dfrac{5}{3}$

b You can subtract the same number from both sides.

$5x + 7 = 10$ take away 7

$5x + 7 - 7 = 10 - 7$

$5x = 3$ $x = \dfrac{3}{5}$

c You can multiply both sides by the same number.

$\dfrac{x}{4} = 3$ multiply by 4

$4\left(\dfrac{x}{4}\right) = 4 \times 3$ $x = 12$

d You can divide both sides by the same number.

$3x = 8$ divide by 3

$\dfrac{3x}{3} = \dfrac{8}{3}$ $x = \dfrac{8}{3} = 2\dfrac{2}{3}$

Exercise 8 Ⓜ

Solve these equations.

1 $x + 7 = 10$	**2** $x + 3 = 20$	**3** $x - 7 = 7$
4 $x - 5 = 11$	**5** $6 + x = 13$	**6** $8 + x = 15$
7 $7 = x + 4$	**8** $7 = x - 6$	**9** $1 = x - 3$
10 $7 + x = 7$	**11** $x - 11 = 20$	**12** $14 = 6 + x$
13 $2x = 20$	**14** $3x = 24$	**15** $5x = 40$
16 $3x + 1 = 10$	**17** $4x + 2 = 22$	**18** $5x + 3 = 18$
19 $2n - 3 = 1$	**20** $3n - 4 = 8$	**21** $6n - 1 = 5$
22 $4a + 3 = 3$	**23** $6a + 7 = 19$	**24** $4 + 2a = 6$
25 $3 + 5y = 4$	**26** $6p - 10 = 32$	**27** $3x - 100 = 500$
28 $11q - 17 = 60$	**29** $3n + 7 = 7$	**30** $8y - 1 = 0$

In question **7**
$7 = x + 4$ is the
same as $x + 4 = 7$.

● **Rule 2** If the x-term is negative, take it to the other side where it becomes positive.

● **Rule 3** If there are x-terms on both sides, collect them on one side.

Solve the equations

a $4 - 3x = 2$

b $2x - 7 = 5 - 3x$

....

a $4 - 3x = 2$ $[+3x]$

       $4 = 2 + 3x$ $[-2]$

       $2 = 3x$ $[\div 3]$

       $\dfrac{2}{3} = x$

b $2x - 7 = 5 - 3x$ $[+3x, +7]$

       $2x + 3x = 5 + 7$

          $5x = 12$ $[\div 5]$

          $x = \dfrac{12}{5} = 2\dfrac{2}{5}$

Exercise 9 Ⓜ

Solve these equations.

1 $2x - 5 = 11$ **2** $3x - 7 = 20$ **3** $2x + 6 = 20$ **4** $5x + 10 = 60$

5 $8 = 7 + 3x$ **6** $12 = 2x - 8$ **7** $-7 = 2x - 10$ **8** $3x - 7 = -10$

9 $12 = 15 + 2x$ **10** $5 + 6x = 7$ **11** $100x - 1 = 98$ **12** $7 = 7 + 7x$

13 $\dfrac{x}{100} + 10 = 20$ **14** $1000x - 5 = -6$ **15** $-4 = -7 + 3x$ **16** $2x + 4 = x - 3$

17 $x - 3 = 3x + 7$ **18** $5x - 4 = 3 - x$ **19** $4 - 3x = 1$ **20** $5 - 4x = -3$

21 $7 = 2 - x$ **22** $3 - 2x = x + 12$ **23** $6 + 2a = 3$ **24** $a - 3 = 3a - 7$

25 $2y - 1 = 4 - 3y$ **26** $7 - 2x = 2x - 7$ **27** $7 - 3x = 5 - 2x$ **28** $8 - 2y = 5 - 5y$

29 $x - 16 = 16 - 2x$ **30** $x + 2 = 3 \cdot 1$ **31** $-x - 4 = -3$ **32** $-3 - x = -5$

● **Rule 4** If there is a fraction in the x-term, multiply out to simplify the equation.

Solve the equations

a $\dfrac{2x}{3} = 10$ **b** $3 = \dfrac{x}{4} - 4$ **c** $\dfrac{a}{3} + 7 = 5$

....

a $\dfrac{2x}{3} = 10$

     $[\times 3]$

     $2x = 30$

     $x = \dfrac{30}{2} = 15$

b $3 = \dfrac{x}{4} - 4$

     $[+4]$

     $7 = \dfrac{x}{4}$

     $[\times 4]$

     $28 = x$

c $\dfrac{a}{3} + 7 = 5$

     $[-7]$

     $\dfrac{a}{3} = -2$

     $[\times 3]$

     $a = -6$

Exercise 10 Ⓜ/Ⓗ

Solve these equations.

1 $\dfrac{x}{5} = 7$ **2** $\dfrac{x}{10} = 13$ **3** $7 = \dfrac{x}{2}$ **4** $\dfrac{x}{2} = \dfrac{1}{3}$

5 $\dfrac{3x}{2} = 5$ **6** $\dfrac{4x}{5} = -2$ **7** $7 = \dfrac{7x}{3}$ **8** $\dfrac{3}{4} = \dfrac{2x}{3}$

9 $\dfrac{5x}{6} = \dfrac{1}{4}$ **10** $-\dfrac{3}{4} = \dfrac{3x}{5}$ **11** $\dfrac{x}{2} + 7 = 12$ **12** $\dfrac{x}{3} - 7 = 2$

13 $\dfrac{x}{5} - 6 = -2$ **14** $4 = \dfrac{x}{2} - 5$ **15** $10 = 3 + \dfrac{x}{4}$ **16** $\dfrac{a}{5} - 1 = -4$

17 $-\dfrac{x}{2} + 1 = -\dfrac{1}{4}$ **18** $-\dfrac{3}{5} + \dfrac{x}{10} = -\dfrac{1}{5} - \dfrac{x}{5}$

● **Rule 5** When an equation has brackets, multiply out the brackets first.

EXAMPLE

Solve the equations

a $x - 2(x - 1) = 1 - 4(x + 1)$ **b** $\dfrac{5}{x} = 2$ **c** $\dfrac{3x - 1}{2} = \dfrac{4x}{3}$

....

a $x - 2(x - 1) = 1 - 4(x + 1)$ **b** $\dfrac{5}{x} = 2$ **c** $\dfrac{3x - 1}{2} = \dfrac{4x}{3}$ [Multiply by 6]

$\quad x - 2x + 2 = 1 - 4x - 4$ $\quad 5 = 2x$ $\quad \dfrac{6(3x - 1)}{2} = \dfrac{6(4x)}{3}$ [Cancel]

$\quad x - 2x + 4x = 1 - 4 - 2$ $\quad \dfrac{5}{2} = x$ $\quad 3(3x - 1) = 2(4x)$

$\quad\quad\quad\quad 3x = -5$ $\quad\quad\quad\quad 9x - 3 = 8x$

$\quad\quad\quad\quad x = -\dfrac{5}{3}$ $\quad\quad\quad\quad x = 3$

Exercise 11 Ⓜ/Ⓗ

Solve these equations.

1 $x + 3(x + 1) = 2x$

2 $1 + 3(x - 1) = 4$

3 $2x - 2(x + 1) = 5x$

4 $2(3x - 1) = 3(x - 1)$

5 $4(x - 1) = 2(3 - x)$

6 $4(x - 1) - 2 = 3x$

7 $4(1 - 2x) = 3(2 - x)$

8 $3 - 2(2x + 1) = x + 17$

9 $4x = x - (x - 2)$

10 $7x = 3x - (x + 20)$

11 $5x - 3(x - 1) = 39$

12 $3x + 2(x - 5) = 15$

13 $7 - (x + 1) = 9 - (2x - 1)$

14 $10x - (2x + 3) = 21$

15 $3(2x + 1) + 2(x - 1) = 23$

16 $5(1 - 2x) - 3(4 + 4x) = 0$

17 $7x - (2 - x) = 0$

18 $3(x + 1) = 4 - (x - 3)$

19 $3y + 7 + 3(y - 1) = 2(2y + 6)$

20 $4(y - 1) + 3(y + 2) = 5(y - 4)$

Exercise 12 Ⓗ

Solve these equations.

1 $\dfrac{7}{x} = 21$

2 $30 = \dfrac{6}{x}$

3 $\dfrac{5}{x} = 3$

4 $\dfrac{9}{x} = -3$

5 $11 = \dfrac{5}{x}$

6 $-2 = \dfrac{4}{x}$

7 $\dfrac{x + 1}{3} = \dfrac{x - 1}{4}$

8 $\dfrac{x + 3}{2} = \dfrac{x - 4}{5}$

9 $\dfrac{2x - 1}{3} = \dfrac{x}{2}$

10 $\dfrac{3x + 1}{5} = \dfrac{2x}{3}$

11 $\dfrac{5}{x - 1} = \dfrac{10}{x}$

12 $\dfrac{12}{2x - 3} = 4$

13 $2 = \dfrac{18}{x + 4}$

14 $\dfrac{5}{x + 5} = \dfrac{15}{x + 7}$

15 $\dfrac{4}{x} + 2 = 3$

16 $\dfrac{6}{x} - 3 = 7$

17 $\dfrac{9}{x} - 7 = 1$

18 $-2 = 1 + \dfrac{3}{x}$

19 $4 - \dfrac{4}{x} = 0$

20 $5 - \dfrac{6}{x} = -1$

21 $\dfrac{x}{3} + \dfrac{x}{4} = 1$

> For question **7**
> see example **c**
> on page 107.

3.2.2 Setting up equations

EXAMPLE

If you multiply a 'mystery' number by 2 and then add 3 the answer is 14. Find the 'mystery' number.

....

Let the mystery number be x.

Then $\quad 2x + 3 = 14$

$\qquad\quad 2x = 11$

$\qquad\quad x = 5\frac{1}{2}$

The 'mystery' number is $5\frac{1}{2}$.

Exercise 13 Ⓜ

Find the 'mystery' number in each question by forming an equation and then solving it.

1 If you multiply the number by 3 and then add 4, the answer is 13.

2 If you multiply the number by 4 and then add 5, the answer is 8.

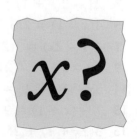

3 If you multiply the number by 2 and then subtract 5, the answer is 4.

4 If you multiply the number by 10 and then add 19, the answer is 16.

5 If you add 3 to the number and then multiply the result by 4, the answer is 10.

6 If you subtract 11 from the number and then treble the result, the answer is 20.

7 If you double the number, add 4 and then multiply the result by 3, the answer is 13.

8 If you treble the number, take away 6 and then multiply the result by 2, the answer is 18.

9 If you double the number and subtract 7 you get the same answer as when you add 5 to the number.

10 If you multiply the number by 5 and subtract 4, you get the same answer as when you add 3 to the number and then double the result.

11 If you multiply the number by 6 and add 1, you get the same answer as when you add 5 to the number and then treble the result.

12 If you add 5 to the number and then multiply the result by 4, you get the same answer as when you add 1 to the number and then multiply the result by 2.

3.2.3 Solving problems using linear equations

So far you have concentrated on solving given equations. Making up your own equations helps you to solve difficult problems.

There are four steps.

i   Call the unknown number x (or any other suitable letter) and state the units where appropriate.

ii Write the problem in the form of an equation.

iii Solve the equation and give the answer in words.

iv Check your solution using the problem and **not** your equation.

EXAMPLE

Find three consecutive even numbers which add up to 792.

· ·

i   Call the smallest number x.
Then the other numbers are $(x + 2)$ and $(x + 4)$ because
they are consecutive **even** numbers.

ii Form an equation
$x + (x + 2) + (x + 4) = 792$

iii   Solve: $3x + 6 = 792$
$$3x = 786$$
$$x = 262$$
The three numbers are 262, 264 and 266.

iv Check $262 + 264 + 266 = 792$ ✓

Exercise 14Ⓜ ⭐Ｆ

Solve each problem by forming an equation. The first questions are
easy but should still be solved using an equation, in order to
practise the method.

1 The length of a rectangle is twice the width.
If the perimeter is 20 cm, find the width.

x

2 The width of a rectangle is one-third of the length. If the perimeter
is 96 cm, find the width.

3 The sum of three consecutive numbers is 276. Find the
numbers.
Let the first number be x.

4 The sum of four consecutive numbers is 90.
Find the numbers.

5 The sum of three consecutive odd numbers is 177.
Find the numbers.

6 Find three consecutive even numbers which add up to 1524.

7 When a number is doubled and then added to 13, the result
is 38. Find the number.

8 If AB is a straight line, find x.

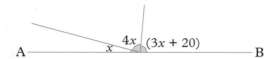

9 The difference between two numbers is 9. Find the numbers, if their sum is 46.

10 The three angles in a triangle are in the ratio $1:3:5$. Find the angles.

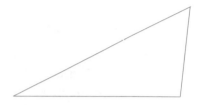

11 The sum of three numbers is 28. The second number is three times the first and the third is 7 less than the second. What are the numbers?

12 If the perimeter of this triangle is 22 cm, find the length of the shortest side.

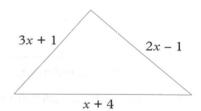

13 David weighs 5 kg less than John, who in turn is 8 kg lighter than Paul. If their total weight is 197 kg, how heavy is each person?

14 The perimeter of the rectangle is 34 cm. Find x.

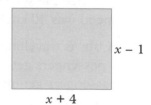

15 The diagram shows a rectangular lawn surrounded by a footpath x m wide.
  a Show that the area of the path is $4x^2 + 14x$.
  b Find an expression, in terms of x, for the distance around the outside edge of the path.
  c Find the value of x when this perimeter is 20 m.

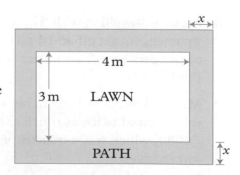

 **Exercise 15** Ⓜ ⭐Ⓕ

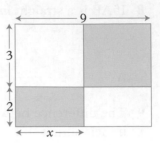

1 Find the value of x so that the areas of the shaded rectangles are equal.

2 Two angles of an isosceles triangle are $a°$ and $(a + 10)°$. Find two possible values of a.

3 The perimeter of the rectangular picture is 37 cm. Find the value of x.

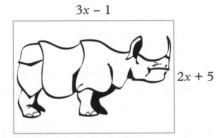

4 A man is 32 years older than his son. Ten years ago he was three times as old as his son was then. Find the present age of each.

5 A man runs to a telephone and back in 900 seconds. His speed on the way to the telephone is 5 m/s and his speed on the way back is 4 m/s. Find the distance to the telephone.

6 A car completes a journey in 10 minutes. For the first half of the distance the speed was 60 km/h and for the second half the speed was 40 km/h. How far is the journey?

7 A bus is travelling with 48 passengers. When it arrives at a stop, x passengers get off and 3 get on. At the next stop half the passengers get off and 7 get on. There are now 22 passengers. Find x.

8 A bus is travelling with 52 passengers. When it arrives at a stop, y passengers get off and 4 get on. At the next stop one-third of the passengers get off and 3 get on. There are now 25 passengers. Find y.

9 Mr Lee left his fortune to his 3 sons, 4 nieces and his wife. Each son received twice as much as each niece and his wife received £6000, which was a quarter of the money. How much did each son receive?

*10 The diagrams show a table with two identical wooden blocks.

Calculate the height of the table, x.

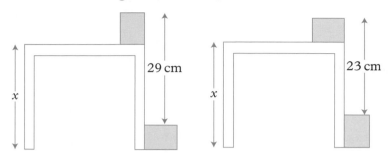

3.2.4 Solving equations involving brackets

Solve the equation $(x + 3)^2 = (x + 2)^2 + 3^2$.

$$(x + 3)^2 = (x + 2)^2 + 3^2$$
$$(x + 3)(x + 3) = (x + 2)(x + 2) + 9$$
$$x^2 + 6x + 9 = x^2 + 4x + 4 + 9$$
$$6x + 9 = 4x + 13 \qquad \text{[subtract } x^2\text{]}$$
$$2x = 4$$
$$x = 2$$

Exercise 16 Ⓜ/Ⓗ

Solve these equations.

1 $x^2 + 4 = (x + 1)(x + 3)$ **2** $x^2 + 3x = (x + 3)(x + 1)$

3 $(x + 3)(x - 1) = x^2 + 5$ **4** $(x + 1)(x + 4) = (x - 7)(x + 6)$

5 $(x - 2)(x + 3) = (x - 7)(x + 7)$ **6** $(x - 5)(x + 4) = (x + 7)(x - 6)$

7 $2x^2 + 3x = (2x - 1)(x + 1)$ **8** $(2x - 1)(x - 3) = (2x - 3)(x - 1)$

9 $x^2 + (x + 1)^2 = (2x - 1)(x + 4)$ **10** $x(2x + 6) = 2(x^2 - 5)$

In questions **11** and **12**, form an equation in x using Pythagoras' theorem, and hence find the length of each side of the triangle. (All the lengths are in cm.)

11

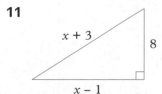

12

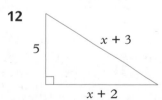

13 The area of this rectangle exceeds the area of the square by $2\,\text{cm}^2$. Find x.

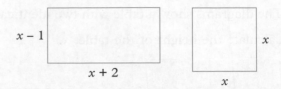

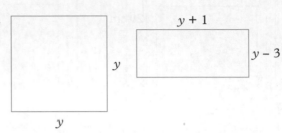

14 The area of the square exceeds the area of the rectangle by $13\,\text{m}^2$. Find y.

3.3 Trial and improvement

3.3.1 Using trial and improvement

Sometimes you cannot write an equation to solve a problem. In such cases, the method of 'trial and improvement' is often a help.

Exercise 17 Ⓜ

1 Think of a rectangle having an area of $72\,\text{cm}^2$ with a base twice its height.
Write the length of the base.

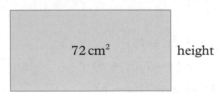

height

2 Find a rectangle of area $75\,\text{cm}^2$ so that its base is three times its height.

3 In each of these rectangles, the base is twice the height. The area is shown inside the rectangle. Find the base and the height.

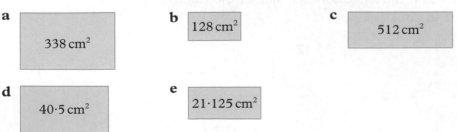

a $338\,\text{cm}^2$

b $128\,\text{cm}^2$

c $512\,\text{cm}^2$

d $40{\cdot}5\,\text{cm}^2$

e $21{\cdot}125\,\text{cm}^2$

4 In each of these rectangles, the base is 1 cm more than the height.
Find each base and height.

a
342 cm²

b
35·75 cm²

c
66·99 cm²

3.3.2 Inexact answers

In some questions it is not possible to find an answer which is precisely
correct. However, you can find answers which are nearer and nearer to the
exact one, perhaps to the nearest 0·1 cm or even to the nearest 0·01 cm.

● You can use a spreadsheet to help you with the calculations.

EXAMPLE

In this rectangle, the base is 1 cm more than the height h cm.
The area is 80 cm². Find the height h, correct to 1 dp.

80 cm² h

$h + 1$

Here you have to solve the equation $h(h + 1) = 80$.

a Try different values for h. **Be systematic**.

$h = 8$	: $8(8 + 1)$	$= 72$	Too small
$h = 9$	: $9(9 + 1)$	$= 90$	Too large
$h = 8·5$	: $(8·5 + 1)$	$= 80·75$	Too large
$h = 8·4$	: $8·4(8·4 + 1)$	$= 78·96$	Too small

b You can now see that the answer is between 8·4 and 8·5 so now
try $x = 8·45$, which is mid-way between 8·4 and 8·5.
$h = 8·45$: $8·45(8·45 + 1) = 79·8525$, which is too small.
So 8·45 is too small for h and the answer is $h = 8·5$, correct to 1 decimal place.

EXAMPLE

Solve the equation $x^3 + 2x = 90$, correct to 1 dp.

Try $x = 4$	: $4^3 + (2 \times 4)$	$= 72$	Too small
$x = 5$	: $5^3 + (2 \times 5)$	$= 135$	Too large
$x = 4·5$	: $4·5^3 + (2 \times 4·5)$	$= 100·125$	Too large
$x = 4·4$	: $4·4^3 + (2 \times 4·4)$	$= 93·984$	Too large
$x = 4·3$	: $4·3^3 + (2 \times 4·3)$	$= 88·107$	Too small
$x = 4·35$	: $4·35^3 + (2 \times 4·35)$	$= 91·013$	Too large

∴ The solution is $x = 4·3$, correct to 1 decimal place.

Exercise 18 Ⓜ

1 In these two rectangles, the base is 1 cm more than the height.

a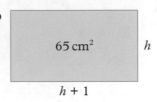

100 cm² h

$h + 1$

b

65 cm² h

$h + 1$

Find the value of h for each one, correct to 1 decimal place.

2 Find a solution to each equation correct to 1 decimal place. Do not use the cube root key on your calculator.
  a $x^3 = 40$ **b** $x^3 = 100$ **c** $x^3 = 300$

3 Find the cube root of each of these numbers correct to 1 decimal place. Do not use the cube root key on your calculator.
  a 4·7   **b** 28   **c** 225

4 Find the positive solutions to these equations, giving the answer correct to 1 decimal place.
  a $x(x - 3) = 11$ **b** $x(x - 2) = 7$
  c $x(x + 1) = 20$ **d** $x^3 + x = 40$

5 An engineer wants to make a solid metal cube of volume 526 cm³. Call the edge of the cube x and write an equation. Find x, giving your answer correct to 1 dp.

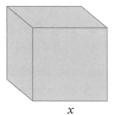

x

6 In this rectangle, the base is 2 cm more than the height.
  If the area is 20 cm² find h correct to 2 dp.

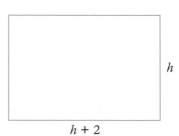

h

$h + 2$

7 For this question you need a calculator with an x^y or a $\wedge$ button.
  Find x correct to 1 decimal place.
  a $x^5 = 313$ **b** $5^x = 77$ **c** $x^x = 100$

8 The number n has two digits after the decimal point and n to the power 10 is 2110 to the nearest whole number. Find n.

*9 The diagram represents a rectangular piece of paper ABCD which has been folded along EF so that C has moved to G.

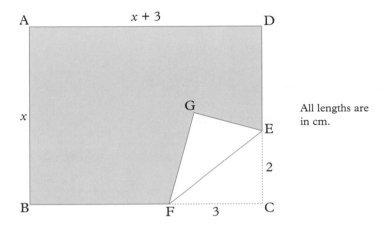

All lengths are in cm.

   a Calculate the area of $\triangle$ECF.
   b Find an expression for the shaded area ABFGED in terms of x.

Given that the shaded area is 20 cm^2, show that $x(x + 3) = 26$
Solve this equation, giving your answer correct to one decimal place.

*10 In the rectangle PQRS, PQ = x cm and QR = 1 cm. The line LM is drawn so that PLMS is a square.
   a Write, in terms of x, the length LQ.
   b If $\dfrac{\text{PQ}}{\text{QR}} = \dfrac{\text{QR}}{\text{LQ}}$, obtain an equation in x.

Hence find x correct to two decimal places.

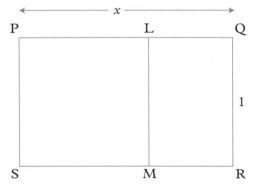

*11 Find the number with a cube 100 times as large as its square root.

$$\square^3 = 100\sqrt{\square}$$

Give your answer correct to 1 decimal place.

*12 In triangle ABC, angle BAC = $x°$, BC = x cm and
AB = $(x + 10)$ cm.

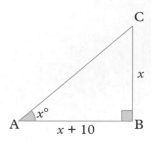

Find the value of x, correct to the nearest 0·5.

*13 A rectangle is drawn inside a semicircle of radius 1 unit.

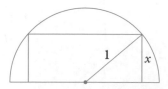

    a Find an expression, in terms of x, for the area of
the rectangle.
    b Use trial and improvement, possibly using a spreadsheet
program, to find the value of x, correct to 2 dp, which
gives the largest value for the area of the rectangle.

*14 The net for an open box is made by cutting corners
from a square card as shown.

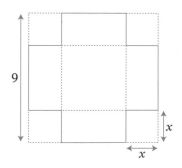

    a Show that the volume of the box is given by the formula
$V = x(9 - 2x)(9 - 2x)$.
    b Find the value of x which gives the maximum volume
of the box. Again you may find it helpful to use a
spreadsheet.

3.4 Sequences

3.4.1 Finding a rule

Here is a sequence of 'houses' made from matches.

   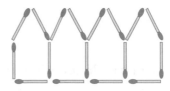

$h = 1, m = 5$ $h = 2, m = 9$ $h = 3, m = 13$

The table on the right records the number of houses h and the number of matches m.

If the number in the h column goes up one at a time, look at the number in the m column. If it goes up (or down) by the same number each time, the function connecting m and h is linear. This means that there are no terms in h^2 or anything more complicated.

h		m
1		5
2		9
3		13
4		17

In this case, the numbers in the m column go up by 4 each time. This suggests that a column for $4h$ might help.

Now it is fairly clear that m is one more than $4h$. So the formula linking m and h is: $\boldsymbol{m = 4h + 1}$

h	4h	m
1	4	5
2	8	9
3	12	13
4	16	17

EXAMPLE

The table shows how r changes with n. What is the formula linking r with n?

n		r
2		3
3		8
4		13
5		18

....

Because r goes up by 5 each time, write another column for $5n$.

n	5n	r
2	10	3
3	15	8
4	20	13
5	25	18

This method only works when the first set of numbers goes up by one each time.

The table shows that r is always 7 less than $5n$, so the formula linking r with n is $r = 5n - 7$.

Exercise 19 Ⓜ

1 This sequence of diagrams shows blue tiles b and white tiles w with the related table.

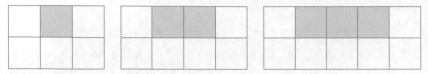

b		w
1		5
2		6
3		7
4		8

What is the formula for w in terms of b? Write it as $w = \ldots$

2 This is a different sequence with blue tiles b and white tiles w and the related table.

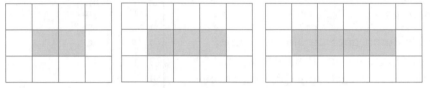

b	$2b$	w
2		10
3		12
4		14
5		16

What is the formula? Write it as $w = \ldots$

3 Here is a sequence of Is.

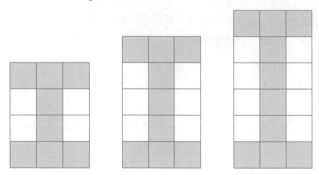

Make your own table for blue tiles b and white tiles w. What is the formula for w in terms of b?

4 This sequence shows matches m arranged in triangles t.

a Make a table for t and m starting like this:

t		m
1		3
2		5
⋮		⋮

b Continue the table and find a formula for m in terms of t.
Write it as $m = \ldots$

5 Here is a different sequence of matches and triangles.

Make a table and find a formula connecting *m* and *t*.

6 In this sequence, there are triangles *t* and squares *s* around the outside.

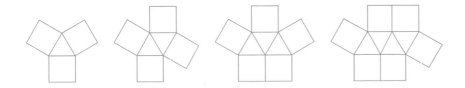

What is the formula connecting *t* and *s*?

7 Look at these tables. In each case, find a formula connecting the two letters.

a

n	p
1	3
2	8
3	13
4	18

b

n	k
2	17
3	24
4	31
5	38

c

n	w
3	17
4	19
5	21
6	23

8 This is one member of a sequence of cubes *c* made from matches *m*.

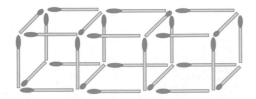

Find a formula connecting *m* and *c*.

9 In these tables, the numbers on the left do not go up by one each time. Try to find a formula in each case.

a

n	y
1	4
3	10
7	22
8	25

b

n	h
2	5
3	9
6	21
10	37

c

n	k
3	14
7	26
9	32
12	41

10 Some attractive loops can be made by fitting pentagon tiles together.

Diagram $n = 1$　　　　Diagram $n = 2$　　　　Diagram $n = 3$

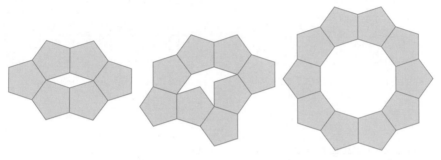

6 tiles t　　　　　8 tiles t　　　　　10 tiles t
14 external edges e　　17 external edges e　　20 external edges e

Make your own tables involving n, t and e.
a Find a formula connecting n and t.
b Find a formula connecting n and e.
c Find a formula connecting t and e.

11 A group of workers were made redundant and the management used a formula to calculate their redundancy money.

The table shows the details.
a Sam has worked for 7 years. How much redundancy money will he receive?
b Write the formula connecting the amount of redundancy paid R and the number of years worked N.

Number of years worked N	Redundancy money £R
1	£5400
2	£5800
3	£6200
4	£6600
10	£9000

12 In each diagram there are blue squares along one diagonal. Find a formula for the number of white squares in the $n \times n$ square.

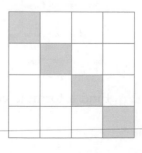

3.4.2 The *n*th term in a sequence

● For the sequence 3, 6, 9, 12, 15 ... the rule is 'add 3'.
Here is the **mapping diagram** for the sequence.
You can find the terms by multiplying the
term number by 3.
So the 10th term is 30, the 33rd term is 99.

A **general** term in the sequence is the *n*th term, where
n stands for any number.
The *n*th term of this sequence is $3n$.

Term number (*n*)		Term
1	→	3
2	→	6
3	→	9
⋮		⋮
10	→	30
⋮		
n	→	$3n$

● Here is a more difficult sequence: 3, 7, 11, 15, ...

The rule is 'add 4' so, in the mapping diagram,
you need a column for 4 times the term number
[that is, $4n$].

You can see that each term is 1 less than $4n$.
So, the 10th term is $(4 \times 10) - 1 = 39$
the *n*th term is $(4 \times n) - 1 = 4n - 1$

Term number (*n*)		4*n*		Term
1	→	4	→	3
2	→	8	→	7
3	→	12	→	11
4	→	16	→	15

Exercise 20 Ⓜ

1 Write each sequence and select the correct expression for the
*n*th term from the list.

  a 10, 20, 30, 40, ...
  b 5, 10, 15, 20, ...
  c 3, 4, 5, 6, ...
  d 3, 5, 7, 9, 11, ...
  e 30, 60, 90, 120, ...
  f 5, 11, 17, 23, ...
  g 1, 4, 7, 10, 13, ...

$5n$ $n + 2$ $6n - 1$ $2n + 1$ $10n$ $30n$ $3n - 2$

2 Copy and complete the mapping diagrams in the tables.
Notice the extra column.

a

Term number (*n*)		2*n*		Term
1	→	2	→	5
2	→	4	→	7
3	→	6	→	9
4	→	8	→	11
⋮				
10	→	☐	→	☐
⋮				
n	→	☐	→	☐

b

Term number (*n*)		3*n*		Term
1	→	3	→	4
2	→	6	→	7
3	→	9	→	10
⋮				
20	→	☐	→	☐
⋮				
n	→	☐	→	☐

3 Here are the first five terms of a sequence: 6, 11, 16, 21, 26, . . .

n	5n	Term
1		6
2		11

> The rule for the sequence is 'add 5' so write a column for '5n' in the table.

a Draw a mapping diagram like those in question **2**.

b Write   **i**   the 10th term

          ii   the nth term.

4 Here are three sequences

A: 1, 4, 7, 10, 13, . . .

B: 6, 10, 14, 18, 22, . . .

C: 5, 12, 19, 26, 33, . . .

For each sequence   **a**   draw a mapping diagram

                      b   find the nth term.

5 Look at the sequences A1, A2, A3, A4 . . . and B1, B2, B3, B4 . . .

Write

a the term A10   **b** the term B10

c the nth term in the 'A' sequence.   **d** the nth term in the 'B' sequence.

6

a Find the nth term in the 'P' sequence.

b Find the nth term in the 'Q' sequence.

7 Here are two more difficult sequences. Copy and complete the mapping diagrams.

a

Term number (n)		Term
1	→ 1×2 →	2
2	→ 2×3 →	6
3	→ 3×4 →	12
4	→ 4×5 →	20
⋮		
10	→ ☐ →	☐
⋮		
n	→ ☐ →	☐

b

Term number (n)		Term
1	→ $1^2 + 1$ →	2
2	→ $2^2 + 1$ →	5
3	→ $3^2 + 1$ →	10
4	→ $4^2 + 1$ →	17
⋮		
10	→ ☐ →	☐
⋮		
n	→ ☐ →	☐

8 Look at the sequence: 5, 8, 13, 20, ...

Decide which of these is the correct expression for the nth term of the sequence.

$4n + 1$		$3n + 2$		$n^2 + 4$

In questions **9** to **13** look carefully at how each sequence is formed. Write

    a the 10th term

    b the nth term.

9 $1^2, 2^2, 3^2, 4^2, \ldots$

10 $(1 \times 2), (2 \times 3), (3 \times 4), (4 \times 5), \ldots$

11 $\dfrac{1}{2}, \dfrac{2}{3}, \dfrac{3}{4}, \dfrac{4}{5}, \ldots$

12 $\dfrac{5}{1^2}, \dfrac{5}{2^2}, \dfrac{5}{3^2}, \dfrac{5}{4^2}, \ldots$

13 $(1 \times 3), (2 \times 4), (3 \times 5), (4 \times 6), \ldots$

3.5 Linear graphs, $y = mx + c$

3.5.1 Drawing linear graphs

A **linear** graph is a graph which is a straight line. The graphs of $y = 3x - 1$, $y = 7 - 2x$, $y = \dfrac{1}{2}x + 100$ are all linear.

Notice that on the right-hand side there are only x-terms and number terms. There are **no** terms involving x^2, x^3, $\dfrac{1}{x}$, etc.

You can use linear graphs to represent a wide variety of practical situations like those in the first exercise below.
Begin by drawing linear graphs **accurately**.

Exercise 21

Draw these graphs, using a scale of 2 cm to 1 unit on the x-axis and 1 cm to 1 unit on the y-axis.

1 $y = 2x + 3$ for x from 0 to 6.

$x \rightarrow \boxed{\times 2} \rightarrow \boxed{+3} \rightarrow y$

x	0	1	2	3	4	5	6
y	3				11		
coordinates					(4,11)		

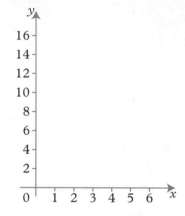

2 $y = 2x - 1$ for x from 0 to 5.

$x \rightarrow \boxed{\times 2} \rightarrow \boxed{-1} \rightarrow y$

x	0	1	2	3	4	5
y	−1					

3 $y = 2x + 1$ for $-3 \leqslant x \leqslant 3$ **4** $y = 3x - 4$ for $-3 \leqslant x \leqslant 3$

5 $y = 8 - x$ for $-2 \leqslant x \leqslant 4$ **6** $y = 10 - 2x$ for $-2 \leqslant x \leqslant 4$

7 $y = \dfrac{x + 5}{2}$ for $-3 \leqslant x \leqslant 3$ **8** $y = 3(x - 2)$ for $-3 \leqslant x \leqslant 3$

9 $y = \dfrac{1}{2}x + 4$ for $-3 \leqslant x \leqslant 3$ **10** $y = 2x - 3$ for $-2 \leqslant x \leqslant 4$

11 $y = 12 - 3x$ for $-2 \leqslant x \leqslant 4$ **12** $y = 8 - 2x$ for $-1 \leqslant x \leqslant 4$

13 Kendal Motors hires out vans.

Copy and complete the table where x is the number of miles travelled and C is the total cost in pounds.

x	0	50	100	150	200	250	300
C	35			65			95

Draw a graph of C against x, using scales of 2 cm for 50 miles on the x-axis and 1 cm for £10 on the C-axis.

a Use the graph to find the number of miles travelled when the total cost was £71.

b What is the formula connecting C and x?

14 Jeff sets up his own business as a plumber.

Copy and complete the table where C stands for his total charge and h stands for the number of hours he works.

h	0	1	2	3
C		33		

Draw a graph with h across the page and C up the page. Use scales of 2 cm to 1 hour for h and 2 cm to £10 for C.

a Use your graph to find how long he worked if his charge was £55·50.

b What is the equation connecting C and h?

15 Some drivers try to estimate their annual cost of repairs £c in relation to their average speed of driving s km/h using the equation $c = 6s + 50$.

Draw the graph for $0 \leqslant s \leqslant 160$. From the graph find

a the estimated repair bill for a man who drives at an average speed of 23 km/h.

b the average speed at which a motorist drives if her annual repair bill is £1000.

3.5.2 Line segment

The graph shows a line segment from A$(1, 2)$ to B$(7, 4)$.

The coordinates of the midpoint M are the mean of the coordinates of A and B, that is M has coordinates

$$\left(\frac{1 + 7}{2}, \frac{2 + 4}{2} \right) = (4, 3)$$

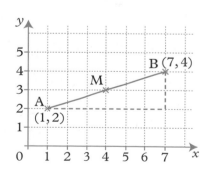

You can find the length of the line segment by using Pythagoras.

$$AB = \sqrt{(7 - 1)^2 + (4 - 2)^2} = \sqrt{40} \text{ units}$$

Exercise 22 Ⓜ/Ⓗ

1 Find the coordinates of the midpoint of the line joining these pairs of points.
   a A (3, 7), B (5, 13)   **b** C (0, 10), D (12, 4)
   c P (−3, 6), Q (5, 0)   **d** X (4, 3), Y (−2, 3)

2 Find the length of the line segment joining these pairs of points.
   a E (7, 4), F (10, 8)   **b** G (0, −2), H (5, 10)
   c S (3, −1), T (10, 4)

3 The line $2x + y = 12$ cuts the axes at the points A and B.

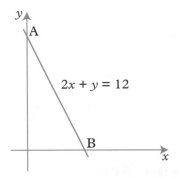

Find the coordinates of A and B and hence find
   a the coordinates of the midpoint of AB
   b the length of the line AB.

> At A, $x = 0$.

4 Find the coordinates of the midpoint of the line joining P $(a, 5a)$ and Q $(7a, −3a)$.

3.5.3 The gradient of a straight line

The gradient of a straight line is a measure of how steep it is.

EXAMPLE

Find the gradient of each line in the triangle.

Gradient of line AB $= \dfrac{3-1}{6-1} = \dfrac{2}{5}$

Gradient of line AC $= \dfrac{6-1}{3-1} = \dfrac{5}{2}$

Gradient of line BC $= \dfrac{6-3}{3-6} = -1$

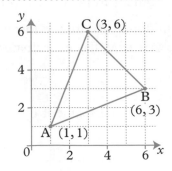

A line which slopes upwards to the right has a **positive** gradient.
A line which slopes upwards to the left has a **negative** gradient.

> Remember:
> '**N** for negative'.

● Gradient = $\dfrac{\text{difference in } y\text{-coordinates}}{\text{difference in } x\text{-coordinates}}$

Exercise 23 Ⓜ

1 Find the gradients of AB, BC and AC.

2 Find the gradients of PQ, PR, QR.

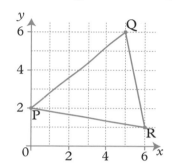

3 Find the gradients of the lines joining these pairs of points.

  a $(5, 2) \rightarrow (7, 8)$
  b $(-1, 3) \rightarrow (1, 6)$
  c $\left(\dfrac{1}{2}, 1\right) \rightarrow \left(\dfrac{3}{4}, 2\right)$
  d $(3{\cdot}1, 2) \rightarrow (3{\cdot}2, 2{\cdot}5)$

4 Find the value of a if the line joining the points $(3a, 4)$ and $(a, -3)$ has a gradient of 1.

5 **a** Write the gradient of the line joining the points $(2m, n)$ and $(3, -4)$.
  b Find the value of n if the line is parallel to the x-axis.
  c Find the value of m if the line is parallel to the y-axis.

6 On the grid 1 square = 1 unit both across and up the page. There are three pairs of intersecting lines. In each case AB is perpendicular to CD.

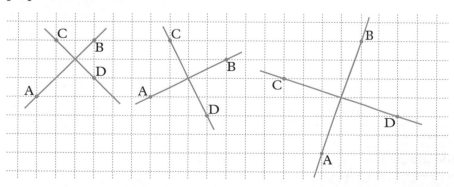

a In each case find the gradient of AB and the gradient of CD.

b In each case work out the product, (gradient of AB) × (gradient of CD).

c Copy and complete this sentence: 'The product of the gradients of lines which are perpendicular is always ☐.'

d A more straightforward result is true for **parallel** lines.
Copy and complete this sentence:
'Parallel lines have _____ gradient.'

7 Copy and complete the table.

Gradient of line 1	Gradient of line 2	Information
5	☐	Lines are parallel
2	☐	Lines are perpendicular
☐	$-\frac{2}{3}$	Lines are parallel
−5	☐	Lines are perpendicular
☐	$\frac{3}{4}$	Lines are perpendicular

> If two lines with gradients m_1 and m_2 are perpendicular, $m_1 \times m_2 = -1$

8 PQR is a right-angled triangle with vertices at P (2, 1), Q (4, 3), and R (8, t). Given that angle PQR is 90°, find t.

3.5.4 Intercept constant c

Here are two straight lines.

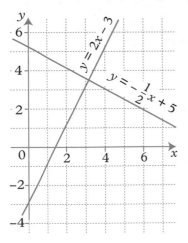

For $y = 2x - 3$, the gradient is 2 and the y-intercept is −3.

For $y = -\frac{1}{2}x + 5$ the gradient is $-\frac{1}{2}$ and the y-intercept is 5.

These two lines illustrate a general rule.

> ● When the equation of a straight line is written in the form
> $y = mx + c$
> the gradient of the line is m and the intercept on the y-axis is c.

EXAMPLE

On a **sketch** graph draw the lines
a $y = 2x + 3$
b $x + 2y - 6 = 0$.

....

a The line $y = 2x + 3$ has a gradient
of 2 and cuts the y-axis at $(0, 3)$.

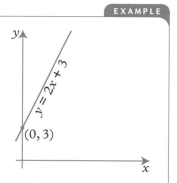

The word 'sketch' implies that you do not plot a series of points but simply show the position and slope of the line.

b Rearrange the equation to make y
the subject.
$x + 2y - 6 = 0$

$$2y = -x + 6$$
$$y = -\frac{1}{2}x + 3$$

The line has a gradient of $-\frac{1}{2}$ and
cuts the y-axis at $(0, 3)$.

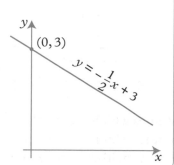

Exercise 24 Ⓜ/Ⓗ

In questions **1** to **20**, find the gradient of the line and the
intercept on the y-axis. Hence draw a small sketch graph of
each line.

1 $y = x + 3$ **2** $y = x - 2$ **3** $y = 2x + 1$ **4** $y = 2x - 5$

5 $y = 3x + 4$ **6** $y = \frac{1}{2}x + 6$ **7** $y = 3x - 2$ **8** $y = 2x$

9 $y = \frac{1}{4}x - 4$ **10** $y = -x + 3$ **11** $y = 6 - 2x$ **12** $y = 2 - x$

13 $y + 2x = 3$ **14** $3x + y + 4 = 0$ **15** $2y - x = 6$ **16** $3y + x - 9 = 0$

17 $4x - y = 5$ **18** $3x - 2y = 8$ **19** $10x - y = 0$ **20** $y - 4 = 0$

21 Find the equations of the lines
A and B.

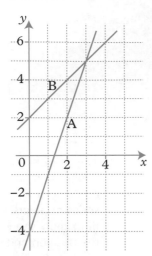

22 Find the equations of the lines
C and D.

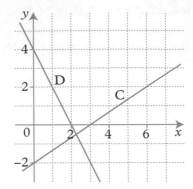

23 Here are the equations of several straight lines.

A $\boxed{y = -x + 7}$ B $\boxed{2y = x - 7}$ C $\boxed{4x + 12y = 5}$

D $\boxed{x + 3y = 10}$ E $\boxed{y = x - 3}$ F $\boxed{y = 5 - 5x}$

G $\boxed{5x + y = 12}$ H $\boxed{y = 3 - 2x}$ I $\boxed{y = 2x + 4}$

a Find **two** pairs of lines which are parallel.
b Find **two** pairs of lines which are perpendicular.
c Find **one** line which is neither parallel nor perpendicular to
   any of the other lines.

*24 The sketch represents a section of the curve $y = x^2 - 2x - 8$.

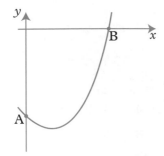

Calculate
a the coordinates of A and of B
b the gradient of the line AB
c the equation of the straight line AB.

3.5.5 Finding the equation of a line

Find the equation of the straight line which passes through $(1, 3)$ and $(3, 7)$.

....

a Let the equation of the line take the form $y = mx + c$.

The gradient, $m = \dfrac{7 - 3}{3 - 1} = 2$

so you can write the equation as $\quad y = 2x + c \quad$ ... [1]

b Since the line passes through $(1, 3)$, substitute 3 for y and 1 for x in [1].

So $\quad 3 = 2 \times 1 + c \qquad$ so $c = 1$.

The equation of the line is $y = 2x + 1$.

Exercise 25 Ⓜ/Ⓗ

In questions **1** to **10** find the equation of the line which

1 passes through $(0, 7)$ at a gradient of 3

2 passes through $(0, -9)$ at a gradient of 2

3 passes through $(0, 5)$ at a gradient of -1

4 passes through $(2, 3)$ at a gradient of 2

5 passes through $(2, 11)$ at a gradient of 3

6 passes through $(4, 3)$ at a gradient of -1

7 passes through $(6, 0)$ at a gradient of $\dfrac{1}{2}$

8 passes through $(2, 1)$ and $(4, 5)$

9 passes through $(5, 4)$ and $(6, 7)$

10 passes through $(0, 5)$ and $(3, 2)$.

*11 Two variables Z and X are thought to be connected by an equation of the form $Z = aX + c$. Here are some values of Z and X.

X	1	2	3·6	4·2	6·4
Z	4	6·5	10·5	12	17·5

Draw a graph with X on the horizontal axis using a scale of 2 cm to 1 unit and Z on the vertical axis with a scale of 1 cm to 1 unit.

Use your graphs to find a and c and hence write the equation relating Z and X.

*12 In an experiment, these measurements of the variables q and t were taken.

q	0·5	1·0	1·5	2·0	2·5	3·0
t	3·85	5·0	6·1	7·0	7·75	9·1

A scientist suspects that q and t are related by an equation of the form $t = mq + c$, (m and c are constants). Plot the values obtained from the experiment and draw the line of best fit through the points. Plot q on the horizontal axis with a scale of 4 cm to 1 unit, and t on the vertical axis with a scale of 2 cm to 1 unit. Find the gradient and intercept on the t-axis and hence estimate the values of m and c.

3.6 Real-life graphs

3.6.1 Line graphs

Exercise 26 Ⓜ

1 The graph shows a return journey by car from Leeds to Scarborough.

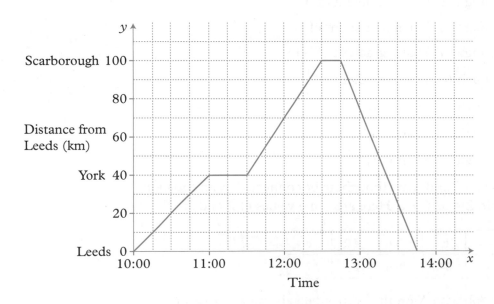

a How far is it from Leeds to York?
b How far is it from York to Scarborough?
c At which two places does the car stop?

    d How long does the car stop at Scarborough?

    e When does the car

       i   arrive in York

       ii arrive back in Leeds?

    f What is the speed of the car

       i   from Leeds to York

       ii from York to Scarborough

       iii from Scarborough to Leeds?

> speed =
> $$\dfrac{\text{distance travelled}}{\text{time taken}}.$$

2 The graph shows the journeys made by a van and a car starting at York, travelling to Durham and returning to York.

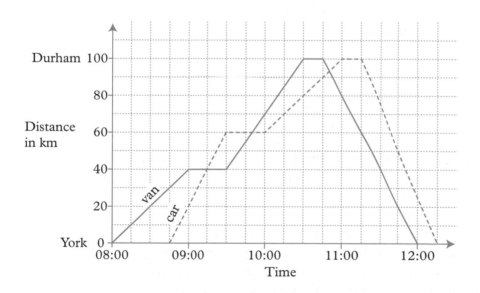

    a For how long was the van stationary during the journey?

    b At what time did the car first overtake the van?

    c At what speed was the van travelling between 09:30 and 10:00?

    d What was the greatest speed attained by the car during the entire journey?

    e What was the average speed of the car over its entire journey?

3 The graph shows the journeys of a bus and a car along the same road. The bus goes from Leeds to Darlington and back to Leeds. The car goes from Darlington to Leeds and back to Darlington.

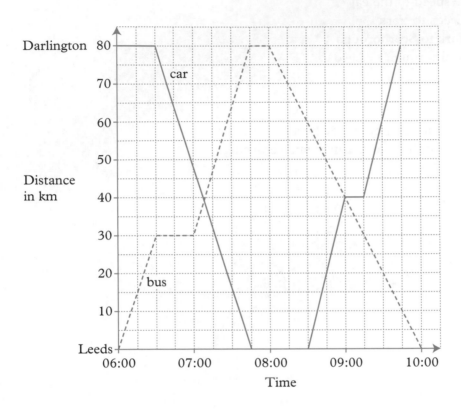

a When did the bus and the car meet for the second time?
b At what speed did the car travel from Darlington to Leeds?
c What was the average speed of the bus over its entire journey?
d Approximately how far apart were the bus and the car at 09:45?
e What was the greatest speed attained by the car during its entire journey?

In questions **4**, **5**, and **6** draw a travel graph to illustrate the journey described. Draw axes with the same scales as in question **3**.

4 Mrs Chuong leaves home at 08:00 and drives at a speed of 50 km/h. After $\frac{1}{2}$ hour she reduces her speed to 40 km/h and continues at this speed until 09:30. She stops from 09:30 until 10:00 and then returns home at a speed of 60 km/h.
Use a graph to find the approximate time at which she arrives home.

5 Mr Coe leaves home at 09:00 and drives at a speed of 20 km/h. After $\frac{3}{4}$ hour he increases his speed to 45 km/h and continues at this speed until 10:45. He stops from 10:45 until 11:30 and then returns home at a speed of 50 km/h.
Use the graph to find the approximate time at which he arrives home.

6 At 10:00 Akram leaves home and cycles to his grandparents' house which is 70 km away. He cycles at a speed of 20 km/h until 11:15, at which time he stops for $\frac{1}{2}$ hour. He then completes the journey at a speed of 30 km/h. At 11:45 Akram's sister, Hameeda, leaves home and drives her car at 60 km/h. Hameeda also goes to her grandparents' house and uses the same road as Akram. At approximately what time does Hameeda overtake Akram?

7 This graph shows the delivery charge for different distances. The delivery charge is made up of a fixed charge and a cost per mile.

   a What is the fixed charge?
   b Work out the cost per mile.

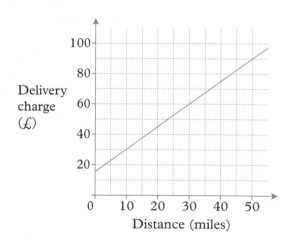

8 **a** Draw axes, as shown, with a scale of 1 cm to 5°. Two equivalent temperatures are 32°F = 0°C and 95°F = 35°C.
   b Draw a line through the points above and use your graph to convert:
     i 40°C into °F
     ii −10°C into °F
     iii 50°F into °C

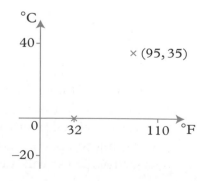

9 The graph shows how the share price of a company varied over a period of weeks. The share price is the price in pence paid for one share in the company.

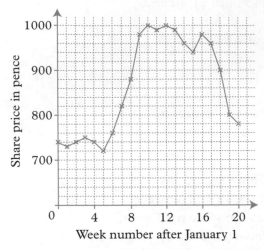

a What was the share price in Week 4?

b Naomi bought 200 shares in Week 6 and sold them all in Week 18. How much profit did she make?

c Mr Gibson can buy (and then sell) 5000 shares.
He consults a very accurate fortune teller who can predict the share price over coming weeks. What is the maximum profit he could make by buying and then selling shares once?

10 In the U.K., petrol consumption for cars is usually quoted in 'miles per gallon'. In other countries the metric equivalent is 'km per litre'.

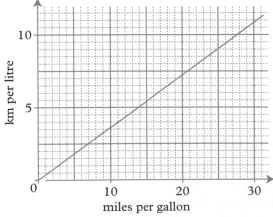

a Convert 14 m.p.g. into km per litre.

b Convert 10 km per litre into m.p.g.

c A car travels 9 km on one litre of petrol. Convert this consumption into miles per gallon. Work out how many gallons of petrol the car will use, if it is driven a distance of 300 miles.

11 A firm makes a profit of P thousand dollars from producing x thousand tiles.

Corresponding values of P and x are given below.

x	0	0·5	1·0	1·5	2·0	2·5	3·0
P	−1·0	0·75	2·0	2·75	3·0	2·75	2·0

Using a scale of 4 cm to one unit on each axis, draw the graph of P against x. [Plot x on the horizontal axis.] Use your graph to find:

a the number of tiles the firm should produce in order to make the maximum profit.

b the minimum number of tiles that should be produced to cover the cost of production.

c the range of values of x for which the profit is more than $2850.

3.6.2 Sketch graphs

Exercise 27 Ⓜ

1 Which of the graphs A to D best fits the statement: 'Unemployment is still rising but by less each month.'?

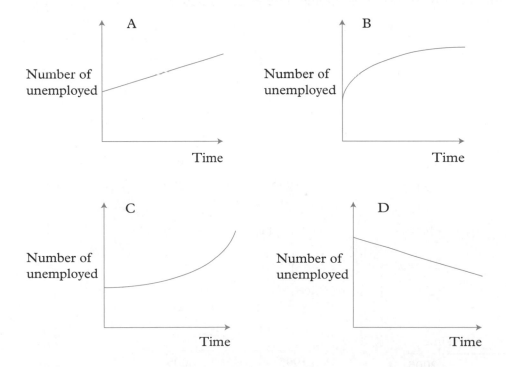

2 Which of the graphs A to D best fits each statement?
  a The birthrate was falling but is now steady.
  b Unemployment, which rose slowly until 2000, is now rising rapidly.
  c Inflation, which has been rising steadily, is now beginning to fall.
  d The price of gold has fallen steadily over the last year.

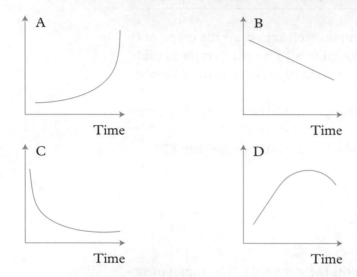

3 Which of the graphs A to D best fits the statement: 'The price of oil was rising more rapidly in 2005 than at any time in the previous ten years.'?

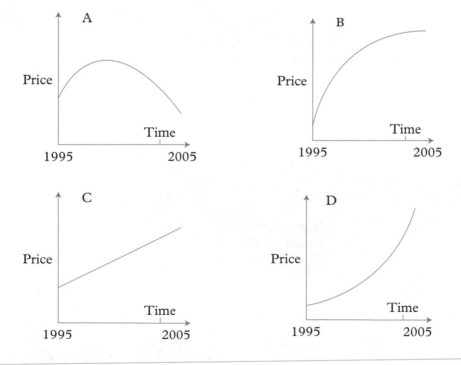

4 A car hire firm charges £30 per day plus 20p per mile. Draw a sketch graph showing 'miles travelled' across the page and 'total cost' up the page. Assume the car is hired for just one day.

5 Water is poured at a constant rate into each of the containers A, B and C.

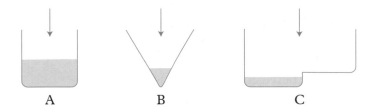

The graphs X, Y and Z show how the water level rises. Decide which graph fits each container.

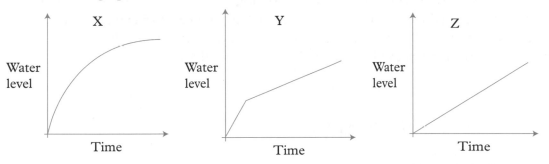

6 The diagrams show three containers P, Q and R with water coming out as they empty.
Sketch three graphs similar to those in question **5** to show how the water levels fall against time.

7 Mr Gibson organises a trip for students from his school. He hires a coach for £100 and decides to divide the cost equally between the students on the trip. Sketch a graph showing the 'number of students' across the page and the 'cost per student' up the page. Is it a linear graph?

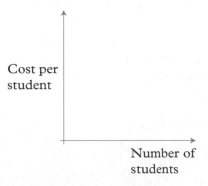

For 1 student, cost per student = £100, For 2 students, cost per student = £$\frac{100}{2}$ = £50 etc.

8 Jan drives along a motorway and the amount of petrol in her tank is monitored as shown on the graph.

a How much petrol did she buy at the first stop?

b What was the petrol consumption in miles per gallon (mpg)
  i before the first stop
  ii between the two stops?

c What was the average petrol consumption over the 200 miles?

After she leaves the second service station, Jan encounters road works and slow traffic for the next 20 miles. The petrol consumption is 20 mpg. After that, the road clears and she travels a further 75 miles during which time the consumption is 30 mpg. Draw the graph and extend it to show the next 95 miles. How much petrol is in the tank at the end of the journey?

3.7 Simultaneous equations

3.7.1 Graphical solution

Louise and Philip are sister and brother. Louise is 5 years older than Philip. The sum of their ages is 12 years. How old is each child?

Let Louise be x years old and Philip be y years old.
The sum of their ages is 12,

so $x + y = 12$

The difference between their ages is 5,

so $x - y = 5$

Because both equations relate to the same information, you can plot both on the same axes.

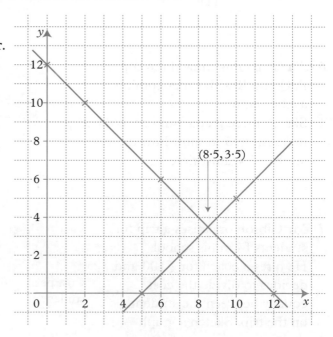

$x + y = 12$ goes through $(0, 12)$, $(2, 10)$, $(6, 6)$, $(12, 0)$.
$x - y = 5$ goes through $(5, 0)$, $(7, 2)$, $(10, 5)$.

You find the values of x and y from the point where the two lines intersect. The point $(8.5, 3.5)$ lies on both lines.

The solution is $x = 8.5$, $y = 3.5$

So Louise is $8\frac{1}{2}$ years old and Philip is $3\frac{1}{2}$ years old.

> Note that only three points are required when you draw a straight-line graph.

Exercise 28 Ⓜ/Ⓗ

1 Use the graphs to solve the equations.
Solve

a $x + y = 10$
$y - 2x = 1$

b $2x + 5y = 17$
$y - 2x = 1$

c $x + y = 10$
$2x + 5y = 17$

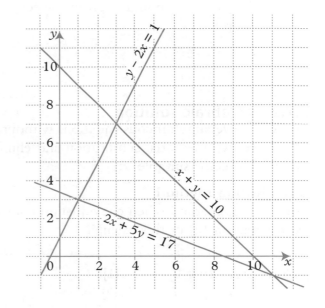

In questions **2** to **6**, solve the simultaneous equations by first drawing graphs.

2 $x + y = 6$
$2x + y = 8$
Draw axes with x and y from 0 to 8.

3 $x + 2y = 8$
$3x + y = 9$
Draw axes with x and y from 0 to 9.

4 $x + 3y = 6$
$x - y = 2$
Draw axes with x from 0 to 8 and y from -2 to 4.

5 $5x + y = 10$
$x - y = -4$
Draw axes with x from -4 to 4 and y from 0 to 10.

6 $a + 2b = 11$
$2a + b = 13$
Here, the unknowns are a and b. Draw the a-axis across the page from 0 to 13 and the b-axis up the page, also from 0 to 13.

7 There are four lines drawn here.
Write the solutions to these pairs of
equations.

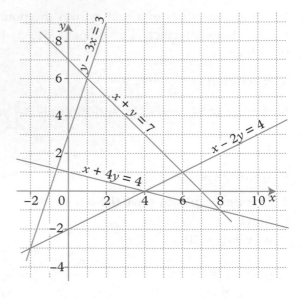

a $x - 2y = 4$
$x + 4y = 4$

b $x + y = 7$
$y - 3x = 3$

c $y - 3x = 3$
$x - 2y = 4$

d $x + 4y = 4$
$x + y = 7$

e $x + 4y = 4$
$y - 3x = 3$

[Here, x and y should be given correct
to 1 dp.]

3.7.2   Algebraic solution

You can solve simultaneous equations without drawing graphs.
There are two methods: substitution and elimination. Choose the
one which seems suitable.

Substitution method

You use this method when one equation contains a single x or y
as in equation [2] of this example.

EXAMPLE

Solve $3x - 2y = 0$ ... [1]
          $2x + y = 7$ ... [2]

a Label the equations so that the working is made clear.
b In **this** case, write y in terms of x from equation [2].
c Substitute this expression for y in equation [1] and solve to
find x.
d Find y from equation [2] using this value of x.

$2x + y = 7$
$\quad y = 7 - 2x$

Substituting for y in [1]
$3x - 2(7 - 2x) = 0$
$\quad 3x - 14 + 4x = 0$
$\qquad\qquad 7x = 14$
$\qquad\qquad\ x = 2$

Substituting for x in [2]
$2 \times 2 + y = 7$
$\qquad\quad y = 3$

The solution is $x = 2$, $y = 3$.

Check your
solution by
substituting your
values for x and y
in equation [1].

Exercise 29 Ⓜ/Ⓗ

Use the substitution method to solve these pairs of simultaneous equations.

1 $2x + y = 5$
$x + 3y = 5$

2 $x + 2y = 8$
$2x + 3y = 14$

3 $3x + y = 10$
$x - y = 2$

4 $2x + y = -3$
$x - y = 2$

5 $4x + y = 14$
$x + 5y = 13$

6 $x + 2y = 1$
$2x + 3y = 4$

7 $2x + y = 5$
$3x - 2y = 4$

8 $2x + y = 13$
$5x - 4y = 13$

9 $7x + 2y = 19$
$x - y = 4$

10 $b - a = -5$
$a + b = -1$

11 $a + 4b = 6$
$8b - a = -3$

12 $a + b = 4$
$2a + b = 5$

13 $3m = 2n - 6\frac{1}{2}$
$4m + n = 6$

14 $2w + 3x - 13 = 0$
$x + 5w - 13 = 0$

15 $x + 2(y - 6) = 0$
$3x + 4y = 30$

16 $2x = 4 + z$
$6x - 5z = 18$

17 $3m - n = 5$
$2m + 5n = 9$

18 $5c - d - 11 = 0$
$4d + 3c = -5$

Elimination method

Use this method when the first method is unsuitable (some people prefer to use it for every question).

> **EXAMPLE**
>
> Solve the equations $\quad 2x + 3y = 5 \qquad \dots [1]$
> $\qquad\qquad\qquad\qquad\quad 5x - 2y = -16 \qquad \dots [2]$
>
> ....
>
> $[1] \times 5 \qquad\qquad 10x + 15y = 25 \qquad \dots [3]$
> $[2] \times 2 \qquad\qquad 10x - 4y = -32 \qquad \dots [4]$
> $[3] - [4] \qquad\quad 15y - (-4y) = 25 - (-32)$
> $\qquad\qquad\qquad\qquad\quad 19y = 57$
> $\qquad\qquad\qquad\qquad\qquad y = 3$
>
> Substitute for y in [1] $\quad 2x + 3 \times 3 = 5$
> $\qquad\qquad\qquad\qquad\quad 2x = 5 - 9 = -4$
> $\qquad\qquad\qquad\qquad\qquad x = -2$
>
> The solution is $x = -2$, $y = 3$.

Remember to check your solution.

Exercise 30 Ⓜ/Ⓗ

Use the elimination method to solve these pairs of simultaneous equations.

1 $2x + 5y = 24$
  $4x + 3y = 20$

2 $5x + 2y = 13$
  $2x + 6y = 26$

3 $3x + y = 11$
  $9x + 2y = 28$

4 $x + 2y = 17$
  $8x + 3y = 45$

5 $3x + 2y = 19$
  $x + 8y = 21$

6 $2a + 3b = 9$
  $4a + b = 13$

7 $2x + 7y = 17$
  $5x + 3y = -1$

8 $5x + 3y = 23$
  $2x + 4y = 12$

9 $3x + 2y = 11$
  $2x - y = -3$

10 $3x + 2y = 7$
  $2x - 3y = -4$

11 $x - 2y = -4$
  $3x + y = 9$

12 $5x - 7y = 27$
  $3x - 4y = 16$

13 $x + 3y - 7 = 0$
  $2y - x - 3 = 0$

14 $3a - b = 9$
  $2a + 2b = 14$

15 $2x - y = 5$
  $\dfrac{x}{4} + \dfrac{y}{3} = 2$

16 $3x - y = 17$
  $\dfrac{x}{5} + \dfrac{y}{2} = 0$

17 $4x - 0.5y = 12.5$
  $3x + 0.8y = 8.2$

18 $0.4x + 3y = 2.6$
  $x - 2y = 4.6$

3.7.3 Solving problems using simultaneous equations

As with linear equations, solving problems involves four steps.

a Call the two unknown numbers x and y.
b Write the problem in the form of two equations.
c Solve the equations and give the answers in words.
d Check your solution using the problem and not your equations.

EXAMPLE

A motorist buys 24 litres of petrol and 5 litres of oil for £26·75, while another motorist buys 18 litres of petrol and 10 litres of oil for £31.
Find the cost of 1 litre of petrol and 1 litre of oil.
....

i Let the cost of 1 litre of petrol be x pence and the cost of 1 litre of oil be y pence.

ii $24x + 5y = 2675$ ... [1]
  $18x + 10y = 3100$ ... [2]

iii Solve the equations $x = 75$, $y = 175$
    1 litre of petrol costs 75 pence.
    1 litre of oil costs 175 pence.

iv Check $24 \times 75 + 5 \times 175 = 2675\text{p} = £26·75$
    $18 \times 75 + 10 \times 175 = 3100\text{p} = £31$

> Notice that, in the equations, you change the units from pounds to pence.

Exercise 31 Ⓜ/Ⓗ

Solve each problem by forming a pair of simultaneous equations.

> Let the numbers be x and y.

1 Find two numbers with a sum of 15 and a difference of 4.

2 Twice the larger number added to three times the smaller number gives 21. Find the numbers, if the difference between them is 3.

3 Twice one number plus the other number add up to 12. The sum of the two numbers is 7. Find the numbers.

4 Double the larger number plus three times the smaller number makes 31. The difference between the numbers is 3. Find the numbers.

5 Here is a puzzle. The ? and * stand for numbers which you have to find. The totals for the rows and columns are given.
Write two equations involving ? and * and solve them to find the values of ? and *.

?	*	?	*	36
?	*	*	?	36
*	?	*	*	33
?	*	?	*	36
39	33	36	33	

6 Angle x is 9° greater than angle y.
Find the angles of the triangle.

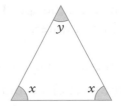

7 The cost of 2 cups and 3 plates is £18.
The cost of 5 cups and 1 plate is £19.
Let the cost of a cup be c and the cost of a plate be p.
Write two simultaneous equations and solve them to find c and p.

8 Shoppers can buy either two televisions and three DVD players for £1750 or four televisions and one DVD player for £1250.
Find the cost of one of each.

9 A pigeon can lay either white or brown eggs. Three white eggs and two brown eggs weigh 13 ounces, while five white eggs and four brown eggs weigh 24 ounces. Find the weight of a brown egg, b, and of a white egg, w.

10 The wage bill for five builders and six joiners is £1340, while the bill for eight builders and three joiners is £1220. Find the wage for b builders and for j joiners.

11 The line $y = mx + c$ passes through the points $(2, 5)$ and $(4, 13)$. Find m and c.

> For the point $(2, 5)$,
> $5 = m \times 2 + c$.
> $(y = mx + c)$

12 A bag contains forty coins, all of them either 2p or 5p coins. If the value of the money in the bag is £1·55, find the number of each kind.

13 A slot machine takes only 10p and 50p coins and contains a total of twenty-one coins altogether. If the value of the coins is £4·90, find the number of coins of each value.

14 Thirty tickets were sold for a concert, some at 60p and the rest at £1. If the total raised was £22, how many had the cheaper tickets?

15 A herring can swim at $14 \, \text{m/s}$ with the current and at $6 \, \text{m/s}$ against it. Find the speed of the current and the speed of the herring in still water.

16 If the numerator and denominator of a fraction are both decreased by one the fraction becomes $\frac{2}{3}$. If the numerator and denominator are both increased by one the fraction becomes $\frac{3}{4}$. Find the original fraction.

17 In three years' time a pet mouse will be as old as his owner was four years ago. Their present ages total 13 years. Find the age of each now.

*18 The curve $y = ax^2 + bx + c$ passes through the points $(1, 8)$, $(0, 5)$ and $(3, 20)$. Find the values of a, b and c and hence the equation of the curve.

*19 The curve $y = ax^2 + bx + c$ passes through $(1, 8)$, $(-1, 2)$ and $(2, 14)$. Find the equation of the curve.

3.7.4 Linear/quadratic simultaneous equations

A pair of simultaneous equations can contain equations which are linear, quadratic, cubic and so on.
Look at the pair of simultaneous equations

$$y = 2x + 2$$
$$\text{and } y = x^2 - 1$$

The diagram shows that the line and the curve intersect at $(-1, 0)$ and $(3, 8)$.

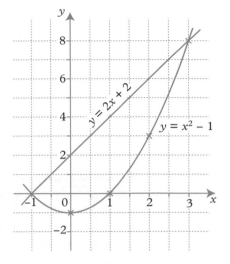

You can solve the equations algebraically using the method of substitution as follows:

$$y = 2x + 2 \quad \dots \text{[A]}$$
$$y = x^2 - 1 \quad \dots \text{[B]}$$

Substitute $y = 2x + 2$ into equation [B]
$$2x + 2 = x^2 - 1$$
$$\Rightarrow \quad x^2 - 2x - 3 = 0$$
$$(x - 3)(x + 1) = 0$$
$$x = 3 \text{ or } x = -1$$
when $x = 3$, $y = 8$
when $x = -1$, $y = 0$ $\Big\}$ This is the solution.

Check the solution by substituting for x and y in equation A.

Exercise 32 Ⓗ

See Section 7·2 page 358.

This work requires a knowledge of the solution of quadratic equations.

1 Solve the simultaneous equations

   a $y = x^2 - 2x$ **b** $y = 7x - 8$
      $y = x + 4$ $y = x^2 - x + 7$

2 Solve the simultaneous equations

   a $y = x^2 - 3x + 7$ **b** $y = 9x - 4$
      $5x - y = 8$ $y = 2x^2$

3 A circle with centre $(0, 0)$ and radius r has equation $x^2 + y^2 = r^2$.
   Find the coordinates of the points of intersection of the line
   $y = x + 1$ and the circle $x^2 + y^2 = 13$.

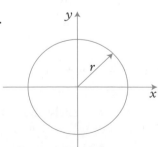

4 Solve the simultaneous equations: $x^2 + y^2 = 20$, $y = x - 2$.

5 Find the coordinates of the two points where the line $y = 4x - 8$ intersects the curve $y^2 = 16x$.

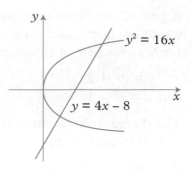

6 Explain why there are no solutions to the simultaneous equations $x^2 + y^2 = 1$ and $y = x + 10$.

7 **a** Draw accurately the graph of the circle $x^2 + y^2 = 25$. Take values of x from -5 to $+5$ and use a scale of 1 cm to 1 unit on both axes.

  b Draw the graph of $y = x + 1$ and find the coordinates of the two points of intersection.

*8 Find the coordinates of the point of intersection of the curve $y = x^3$ and the line $x + y = 10$.

Test yourself

1 **a** Solve the equation $\dfrac{x + 2}{2} = 2x$

  b Factorise $n^2 - 3n$

  c Expand and simplify $(x + 5)(2x - 1)$

  d Expand $m(m^3 - 2m + 1)$

2 **a** Factorise completely
$$3x^2 - 12y^2$$

  b Simplify
$$\frac{x^2 - 3x}{x^2 - 9}$$

(OCR, 2005)

3 $P = x(y + 2)$
$Q = xy + 2$

Show clearly that $P - Q = 2(x - 1)$.

(AQA, 2007)

4 a Simplify $\quad 5bc + 2bc - 4bc$
  b Simplify $\quad 4x + 3y - 2x + 2y$
  c Simplify $\quad m \times m \times m$
  d Simplify $\quad 3n \times 2p$

(Edexcel, 2008)

5 Use trial and improvement to solve the equation $3^x = 12$
Give your answer correct to 2 decimal places.

6 The equation $x^4 + x = 300$ has a solution between 4 and 5.
Use trial and improvement to find this solution, giving your
answer correct to one decimal place.

7 A pattern using
pentagons is made
of sticks.
  a Write down an
    expression for the
    number of sticks in
    Diagram n.
  b Which Diagram
    uses 201 sticks?

  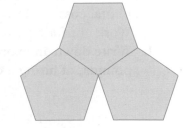

Diagram 1   Diagram 2   Diagram 3
5 sticks   9 sticks   13 sticks

(AQA, 2003)

8 Find the nth term of the following sequences.
  a 1, 4, 9, 16, 25, ...

  b −2, 5, 12, 19

9 The diagram shows three points $A(-1, 5)$, $B(2, -1)$ and $C(0, 5)$.
The line **L** is parallel to AB and passes through C.

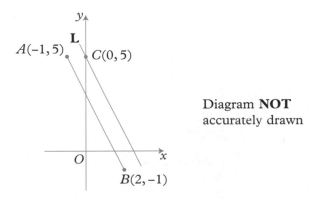

Diagram **NOT**
accurately drawn

Find the equation of the line **L**.

(Edexcel, 2005)

10 Solve algebraically these simultaneous equations.
$$3x + 2y = 7$$
$$y = x^2 - 2x + 3$$

(OCR, 2009)

11 This rule is used to work out the total cost, in pounds, of hiring a carpet cleaner.

> Multiply the number of days' hire by 4.
> Add 6 to your answer.

Peter hires a carpet cleaner.
The total cost is £18.
a Work out for how many days he hires the carpet cleaner.
b Write down an expression, in terms of n, for the total cost, in pounds, of hiring a carpet cleaner for n days.

(Edexcel, 2008)

12 The nth term of a sequence is $4n - 9$.
a Work out the first four terms.
b What is the difference between the 74th term and the 73rd term?
c The last term of this sequence is 391.
How many terms are there in this sequence?

(AQA, 2007)

13 ABC is an isosceles triangle.

$AB = AC$
Angle $A = x°$
a Find an expression, in terms of x, for the size of angle B.
b Solve the simultaneous equations
$$3p + q = 11$$
$$p + q = 3$$

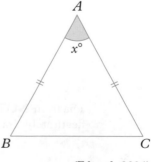

(Edexcel, 2004)

14 Solve the simultaneous equations
$$x^2 + y^2 = 34$$
$$y - x = 8$$

15 x, $2x + y$, $3x + 2y$, $4x + 3y$ are the first four terms of a sequence.
    a Write down the fifth term of the sequence.
    The sum of the first four terms is 5 and the fifth term has the value 1.
    b Find the values of x and y.
    c **i** In terms of n, x, y, write down the nth term of the sequence.
        ii Using the values of x, y found in (b), find the value of n
           for which the nth term is 0.

(CCEA)

16 **a** Solve the following equation.
        $5x - 6 = 3(10 - x)$
    b Expand and simplify:
        $4(2y - 3) - 3(y + 5)$

(WJEC, 2003)

17 **a** Simplify.
        $2xy - 3xy + 4xy$
    b Find an expression for the perimeter of this shape.
    Give your answer as simply as possible in terms of a and b.

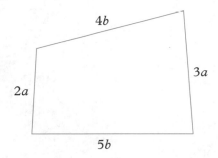

(OCR, 2009)

Functional task 3

Exercise to lose weight

Stuart thinks he may be a little overweight but how is not sure by how much. He uses a formula to calculate his Body Mass Index (BMI) which will tell him what is a healthy weight for his height.

$$BMI = \frac{\text{Weight (in kg)}}{\text{Height}^2 \text{ (in m)}}$$

> This formula only works for adults. Children's bodies are still growing.

For example, if you weigh 78 kg and your height is 1·85 m,

$$\text{your BMI} = \frac{78}{1·85^2} = 22·8$$

This chart is used to tell someone if their weight is healthy (acceptable) for their height.

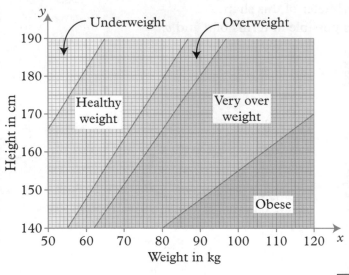

Task 1

Which category is each of Stuart's friends in [e.g. underweight, obese]?

	Weight (kg)	Height (m)
Phil	65	1·78
Annie	72	1·60
Kate	50	1·73
Jack	80	1·86
Stuart	75	1·55
Hans	104	1·55

Task 2

Stuart wants to lose weight by taking more exercise. He is told that to lose **500 g** of fat from exercise alone means burning about **3500** extra calories.

This table shows the number of calories burned per hour for a man who weighs the same as Stuart.

Exercise Table

Aerobics	518	Fishing	345
Bicycling (fast)	1380	Football	776
Circuit training	690	Mowing lawn	474
House cleaning	302	Rowing machine	1035
Playing darts	216	Running (10 min mile)	863

Stuart makes an exercise program:

Monday	Rowing machine $\frac{1}{2}$h, Play darts 1h
Tuesday	Mow lawn $\frac{1}{2}$h, Fishing 3h
Wednesday	House cleaning 1h
Thursday	Aerobics 1h
Friday	Running $\frac{1}{2}$h
Saturday	Bicycling (fast) 1h
Sunday	Fishing 6h

1 How many weeks of exercise on this program will it take for him to lose 20 kg?

2 If Hans did the same exercise, how long would it take to get his weight into the 'acceptable' range?

4 Shape, space and measures 1

In this unit you will:
- revise angle facts
- learn about congruent shapes
- construct and use loci
- use Pythagoras' theorem in 2-D and 3-D problems
- find areas of simple and composite plane shapes
- find the area and circumference of a circle, sector and segment
- find the volume and surface area of prisms, pyramids, spheres and cones
- use the properties of similar shapes.

Functional skills coverage and range:
- Find area, perimeter and volume of common shapes.

Links
Knowing how to draw accurate shapes and constructions is important for everything from computer design, building large buildings and bridges, or designing a space shuttle to drawing out a vegetable plot for an allotment.

4.1 Angles

4.1.1 Angle facts

- The angles at a point add up to 360°.
- The angles on a straight line add up to 180°.

EXAMPLE

Find the missing angles.

a

b

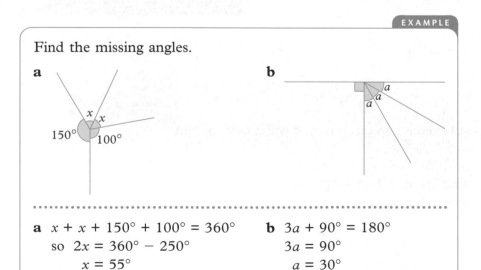

a $x + x + 150° + 100° = 360°$
so $2x = 360° - 250°$
$x = 55°$

b $3a + 90° = 180°$
$3a = 90°$
$a = 30°$

● The angles in a triangle add up to 180°.

Find the angles a and x.

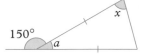

$a = 180° - 150° = 30°$

The triangle is isosceles

$\therefore \quad 2x + 30° = 180°$

$\qquad 2x = 150°$

$\qquad \ \ x = 75°$

● When a line cuts a pair of parallel lines all the acute angles are equal and all the obtuse angles are equal.

Corresponding angles   Alternate angles

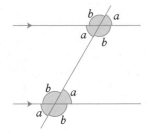

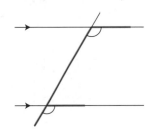

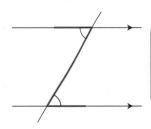

Some people remember 'F angles' and 'Z angles'.

Exercise 1 Ⓜ

Find the angles marked with letters. The lines AB and CD are straight.

1

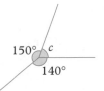

2

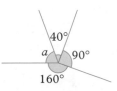

3

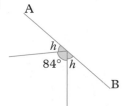

4

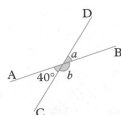

5

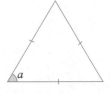

6

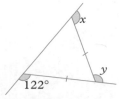

7

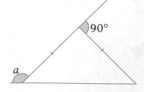

8

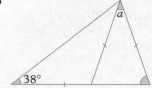

9

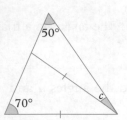

10

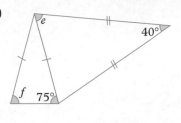

11

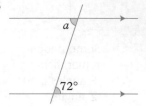

12

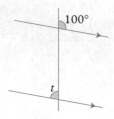

13

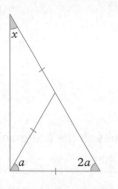

14

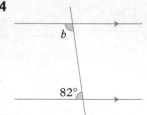

15

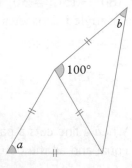

16

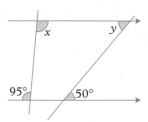

17

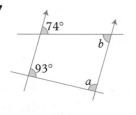

18

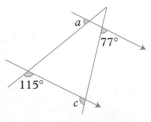

19

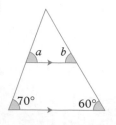

20

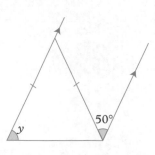

21

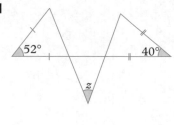

22 Here is a triangle.   A line is drawn parallel to one side.

    a What is angle x? What is angle y?
    b Prove that the sum of the angles in a triangle is 180°.

23 a Write the sum of the angles p, q and r.
    b Write the sum of the angles q and x.
    c Hence show that the exterior angle of a triangle is equal
        to the sum of the interior angles at the other two vertices.

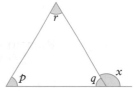

24 The diagram shows two equal squares joined to
    a triangle.
    Find the angle x.

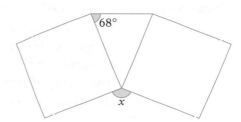

25 Find the angle a between the
    diagonals of the parallelogram.

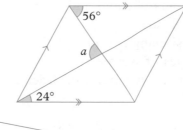

26 The diagram shows a
    series of isosceles triangles
    drawn between two lines.
    Find the value of x.

27 In the diagram AB = BD = DC
    and $\angle ADC - \angle DAC = 70°$
    Find $\angle ADB$.

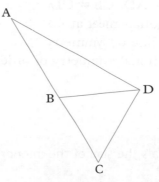

4.1.2 Quadrilaterals

Square
Four equal sides
All angles 90°
Four lines of symmetry
Rotational symmetry
of order 4

Rectangle
(not square)
Two pairs of equal and parallel sides
All angles 90°
Two lines of symmetry
Rotational symmetry of order 2

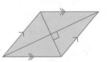

Rhombus
Four equal sides
Opposite sides parallel
Diagonals bisect at right angles
Diagonals bisect angles of rhombus
Two lines of symmetry
Rotational symmetry of order 4

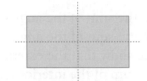

Parallelogram
Two pairs of equal and
parallel sides
Opposite angles equal
No lines of symmetry
Rotational symmetry of order 2

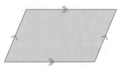

Trapezium
One pair of parallel sides
Rotational symmetry of order 1

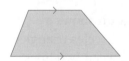

Kite
AB = AD, CB = CD
Diagonals meet at 90°
One line of symmetry
Rotational symmetry of order 1

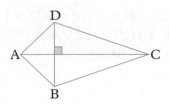

● For all quadrilaterals the sum of the interior angles is 360°.

Exercise 2 Ⓜ

1 Copy the table and fill all the boxes with either ticks or crosses.

	Diagonals always equal	Diagonals always perpendicular	Diagonals always bisect the angles	Diagonals always bisect each other
Square				
Rectangle				
Parallelogram				
Rhombus				
Kite				

2 Find the angle *x*.

a
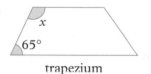
65°

trapezium

b
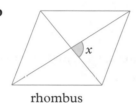
x

rhombus

c
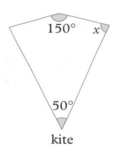
150°   *x*

50°

kite

3 Copy each diagram on square grid paper and mark with a cross the fourth vertex.

a
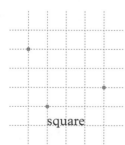
square

b
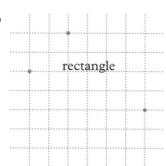
rectangle

c
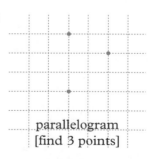
parallelogram
[find 3 points]

4 Which quadrilateral has all four sides the same length but only opposite angles equal?

5 True or false: 'All squares are rectangles.'

6 True or false: 'Any quadrilateral can be cut into two equal triangles.'

7 Name each of these shapes.
  a ABFG
  b CEFI
  c ABEH
  d ABDI

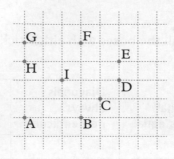

8 Name each of these shapes.
  a BIGE
  b ABEH
  c BCDF
  d CJGD
  e CJE

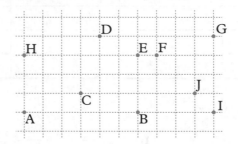

9 On square grid paper draw a quadrilateral with just two right angles and only one pair of parallel sides.

Exercise 3 Ⓜ

1 ABCD is a rhombus with diagonals that intersect at M. Find the coordinates of C and D.

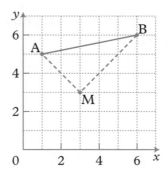

2 Quadrilateral PQRS is split into two triangles.
  a Write the sum of the angles *a*, *b* and *c*.
  b Write the sum of the angles *d*, *e* and *f*.
  c Show that the sum of the angles in the quadrilateral is 360°.

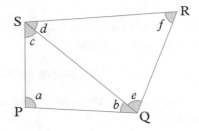

In questions **3** to **11** begin by drawing a diagram and remember to put the letters around the shape in alphabetical order.

3 In a parallelogram WXYZ, ∠WXY = 72°, ∠ZWY = 80°. Calculate
  a ∠WZY   **b** ∠XWZ   **c** ∠WYX

4 In a kite ABCD, AB = AD, BC = CD, ∠CAD = 40° and ∠CBD = 60°. Calculate
  a ∠BAC   **b** ∠BCA   **c** ∠ADC

5 In a rhombus ABCD, ∠ABC = 64°. Calculate
  a ∠BCD   **b** ∠ADB   **c** ∠BAC

6 In the parallelogram PQRS line QA bisects (cuts in half) angle PQR. Calculate the size of angle RAQ.

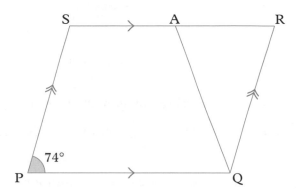

7 In a trapezium ABCD, AB is parallel to DC, AB = AD, BD = DC and ∠BAD = 128°. Find
  a ∠ABD   **b** ∠BDC   **c** ∠BCD

8 ABE is an equilateral triangle drawn inside square ABCD. Calculate the size of angle DEC.

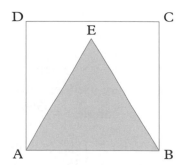

9 In a kite PQRS with PQ = PS and RQ = RS, ∠QRS = 40° and ∠QPS = 100°. Find ∠PQR.

10 In a rhombus PQRS, ∠RPQ = 54°. Find
  a ∠PRQ   **b** ∠PSR   **c** ∠RQS

11 In a kite PQRS, ∠RPS = 2∠PRS, PQ = QS = PS and QR = RS. Find
  a ∠QPS   **b** ∠PRS   **c** ∠QSR

4.1.3 Angles in polygons

Exterior angles of a polygon

The exterior angle of a polygon is the angle between a produced side and the adjacent side of the polygon.
The word 'produced' in this context means 'extended'.

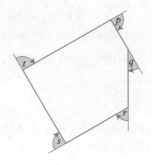

If you put all the exterior angles together you can see that the sum of the angles is 360°. This is true for any polygon.

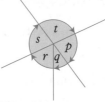

- The sum of the **exterior** angles of a polygon = 360°.

- In a **regular** polygon all exterior angles are equal.

- For a regular polygon with n sides, each exterior angle = $\dfrac{360°}{n}$.

EXAMPLE

a Calculate the size of each exterior angle (marked e) in this regular octagon.

b Calculate the size of each interior angle (marked i).

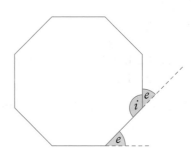

....

a There are 8 exterior angles and the sum of these angles is 360°.

so angle $e = \dfrac{360}{8} = 45°$

b $e + i = 180°$ (angles on a straight line)
so $i = 135°$

Interior angles of a polygon

Interior angle = 180° − exterior angle
            (angles on a straight line)

- In a **regular** polygon all interior angles are equal.

See Question **8** for a general formula.

Exercise 4 Ⓜ

1 Look at the polygon.
   a Calculate each exterior angle.
   b Check that the total of the exterior angles is 360°.

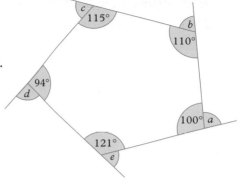

2 When shapes tessellate, you can add up the angles at a point to make 360°.
   Explain why regular hexagons tessellate and why regular pentagons do not.

3 The diagram shows a regular decagon.
   a Calculate the angle *a*.
   b Calculate the interior angle of a regular decagon.
   c Do regular decagons tessellate?

4 Find **a** the exterior angle
         b the interior angle of a regular polygon with
            i 9 sides **ii** 18 sides **iii** 45 sides **iv** 60 sides.

5 Find the angles marked with letters.

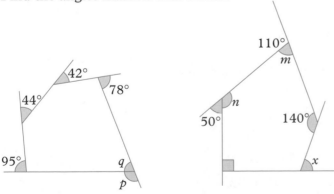

6 Each exterior angle of a regular polygon is 15°.
   How many sides has the polygon?

7 Each interior angle of a regular polygon is 140°.
   How many sides has the polygon?

> Do these polygons in questions 6 and 7 tessellate?

8 Each exterior angle of a regular polygon is 18°.
How many sides has the polygon?

9 a Look at these diagrams.

pentagon, 5 sides

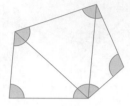

hexagon, 6 sides

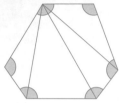

There are 3 triangles.
Sum of interior angles = 3 × 180°
                        = (5 − 2) × 180°

There are 4 triangles.
Sum of interior angles = 4 × 180°
                        = (6 − 2) × 180°

● In general: The sum of the interior angles of a polygon
  with n sides is $(n − 2) × 180°$ or $(2n − 4)$ right angles.

b Use the formula in part **a** to find the sum of the interior
angles in
i an octagon (8 sides)
ii a decagon (10 sides).

10 A regular dodecagon has 12 sides.
a Use the formula from question **8** to calculate the size
of each interior angle, i.
b Use your answer to find the size of each exterior angle, e.
c Check your answer for angle e by using the fact that the
sum of the exterior angles is always 360°.

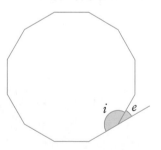

11 The sides of a regular polygon
subtend angles of 18° at the centre
of the polygon.
How many sides has the polygon?

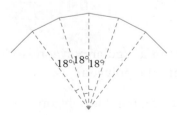

'Subtends' means
that the side of
the polygon joins
the ends of the
arms of the angle.

4.2 Congruent shapes

4.2.1 Finding congruent shapes

● **Congruent** shapes are exactly the same in shape and size.
Shapes are congruent if one shape can be fitted exactly over
the other.

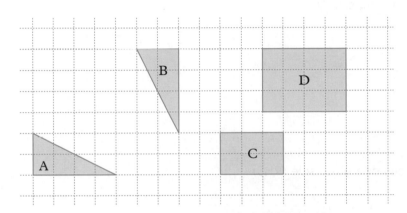

A and B are congruent.   C and D are not congruent.

Exercise 5 Ⓜ

1 Make a list of the pairs of congruent shapes.

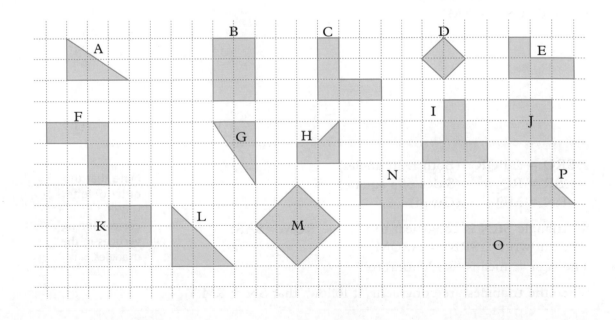

4.2.2 Congruent triangles

There are four types of congruence for triangles.

● Two sides and the included angle (SAS)

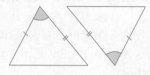

● Two angles and a corresponding side (AAS)

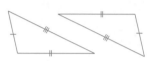

● Three sides (SSS)

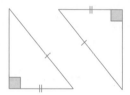

● Right angle, hypotenuse and one other side (RHS)

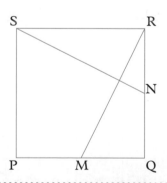

Proof using congruent triangles

PQRS is a square. M is the midpoint of PQ,
N is the midpoint of RQ.
Prove that SN = RM.

$\angle$SRQ = $\angle$PQR (Both right angles)
SR = RQ (Sides of a square)
RN = MQ (Given)

So triangle SRN is congruent to triangle RQM. (SAS)
You can write $\triangle$SRN $\equiv$ $\triangle$RQM
The sign $\equiv$ means 'is identical to'.

Since the triangles are congruent, it follows that SN = RM.

Note that the order of letters shows corresponding vertices in the triangles.

Exercise 6 Ⓗ

For questions **1** to **6**, decide whether the triangles in each pair of triangles are congruent. If they are congruent, state which conditions for congruency are satisfied.

1 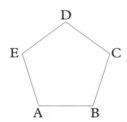   **2** 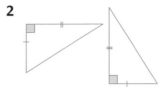   **3**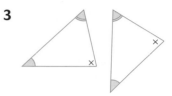

4 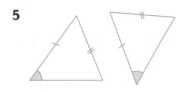   **5** 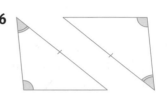   **6**

7 ABCDE is a regular pentagon. List all the triangles which are congruent to triangle BCE.

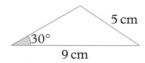

8 By construction show that it is possible to draw two **different** triangles with the angle and sides shown.

This shows that 'SSA' is not a condition for triangles to be congruent.

9 Draw triangle ABC with AB = BC and point D at the midpoint of AC.
Prove that triangles ABD and CBD are congruent and state which case of congruency applies.
Hence prove that angles A and C are equal.

10 Tangents TA and TB touch the circle with centre O.
Use congruent triangles to prove that the two tangents are the same length.

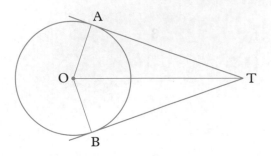

> A tangent is perpendicular to the radius at the point where it touches the circle.

11 Triangle LMN is isosceles with LM = LN; X and Y are points on LM, LN respectively such that LX = LY. Prove that triangles LMY and LNX are congruent.

> Draw diagrams for questions **11**, **12** and **13**.

12 ABCD is a quadrilateral and a line through A parallel to BC meets DC at X. If ∠D = ∠C, prove that △ADX is isosceles.

13 XYZ is a triangle with XY = XZ. The bisectors of angles Y and Z meet the opposite sides in M and N respectively. Prove that YM = ZN.

14 In the diagram, DX = XC, DV = ZC and the lines AB and DC are parallel. Prove that
a AX = BX
b AC = BD
c triangles DBZ and CAV are congruent.

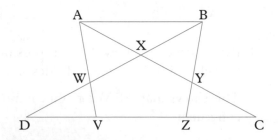

15 Draw a parallelogram ABCD and prove that its diagonals bisect each other.

16 Draw a square and label it PQRS. Draw equilateral triangles PQA and RQB so that A is inside and B is outside the square.

By considering triangles PQR and AQB, prove that AB is equal to PR.

4.3 Locus and constructions

In mathematics, the word **locus** describes the position of points which obey a certain rule. The locus can be the path traced out by a moving point.

For example, the locus of points which are equidistant from a fixed point O is a **circle** with centre O.

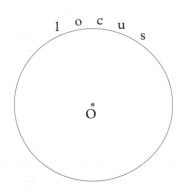

4.3.1 Constructions with a ruler and compasses

A Perpendicular bisector of a line segment AB
With centres A and B, draw two arcs.
Join the points where these arcs intersect.
The perpendicular bisector is the broken line.

Note that the perpendicular bisector is the locus of points which are equidistant from points A and B.

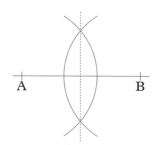

B Perpendicular from point P to a line

With centre P draw an arc to cut the line at A and B.

Construct the perpendicular bisector of AB.

C Bisector of an angle
With centre A draw arc PQ.
With centres at P and Q draw two more arcs.
Join the point where the arcs intersect to A.
This is the angle bisector that is then drawn.

Note that the angle bisector is the locus of points which are equidistant from the lines AP and AQ.

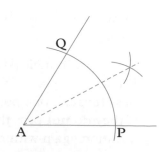

Exercise 7 Ⓜ [Use plain unlined paper]

Use only a pencil, a straight edge and a pair of compasses.

1 Draw a horizontal line AB of the length 6 cm. Construct the perpendicular bisector of AB.

2 Draw a vertical line CD of length 8 cm. Construct the perpendicular bisector of CD.

3 Draw a line and a point P about 4 cm from the line. Construct the line which passes through P which is perpendicular to the first line.

See construction B.

4 a Using a set square, draw a right-angled triangle ABC as shown. For greater accuracy draw lines slightly longer than 8 cm and 6 cm and then mark the points A, B and C.
   b Construct the perpendicular bisector of AB.
   c Construct the perpendicular bisector of AC.
   d If you do this accurately, your two lines from **b** and **c** should cross exactly on the line BC.

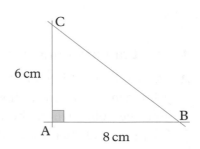

5 This is the construction of a perpendicular from a point P on a line, using ruler and compasses.
   a With centre P, draw arcs to cut the line at A and B.
   b Now construct the perpendicular bisector of AB.

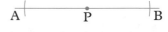

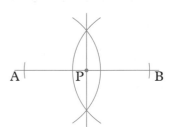

6 Draw an angle of about 60°. Construct the bisector of the angle.

See construction C.

7 Draw an angle of about 80°. Construct the bisector of the angle.

8 Draw any triangle ABC and then construct the bisector of angles A, B and C. If you do this accurately the three bisectors should all pass through one point.
   If they do **not** pass through one point (or very nearly), do this question again with a new triangle ABC.

Exercise 8 Ⓜ

1 Draw the locus of a point P which moves so that it is always 3 cm
from a fixed point X.

•X

2 Mark two points P and Q which are 10 cm apart. Draw the locus of
points which are equidistant from P and Q.

3 Draw two lines AB and AC of length 8 cm, where ∠BAC = 40°.
Draw the locus of points which are equidistant from AB
and AC.

4 Draw three copies of square PQRS and show the locus of points
inside the square which are
  a equidistant from P and R
  b nearer to P than to R
  c less than 3 cm from S.

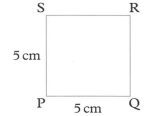

5 A sphere rolls along a surface from A to B. Sketch the locus of the
centre of the sphere in each case.

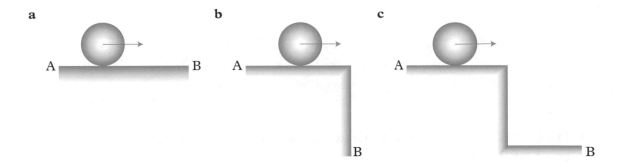

a

b

c

6 A rectangular slab is rotated
around corner B from position 1 to
position 2. Draw a diagram,
on square grid paper, to show
  a the locus of corner A
  b the locus of corner C.

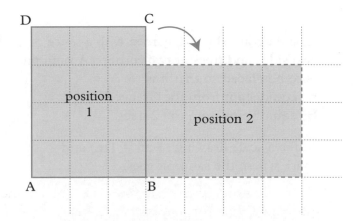

7 The diagram shows a section of coastline with a lighthouse L and coastguard station C. A sinking ship sends a distress signal. The ship appears to be up to 40 km from L and up to 20 km from C.

Copy the diagram and show the region in which the sinking ship could be.
(One grid square represents 100 km².)

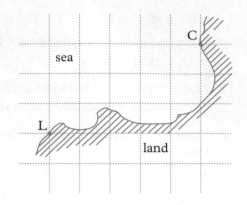

8 a Draw the triangle LMN full size.
  b Draw the locus of the points which are
    i equidistant from L and N
    ii equidistant from LN and LM
    iii 4 cm from M.
  (Draw the three loci in different colours.)

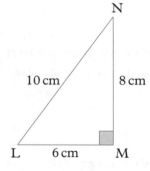

9 Draw a line AB of length 6 cm. Draw the locus of a point P so that angle ABP = 90°.

10 The diagram shows a garden with a fence on two sides and trees at two corners. A sand pit is to be placed so that it is
  a equidistant from the 2 fences
  b equidistant from the 2 trees.

Make a scale drawing (1 cm = 1 m) and mark where the sand pit goes.

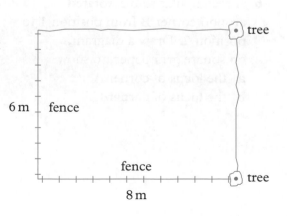

Exercise 9 Ⓜ

1 The line segment AB is 4 cm long. Draw AB and
sketch the locus of a point P which moves so that P is
always within 2 cm of the line segment.

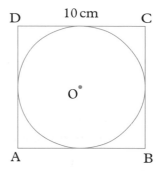

2 A circle, centre O, radius 5 cm, is inscribed inside
a square ABCD. Point P moves so that OP ≤ 5 cm and
BP ≤ 5 cm. Copy the diagram and shade the set
of points indicating where P can be.

3 Construct a triangle ABC where AB = 9 cm, BC = 7 cm and
AC = 5 cm.
   a Sketch and describe the locus of points within the triangle
   which are equidistant from AB and AC.
   b Shade the set of points within the triangle which are less than
   5 cm from B and are also nearer to AC than to AB.

4 A goat is tied to one corner on the outside of a barn.
The diagram shows a plan view.
Sketch two plan views of the barn and show the locus of points
where the goat can graze if
   a the rope is 4 m long
   b the rope is 7 m long.

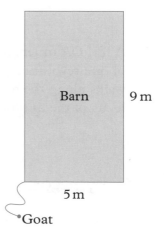

5 Draw two points M and N, 16 cm apart. Draw the locus of a
point P which moves so that the area of triangle MNP is 80 cm^2.

6 Describe the locus of a point which moves in three-dimensional
space and is equidistant from two fixed points.

7 Draw two points A and B 10 cm apart.

Place the corner of a piece of paper (or a set square) so that the
edges of the paper pass through A and B.
Mark the position of corner C.
Slide the paper around so the edge still passes through A and B
and mark the new position of C. Repeat several times and describe
the locus of the point C which moves so that angle ACB is
always 90°.

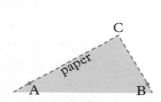

8 Draw any triangle ABC and construct the bisectors of angles B and C to meet at point Y.

With centre at Y draw a circle which just touches the sides of the triangle. This is the **inscribed** circle of the triangle.

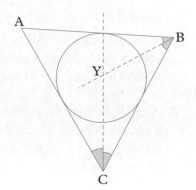

9 a Draw axes with x and y from 0 to 10.
   b Draw the locus of points which are equidistant from the points $(1, 9)$ and $(5, 5)$.
   c Draw the locus of points which are 4 units from the point $(6, 6)$.

10 Copy the diagram shown. OC = 4 cm.
   Sketch the locus of P which moves so that PC is equal to the perpendicular distance from P to the line AB.
   This locus is called a **parabola**.

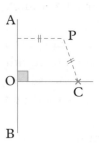

11 A rod OA of length 60 cm rotates about O at a constant rate of one revolution per minute. An ant, with good balance, walks along the rod at a speed of 1 cm per second. Sketch the locus of the ant for one minute after it leaves O.

4.4 Pythagoras' theorem

Pythagoras (569–500 BC) was one of the first of the great mathematical names in Greek antiquity. He settled in southern Italy and formed a mysterious brotherhood with his students who were bound by an oath not to reveal the secrets of numbers and who exercised great influence. They laid the foundations of arithmetic through geometry and were among the first mathematicians to develop the idea of proof.

● In a right-angled triangle the square on the hypotenuse is equal to the sum of the squares on the other two sides.

$a^2 + b^2 = c^2$

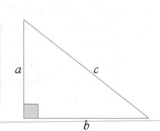

4.4.1 Using Pythagoras' theorem in two dimensions

EXAMPLE

a Find the length x.

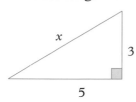

b Find the length y.

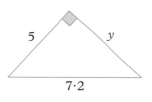

The side on its own in the equation is the hypotenuse.

....

a $x^2 = 3^2 + 5^2$
   $x^2 = 9 + 25$
   $x^2 = 34$
   $x = \sqrt{34}$ [This is the *exact* answer]
   $x = 5 \cdot 8$ (2 sf)

b $y^2 + 5^2 = 7 \cdot 2^2$
   $y^2 + 25 = 51 \cdot 84$
   $y^2 = 26 \cdot 84$
   $y = \sqrt{26 \cdot 84}$
   $y = 5 \cdot 2$ (2 sf)

Exercise 10 Ⓜ

In questions **1** to **8**, find x correct to 2 significant figures. All the lengths are in cm.

1
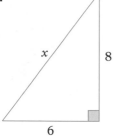

2
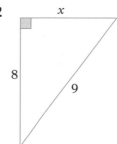

3
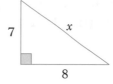

4
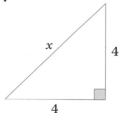

5
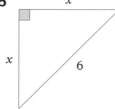

6
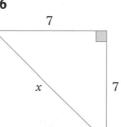

7
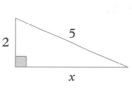

8
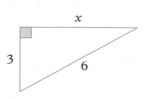

9 Find the length of a diagonal of a rectangle of length 9 cm and width 4 cm. Give the exact answer as a surd.

A surd is a number like $\sqrt{11}$ or $\sqrt{97}$.

10 An isosceles triangle has sides 10 cm, 10 cm and 4 cm.
Find the height of the triangle.

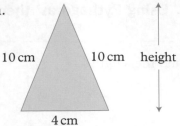

10 cm   10 cm   height

4 cm

11 A 4 m ladder rests against a vertical wall with its foot 2 m from the wall. How far up the wall does the ladder reach?

12 A ship sails 20 km due north and then 35 km due east. How far is it from its starting point?

13 Find the length of a diagonal of a square of side 9 cm.

14 The square and the rectangle have the same length diagonal. Find x.

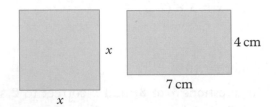

x 4 cm

7 cm

x

15 a Find the height of the triangle, h.
  b Find the area of the triangle ABC.

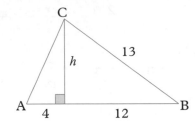

C

h 13

A   4   12   B

16 A thin wire of length 18 cm is bent in the shape shown.

Calculate the length from A to B.

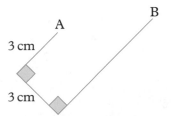

B

A

3 cm

3 cm

17 A paint tin is a cylinder of radius 12 cm and height 22 cm. Leonardo, the painter, drops his stirring stick into the tin and it disappears.
Work out the maximum length of the stick.

18 A square of side 10 cm is drawn inside a circle.
Find the shaded area.

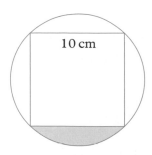

*19 Calculate the area of a regular hexagon of side 12 cm.

4.4.2 Three-dimensional problems

Find the length of the diagonal BH in the cuboid.

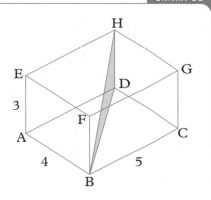

- -

Redraw the base ABCD.

$BD^2 = 5^2 + 4^2$
$BD^2 = 25 + 16$
$BD = \sqrt{41}$

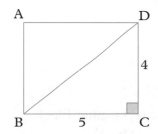

Redraw triangle BHD.

$BH^2 = 3^2 + (\sqrt{41})^2$
$BH^2 = 9 + 41$
$BH = \sqrt{50}$

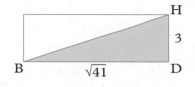

Exercise 11 Ⓗ

1 In the diagram A is (1, 2) and B is (6, 4).
Work out the length AB. (First find the lengths
of AN and BN.)

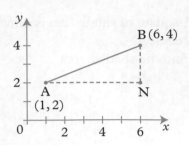

2 On square grid paper plot P(1, 3), Q(6, 0), R(6, 6). Find the
lengths of the sides of triangle PQR. Is the triangle isosceles?

In questions **3** to **8** find x.

3

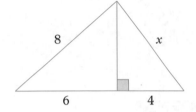

4

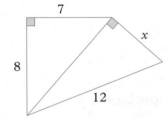

5

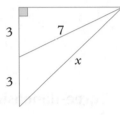

6

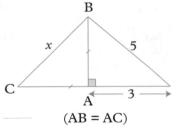

(AB = AC)

7

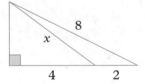

8

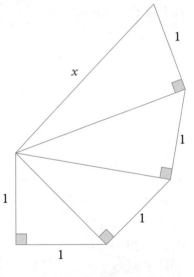

9 The diagram shows a rectangular block.
Calculate **a** AC **b** AY.

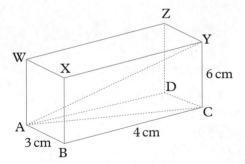

10 The diagram shows a cuboid 5 cm by 4 cm by 12 cm.
Calculate **a** AC **b** AD.

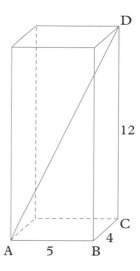

11 Find the length of a diagonal of a rectangular room of length 5 m, width 3 m and height 2·5 m.

12 Find the height of a rectangular box of length 8 cm, width 6 cm where the length of a diagonal is 11 cm.

13 TC is a vertical pole whose base lies at a corner of the horizontal rectangle ABCD.
The top of the pole T is connected by straight wires to points A, B and D.

Calculate **a** TC
        b TD
       *c TA.

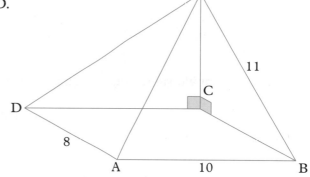

14 The diagonal of a rectangle exceeds the length by 2 cm.
If the width of the rectangle is 10 cm, find the length.

15 A ladder reaches H when held vertically against a wall.
When the base is 6 feet from the wall, the top of the ladder is 2 feet lower than H.
How long is the ladder?

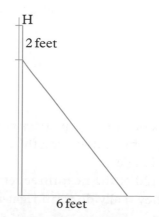

16 a Sketch axes in three dimensions as shown in the diagram.
  b Find the length of the line from O $(0, 0, 0)$ to A $(3, 4, 7)$.
  c Find the length of the line from O $(0, 0, 0)$ to B $(1, 10, 13)$.

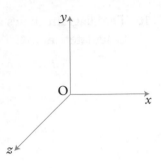

17 The diagram represents the starting position (AB) and the finishing position (CD) of a ladder as it slips. The ladder is leaning against a vertical wall.

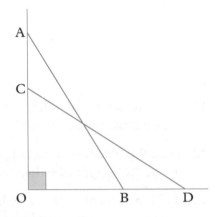

AC = x, OC = 4AC, BD = 2AC and OB = 5 m.
Form an equation in x, find x and hence find the length of the ladder.

18 The best known right-angled triangle is the 3, 4, 5 triangle $[3^2 + 4^2 = 5^2]$.
It is interesting to look at other right-angled triangles where all the sides are whole numbers.
  a i Find c if $a = 5$, $b = 12$.
   ii Find c if $a = 7$, $b = 24$.
  iii Find a if $c = 41$, $b = 40$.

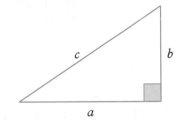

  b Write the results in a table.

a	b	c
3	4	5
5	12	?
7	24	?
?	40	41

  c Look at the sequences in the 'a' column and in the 'b' column. Also write the connection between b and c for each triangle.
  d Predict the next three sets of values of a, b, c. Check to see if they really do form right-angled triangles.

19 The diagram shows the net of a square-based pyramid. The base is 8 cm × 8 cm and the vertical height is 10 cm. Find x.

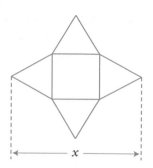

4.5 Three-dimensional coordinates

In three-dimensional space, we need three coordinates to describe how the position of a point relates to the common origin of three axes. In this diagram, O is the origin for the three axes. For A, $x = 5$, $y = 4$, $z = 2$, so A is written (5, 4, 2).

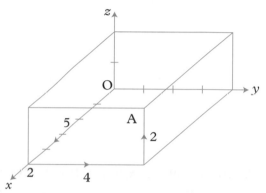

Exercise 12 Ⓜ/Ⓗ

1 Write down the coordinates of the points A, B, C, D.

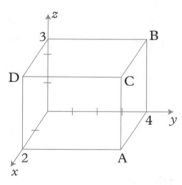

2 a Write down the coordinates of B, C, Q, R.
 b Write down the coordinates of the midpoints of
  i AD **ii** DS **iii** DC.
 c Write down the coordinates of the centre of the face
  i ABCD **ii** PQRS **iii** RSDC.
 d Write down the coordinates of the centre of the box.

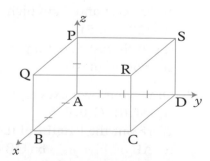

3 a Write down the coordinates of C, R, B, P, Q.
  b Write down the coordinates of the midpoints of
    i QB   **ii** PQ.

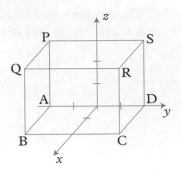

4 Use Pythagoras' theorem to calculate the lengths of
  i LM   **ii** LN   **iii** NM.

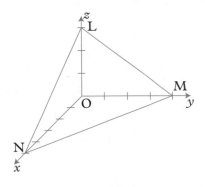

5 In the cuboid below, OP = 4, OQ = 7, OR = 5. Write down the
coordinates of the centre of the face
  i ABCR   **ii** BCQD   **iii** OPDQ.

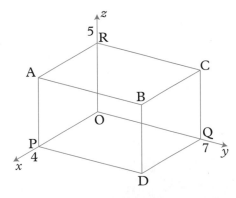

6 Measured from a control tower O, an aircraft is 20 km north,
30 km east and 5 km high. So with origin at O, its coordinates
are (20, 30, 5).
It travels east at 3 km per minute. What are its coordinates
after 5 minutes?

7 Draw your own axes and plot the points $O(0, 0, 0)$, $A(3, 0, 0)$,
$B(3, 4, 0)$, $C(0, 0, 5)$.
Work out the lengths of these lines.
  a AB   **b** OB   **c** CB

8 A solid object has vertices at $(0, 0, 0)$, $(6, 0, 0)$, $(6, 6, 0)$, $(0, 6, 0)$ and $(3, 3, 4)$.
Plot the points on a diagram and name the object.

*9 Plot the points O$(0, 0, 0)$, A$(6, 8, 0)$, B$(6, 8, 10)$.
Work out   **a**   OA
           b   angle BOA.

*10 A prisoner starts his escape tunnel at $(0, 0, 0)$. He digs in a straight line to a sewer $(5, 3, -2)$. He then crawls along the sewer to $(5, 20, -2)$. Finally he digs in a straight line to escape at $(0, 24, 0)$. Work out the total length of the three sections of his escape route. The units are given in metres.

4.6   Area

4.6.1   Rectangles, triangles and composite shapes

- For a **rectangle**,
  area = base × height

- For a **triangle**,
  area = $\dfrac{1}{2}$ × base × height

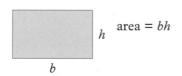

area = bh

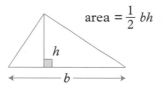

area = $\dfrac{1}{2} bh$

Exercise 13 Ⓜ

The shapes in questions **1** to **11** consist of rectangles joined together. Find the area of each shape. All lengths are in cm.

1

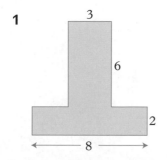

2

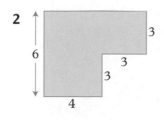

3

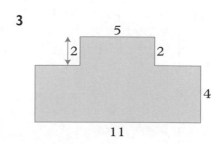

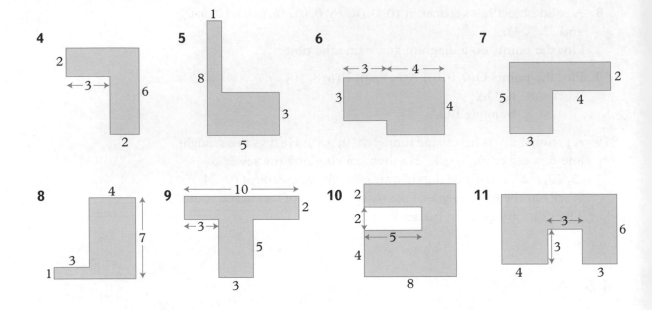

4
5
6
7

8
9
10
11

12 Find the height of each rectangle.

a area = 32 cm² 8 cm ?

b area = 33 cm² 6 cm ?

c area = 43 cm² 10 cm ?

13 Calculate the total surface area of this cuboid.

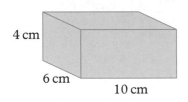

4 cm
6 cm
10 cm

Find the total area of each shape consisting of a rectangle and a triangle. All lengths are in cm.

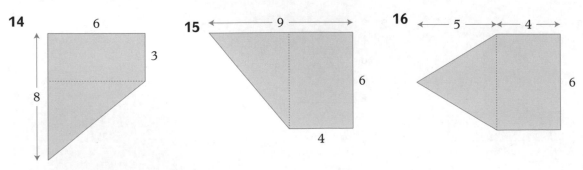

14 6 3 8

15 9 6 4

16 5 4 6

17 Find the height of each of these triangles.

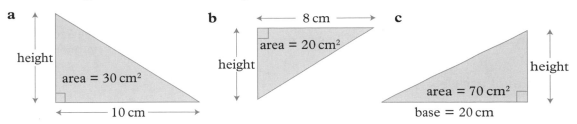

18 This triangle has an area of $2\,\text{cm}^2$.

    a On square dotty paper draw a triangle
       with an area of $3\,\text{cm}^2$.

    b Draw a triangle, different to the one shown,
       with an area of $2\,\text{cm}^2$.

19 Joe said 'There are $100\,\text{cm}^2$ in $1\,\text{m}^2$.
    Mark said 'That's not right because one square metre is
    $100\,\text{cm}$ by $100\,\text{cm}$ so there are $10\,000\,\text{cm}^2$ in $1\,\text{m}^2$.'
    Who is right?

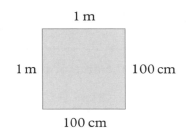

20 Here are three shapes made with centimetre squares.

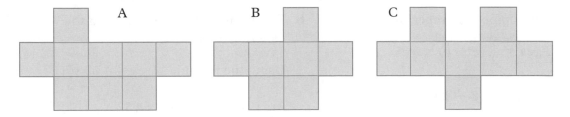

    a Find the perimeter of each shape.

    b Draw a shape of your own design with a perimeter of $14\,\text{cm}$.
       Ask another student to check your shape.

4.6.2 Triangle, trapezium and parallelogram

Triangle

For a triangle,   area $= \frac{1}{2} \times$ base $\times$ height **or**

See Section 6·2 for more on $\sin C$

- area of triangle $= \frac{1}{2} ab \sin C$

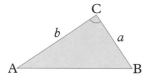

Use the second formula when you know two sides and the included angle. The **included** angle is the angle between the two sides you know.

Trapezium

A trapezium has two parallel sides.

- area of trapezium $= \frac{1}{2}(a + b) \times h$

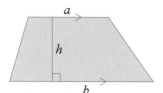

Parallelogram

- area of parallelogram $= b \times h$
  or
  area of parallelogram $= ab \sin \theta$

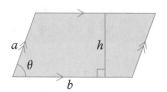

EXAMPLE

Find the area of each shape.

a

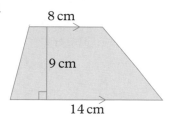

8 cm

9 cm

14 cm

b

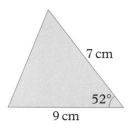

7 cm

52°

9 cm

....

a The shape is a trapezium.

Area $= \frac{1}{2}(a + b) \times h$

$= \frac{1}{2}(8 + 14) \times 9$

$= 99 \text{ cm}^2$

b The triangle has two sides with the included angle.

Using the formula 'Area $= \frac{1}{2} ab \sin C$'

Area $= \frac{1}{2} \times 7 \times 9 \times \sin 52°$

$= 24·8 \text{ cm}^2$ (3 sf)

- To use a calculator to find a sine: | sin | 5 | 2 | = |

Exercise 14 Ⓜ

For questions **1** to **6**, find the area of each shape. All lengths are in cm.

1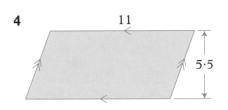

2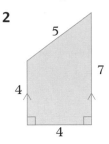

3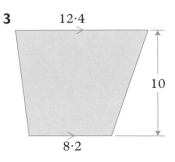

4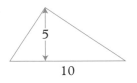

5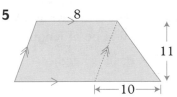

6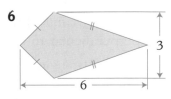

7 A rectangle has an area of $117\,\text{m}^2$ and a width of 9 m. Find its length.

8 A trapezium of area $105\,\text{cm}^2$ has parallel sides of length 5 cm and 9 cm. How far apart are the parallel sides?

9 A floor 5 m by 20 m is covered by square tiles of side 20 cm. How many tiles are needed?

10 Find the area of each triangle. All lengths are in cm.

a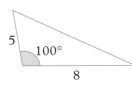

b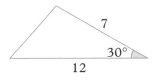

c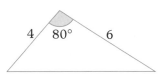

d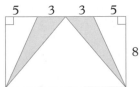

e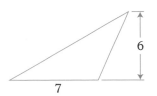

f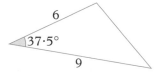

In questions **11** to **16**, find the area shaded.

11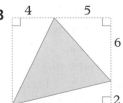

12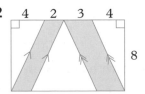

13

14
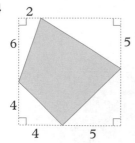

15
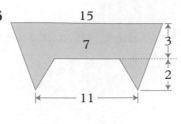

16
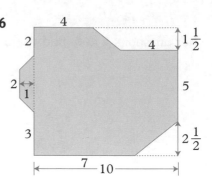

*17 Copy and complete a derivation of the formula, area $= \frac{1}{2} ab \sin C$.

a Draw a triangle ABC and draw a line through A perpendicular to BC. Let this line be of length h.

Area of triangle ABC $= \frac{1}{2} \times$ base $\times$ height

$\qquad\qquad\qquad\quad = \frac{1}{2} ah$

From triangle ADC, $\sin C = \dfrac{h}{\square}$

So $h = \square \times \sin C$

Area of triangle ABC $= \frac{1}{2} a \square \sin C$.

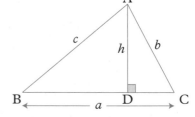

Exercise 15 Ⓜ ⭐Ⓕ

1 A rectangular pond measures 10 m by 6 m.
The path surrounding the pond is 2 m wide.

10 m

pond   6 m

path

Find the area of the path.

2 This shape is made with centimetre squares.
   a Write its perimeter.
   b Write its area.
   c Draw a shape of your own design with area 8 cm^2 and perimeter 14 cm.

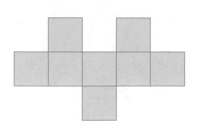

3 Find the length x. The area is shown inside the shape.

a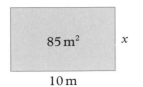

85 m² x

10 m

b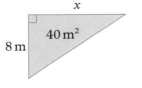

x

8 m   40 m²

c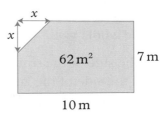

x

x 62 m²   7 m

10 m

4 Work out the area of this shape.
All lengths are in cm.

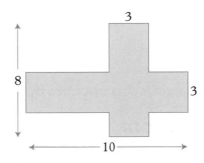

3

8

3

10

5 The arrowhead has an area of $3·6$ cm².
Find the length x.

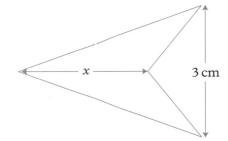

x 3 cm

6 The diagram shows a square ABCD in which
DX = XY = YC = AW. The area of the square is 45 cm².

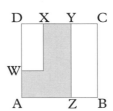

D   X   Y   C

W

A   Z   B

  a What is the fraction $\dfrac{DX}{DC}$?

  b What fraction of the square is shaded?

  c Find the area of the unshaded part.

7 On square grid paper draw a 7×7 square. Design a pattern which
divides it up into nine smaller squares.

8 A rectangular field, 400 m long, has an area of 6 hectares.
Calculate the perimeter of the field. [1 hectare $= 10\,000$ m²]

9 On square grid paper draw a triangle with vertices at $(1, 1)$, $(5, 3)$,
$(3, 5)$. Find the area of the triangle.

10 Draw a quadrilateral with vertices at $(1, 1)$, $(6, 2)$, $(5, 5)$, $(3, 6)$.
Find the area of the quadrilateral.

11 A square wall is covered with square tiles. There are 85 tiles altogether along the two diagonals. How many tiles are there on the whole wall?

12 The side of the small square is half the length of the side of the large square. The L-shape has an area of $75\,\text{cm}^2$. Find the side length of the large square.

13 A regular hexagon is circumscribed by a circle of radius 3 cm with centre O.
  a What is angle EOD?
  b Find the area of triangle EOD and hence find the area of the hexagon ABCDEF.

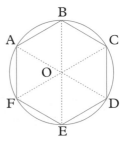

14 Find the area of a parallelogram ABCD with AB = 7 m, AD = 20 m and $\angle BAD = 62°$.

15 In the diagram, AE = $\frac{1}{3}$ AB. Find the area shaded.

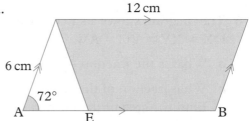

16 The area of an equilateral triangle ABC is $50\,\text{cm}^2$. Find AB.

17 The area of a triangle XYZ is $11\,\text{m}^2$. If YZ = 7 m and $\angle XYZ = 130°$, find XY.

18 Find the length of the side of an equilateral triangle of area $10\cdot2\,\text{m}^2$.

19 A rhombus has an area of $40\,\text{cm}^2$ and adjacent angles of $50°$ and $130°$. Find the length of a side of the rhombus.

*20 The diagram shows a part of the perimeter of a regular polygon with n sides.
The centre of the polygon is at O and
OA = OB = 1 unit.

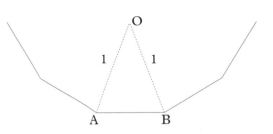

  a What is the angle AOB in terms of n?
  b Work out an expression in terms of n for
    i the area of triangle OAB
    ii the area of the whole polygon.
  c Find the area of the polygons where $n = 100$ and $n = 1000$. What do you notice?

*21 The area of a regular pentagon is $600\,\text{cm}^2$.
Calculate the length of one side of the pentagon.

4.7 Circles, arcs, sectors and segments

4.7.1 Area and circumference

- For any circle, the ratio $\left(\dfrac{\text{circumference}}{\text{diameter}}\right)$ is equal to π.

The value of π is usually taken to be $3 \cdot 14$, but this is not an exact value. Through the centuries, mathematicians have been trying to obtain a better value for π.

For example, in the third century AD, the Chinese mathematician Liu Hui obtained the value $3 \cdot 14159$ by considering a regular polygon having 3072 sides! Ludolph van Ceulen (1540–1610) worked even harder to produce a value correct to 35 significant figures. He was so proud of his work that he had this value of π engraved on his tombstone. Electronic computers are now able to calculate the value of π to many millions of figures, but its value is still not exact. It was shown in 1761 that π is an **irrational number** which, like $\sqrt{2}$ or $\sqrt{3}$ cannot be expressed exactly as a fraction.

The first fifteen significant figures of π can be remembered from the number of letters in each word of this sentence.

How I need a drink, cherryade of course, after the silly lectures involving Italian kangaroos.

There remain a lot of unanswered questions concerning π, and many mathematicians today are still working on them.

Learn these formulae.

- Circumference of a circle $= \pi d$ d = diameter
  $= 2\pi r$ r = radius
- Area of a circle $= \pi r^2$

Find the circumference and area of a circle of diameter 8 cm.
Take π from a calculator.

Circumference $= \pi d$
$= \pi \times 8$
$= 25 \cdot 1$ cm (3 sf)
Area $= \pi r^2$
$= \pi \times 4^2$
$= 50 \cdot 3$ cm^2 (3 sf)

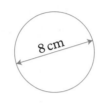

These answers are not exact. You can write the exact answers
by leaving π in the answer. So the exact answers above are:
circumference $= 8\pi$ cm, area $= 16\pi$ cm^2.

Exercise 16 Ⓜ ⭐

1 Find the circumference. For parts **a** to **d**, use the π button on a
calculator or take $\pi = 3 \cdot 142$. Give the answers correct to
3 significant figures. For parts **e** to **h** give the exact answer using π.

a

11 cm

b

8 cm

c

6 cm

d

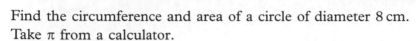

5 cm

e

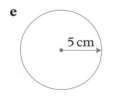

5 cm

f

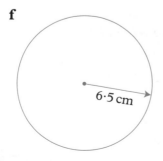

6·5 cm

g

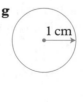

1 cm

h

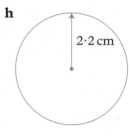

2·2 cm

2 The current 10p coin has a diameter of 2·4 cm, but the original 10p coin had a diameter of 2·8 cm.

How much longer, to the nearest mm, was the circumference of the original coin?

3 A circular pond has a diameter of 2·7 m. Calculate the length of the perimeter of the pond.

4 A running track has two semicircular ends of radius 34 m and two straights of 93·2 m as shown in the diagram.

Calculate the total distance around the track to the nearest metre.

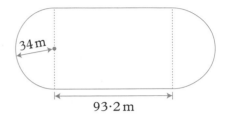

5 A fly, perched on the tip of the minute hand of a grandfather clock, is 14·4 cm from the centre of the clock face.
How far does the fly move between 12:00 and 12:30?

6 Sue cycles for 400 m. Her cycle wheels have a diameter of 60 cm.
How many complete revolutions does her front wheel make?

7 The diagram shows a framework for a target, consisting of 2 circles of wire and 6 straight pieces of wire. The radius of the outer circle is 30 cm and the radius of the inner circle is 15 cm.
Calculate the exact total length of wire needed for the whole framework. Leave π in your answer.

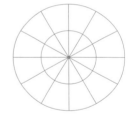

8 Calculate the area of each circle in question **1**.

9 A spinner of radius 7·5 cm is divided into six equal sectors.
Calculate the area of each sector.

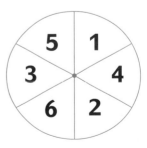

10 A circular swimming pool of diameter 12·6 m is to be covered by a plastic sheet to keep out leaves and insects.
Work out the area of the pool.

11 A circle of radius 5 cm is inscribed inside a square as shown in the diagram.
Find the shaded area.
Give the exact answer by leaving π in your answer.

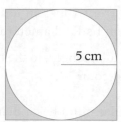

5 cm

12 A large circular lawn is sprayed with weedkiller. Each square metre of grass requires 2 g of weedkiller. How much weedkiller is needed for a lawn of radius 27 m?

EXAMPLE

Find the perimeter and area of this shape.

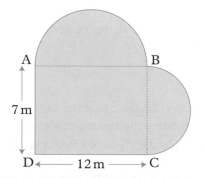

A   B

7 m

D $\longleftarrow$ 12 m $\longrightarrow$ C

Perimeter = arc AB + arc BC + CD + DA

$$= \left(\frac{\pi \times 12}{2}\right) + \left(\frac{\pi \times 7}{2}\right) + 12 + 7$$

$$= 9\frac{1}{2}\pi + 19 = 48{\cdot}8 \text{ m (3 sf)}$$

Area = large semicircle + small semicircle + rectangle

$$= \left(\frac{\pi \times 6^2}{2}\right) + \left(\frac{\pi \times 3{\cdot}5^2}{2}\right) + (12 \times 7)$$

$$= 160 \text{ m}^2 \text{ (3 sf)}$$

Do not approximate **early** in questions of this type. If your final answer is given to 3 sf, work to 5 sf in the working. Better still, leave the working in the calculator memory.

Exercise 17 Ⓜ

For each shape find **a** the perimeter   **b** the area.
All lengths are in cm unless otherwise stated. All the arcs are either semicircles or quarter circles.

1
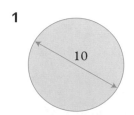
10

2

6

3
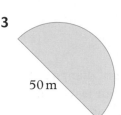
50 m

4
22

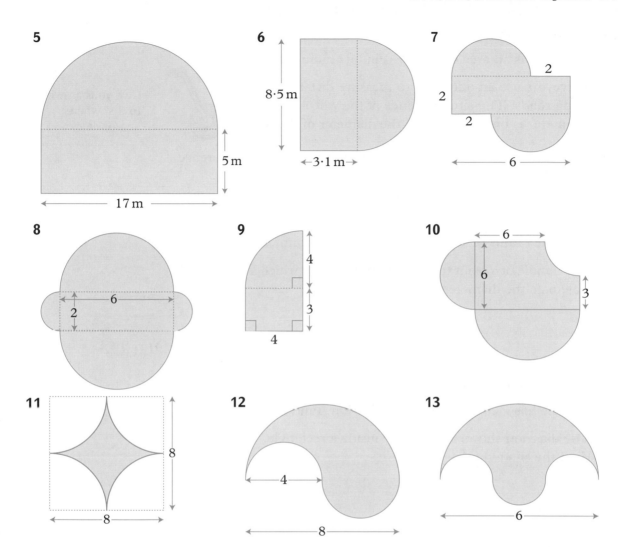

4.7.2 Finding the radius

a A circle has a circumference of 20 m. Find the radius of the circle.

b A circle has an area of 45 cm^2. Find the radius of the circle.

. . . .

a Let the radius of the circle be r m.

Circumference = $2\pi r$

$$2\pi r = 20$$

So $$r = \frac{20}{2\pi}$$

$$r = 3 \cdot 18$$

The radius of the circle is $3 \cdot 18$ m (3 sf).

b Let the radius of the circle be r cm.

$$\pi r^2 = 45$$

$$r^2 = \frac{45}{\pi}$$

$$r = \sqrt{\left(\frac{45}{\pi}\right)} = 3 \cdot 78$$

The radius of the circle is $3 \cdot 78$ cm (3 sf).

Exercise 18 Ⓜ

1 A circle has an area of 15 cm². Find its radius.

2 Surveyors use an odometer to measure distances along roads. The circumference of the wheel of an odometer is one metre. Find the diameter of the wheel.

> Give your answers to 3 sf where appropriate.

3 Find the radius of each of these circles from their circumference, C, or their area, A.
   a $C = 72$ m   **b** $C = 260$ km   **c** $A = 44$ m²
   d $A = 18 \cdot 5$ cm²   **e** $C = 10^8$ km

4 Find the radius of a circular saucepan of circumference 58·6 cm.

5 The handle of a paint tin is a semicircle of wire which is 28 cm long. Calculate the diameter of the tin.

6 A circle has an area of 16 mm². Find its circumference.

7 A circle has a circumference of 2500 km. Find its area.

8 The diagram shows a semicircle inside a rectangle. Find the shaded area.

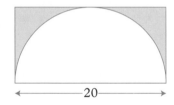

9 Discs of radius 4 cm are cut from a rectangular plastic sheet of length 84 cm and width 24 cm.

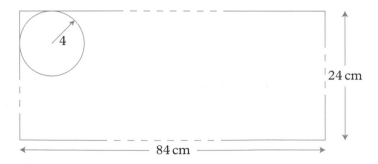

How many complete discs can be cut out?
Find
   a the total area of the discs cut
   b the area of the sheet wasted.

10 The tyre of a car wheel has an outer diameter of 30 cm. How many times will the wheel rotate on a journey of 5 km?

11 A golf ball of diameter 1·68 inches rolls a distance of 4 m in a straight line. How many times does the ball rotate completely? (1 inch = 2·54 cm)

12 A circular pond of radius 6 m is surrounded by a path of width 1 m. Find the area of the path.

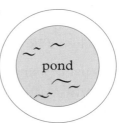

13 A rectangular metal plate has a length of 65 cm and a width of 35 cm.
It is melted down and recast into circular discs of the same thickness. How many complete discs can be formed if
a the radius of each disc is 3 cm?
b the radius of each disc is 10 cm?

14 Calculate the radius of a circle with an area equal to the sum of the areas of three circles of radii 2 cm, 3 cm and 4 cm respectively.

15 The diagram shows a lawn (unshaded) surrounded by a path of uniform width (shaded). The curved end of the lawn is a semicircle of diameter 10 m.

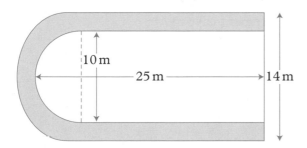

Calculate the total area of the path.

16 The diameter of a circle is given as 10 cm, correct to the nearest cm. Calculate
a the maximum possible circumference
b the minimum possible area of the circle consistent with this data.

17 A square is inscribed in a circle of radius 7 cm. Find
    a the area of the square
    b the area shaded.

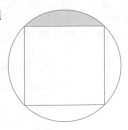

18 The governor of a prison has 100 m of wire fencing. What area can he enclose if he makes a circular compound?

19 In 'equable shapes' the numerical value of the area is equal to the numerical value of the perimeter.
Find the dimensions of these equable shapes
    a square   **b** circle   **c** equilateral triangle.

*20 The semicircle and the isosceles triangle have the same base AB and the same area.
Find the angle x.

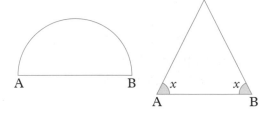

*21 Mr Gibson decided to measure the circumference of the earth using a very long tape measure. For reasons best known to himself he held the tape measure 1 m from the surface of the (perfectly spherical) earth all the way round. When he had finished Mrs Gibson told him that his measurement gave too large an answer. She suggested taking off 6 m. Was she correct? (Take the radius of the earth to be 6400 km (if you need it).)

*22 The large circle has a radius of 10 cm. Find the radius of the largest circle which will fit in the middle.

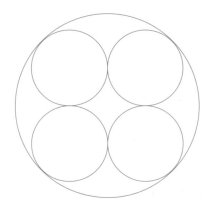

Exercise 19 Ⓜ (Mixed questions)

1 The sloping line divides the area of the square in the ratio $1:5$. What is the ratio $a:b$?

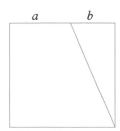

2 ABCD is a square of side 14 cm. P, Q, R, S are the midpoints of the sides. Semicircles are drawn with centres P, Q, R, S as shown on the diagram.

Find the shaded area, using $\pi = \dfrac{22}{7}$.

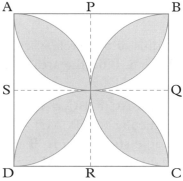

3 What fraction of the area of a circle of radius 4 cm is more than 3 cm from the centre?

4 Two equal circles are cut from a square piece of paper of side 2 m. Calculate the radius of the largest possible circles. Give your answer correct to 3 significant figures.

5 These circles have radii 2, 3, 5 cm. Express the shaded area as a percentage of the area of the largest circle.

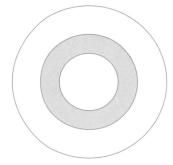

6 A square of side x cm is drawn in the middle of a square of side 10 cm. The corners of the smaller square are 2 cm from the corners of the larger square and are on the diagonals of the larger square. Find x correct to 4 sf.

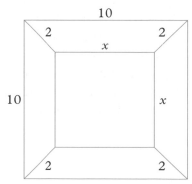

7 ABCDEFGH is a regular octagon. Calculate angle ACD.

8 The diagram shows two diagonals
drawn on the faces of a cube.
Find the angle between the diagonals.

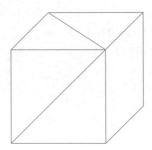

*9 The radius of the circle inscribed in a regular hexagon has
length 4 cm. Find the area of the hexagon, correct to 3 sf.

*10 In △ABC, AC = BC = 12 cm. CD is
perpendicular to AB. MD = 1 cm.
M is equidistant from A, B and C.
Calculate the length CM.

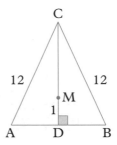

*11 The circumcircle and the inscribed circle of an equilateral triangle
are shown. Calculate the ratio of the area of the circumcircle to the
area of the inscribed circle.

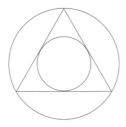

*12 ABCDEFGH is a regular octagon of side 1 unit.
  a Calculate the length AF.
  b Calculate the length AC.
  *c You probably used a calculator to work out AF.
     Can you calculate the **exact** value of AF? Leave your
     answer in a form involving a square root.

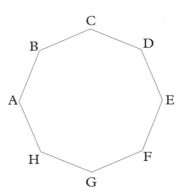

4.7.3 Arcs and sectors

Arc length

- Arc length, $l = \dfrac{\theta}{360} \times 2\pi r$

  The arc length is a fraction of the whole circumference depending on the angle at the centre of the circle.

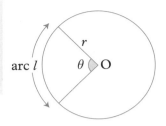

Sector area

- Sector area, $A = \dfrac{\theta}{360} \times \pi r^2$

The sector area is a fraction of the whole area depending on the angle at the centre of the circle.

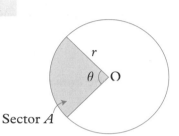

Sector A

EXAMPLE

a Find the length of an arc which subtends an angle of 140° at the centre of a circle of radius 12 cm. (Take $\pi = \dfrac{22}{7}$)

Arc length $= \dfrac{140}{360} \times 2 \times \dfrac{22}{7} \times 12$

$= \dfrac{88}{3}$

$= 29\dfrac{1}{3}$ cm

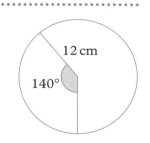

12 cm

140°

b A sector of a circle of radius 10 cm has an area of 25 cm^2. Find the angle at the centre of the circle.

Let the angle at the centre of the circle be θ.

$\dfrac{\theta}{360} \times \pi \times 10^2 = 25$

$\therefore \qquad \theta = \dfrac{25 \times 360}{\pi \times 100}$

$\theta = 28\cdot6°$ (1 dp)

The angle at the centre of the circle is $28\cdot6°$.

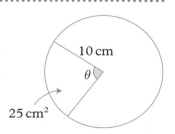

10 cm

θ

25 cm^2

Exercise 20 (H)

1 Find the length of the minor arc AB.

a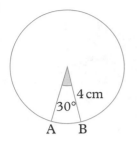

b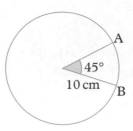

c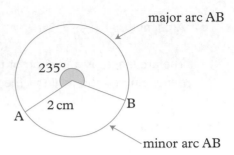

2 A pendulum of length 55 cm swings through an angle of 16°. Through what distance does the tip of the pendulum swing?

3 Find the area of the shaded sector.

a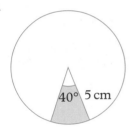

b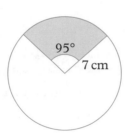

c

4 This question refers to the diagram.
Find the quantities marked with a ✓.

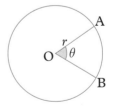

	r	θ	arc AB	area of sector AOB
a	8 cm	45°	✓	
b	10 cm	70°		✓
c	5·8 cm	115°	✓	
d	28 cm	22°		✓
e	65 cm	107°		✓

5 Find the shaded area in each diagram.

a

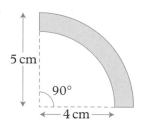

b

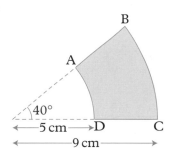

6 Find the length of the perimeter of the shaded area in question **5b**.

7 A sector is cut from a round cheese as shown. Calculate the volume of the piece of cheese.

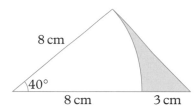

8 Work out the shaded area.

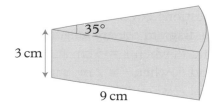

9 The lengths of the minor and major arcs of a circle are 5·2 cm and 19·8 cm respectively. Find
 a the radius of the circle
 b the angle subtended at the centre by the minor arc.

10 The length of the minor arc AB of a circle, centre O, is 2π cm and the length of the major arc is 22π cm. Find
 a the radius of the circle
 b the acute angle AOB.

11 A wheel of radius 10 cm is turning at a rate of 5 revolutions per minute. Calculate
 a the angle through which the wheel turns in 1 second
 b the distance moved by a point on the rim in 2 seconds.

The diagram shows an arc of length 10 cm on a circle of radius r. Find r.

$$\text{Arc length} = \frac{42}{360} \times 2 \times \pi \times r$$

$$\therefore \quad \frac{42}{360} \times 2 \times \pi \times r = 10$$

$$r = \frac{10 \times 360}{42 \times 2 \times \pi}$$

$$r = 13\cdot6 \text{ cm (3 sf)}$$

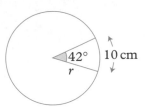

Exercise 21 Ⓗ

These questions cover extended topics.

1 In the diagram the arc length is l and the sector area is A.
  a Find θ, when $r = 5$ cm and $l = 7\cdot5$ cm.
  b Find θ, when $r = 2$ m and $A = 2$ m^2.
  c Find r, when $\theta = 55°$ and $l = 6$ cm.

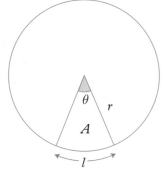

2 Two parallel lines are drawn 2 cm from the centre of a circle of radius 4 cm. Calculate the area shaded.

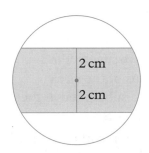

2 cm

2 cm

3 Find the angle θ so that the arc length is equal to the radius. The angle you have found is called a radian. Your calculator may have a ⎸RAD⎸ mode, where ⎸RAD⎸ stands for radian.

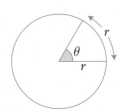

4 a ABCDE is a regular pentagon.
Calculate the size of angle x.

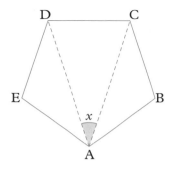

See Section 6.8
(page 333).

b In this diagram, arcs are drawn with centres at the
opposite corners of the pentagon, for example,
arc CD has centre at A. The radius of each arc is
5 cm. Show that the perimeter of the shape is 5π cm.

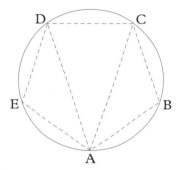

5 In the diagram, the arc length is l and the sector area is A.
  a Find l, when $\theta = 72°$ and $A = 15\,\text{cm}^2$.
  b Find l, when $\theta = 135°$ and $A = 162\,\text{m}^2$.
  c Find A, when $l = 11\,\text{cm}$ and $r = 5\cdot2\,\text{cm}$.

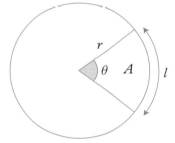

6 An arc of length one nautical mile subtends an angle of one minute
at the centre of the circle around the equator of the earth.
Calculate the length of a nautical mile in metres, given that
60 minutes = 1 degree and the radius of the earth is about
6370 km.

7 The length of an arc of a circle is 12 cm. The corresponding sector
area is 108 cm². Find
  a the radius of the circle
  b the angle subtended at the centre of the circle by the arc.

8 The length of an arc of a circle is 7·5 cm. The corresponding sector
area is 37·5 cm². Find
  a the radius of the circle
  b the angle subtended at the centre of the circle by the arc.

9 In the diagram, AB is a tangent to the circle at A. Straight lines BCD and ACE pass through the centre of the circle C. Angle ACB = x and the radius of the circle is 1 unit.

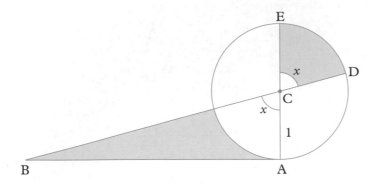

The two shaded areas are equal.

a Show that x satisfies the equation $x = \left(\dfrac{90}{\pi}\right)\tan x$.

b Use trial and improvement to find a solution for x correct to 1 decimal place.

*10 In the diagram, BC is a diameter of the semicircle, $\angle ABC = 90°$,

shaded area ① = shaded area ② and angle ACB = x.
Show that $\tan x = \dfrac{\pi}{4}$.
Hence find angle x.

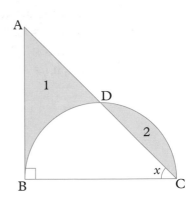

> See Section 6.2 for more on tan x.

4.7.4 Segments

The line AB is a chord. The area of a circle cut off by a chord is called a **segment**. In the diagram the **minor** segment is shaded and the **major** segment is unshaded.

- The line from the centre of a circle to the midpoint M of a chord **bisects** the chord at right angles.

- The line from the centre of a circle to the midpoint of a chord bisects the angle subtended by the chord at the centre of the circle.

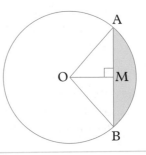

XY is a chord of length 12 cm of a circle of radius 10 cm, centre O. Calculate
a the angle XOY
b the area of the minor segment cut off by the chord XY.

Let the midpoint of XY be M.

$$MY = 6\,cm$$

$$\sin \angle MOY = \frac{6}{10}$$

$$\angle MOY = 36{\cdot}87°$$

$$\angle XOY = 2 \times 36{\cdot}87$$

$$= 73{\cdot}74°$$

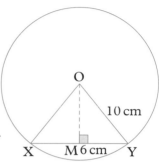

area of minor segment = area of sector XOY − area of △XOY

$$\text{area of sector XOY} = \frac{73{\cdot}74}{360} \times \pi \times 10^2$$

$$= 64{\cdot}35\,cm^2$$

$$\text{area of } \triangle XOY = \frac{1}{2} \times 10 \times 10 \times \sin 73{\cdot}74° = 48{\cdot}00\,cm^2$$

> Using the formula
> Area of triangle
> $= \frac{1}{2}\,ab \sin C$

So area of minor segment = 64·35 − 48·00
$$= 16{\cdot}4\,cm^2 \text{ (3 sf)}$$

Exercise 22 Ⓗ

Use the π button on your calculator.

1 The chord AB subtends an angle of 130° at the centre O.
   The radius of the circle is 8 cm. Find
   a the area of sector OAB
   b the area of triangle OAB
   c the area of the minor segment (shown shaded).

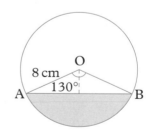

2 Find the shaded area when
   a $r = 6\,cm, \ \theta = 70°$
   b $r = 14\,cm, \ \theta = 104°$
   c $r = 5\,cm, \ \theta = 80°$.

3 Find θ and hence the shaded area when
   a AB = 10 cm, $r = 10$ cm
   b AB = 8 cm, $r = 5$ cm.

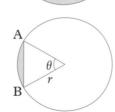

4 How far is a chord of length 8 cm from the centre of a circle of radius 5 cm?

5 How far is a chord of length 9 cm from the centre of a circle of radius 6 cm?

6 The diagram shows the cross-section of a cylindrical pipe with water lying in the bottom.
   a If the maximum depth of the water is 2 cm and the radius of the pipe is 7 cm, find the shaded area.
   b What is the **volume** of water in a length of 30 cm of the pipe?

7 An equilateral triangle is inscribed in a circle of radius 10 cm. Find
   a the area of the triangle
   b the shaded area.

8 A regular hexagon is circumscribed by a circle of radius 6 cm.
   Find the shaded area.

9 A regular octagon is circumscribed by a circle of radius r cm. Find the area enclosed between the circle and the octagon. (Give the answer in terms of r.)

*10 Find the radius of the circle
   a when $\theta = 90°$, $A = 20 \text{ cm}^2$
   b when $\theta = 30°$, $A = 35 \text{ cm}^2$
   c when $\theta = 150°$, $A = 114 \text{ cm}^2$.

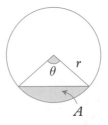

*11 The diagram shows a regular pentagon of side 10 cm with a star inside.
   Calculate the area of the star.

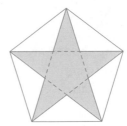

4.8 Volume and surface area

4.8.1 Volume and surface area of a cuboid

● The amount of space which an object occupies is called its volume.

A cube with edges 1 cm long has a volume of one cubic centimetre. You write this as 1 cm^3.

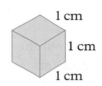

A cuboid is a solid object with faces that are all rectangles. Shapes A and B are cuboids. Do not confuse a cuboid with a **cube**, where all the edges are the same length. A cube is a special kind of cuboid.

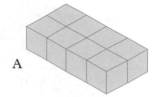

A

● The volume of a cuboid is given by
Volume = length × width × height

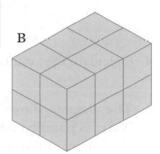

B

For shape A, volume = 4 × 2 × 1 = 8 cm^3
For shape B, volume = 2 × 3 × 2 = 12 cm^3

Faces, edges and vertices
Many three-dimensional shapes have **faces, edges** and **vertices** (plural of **vertex**). The diagram shows a cuboid.

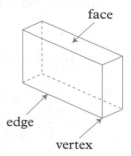

● The faces of the cuboid are the flat surfaces of the shape. There are 6 faces on a cuboid.

● The edges of the cuboid are the lines that make up the shape. There are 12 edges on a cuboid.

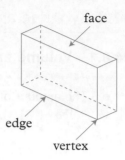
face

● The **vertices** of the cuboid are where the edges meet at a point.
There are 8 vertices on a cuboid.

● The **surface area** of a shape is the sum of the area of all its faces.

edge

vertex

Exercise 23

In questions **1** to **3** work out the volume and surface area of each
cuboid. All lengths are in cm.

1

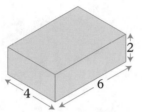

2

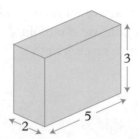

3

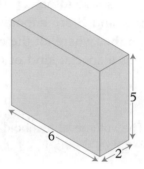

4 Calculate the volume of a cube of side 5 cm.

5 The diagram shows an empty swimming pool.
Water is pumped into the pool at a rate
of 2 m³ per minute.
How long will it take to fill the pool?

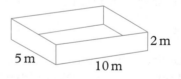

2 m
5 m
10 m

6 Find the missing side *x* in each case. All lengths are in cm.

a

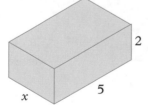

Volume = 30 cm³

b

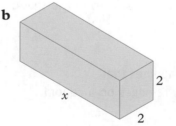

Volume = 22 cm³

c

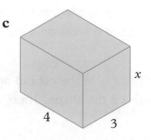

Volume = 30 cm³

7 **a** Sketch a diagram of a cube of volume 1 m³.
b How many cubic centimetres (cm³) are there in 1 m³?
c What is the surface area of the cube?

4.8.2 Volume and surface area of a prism and volume of a cylinder

- A prism is an object with a uniform cross-section.

- A cylinder is a prism with a circular cross-section.

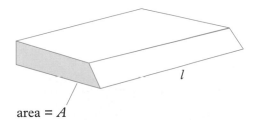

area = A

Volume of prism
= (area of cross-section) × length
= $A \times l$

Surface area of prism
= sum of areas of all faces

radius = r
height = h
Volume of cylinder = area of cross-section
× length
Volume = $\pi r^2 h$

a Find the volume and surface area of the prism.

b Find the volume of the cylinder.

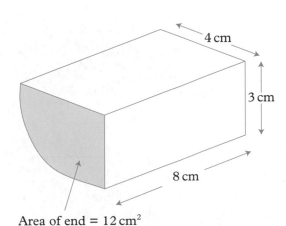

3 cm

4 cm

8 cm

Area of end = 12 cm^2

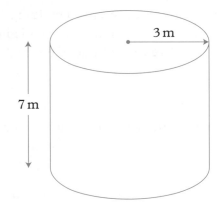

3 m

7 m

a Volume = $A \times l$
  = 12 × 8
  = 96 cm^3

Surface area =
(2 × 4 × 8) + (2 × 8 × 3) +
(2 × 12) = 136 cm^2

b Volume = $\pi \times 3^2 \times 7$
  = 198 m^3 (3 sf)

Exercise 24 Ⓜ ⭐Ⓕ

1 Calculate the volume of each prism. All lengths are in cm.

a Area of end = 15 cm²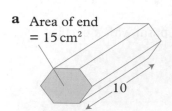

b Area of end = 5 cm²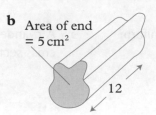

c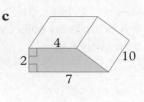

d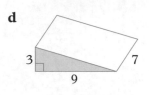

e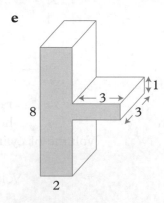

f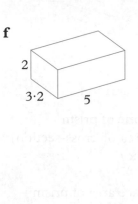

2 Calculate the surface area of shapes **c**, **d**, **e** and **f**.

3 A cylindrical bar has a cross-sectional area of 12 cm² and a length of two **metres**. Calculate the volume of the bar in cm³.

4 The diagram represents a building.
 a Calculate the area of the shaded end.
 b Calculate the volume of the building.

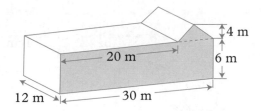

5 Calculate the volume of each cylinder.

a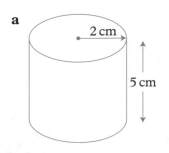

b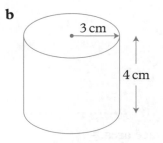

c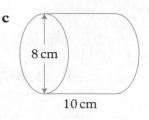

6 Calculate the volumes of these cylinders.
 a $r = 4\,\text{cm}$, $h = 10\,\text{cm}$
 b $r = 11\,\text{m}$, $h = 2\,\text{m}$

7 A gas cylinder has diameter 18 cm and length 40 cm.
Calculate the capacity of the cylinder, correct to the nearest litre.

8 A solid cylinder of radius 5 cm and length 15 cm is made
from material of density 6 g/cm^3. Calculate the mass of the
cylinder.

9 Cylinders are cut along the axis of symmetry to form these objects.
Find the volume of each object.

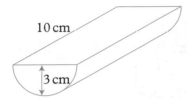
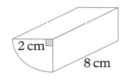

10 A rectangular block has dimensions 20 cm × 7 cm × 7 cm. Find
the volume of the largest solid cylinder which can be cut from this
block.

11 The two solid cylinders shown have the same mass.
Calculate the density, x g/cm^3, of cylinder B.

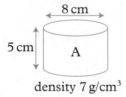
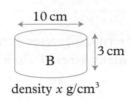

density 7 g/cm^3 density x g/cm^3

EXAMPLE

Calculate the height of a cylinder of volume 500 cm^3 and
base radius 8 cm.

....

Let the height of the cylinder be h cm.

$$\pi r^2 h = 500$$

$$3 \cdot 14 \times 8^2 \times h = 500$$

$$h = \frac{500}{3 \cdot 14 \times 64}$$

$$h = 2 \cdot 49 \ (3 \text{ sf})$$

The height of the cylinder is 2·49 cm.

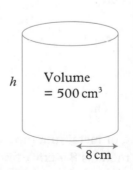

h Volume
= 500 cm^3

8 cm

Exercise 25 Ⓜ ⭐

1 Find the height of a cylinder of volume 200 cm³ and radius 4 cm.

2 Find the length of a cylinder of volume 2 litres and radius 10 cm.

3 Find the radius of a cylinder of volume 45 cm³ and length 4 cm.

4 When 3 litres of oil is removed from an upright cylindrical can, the level falls by 10 cm. Find the radius r of the can.

5 A solid cylinder of radius 4 cm and length 8 cm is melted down and recast into a solid cube. Find the length of the side of the cube.

6 A solid rectangular block of copper 5 cm by 4 cm by 2 cm is drawn out to make a cylindrical wire of diameter 2 mm. Calculate the length of the wire.

7 Water flows through a circular pipe of internal diameter 3 cm at a speed of 10 cm/s. If the pipe is full, how many litres of water issue from the pipe in one minute?

8 Water issues from a hose-pipe of internal diameter 1 cm at a rate of 5 litres per minute. At what speed is the water flowing through the pipe?

9 A cylindrical metal pipe has external diameter of 6 cm and internal diameter of 4 cm.

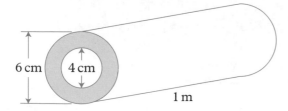

Calculate the volume of metal in a pipe of length 1 m. If the density of the metal is 8 g/cm³, find the weight of the pipe.

10 Henry decided to build a garage and began by calculating the number of bricks required. The garage was to be 6 m by 4 m and 2·5 m in height. Each brick measures 22 cm by 10 cm by 7 cm. Henry estimated that he would need about 40 000 bricks. Is this a reasonable estimate? Explain your answer.

11 A cylindrical can of internal radius 20 cm stands upright on a flat surface. It contains water to a depth of 20 cm. Calculate the rise h in the level of the water when a brick of volume 1500 cm^3 is immersed in the water.

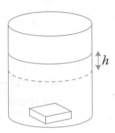

12 A cylindrical tin of height 15 cm and radius 4 cm is filled with sand from a rectangular box. How many times can the tin be filled if the dimensions of the box are 50 cm by 40 cm by 20 cm?

13 Rain which falls onto a flat rectangular surface of length 6 m and width 4 m is collected in a cylinder of internal radius 20 cm. What is the depth of water in the cylinder after a storm in which 1 cm of rain fell?

4.8.3 Volume of pyramids, spheres and cones

Pyramid Volume	Sphere Volume	Cone Volume
$= \frac{1}{3}$ (base area) × height	$= \frac{4}{3}\pi r^3$	$= \frac{1}{3}\pi r^2 h$

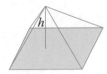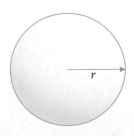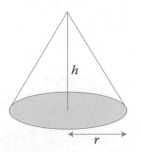

The formula for the volume of a pyramid is demonstrated below.

Figure 1 shows a cube of side $2a$ broken down into six pyramids of height a as shown in Figure 2. If the volume of each pyramid is V, then

$6V = 2a \times 2a \times 2a$

$V = \frac{1}{6} \times (2a)^2 \times 2a$

so $V = \frac{1}{3} \times (2a)^2 \times a$

$V = \frac{1}{3}$ (base area) $\times$ height.

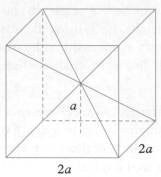

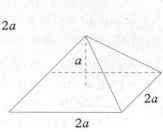

Figure 1   Figure 2

EXAMPLE

a Calculate the volume of a cone of radius 2 cm and height 5 cm.

b A solid sphere of radius 4 cm is made of metal of density 8 g/cm³. Calculate the mass of the sphere.

· ·

a Volume $= \frac{1}{3} \times \pi \times 2^2 \times 5$

$= \frac{\pi \times 20}{3}$

$= 20 \cdot 9 \text{ cm}^3$ (3 sf)

b Volume $= \frac{4}{3} \times \pi \times 4^3 \text{ cm}^3$

Mass $= \frac{4}{3} \times \pi \times 4^3 \times 8 = 2140 \text{ g}$ (3 sf)

Exercise 26 Ⓗ

In questions **1** to **6** find the volume of each object. All lengths are in cm.

1
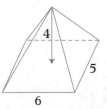

2

3
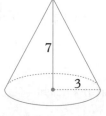

4
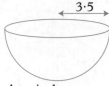
hemisphere

5
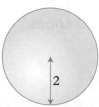

6
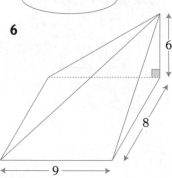

7 A solid sphere is made of metal of density $9\,\text{g/cm}^3$. Calculate the mass of the sphere if the radius is $5\,\text{cm}$.

8 Find the volume of a hemisphere of radius $5\,\text{cm}$.

9 Calculate the volume of a cone of height and radius $7\,\text{m}$.

10 A cone is attached to a hemisphere of radius $4\,\text{cm}$. If the total length of the object is $10\,\text{cm}$, find its volume.

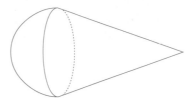

11 Find the height of a pyramid of volume $20\,\text{m}^3$ and base area $12\,\text{m}^2$.

12 A single drop of oil is a sphere of radius $3\,\text{mm}$. The drop of oil falls on water to produce a thin circular film of radius $100\,\text{mm}$. Calculate the thickness of this film in mm.

13 Gold is sold in solid spherical balls. Which is worth more: 10 balls of radius $2\,\text{cm}$ or 1 ball of radius $4\,\text{cm}$?

r = 2 cm   r = 4 cm

14 A toy consists of a cylinder of diameter $6\,\text{cm}$ 'sandwiched' between a hemisphere and a cone of the same diameter. If the cone is of height $8\,\text{cm}$ and the cylinder is of height $10\,\text{cm}$, find the total volume of the toy.

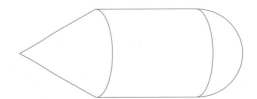

15 Water is flowing into an inverted cone, of diameter and height $30\,\text{cm}$, at a rate of 4 litres per minute. How long, in seconds, will it take to fill the cone?

16 Metal spheres of radius $2\,\text{cm}$ are packed into a rectangular box of internal dimensions $16\,\text{cm} \times 8\,\text{cm} \times 8\,\text{cm}$. When 16 spheres are packed the box is filled with a preservative liquid. Find the volume of this liquid.

17 The Great Pyramids in Egypt are seriously large! One Pyramid has a square base of side $100\,\text{m}$ and height $144\,\text{m}$. An average slave could just manage to carry a brick measuring $40\,\text{cm}$ by $20\,\text{cm}$ by $16\,\text{cm}$. Assuming no spaces were left for mummies or treasure, work out how many bricks would be needed to make this Pyramid.

18 The cylindrical end of a pencil is sharpened to produce a perfect cone with no overall loss of length. If the diameter of the pencil is 1 cm, and the cone is of length 2 cm, calculate the volume of the shavings.

19 One corner of a solid cube of side 8 cm is removed by cutting through the midpoints of three adjacent sides. Calculate the volume of the piece removed.

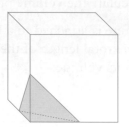

20 A hollow spherical vessel has internal and external radii of 6 cm and 6·4 cm respectively. Calculate the weight of the vessel if it is made of metal of density 10 g/cm³.

Finding the radius

EXAMPLE

Calculate the radius of a sphere of volume 500 cm³.

· ·

Let the radius of the sphere be r cm.

$$\frac{4}{3}\pi r^3 = 500$$

$$r^3 = \frac{3 \times 500}{4\pi}$$

$$r = \sqrt[3]{\left(\frac{3 \times 500}{4\pi}\right)} = 4\cdot92 \ (3 \ \text{sf})$$

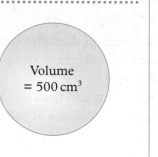

Volume
= 500 cm³

The radius of the sphere is 4·92 cm.

Exercise 27 Ⓗ

1 Calculate the radius of each sphere.

a

Volume
= 250 cm³

b

Volume
= 60 cm³

c

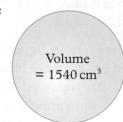

Volume
= 1540 cm³

2 A solid metal sphere of radius 4 cm is recast into a solid cube. Find the length of a side of the cube.

3 Find the height of a cone of volume 2500 cm^3 and radius 10 cm.

4 A solid metal sphere is recast into many smaller spheres. Calculate the number of the smaller spheres if the initial and final radii are

    a initial radius $= 10 \text{ cm}$, final radius $= 2 \text{ cm}$

    b initial radius $= 1 \text{ m}$,   final radius $= \frac{1}{3} \text{ cm}$.

5 A spherical ball is immersed in water contained in a vertical cylinder.

    Assuming the water covers the ball, calculate the rise in the water level h, if

    a sphere radius $= 3 \text{ cm}$, cylinder radius $= 10 \text{ cm}$

    b sphere radius $= 2 \text{ cm}$, cylinder radius $= 5 \text{ cm}$.

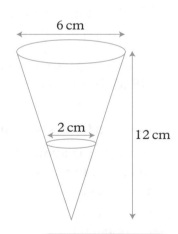

6 A spherical ball is immersed in water contained in a vertical cylinder. The rise in water level is measured in order to calculate the radius of the spherical ball. Calculate the radius of the ball in these cases

    a cylinder of radius 10 cm, water level rises 4 cm

    b cylinder of radius 100 cm, water level rises 8 cm.

7 The diagram shows the cross-section of an inverted cone of height $MC = 12 \text{ cm}$.

If $AB = 6 \text{ cm}$ and $XY = 2 \text{ cm}$, use similar triangles to find the length NC. Hence find the volume of the cone of height NC.

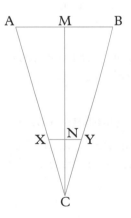

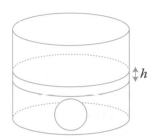

> See section 4.9, page 225, for similar triangles.

8 An inverted cone of height 10 cm and base radius $6\cdot4 \text{ cm}$ contains water to a depth of 5 cm, measured from the vertex. Calculate the volume of water in the cone.

9 An inverted cone of height 15 cm and base radius 4 cm contains water to a depth of 10 cm. Calculate the volume of water in the cone.

10 A frustum is a cone with 'the end chopped off'. A bucket in the shape of a frustum as shown in the diagram has diameters of 10 cm and 4 cm at its ends and a depth of 3 cm.
Calculate the volume of the bucket.

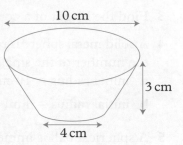

11 Find the volume of a frustum with end diameters of 60 cm and 20 cm and a depth of 40 cm.

12 Builders' sand is tipped into the shape of a cone where the height and radius are both 1·4 m.
   a Calculate the volume of the sand, correct to 1 dp.
   b A further 2 m^3 is added to the pile. This now forms a larger cone, but the height still equals the radius. Calculate the height of this larger pile correct to 1 dp.

*13 The diagram shows the major sector of a circle of radius 10 cm.
   a Find, as a multiple of π, the arc length of the sector.
The straight edges are brought together to make a cone.
Calculate
   b the radius of the base of the cone
   c the vertical height of the cone.

*14 A sphere passes through the eight vertices of a cube of side 10 cm. Find the volume of the sphere.

4.8.4 Surface areas of cylinders, cones and spheres

This section is about the surface areas of the **curved** parts of cylinders, spheres and cones. The areas of the plane faces are easier to find.

- **Cylinder**
  Curved surface area = 2πrh

- **Sphere**
  Surface area = 4πr^2

- **Cone**
  Curved surface area = πrℓ, where ℓ is the slant height

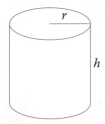

EXAMPLE

Calculate the **total** surface area of a solid cylinder of radius 3 cm and height 8 cm.

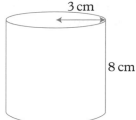

Curved surface area $= 2\pi rh$
$$= 2 \times \pi \times 3 \times 8$$
$$= 48\pi \text{ cm}^2$$

Area of two ends $= 2 \times \pi r^2$
$$= 2 \times \pi \times 3^2$$
$$= 18\pi \text{ cm}^2$$

Total surface area $= (48\pi + 18\pi) \text{ cm}^2$
$$= 207 \text{ cm}^2 \text{ (3 sf)}$$

Exercise 28 Ⓗ

1 Work out the **curved** surface area of these objects. Leave π in your answers. All lengths are in cm.

a

$r = 3$

b

$r = 4$
$h = 5$

c

$l = 10$

$r = 6$

d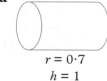

$r = 0\cdot7$
$h = 1$

2 Work out the **total** surface area of these objects. Leave π in your answers. All lengths are in cm.

a

$l = 6$

$r = 4$

b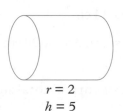

$r = 2$
$h = 5$

c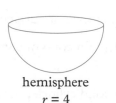

hemisphere
$r = 4$

d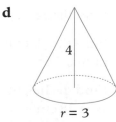

4

$r = 3$
vertical height $= 4$

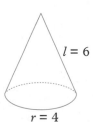 **3** A solid cylinder of height 10 cm and radius 4 cm is to be plated with material costing £11 per cm^2. Find the cost of the plating.

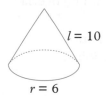 **4** A tin of paint covers a surface area of 60 m^2 and costs £4·50. Find the cost of painting the outside surface of a cylindrical gas holder of height 30 m and radius 18 m. The top of the gas holder is a flat circle.

5 A solid wooden cylinder of height 8 cm and radius 3 cm is cut in two along a vertical axis of symmetry. Calculate the total surface area of the two pieces.

6 Calculate the total surface area of the combined cone/cylinder/hemisphere.

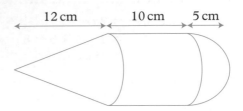

7 A man is determined to spray the entire surface of the earth (including the oceans) with a revolutionary new weed killer. If it takes him 10 seconds to spray 1 m^2, how long will it take to spray the whole world?
(radius of the earth = 6370 km; ignore leap years)

8 A rectangular piece of card 10 cm by 30 cm is rolled up to make a tube (with no overlap). Find the radius of the tube if
a the long sides are joined
b the short sides are joined.

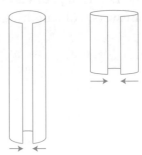

9 Find the radius of a sphere of surface area 34 cm^2.

10 Find the slant height of a cone of curved surface area 20 cm^2 and radius 3 cm.

11 Find the height of a solid cylinder of radius 1 cm and **total** surface area 28 cm^2.

12 Find the volume of a sphere of surface area 100 cm^2.

13 Find the surface area of a sphere of volume 28 cm^3.

14 An inverted cone of vertical height 12 cm and base radius 9 cm contains water to a depth of 4 cm. Find the area of the interior surface of the cone not in contact with the water.

15 A circular paper of radius 20 cm is cut in half and each half is made into a hollow cone by joining the straight edges. Find the slant height and base radius of each cone.

16 A solid metal cube of side 6 cm is recast into a solid sphere.
   a Find the radius of the sphere.
   b By how much is the surface area of the original cube greater than the surface area of the new sphere?

17 A solid cuboid, measuring 5 cm × 5 cm × 3 cm, has a hole of radius 2 cm drilled right through.
Calculate the surface area of the object.

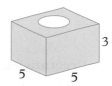

18 A cone of radius 6 cm has a **total** surface area of 300 cm^2.
Calculate the volume of the cone.

4.9 Similar shapes

4.9.1 Properties of similar shapes

- Shapes which are mathematically **similar** have the same shape. All corresponding angles are equal, and corresponding lengths are in the same ratio.

If two shapes are similar one shape is an enlargement of the other.

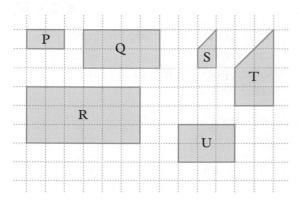

P, Q and R are similar   S and T are similar

U is **not** similar to any of P, Q and R.

● Notice that all **circles** are similar to each other
and all **squares** are similar to each other.

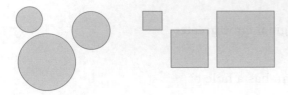

In general **rectangles** are **not** similar to each other.

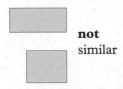

not
similar

The two triangles A and B are similar if they have the same
angles.

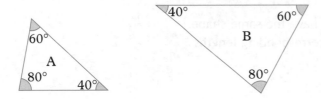

EXAMPLE

The two triangles are similar.
Find x.

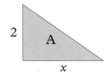

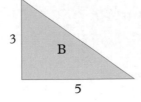

Triangle B is an enlargement of
triangle A.
Corresponding sides are in the
same ratio.

$$\frac{x}{5} = \frac{2}{3}$$

$$x = \frac{2}{3} \times 5$$

$$x = 3\frac{1}{3}$$

Exercise 29 Ⓜ/Ⓗ

1 Look at the diagram and use the letters to make a list of pairs of shapes which are similar.

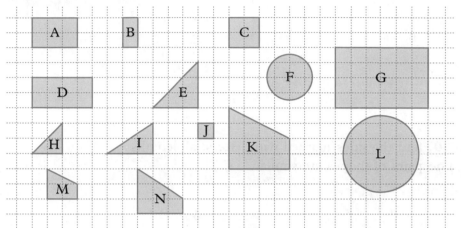

2 Copy each shape and draw a new shape which is similar in each case.

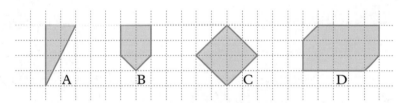

3 Decide which of the triangles **a, b** and **c** are similar to the blue triangle.

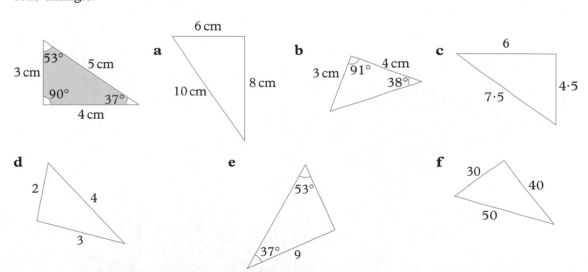

4 Which of the shapes B, C, D is/are similar to shape A?

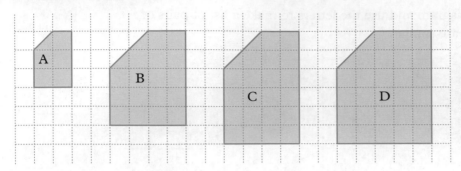

In questions **5** to **10**, find the sides marked with letters; all lengths are given in cm. The pairs of shapes are similar.

5

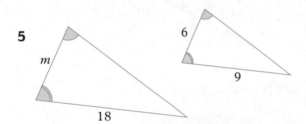

6

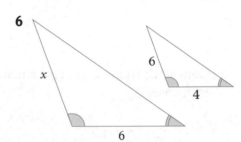

7

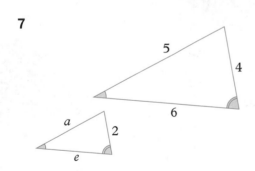

8

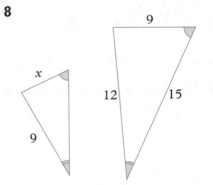

9

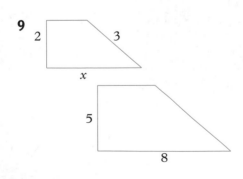

10

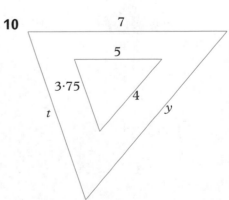

11 Picture B is an enlargement of picture A. Calculate the length x.

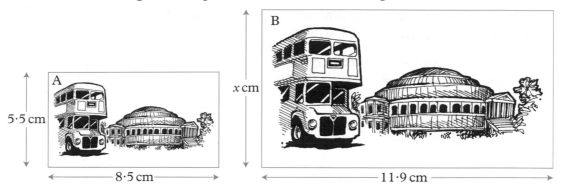

12 The drawing shows a rectangular picture 16 cm × 8 cm surrounded by a border of width 4 cm.
Are the two rectangles similar?

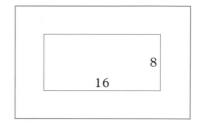

13 Which of the following **must** be similar to each other?
   a two equilateral triangles   **b** two rectangles
   c two isosceles triangles   **d** two squares
   e two regular pentagons   **f** two kites
   g two rhombuses   **h** two circles

14 a Explain why triangles ABC and EBD are similar.
   b Given that EB = 7 cm, calculate the length AB.
   c Write the length AE.

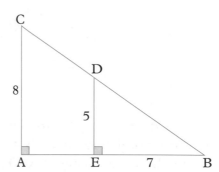

In questions **15**, **16** and **17** use similar triangles to find the sides marked with letters. All lengths are in cm.

15

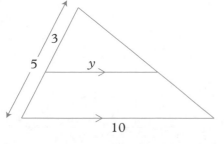

16

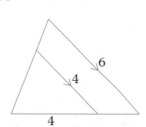

17

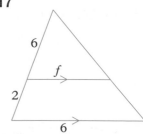

Exercise 30 Ⓗ (More difficult)

1 A tree of height 4 m casts a shadow of length 6·5 m. Find the height of a house casting a shadow 26 m long.

2 A small cone is cut from a larger cone. Find the radius of the smaller cone.

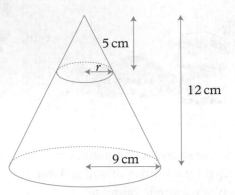

3 The diagram shows the side view of a swimming pool being filled with water. Calculate the length x.

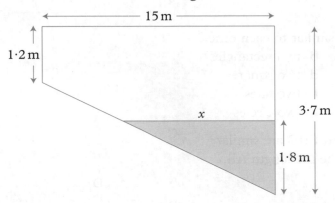

4 The diagonals of a trapezium ABCD intersect at O. AB is parallel to DC, AB = 3 cm and DC = 6 cm.
   a Show that triangles ABO and CDO are similar.
   b If CO = 4 cm and OB = 3 cm, find AO and DO.

In questions **5** and **6** find the sides marked with letters.

5 ∠BAC = ∠DBC   **6**

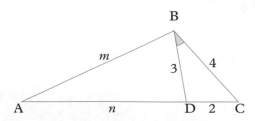

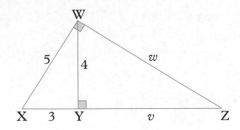

7 A rectangle 11 cm by 6 cm is similar to a rectangle 2 cm by x cm. Find the two possible values of x.

8 From the rectangle ABCD a square is cut off to leave rectangle BCEF. Rectangle BCEF is similar to ABCD. Find x and hence state the ratio of the sides of rectangle ABCD.

> ABCD is called the Golden Rectangle and is an important shape in architecture.

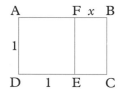

9 Triangles ABC and EBD are similar but DE is **not** parallel to AC. Work out the length x.

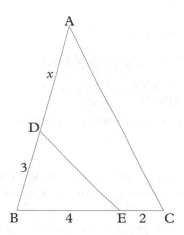

10 In the diagram $\angle ABC = \angle ADB = 90°$, AD = p and DC = q.
   a Use similar triangles ABD and ABC to show that $x^2 = pz$.
   b Find a similar expression for y^2.
   c Add the expressions for x^2 and y^2 and hence prove Pythagoras' theorem.

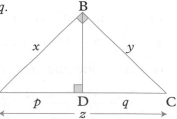

*11 In a triangle ABC, a line is drawn parallel to BC to meet AB at D and AC at E. DC and BE meet at X. Prove that
   a the triangles ADE and ABC are similar
   b the triangles DXE and BXC are similar
   c $\dfrac{AD}{AB} = \dfrac{EX}{XB}$

4.9.2 Areas of similar shapes

The two rectangles, ABCD and WXYZ, are similar.
The ratio of their corresponding sides is k.

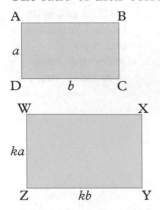

area of ABCD = ab

area of WXYZ = $ka \times kb = k^2 ab$

So $\dfrac{\text{area of WXYZ}}{\text{area ABCD}} = \dfrac{k^2 ab}{ab} = k^2$

This illustrates an important general rule for all similar shapes.

- If two shapes are similar and the ratio of corresponding sides is k, then the ratio of their areas is k^2.

k is called the linear scale factor and k^2 is called the area scale factor.

This result also applies for the surface areas of similar three-dimensional objects.

EXAMPLE

In triangle ABC, XY is parallel to BC and $\dfrac{\text{AB}}{\text{AX}} = \dfrac{3}{2}$. If the area of triangle AXY is 4 cm², find the area of triangle ABC.

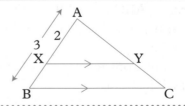

The triangles ABC and AXY are similar.

Ratio of corresponding sides $(k) = \dfrac{3}{2}$

Ratio of areas $(k^2) = \dfrac{9}{4}$

So $\qquad$ Area of $\triangle$ABC $= \dfrac{9}{4} \times$ (area of $\triangle$AXY)

$$= \dfrac{9}{4} \times (4) = 9 \text{ cm}^2$$

Two similar triangles have areas of $18\,\text{cm}^2$ and $32\,\text{cm}^2$ respectively.
If the base of the smaller triangle is $6\,\text{cm}$, find the base of the larger triangle.

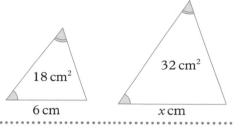

$$\text{Ratio of areas } (k^2) = \frac{32}{18} = \frac{16}{9}$$

$$\text{Ratio of corresponding sides } (k) = \sqrt{\left(\frac{16}{9}\right)} = \frac{4}{3}$$

$$\text{So base of larger triangle} = 6 \times \frac{4}{3} = 8\,\text{cm}$$

Exercise 31 Ⓜ/Ⓗ

In this exercise, a number written inside a figure represents the area of the shape in cm^2. Numbers on the outside give linear dimensions in cm.
In each case the shapes are similar. In questions **1** to **6**, find the unknown area A.

> Remember: If linear scale factor = k, area scale factor = k^2.

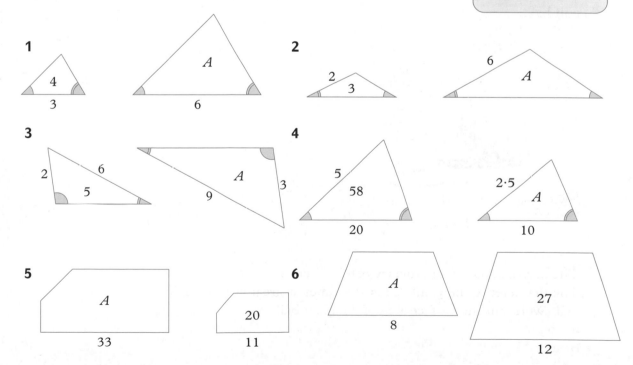

In questions **7** to **10**, find the lengths marked for each pair of similar shapes.

7

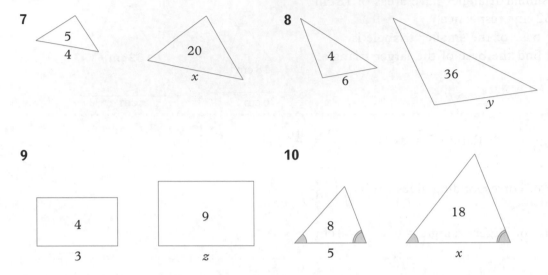

5
4

20
x

8

4
6

36
y

9

4
3

9
z

10

8
5

18
x

11 A triangle with sides 3, 4 and 5 cm has an area of 6 cm².
A similar triangle has an area of 54 cm². Find the sides of the larger triangle.

12 A giant ball is made to promote the sales of a new make of golf ball.

The surface area of an ordinary ball is 50 cm².
The diameter of the giant ball is 100 times as great as a normal ball. Work out the surface area of the giant ball
a in cm²
b in m².

13 It takes 30 minutes to cut the grass in a square field of side 20 m.
How long will it take to cut the grass in a square field of side 60 m?

14 A floor is covered by 600 tiles which are 10 cm by 10 cm. How many 20 cm by 20 cm tiles are needed to cover the same floor?

15 A wall is covered by 160 tiles which are 15 cm by 15 cm. How many 10 cm by 10 cm tiles are needed to cover the same wall?

16 A ball of radius r cm has a surface area of 20 cm^2. Find, in terms of r, the radius of a ball with a surface area of 500 cm^2.

17 AD = 3 cm, AB = 5 cm and area of $\triangle$ADE = 6 cm^2.

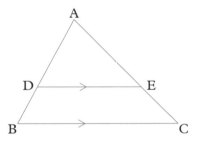

Find
a the area of $\triangle$ABC
b the area of DECB.

18 XY = 5 cm, MY = 2 cm and area of $\triangle$MYN = 4 cm^2.

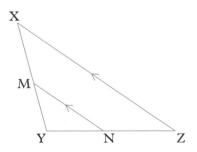

Find
a the area of $\triangle$XYZ
b the area of MNZX.

19 The triangles ABC and EBD are similar.
(AC and DE are **not** parallel.)

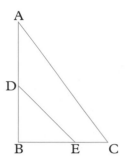

If AB = 8 cm, BE = 4 cm and the area of △DBE = 6 cm²,
find the area of △ABC.

20 When potatoes are peeled do you lose more of the potatoes or less
when big potatoes are used as opposed to small ones?

21 A supermarket offers cartons of cheese with 10%
extra for the same price.

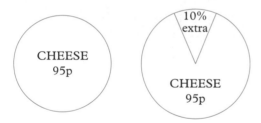

The old carton has a radius of 6 cm and a height of 1·2 cm. The
new carton has the same height. Calculate its radius.

22 An A3 sheet of paper can be cut into two sheets of A4.
The A3 and A4 sheets are mathematically similar.

Find the scale factor: $\left(\dfrac{\text{long side of A3 sheet}}{\text{long side of A4 sheet}}\right)$ $\left[\text{that is, } \dfrac{x}{y}\right]$

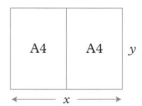

*23 Rectangle ABCD is inscribed in △PQR. The length of AD is
one-third the length of the perpendicular from R to PQ.

Calculate the ratio $\left(\dfrac{\text{area ABCD}}{\text{area PQR}}\right)$.

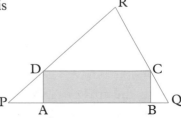

4.9.3 Volumes of similar objects

When solid objects are similar, one is an accurate enlargement of the other.

● If two objects are similar and the ratio of corresponding sides is
  k, then the ratio of their volumes is k^3.

A line has one dimension, and the scale factor is used once.
An area has two dimensions, and the scale factor is used twice.
A volume has three dimensions, and the scale factor is used three times.

EXAMPLE

Two similar cylinders have heights of 3 cm and 6 cm respectively.
If the volume of the smaller cylinder is 30 cm³, find the volume of the
larger cylinder.

$$\text{ratio of heights } (k) = \frac{6}{3} \text{ (linear scale factor)}$$

$$= 2$$

$$\text{So, ratio of volumes } (k^3) = 2^3 \text{ (volume scale factor)}$$

$$= 8$$

$$\text{and volume of larger cylinder} = 8 \times 30$$

$$= 240 \text{ cm}^3$$

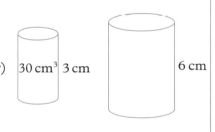

30 cm³ 3 cm   6 cm

EXAMPLE

Two similar spheres made of the same
material have masses of 32 kg and 108 kg
respectively. If the radius of the larger
sphere is 9 cm, find the radius of the
smaller sphere.

Take the ratio of masses to be the same as
the ratio of volumes.

$$\text{ratio of volumes } (k^3) = \frac{32}{108} = \frac{8}{27}$$

$$\text{ratio of corresponding lengths } (k) = \sqrt[3]{\left(\frac{8}{27}\right)} = \frac{2}{3}$$

$$\text{So, radius of smaller sphere} = \frac{2}{3} \times 9 = 6 \text{ cm}$$

108 kg

32 kg

$r = ?$

$r = 9$ cm

Exercise 32 Ⓜ/Ⓗ

In this exercise, the objects are similar and a number written inside a figure represents the volume of the object in cm³.

Numbers on the outside give linear dimensions in cm. In questions **1** to **8**, find the unknown volume *V*.

> Remember:
> If linear scale factor = *k*, volume scale factor = k^3.

1

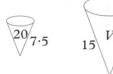 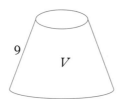

60   *V*

5   10

2

5 | 20   15 | *V*

3

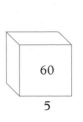 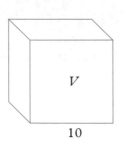

20/7·5   15\ *V*

4

(4·5)
radius = 1·2   *V*

radius = 12

5

24   6   9   *V*

6

88   *V*

6·2   3·1

7

 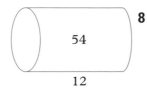

V   54

8   12

8

36   9   *V*   12

In questions **9** to **14**, find the lengths marked by a letter.

9

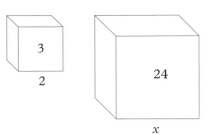

3

2

24

x

10

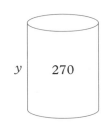

7 10

y 270

11

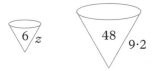

6 z

48 9·2

12

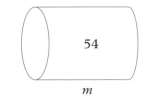

16

6

54

m

13

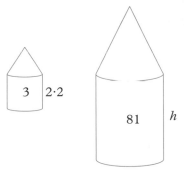

3 2·2

81 h

14

3 d

80 270

15 Two similar jugs have heights of 4 cm and 6 cm respectively. If the capacity of the smaller jug is 50 cm^3, find the capacity of the larger jug.

16 Two similar cylindrical tins have base radii of 6 cm and 8 cm respectively. If the capacity of the larger tin is 252 cm^3, find the capacity of the small tin.

17 Two solid metal spheres have masses of 5 kg and 135 kg respectively. If the radius of the smaller one is 4 cm, find the radius of the larger one.

18 Two similar cones have surface areas in the ratio 4 : 9. Find the ratio of
a their lengths   **b** their volumes.

19 The areas of the bases of two similar glasses are in the ratio 4 : 25. Find the ratio of their volumes.

20 A model of a treasure chest is 3 cm long and has a volume of $11\,\text{cm}^3$. Work out the length of the real treasure chest if its volume is $88\,000\,\text{cm}^3$.

21 The ingredients of a standard chocolate bar cost 20p.
A giant-sized bar is similar in shape and four times as long.
What should be the cost of the ingredients of the giant bar?

22 Sam has a model of a pirate ship which is made
to a scale of $1:50$.
Copy and complete the table using the
given units.

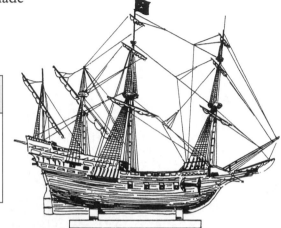

	On model	On actual ship
Length	42 cm	_____ m
Capacity of hold	500 cm³	_____ m³
Area of sails	_____ cm²	175 m²
Number of cannon	12	_____
Deck area	370 cm²	_____ m²

23 Two solid spheres have surface areas of $5\,\text{cm}^2$ and $45\,\text{cm}^2$ respectively and the mass of the smaller sphere is 2 kg. Find the mass of the larger sphere.

24 The masses of two similar objects are 24 kg and 81 kg respectively. If the surface area of the larger object is $540\,\text{cm}^2$, find the surface area of the smaller object.

25 A cylindrical can has a circumference of 40 cm and a capacity of 4·8 litres. Find the capacity of a similar cylinder of circumference 50 cm.

26 A container has a surface area of $5000\,\text{cm}^2$ and a capacity of 12·8 litres. Find the surface area of a similar container which has a capacity of 5·4 litres.

27 A full size snooker ball has a diameter of $2\frac{1}{16}$ inches and weighs 133·1 g. Calculate the weight of a snooker ball of diameter $1\frac{7}{8}$ inches, assuming that both balls are made of the same material.

Test yourself

1 The diagram shows the angles, in degrees, of a quadrilateral. Write an equation in terms of x and solve it to find the four angles of the quadrilateral.

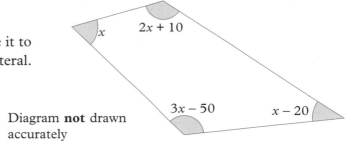

Diagram **not** drawn accurately

2 The diagram shows part of a regular 10-sided polygon. Work out the size of the angle marked x.

Diagram **NOT** accurately drawn

(Edexcel, 2008)

3 Work out the area of the shape.

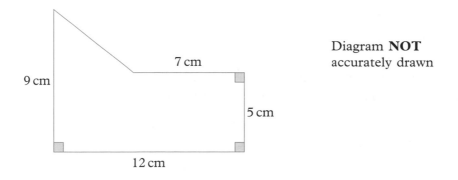

Diagram **NOT** accurately drawn

(Edexcel, 2008)

4 A, B, C and D lie on the circumference of a circle.
AB is parallel to CD.
BC is a diameter.
Prove that the triangles ABC and DCB are congruent.

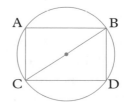

(OCR, 2003)

5 Use ruler and compasses to **construct** an angle of 30°. You must show all construction lines.

6 Use ruler and compasses to **construct** the perpendicular to the line segment AB that passes through the point P.
You must show all construction lines.

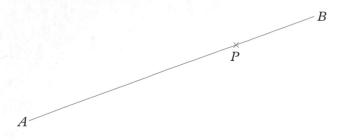

(Edexcel, 2004)

7

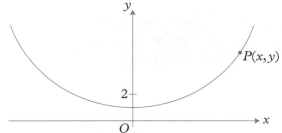

The diagram shows a sketch of a curve. The point $P(x, y)$ lies on the curve. The locus of P has the following property:

The distance of the point P from the point $(0, 2)$ is the same as the distance of the point P from the x-axis.

Show that $y = \frac{1}{4}x^2 + 1$

(Edexcel, 2004)

8 The length of a rectangle is twice the width of the rectangle.
The length of a diagonal of the rectangle is 25 cm.
Work out the area of the rectangle.
Give your answer as an integer.

(Edexcel, 2004)

9 Calculate the length of the longest thin straight rod that will fit inside this cuboid.

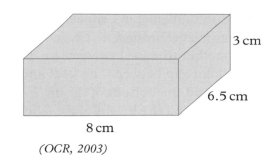

3 cm

6.5 cm

8 cm

(OCR, 2003)

10 *ABC* is a triangle.
AB = 8 cm
BC = 14 cm
Angle *ABC* = 106°
Calculate the area of the triangle.
Give your answer correct to 3 significant figures.

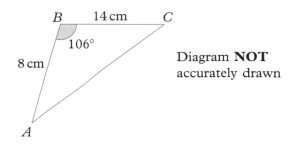

B   14 cm   *C*

106°

8 cm

A

Diagram **NOT** accurately drawn

(Edexcel, 2005)

11 Calculate the perimeter of the semicircle shown.

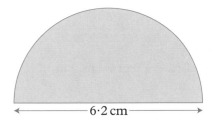

6·2 cm

12 A circular hole has diameter 38 mm. A square peg will just fit into the hole. Find the length of a side of the square.

(OCR, 2004)

13 The diagram shows a sector of a circle of radius 9 centimetres.
Find the perimeter of the sector.
Give your answer in terms of π.

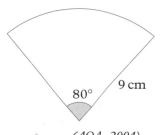

Not drawn accurately

80°   9 cm

(AQA, 2004)

14 A tin of diameter 7 cm and height 12 cm has a label around it. The label is glued together using a 1 cm overlap. There is a 1 cm gap between the label and the top and the bottom of the tin. Find the length and the height of the label.

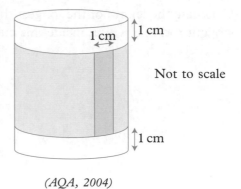

Not to scale

(AQA, 2004)

15 The diagram shows the garden of a house.
There is a security light, S, in the garden and two security lights, A and B, on the house wall.
The lights are at ground level.

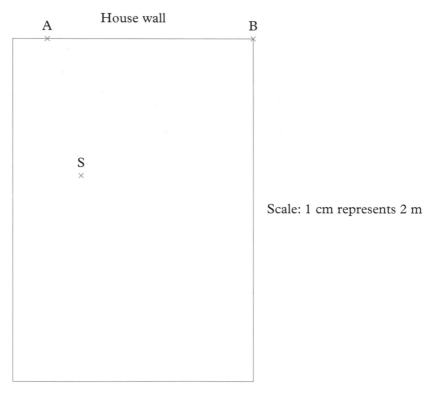

Scale: 1 cm represents 2 m

The security light in the garden comes on when it detects movement within 7 m.
Each security light on the house wall comes on when it detects movement within 4 m.
A fox is in the garden.
Copy the diagram and indicate clearly the region where the fox can move **without** making any of the lights come on.

(OCR, 2008)

16 a A cylinder has base radius 4 cm and height 10 cm.
Work out the volume of the cylinder.
Leave your answer as a multiple of π.
Give the units of your answer.

b Two cones are mathematically similar.
The height of the smaller cone is exactly half the
height of the larger cone.
The surface area of the **larger cone** is $112\pi \text{ cm}^2$.
What is the surface area of the smaller cone?

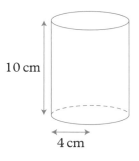
10 cm

4 cm

(OCR, 2008)

17 Two similar pentagons are shown below. Find the lengths a and b.

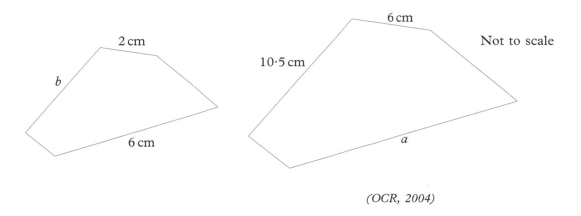

2 cm

b

6 cm

6 cm

10·5 cm

Not to scale

a

(OCR, 2004)

18 A child's rugby ball is 10 cm long and has a volume of 200 cm^3. It is
similar in shape to a full-size rugby ball. A full-size rugby ball is
22 cm long.

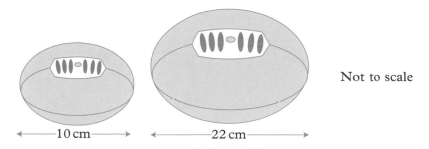

10 cm

22 cm

Not to scale

Find the volume of the full-size ball.

(AQA, 2004)

19 Triangles *ABC* and *ACD* are similar.
   AB = 5 cm and *AC* = 6 cm.

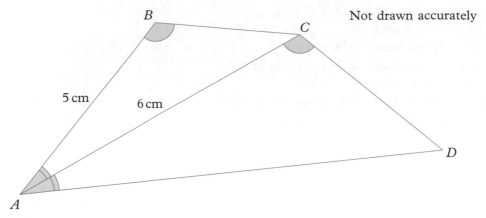

Not drawn accurately

Calculate the length of *AD*.

(AQA, 2004)

20 The diagram shows a cylinder.
   The diameter of the cylinder is 20 cm.
   The height of the cylinder is 2 cm.

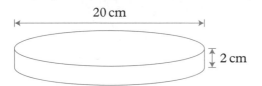

Not drawn accurately

  a Work out the volume of the cylinder.
   Use π = 3.14
  b Write your answer to part (**a**) in litres.

(AQA, 2007)

5 Algebra 2

In this unit you will:
- learn how to change the subject of a formula
- learn how to solve inequalities and represent them as regions on a graph
- solve problems in direct and indirect proportion
- identify and plot curved graphs
- use graphs to solve equations.

Functional skills coverage and range:
- Understand, use, and calculate ratio and proportion, including problems involving scale
- Understand and use simple equations and simple formulae involving one- or two-step operations.

Links
If you have an idea of the shape of a graph then you can model a situation to find out the exact equation, for example, how fast a radioactive isotope is decaying. This could tell you exactly what the isotope is.

5.1 Changing the subject of a formula

The operations that you use in solving ordinary linear equations are exactly the same as the operations you need to change the subject of a formula.

5.1.1 Simple formulae

EXAMPLE

Make x the subject in these formulae.

a $ax - p = t$ **b** $y(x + y) = v^2$

....

a $ax - p = t$ **b** $y(x + y) = v^2$

$ax = t + p$ $yx + y^2 = v^2$

$x = \dfrac{t + p}{a}$ $yx = v^2 - y^2$

$x = \dfrac{v^2 - y^2}{y}$

Exercise 1 Ⓜ

Make x the subject.

1 $x + b = e$	**2** $x - t = m$	**3** $x - f = a + b$
4 $x + h = A + B$	**5** $x + t = y + t$	**6** $a + x = b$
7 $k + x = m$	**8** $v + x = w + y$	**9** $ax = b$

10 $hx = m$

11 $mx = a + b$

12 $kx = c - d$

13 $vx = e + n$

14 $3x = y + z$

15 $xp = r$

16 $xm = h - m$

17 $ax + t = a$

18 $mx - e = k$

19 $ux - h = m$

20 $ex + q = t$

21 $kx - u^2 = v^2$

Now do these.

22 $gx + t^2 = s^2$

23 $xa + k = m^2$

24 $xm - v = m$

25 $a + bx = c$

26 $t + sx = y$

27 $y + cx = z$

28 $a + hx = 2a$

29 $mx - b = b$

30 $kx + ab = cd$

31 $a(x - b) = c$

32 $c(x - d) = e$

33 $m(x + m) = n^2$

34 $k(x - a) = t$

35 $h(x - h) = k$

36 $m(x + b) = n$

37 $a(x - a) = a^2$

38 $c(a + x) = d$

39 $m(b + x) = e$

> In questions 31–39 multiply out the brackets first.

5.1.2 Formulae involving fractions

> **EXAMPLE**

Make x the subject in these formulae.

a $\dfrac{x}{a} = p$

b $\dfrac{m}{x} = t$

c $\dfrac{a^2}{m} = \dfrac{d}{x}$

....

a $\dfrac{x}{a} = p$

$\quad x = ap$

b $\dfrac{m}{x} = t$

$\quad m = xt$

$\quad \dfrac{m}{t} = x$

c $\dfrac{a^2}{m} = \dfrac{d}{x}$

$\quad xa^2 = dm$

$\quad x = \dfrac{dm}{a^2}$

> Notice in parts **b** and **c** that when x is on the bottom you start by multiplying both sides by x.

Exercise 2 Ⓜ

Make x the subject.

1 $\dfrac{x}{t} = m$

2 $\dfrac{x}{e} = n$

3 $\dfrac{x}{p} = a$

4 $am = \dfrac{x}{t}$

5 $bc = \dfrac{x}{a}$

6 $e = \dfrac{x}{y^2}$

7 $\dfrac{x}{a} = (b + c)$

8 $\dfrac{x}{t} = (c - d)$

9 $\dfrac{x}{m} = s + t$

10 $\dfrac{x}{k} = h + i$

11 $\dfrac{x}{b} = \dfrac{a}{c}$

12 $\dfrac{x}{m} = \dfrac{z}{y}$

13 $\dfrac{x}{h} = \dfrac{c}{d}$

14 $\dfrac{m}{n} = \dfrac{x}{e}$

15 $\dfrac{b}{e} = \dfrac{x}{h}$

16 $\dfrac{x}{(a + b)} = c$

17 $\dfrac{x}{(h + k)} = m$

18 $\dfrac{x}{u} = \dfrac{m}{y}$

19 $\dfrac{x}{(h - k)} = t$

20 $\dfrac{x}{(a + b)} = (z + t)$

21 $t = \dfrac{e}{x}$

22 $a = \dfrac{e}{x}$

23 $m = \dfrac{h}{x}$

24 $\dfrac{a}{b} = \dfrac{c}{x}$

> When x is on the bottom, multiply both sides by x.

25 $\dfrac{u}{x} = \dfrac{c}{d}$ **26** $\dfrac{m}{x} = t^2$ **27** $\dfrac{h}{x} = \sin 20°$

28 $\dfrac{e}{x} = \cos 40°$ **29** $\dfrac{m}{x} = \tan 46°$ **30** $\dfrac{a^2}{b^2} = \dfrac{c^2}{x}$

> You will meet sine, cos and tan in chapter 6.

5.1.3   Formulae with negative *x*-terms

EXAMPLE

Make x the subject of these formulae.

a $t - x = a^2$ **b** $h - bx = m$ **c** $a(m - x) = h$

. . . .

a $t - x = a^2$ **b** $h - bx = m$ **c** $a(m - x) = h$
$\qquad t = a^2 + x \qquad\qquad h = m + bx \qquad\qquad am - ax = h$
$\quad t - a^2 = x \qquad\qquad h - m = bx \qquad\qquad\qquad am = h + ax$
$\quad\text{or } x = t - a^2 \qquad\quad \dfrac{h - m}{b} = x \qquad\qquad am - h = ax$
$\qquad\qquad\qquad\qquad\qquad\qquad\qquad\qquad\qquad \dfrac{am - h}{a} = x$

> Notice that in each question the first step is to make the x-term positive by taking it to the other side.

Exercise 3 Ⓜ

Make x the subject.

1 $a - x = y$ **2** $h - x = m$ **3** $z - x = q$ **4** $v = b - x$

5 $m = k - x$ **6** $h - cx = d$ **7** $y - mx = c$ **8** $k - ex = h$

9 $a^2 - bx = d$ **10** $m^2 - tx = n^2$ **11** $v^2 - ax = w$ **12** $y - x = y^2$

13 $k - t^2x = m$ **14** $e = b - cx$ **15** $z = h - gx$ **16** $a + b = c - dx$

17 $y^2 = v^2 - kx$ **18** $h = d - fx$ **19** $a(b - x) = c$ **20** $h(m - x) = n$

Make a the subject.

21 $m(c - a) = t$ **22** $v(p - a) = w$ **23** $e = d(q - a)$ **24** $b^2 - a = r^2$

25 $\dfrac{x - a}{f} = 2f$ **26** $\dfrac{B - Aa}{D} = E$ **27** $\dfrac{D - Ea}{N} = B$ **28** $\dfrac{h - fa}{b} = x$

29 $\dfrac{v^2 - ha}{C} = d$ **30** $\dfrac{M(a + B)}{N} = T$ **31** $\dfrac{f(Na - e)}{m} = B$ **32** $\dfrac{T(M - a)}{E} = F$

Make x the subject (more difficult).

33 $\dfrac{2}{x} + 1 = 3y$ **34** $\dfrac{5}{x} - 2 = 4z$ **35** $\dfrac{A}{x} + B = C$ **36** $\dfrac{V}{x} + G = H$

37 $\dfrac{r}{x} - t = n$ **38** $q = \dfrac{b}{x} + d$ **39** $t = \dfrac{m}{x} - n$ **40** $h = d - \dfrac{b}{x}$

41 $C - \dfrac{d}{x} = e$ **42** $r - \dfrac{m}{x} = e^2$ **43** $t^2 = b - \dfrac{n}{x}$ **44** $\dfrac{d}{x} + b = mn$

45 $3M = M + \dfrac{N}{P + x}$ **46** $A = \dfrac{B}{c + x} - 5A$ **47** $\dfrac{m^2}{x} - n = -p$ **48** $t = w - \dfrac{q}{x}$

5.1.4 Formulae with squares and square roots

Make x the subject in these formulae.

a $mx^2 = b$ **b** $x^2 + h = k$ **c** $ax^2 + b = c$

d $\sqrt{x} = u$ **e** $\sqrt{x - e} = t$ **f** $\sqrt{x^2 + A} = B$

....

a $mx^2 = b$

$x^2 = \dfrac{b}{m}$

$x = \pm\sqrt{\dfrac{b}{m}}$

b $x^2 + h = k$

$x^2 = k - h$

$x = \pm\sqrt{(k - h)}$

c $ax^2 + b = c$

$ax^2 = c - b$

$x^2 = \dfrac{c - b}{a}$

$x = \pm\sqrt{\dfrac{c - b}{a}}$

d $\sqrt{x} = u$

$x = u^2$ (square both sides)

e $\sqrt{x - e} = t$

$x - e = t^2$

$x = t^2 + e$

f $\sqrt{x^2 + A} = B$

$x^2 + A = B^2$

$x^2 = B^2 - A$

$x = \pm\sqrt{(B^2 - A)}$

Exercise 4 Ⓜ/Ⓗ

Make x the subject.

1 $cx^2 = h$ **2** $bx^2 = f$ **3** $x^2t = m$ **4** $x^2y = (a + b)$

5 $mx^2 = (t + a)$ **6** $x^2 - a = b$ **7** $x^2 + c = t$ **8** $x^2 + y = z$

9 $x^2 - a^2 = b^2$ **10** $x^2 + t^2 = m^2$ **11** $x^2 + n^2 = a^2$ **12** $ax^2 = c$

13 $hx^2 = n$ **14** $cx^2 = z + k$ **15** $ax^2 + b = c$ **16** $dx^2 - e = h$

17 $gx^2 - n = m$ **18** $x^2m + y = z$ **19** $a + mx^2 = f$ **20** $a^2 + x^2 = b^2$

Make x the subject.

21 $\sqrt{x} = 2z$ **22** $\sqrt{(x - 2)} = 3y$ **23** $\sqrt{(x + C)} = D$ **24** $\sqrt{(ax + b)} = c$

25 $b = \sqrt{(gx - t)}$ **26** $\sqrt{(d - x)} = t$ **27** $c = \sqrt{(n - x)}$ **28** $g = \sqrt{(c - x)}$

29 $\sqrt{(Ax + B)} = \sqrt{D}$ **30** $x^2 = g$ **31** $x^2 = B$ **32** $x^2 - A = M$

Make k the subject.

33 $C - k^2 = m$ **34** $mk^2 = n$ **35** $\dfrac{kz}{a} = t$ **36** $n = a - k^2$

37 $\sqrt{(k^2 - A)} = B$ **38** $t = \sqrt{(m + k^2)}$ **39** $A\sqrt{(k + B)} = M$ **40** $\sqrt{\left(\dfrac{N}{k}\right)} = B$

41 $\sqrt{(a^2 - k^2)} = t$ **42** $2\pi\sqrt{(k + t)} = 4$ **43** $\sqrt{(ak^2 - b)} = C$ **44** $k^2 + b = x^2$

5.1.5 Formulae with *x* on both sides

EXAMPLE

Make *x* the subject of each formula.

a $Ax - B = Cx + D$

b $x + a = \dfrac{x + b}{c}$

....

a $Ax - B = Cx + D$

$Ax - Cx = D + B$ (*x*-terms on one side)

$x(A - C) = D + B$ (factorise)

$x = \dfrac{D + B}{A - C}$

b $x + a = \dfrac{x + b}{c}$

$c(x + a) = x + b$

$cx + ca = x + b$

$cx - x = b - ca$ (*x*-terms on one side)

$x(c - 1) = b - ca$ (factorise)

$x = \dfrac{b - ca}{c - 1}$

Exercise 5 Ⓜ/Ⓗ

Make *y* the subject.

1 $5(y - p) = 2(y + x)$

2 $x(y - 3) = p(3 - y)$

3 $Ny + B = D - Ny$

4 $My - D = E - 2My$

5 $ay + b = 3b + by$

6 $my - c = e - ny$

7 $xy + 4 = 7 - ky$

8 $Ry + D = Ty + C$

9 $ay - x = z + by$

10 $m(y + a) = n(y + b)$

11 $x(y - b) = y + d$

12 $\dfrac{a - y}{a + y} = b$

13 $\dfrac{1 - y}{1 + y} = \dfrac{c}{d}$

14 $\dfrac{M - y}{M + y} = \dfrac{a}{b}$

15 $m(y + n) = n(n - y)$

16 $y + m = \dfrac{2y - 5}{m}$

17 $y - n = \dfrac{y + 2}{n}$

18 $y + b = \dfrac{ay + e}{b}$

19 $\dfrac{ay + x}{x} = 4 - y$

20 $c - dy = e - ay$

21 $y(a - c) = by + d$

22 $y(m + n) = a(y + b)$

23 $t - ay = s - by$

24 $\dfrac{y + x}{y - x} = 3$

25 $\dfrac{v - y}{v + y} = \dfrac{1}{2}$

26 $y(b - a) = a(y + b + c)$

27 $\sqrt{\left(\dfrac{y + x}{y - x}\right)} = 2$

28 $\sqrt{\left(\dfrac{z + y}{z - y}\right)} = \dfrac{1}{3}$

29 $\sqrt{\left[\dfrac{m(y + n)}{y}\right]} = p$

30 $n - y = \dfrac{4y - n}{m}$

Exercise 6 Ⓜ/Ⓗ Ⓕ⭐

1 A formula for calculating velocity is $v = u + at$.

    a Rearrange the formula to express *a* in terms of *v*, *u* and *t*.

    b Calculate *a* when $v = 20$, $u = 4$, $t = 8$.

2 The area of a sector of a circle is given

    by the formula $A = \dfrac{x\pi r^2}{360}$.

    Express *x* in terms of *A*, π and *r*.

3 a Express k in terms of P, m and y, when $P = \dfrac{mk}{y}$.
   b Express y in terms of P, m and k.

4 A formula for calculating repair bills, R, is $R = \dfrac{n-d}{p}$.

   a Express n in terms of R, p and d.
   b Calculate n when $R = 400$, $p = 3$ and $d = 55$.

5 The formula for the area of a circle in $A = \pi r^2$.
   Express r in terms of A and π.

6 The volume of a cylinder is given by $V = \pi r^2 h$.
   Express h in terms of V, π and r.

7 The surface area, A, and volume, V, of a sphere are given by the
   formulae $A = 4\pi r^2$ and $V = \dfrac{4}{3}\pi r^3$. Make r the subject of each formula.

Exercise 7 Ⓜ/Ⓗ

Make the letter in brackets the subject.

1 $ax - d = h$ $[x]$	**2** $zy + k = m$ $[y]$	**3** $d(y + e) = f$ $[y]$
4 $m(a + k) = d$ $[k]$	**5** $a + bm = c$ $[m]$	**6** $ae^2 = b$ $[e]$
7 $yt^2 = z$ $[t]$	**8** $x^2 - c = e$ $[x]$	**9** $my - n = b$ $[y]$
10 $a(z + a) = b$ $[z]$	**11** $\dfrac{a}{x} = d$ $[x]$	**12** $\dfrac{k}{m} = t$ $[k]$
13 $\dfrac{u}{m} = n$ $[u]$	**14** $\dfrac{y}{x} = d$ $[x]$	**15** $\dfrac{a}{m} = t$ $[m]$
16 $\dfrac{d}{g} = n$ $[g]$	**17** $\dfrac{t}{k} = (a + b)$ $[t]$	**18** $y = \dfrac{v}{e}$ $[e]$
19 $c = \dfrac{m}{y}$ $[y]$	**20** $\dfrac{a^2}{m} = b$ $[a]$	**21** $g(m + a) = b$ $[m]$
22 $h(h + g) = x^2$ $[g]$	**23** $y - t = z$ $[t]$	**24** $me^2 = c$ $[e]$
25 $a(y + x) = t$ $[x]$	**26** $uv - t^2 = y^2$ $[v]$	**27** $k^2 + t = c$ $[k]$
28 $k - w = m$ $[w]$	**29** $b - an = c$ $[n]$	**30** $m(a + y) = c$ $[y]$
31 $pq - x = ab$ $[x]$	**32** $a^2 - bk = t$ $[k]$	**33** $v^2 z = w$ $[z]$
34 $c = t - u$ $[u]$	**35** $xc + t = 2t$ $[c]$	**36** $m(n + w) = k$ $[w]$
37 $v - mx = t$ $[m]$	**38** $c = a(y + b)$ $[y]$	**39** $m(a - c) = e$ $[c]$
40 $ba^2 = c$ $[a]$	**41** $\dfrac{a}{p} = q$ $[p]$	**42** $\dfrac{a}{n^2} = e$ $[n]$
43 $\dfrac{h}{f^2} = m$ $[f]$	**44** $\dfrac{v}{x^2} = n$ $[x]$	**45** $v - ac = t^3$ $[c]$
46 $a(a^2 + y) = b^3$ $[y]$	**47** $ah^2 - d = b$ $[h]$	**48** $h(h + k) = bc$ $[k]$
49 $u^2 - n^2 = v^2$ $[n]$	**50** $m(b - z) = b^3$ $[z]$	

5.2 Inequalities and regions

5.2.1 Inequality symbols

There are four inequality symbols.

- $x < 4$ means 'x is **less than** 4'
- $y > 7$ means 'y is **greater than** 7'
- $z \leqslant 10$ means 'z is **less than or equal to** 10'
- $t \geqslant -3$ means 't is **greater than or equal to** -3'

When there are two symbols in one statement look at each part separately.

For example, if n is an **integer** and $3 < n \leqslant 7$,

n has to be greater than 3 but at the same time it has to be less than or equal to 7.

So n could be 4, 5, 6 or 7 only.

EXAMPLE

Illustrate on a number line the range of values of x stated.

a $x > 1$

$\qquad$ 1

The circle at the left-hand end
of the range is open.
This means that 1 is not included.

b $x \leqslant -2$

$\qquad$ -2

The circle at -2 is filled in to
indicate that -2 is included.

c $1 \leqslant x < 4$

$\qquad$
1 $\qquad$ 4

Exercise 8 Ⓜ

1 Write each statement with either $>$ or $<$ in the box.

a 3 ☐ 7 **b** 0 ☐ -2

c 3·1 ☐ 3·01 **d** -3 ☐ -5

e 100 mm ☐ 1 m **f** 1 kg ☐ 1 lb

2 Write the inequality displayed. Use x for the variable.

3 Draw a number line to display these inequalities.

   a $x \geqslant 7$ **b** $x < 2 \cdot 5$ **c** $1 < x < 7$

   d $0 \leqslant x \leqslant 4$ **e** $-1 < x \leqslant 5$

4 Write an inequality for each statement.

   a You must be at least 16 to get married. [Use A for age.]

   b Vitamin J1 is not recommended for people over 70 or for children 3 years or under. [Use A for age recommended.]

   c To cook beef the oven temperature should be between $150\,°C$ and $175\,°C$. [Use T for temperature.]

   d Applicants for training as paratroopers must be at least $1 \cdot 75\,m$ tall. [Use h for height.]

5 Answer 'true' or 'false':

   a n is an integer and $1 < n \leqslant 4$, so n can be 2, 3 or 4.

   b x is an integer and $2 \leqslant x < 5$, so x can be 2, 3 or 4.

   c p is an integer and $p \geqslant 10$, so p can be 10, 11, 12, 13 ...

6 Which of the numbers x, below, satisfy $x^2 < 90$?

   $7, \, -6, \, 10, \, 8 \cdot 5, \, \sqrt{95}$

7 Write one inequality to show the values of x which satisfy both of these inequalities.

   $\boxed{x \leqslant 7}$ $\boxed{x > 2}$

8 Write one inequality to show the values of x which satisfy all three of these inequalities.

   $\boxed{x < 5}$ $\boxed{0 < x < 6}$ $\boxed{3 \leqslant x < 10}$

5.2.2 Solving inequalities

Follow the same procedure that you use for solving equations except that when you multiply or divide by a **negative** number you **reverse** the inequality.

For example, $4 > -2$ but if you multiply by -2 then $-8 < 4$

It is best to avoid dividing by a negative number as in the following example, part **b**.

EXAMPLE

Solve these inequalities.

a $x + 11 < 4$ **b** $8 > 13 - x$
c $2x - 1 > 5$ **d** $x + 1 < 2x < x + 3$

....

a $x + 11 < 4$ **b** $8 > 13 - x$
   $x < -7$ (subtract 11) $8 + x > 13$ (add x)
                                                      $x > 5$ (subtract 8)

c $2x - 1 > 5$ **d** $x + 1 < 2x < x + 3$
      $2x > 5 + 1$ (add 1)   Solve the two inequalities separately.
      $x > \dfrac{6}{2}$ (divide by 2) $x + 1 < 2x \qquad 2x < x + 3$
                                                      $1 < x \qquad\quad x < 3$
      $x > 3$ The solution is $1 < x < 3$.

Exercise 9 Ⓜ

Solve these inequalities.

1 $x - 3 > 10$ **2** $x + 1 < 0$ **3** $5 > x - 7$

4 $2x + 1 \leqslant 6$ **5** $3x - 4 > 5$ **6** $10 \leqslant 2x - 6$

7 $5x < x + 1$ **8** $2x \geqslant x - 3$ **9** $4 + x < -4$

10 $3x + 1 < 2x + 5$ **11** $2(x + 1) > x - 7$ **12** $7 < 15 - x$

13 $9 > 12 - x$ **14** $4 - 2x \leqslant 2$ **15** $3(x - 1) < 2(1 - x)$

16 $7 - 3x < 0$ **17** $\dfrac{x}{3} < -1$ **18** $\dfrac{2x}{5} > 3$

19 $2x > 0$ **20** $\dfrac{x}{4} < 0$

21 The height of this picture has to be
greater than the width.
Find the range of possible values of x.

height
$2(x + 1)$

width
$(x + 7)$

22 $10 \leqslant 2x \leqslant x + 9$ **23** $x < 3x + 2 < 2x + 6$

24 $10 \leqslant 2x - 1 \leqslant x + 5$ **25** $3 < 3x - 1 < 2x + 7$

26 $x - 10 < 2(x - 1) < x$ **27** $4x + 1 < 8x < 3(x + 2)$

In questions **22** to
27, solve the two
inequalities
separately.

28 Sumitra said 'I think of an integer.
   I subtract 14.
   I multiply the result by 5.
   I divide by 2.
   The answer is greater than the number I
   thought of.'

Write an inequality and solve it to find the smallest number
Sumitra could have thought of.

Take care when there are squares and square roots in inequalities.

The equation $x^2 = 4$ has solutions $x = \pm 2$, which is correct.

For the inequality $x^2 < 4$, you might wrongly write $x < \pm 2$.
Consider $x = -3$, say.
   -3 is less than -2 and is also less than $+2$.
   But $(-3)^2$ is not less than 4 and so
   $x = -3$ does not satisfy the inequality $x^2 < 4$.
The correct solution for $x^2 < 4$ is $-2 < x < 2$.

$(-3)^2 = 9$

EXAMPLE

a Solve the inequality
$2x^2 - 1 > 17$.

b List the solutions which satisfy
$2 \le n < 14$; n is a prime number

. . . .

a $2x^2 - 1 > 17$
$\quad 2x^2 > 18$
$\quad\ \ x^2 > 9$
$\quad x > 3$ or $x < -3$ (Avoid the temptation
$\qquad\qquad\qquad\qquad$ to write $x > \pm 3$!)

b The prime numbers in the range
specified are 2, 3, 5, 7, 11, 13.

Exercise 10 Ⓜ

1 The area of the rectangle must be greater than the area of the
triangle.
Find the range of possible values of x.

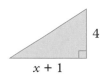

4

$x + 1$

3

$x - 2$

For questions **2** to **8**, list the solutions which satisfy the given condition.

2 $3a + 1 < 20$; a is a positive integer.

3 $b - 1 \geqslant 6$; b is a prime number less than 20.

4 $1 < z < 50$; z is a square number.

5 $2x > -10$; x is a negative integer.

6 $x + 1 < 2x < x + 13$; x is an integer.

7 $0 \leqslant 2z - 3 \leqslant z + 8$; z is a prime number.

8 $\frac{a}{2} + 10 > a$; a is a positive even number.

9 Given that $4x > 1$ and $\frac{x}{3} \leqslant 1\frac{1}{3}$, list the possible integer values of x.

10 State the smallest integer n for which $4n > 19$.

11 Given that $-4 \leqslant a \leqslant 3$ and $-5 \leqslant b \leqslant 4$, find
   a the largest possible value of a^2
   b the smallest possible value of ab
   c the largest possible value of ab
   d the value of b if $b^2 = 25$.

12 For any shape of triangle ABC, complete the statement
   AB + BC $\Box$ AC, by writing $<$, $>$ or $=$ inside the box.

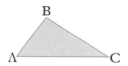

13 Find a simple fraction r such that $\frac{1}{3} < r < \frac{2}{3}$.

14 Find the largest prime number p such that $p^2 < 400$.

15 Find the integer n such that $n < \sqrt{300} < n + 1$.

16 If $f(x) = 2x - 1$ and $g(x) = 10 - x$ for what values of x is $f(x) > g(x)$?

17 a The solution of $x^2 < 9$ is $-3 < x < 3$.
   [The square roots of 9 are -3 and 3.]
   b Copy and complete.
      i If $x^2 < 100$, then $\Box < x < \Box$
      ii If $x^2 < 81$, then $\Box < x < \Box$
      iii If $x^2 > 36$, then $x > \Box$ or $x < \Box$

Solve the inequalities.

18 $x^2 < 25$ **19** $x^2 \leqslant 16$ **20** $x^2 > 1$
21 $2x^2 \geqslant 72$ **22** $3x^2 + 5 > 5$ **23** $5x^2 - 2 < 18$

24 Given $2 \leqslant p \leqslant 10$ and $1 \leqslant q \leqslant 4$, find the range of values of
   a pq **b** $\frac{p}{q}$ **c** $p - q$ **d** $p + q$

25 If $2^r > 100$, what is the smallest integer value of r?

26 Given $\left(\frac{1}{3}\right)^x < \frac{1}{200}$, what is the smallest integer value of x?

27 Find the smallest integer value of x which satisfies $x^x > 10\,000$.

28 What integer values of x satisfy $100 < 5^x < 10\,000$?

*29 If x is an acute angle and $\sin x > \frac{1}{2}$, write the range of values that x can take.

*30 If x is an acute angle and $\cos x > \frac{1}{4}$, write the range of values that x can take.

5.2.3 Shading regions

You can represent inequalities on a graph, particularly where two variables (x and y) are involved.

EXAMPLE

Draw a sketch graph and shade the area which represents the set of points that satisfy each of these inequalities.

a $x > 2$ **b** $1 \leqslant y \leqslant 5$ **c** $x + y \leqslant 8$

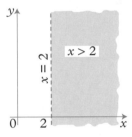 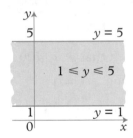 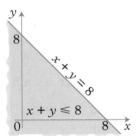

In each graph, the required region is shaded.

In **a**, the line $x = 2$ is shown as a broken line to indicate that the points on the line are not included.

In **b** and **c** points on the line **are** included 'in the region' and the lines are drawn unbroken.

To decide which side to shade when the line is sloping, take a **trial point**. This can be any point which is not actually on the line.

In **c** above, the trial point could be $(1, 1)$.

Is $(1, 1)$ in the region $x + y \leqslant 8$?
It satisfies $x + y < 8$ because $1 + 1 = 2$, which is less than 8.
So below the line is $x + y < 8$ which is in the shaded region.

Exercise 11 Ⓜ/Ⓗ

In questions **1** to **6**, describe the region which is shaded.

1

2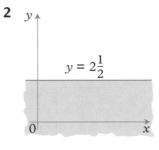

3

4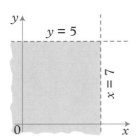

5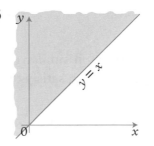

6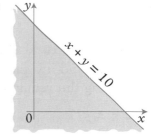

7 The point $(1, 1)$, marked $\star$, lies in the shaded region.
Use this as a trial point to describe the shaded region as follows:
Is the shaded region $2x - y > 3$?
Try $x = 1$, $y = 1$. Is $2 - 1 > 3$? No.
Copy and complete 'So the shaded region is $2x - y \,\square\, 3$.'

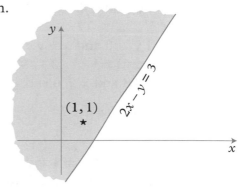

8 The point $(3, 1)$, marked $\star$, lies in the shaded triangle.
Use this as a trial point to write the three inequalities which describe the shaded region.

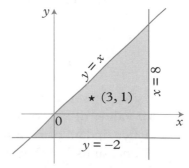

9 A trial point $(1, 2)$ lies inside the shaded triangles. Write the three inequalities which describe each shaded region.

a

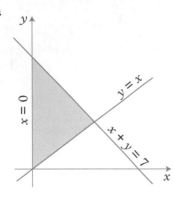

b

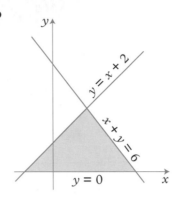

For questions **10** to **27**, draw a sketch graph similar to those in question **9** and indicate the set of points which satisfy the inequalities by shading the required region.

10 $2 < x < 7$

11 $0 < y < 3\frac{1}{2}$

12 $-2 < x < 2$

13 $x < 6$ and $y < 4$

14 $0 < x < 5$ and $y < 3$

15 $1 < x < 6$ and $2 < y < 8$

16 $-3 < x < 0$ and $-4 < y < 2$

17 $y < x$

18 $x + y < 5$

19 $y > x + 2$ and $y < 7$

20 $x > 0$ and $y > 0$ and $x + y < 7$

21 $x > 0$ and $x + y < 10$ and $y > x$

22 $8 > y > 0$ and $x + y > 3$

23 $x + 2y < 10$ and $x > 0$ and $y > 0$

24 $3x + 2y < 18$ and $x > 0$ and $y > 0$

25 $x > 0$, $y > x - 2$, $x + y < 10$

26 $3x + 5y < 30$ and $y > \dfrac{x}{2}$

27 $y > \dfrac{x}{2}$, $y < 2x$ and $x + y < 8$

28 The two lines $y = x + 1$ and $x + y = 5$ divide the graph into four regions A, B, C, D. Write the two inequalities which describe each of the regions A, B, C, D.

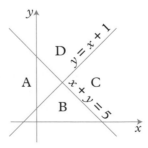

*29 Using the same axes, draw the graphs of $xy = 10$ and $x + y = 9$ for values of x from 1 to 10. Hence find all pairs of positive integers with products greater than 10 and sums less than 9.

5.3 Direct and inverse proportion

5.3.1 Direct proportion

a When you buy petrol, the more you buy the more money you have to pay. So if 2·2 litres costs 198p, then 4·4 litres will cost 396p. The cost of petrol is **directly proportional** to the quantity bought. To show that quantities are proportional, you use the symbol '$\propto$'. So in the example if the cost of petrol is c pence and the number of litres of petrol is l, you write

$c \propto l$

The '$\propto$' sign can always be replaced by ' $= k$' where k is a constant.

So $c = kl$

From above, if $c = 198$ when $l = 2·2$
then $198 = k \times 2·2$

$$k = \frac{198}{2·2} = 90$$

You can then write $c = 90l$, and this allows you to find the value of c for any value of l, and **vice versa**.

b If a quantity z is proportional to a quantity x, you can write

$z \propto x$ or $z = kx$

Two other expressions are sometimes used when quantities are directly proportional. You can say

    'z varies as x'
or 'z varies directly as x'.

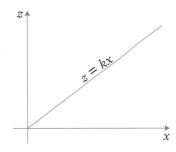

- When z and x are directly proportional the graph connecting z and x is a straight line which passes through the origin.

If y varies as z, and $y = 2$ when $z = 5$, find

a the value of y when $z = 6$
b the value of z when $y = 5$.

Because $y \propto z$, then $y = kz$ where k is a constant.

$$y = 2 \text{ when } z = 5$$
So $\quad 2 = k \times 5$
$$k = \frac{2}{5}$$
So $\quad y = \frac{2}{5}z$

a When $z = 6$, $y = \frac{2}{5} \times 6 = 2\frac{2}{5}$.

b When $y = 5$, $5 = \frac{2}{5}z$; $z = \frac{25}{2} = 12\frac{1}{2}$.

The value V of a diamond is proportional to the square of its weight W.
If a diamond weighing 10 grams is worth £200, find

a the value of a diamond weighing 30 grams
b the weight of a diamond worth £5000.

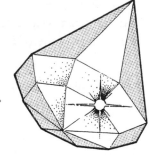

$$V \propto W^2$$
or $\quad V = kW^2$ where k is a constant.
$$V = 200 \text{ when } W = 10$$
So $\quad 200 = k \times 10^2$
$$k = 2$$
So $\quad V = 2W^2$

a When $W = 30$,
$$V = 2 \times 30^2 = 2 \times 900$$
$$V = £1800$$

So a diamond of weight 30 grams is worth £1800.

b When $\quad V = 5000$,
$$5000 = 2 \times W^2$$
$$W^2 = \frac{5000}{2} = 2500$$
$$W = \sqrt{2500} = 50$$

So a diamond of value £5000 weighs 50 grams.

Exercise 12 Ⓗ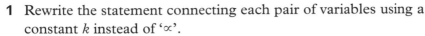

1 Rewrite the statement connecting each pair of variables using a constant k instead of '$\propto$'.

   a $S \propto e$ **b** $v \propto t$ **c** $x \propto z^2$

   d $y \propto \sqrt{x}$ **e** $T \propto \sqrt{L}$

> Reminder: If $x \propto d$, then $x = kd$.

2 y is proportional to t so that $y = kt$. If $y = 6$ when $t = 4$, calculate the value of k and hence find
   a the value of y when $t = 6$
   b the value of t when $y = 4$.

3 z is proportional to m. If $z = 20$ when $m = 4$, calculate
   a the value of z when $m = 7$
   b the value of m when $z = 55$.

4 A varies directly as r^2. If $A = 12$, when $r = 2$, calculate
   a the value of A when $r = 5$
   b the value of r when $A = 48$.

5 Given that $z \propto x$, copy and complete the table.

x	1	3		$5\frac{1}{2}$
z	4		16	

6 Given that $V \propto r^3$, copy and complete the table.

r	1	2		$1\frac{1}{2}$
V	4		256	

7 The pressure of the water, P, at any point below the surface of the sea varies as the depth of the point below the surface, d. If the pressure is 200 newtons/cm^2 at a depth of 3 m, calculate the pressure at a depth of 5 m.

8 The distance d through which a stone falls from rest is proportional to the square of the time taken, t. If a stone falls 45 m in 3 seconds, how far will it fall in 6 seconds? How long will it take to fall 20 m?

9 The energy, E, stored in an clastic band is proportional to the square of the extension, x. When the elastic is extended by 3 cm, the energy stored is 243 joules.
   a What is the energy stored when the extension is 5 cm?
   b What is the extension when the stored energy is 36 joules?

10 The resistance to motion of a car is proportional to the square of the speed of the car.

    a If the resistance is 4000 newtons at a speed of 20 m/s, what is the resistance at a speed of 30 m/s?

    b At what speed is the resistance 6250 newtons?

11 In an experiment, Julie made measurements of w and p.

w	2	5	7
p	1·6	25	68·6

Which of these laws fits the results?

$p \propto w$, $p \propto w^2$, $p \propto w^3$.

12 A road research organisation recently claimed that the damage to road surfaces was proportional to the fourth power of the axle load. The axle load of a 44-ton HGV is about 15 times that of a car. Calculate the ratio of the damage to road surfaces made by a 44-ton HGV and a car.

5.3.2 Inverse proportion

If you travel a distance of 200 m at 10 m/s, the time taken is 20 s.
If you travel the same distance at 20 m/s, the time taken is 10 s.
As you **double** the speed, you **halve** the time taken.
For a fixed journey, the time taken is **inversely proportional** to the speed at which you travel.
If s is inversely proportional to t, you write

$$s \propto \frac{1}{t}$$

or $s = k \times \dfrac{1}{t}$

Notice that the product $s \times t$ is constant.
The graph connecting s and t is a curve.
The shape of the curve is similar to that of $y = \dfrac{1}{x}$.

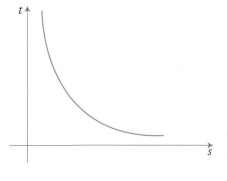

> Sometimes you will see 'x varies inversely as y'. It means the same as 'x is inversely proportional to y'.

z is inversely proportional to t^2 and $z = 4$ when $t = 1$.
Calculate z when $t = 2$.

··

$$z \propto \frac{1}{t^2} \qquad \text{or} \qquad z = k \times \frac{1}{t^2} \ (k \text{ is a constant})$$

$z = 4$ when $t = 1$

So $\quad 4 = k\left(\dfrac{1}{1^2}\right)$

so $\quad k = 4$

So $\quad z = 4 \times \dfrac{1}{t^2}$

When $t = 2$, $\quad z = 4 \times \dfrac{1}{2^2} = 1$

Exercise 13 Ⓗ

1 Rewrite the statements connecting the variables using a constant
of variation, k.

  a $x \propto \dfrac{1}{y}$ **b** $s \propto \dfrac{1}{t^2}$ **c** $t \propto \dfrac{1}{\sqrt{q}}$

  d m varies inversely as w

  e z is inversely proportional to t^2.

2 T is inversely proportional to m. If $T = 12$ when $m = 1$, find

  a T when $m = 2$ **b** T when $m = 24$.

> Start by writing
> $T = \dfrac{k}{m}$ and then
> find k.

3 L is inversely proportional to x. If $L = 24$ when $x = 2$, find

  a L when $x = 8$ **b** L when $x = 32$.

4 b varies inversely as e. If $b = 6$ when $e = 2$, calculate

  a the value of b when $e = 12$

  b the value of e when $b = 3$.

5 x is inversely proportional to y^2. If $x = 4$ when $y = 3$, calculate

  a the value of x when $y = 1$

  b the value of y when $x = 2\frac{1}{4}$.

6 p is inversely proportional to $\sqrt{y}$. If $p = 1\cdot2$ when $y = 100$,
calculate

  a the value of p when $y = 4$

  b the value of y when $p = 3$.

7 Given that $z \propto \dfrac{1}{y}$, copy and complete the table.

y	2	4		$\frac{1}{4}$
z	8		16	

8 Given that $v \propto \dfrac{1}{t^2}$, copy and complete the table.

t	2	5		10
v	25		$\frac{1}{4}$	

9 e varies inversely as $(y - 2)$. If $e = 12$ when $y = 4$, find

   a e when $y = 6$ **b** y when $e = \dfrac{1}{2}$.

10 The volume, V, of a given mass of gas varies inversely as the pressure, P. When $V = 2\,\text{m}^3$, $P = 500\,\text{N/m}^2$.
   a Find the volume when the pressure is $400\,\text{N/m}^2$.
   b Find the pressure when the volume is $5\,\text{m}^3$.

11 The number of hours, N, required to dig a certain hole is inversely proportional to the number of people available, x.

   When 6 people are digging, the hole takes 4 hours.
   a Find the time taken when 8 people are available.
   b If it takes $\dfrac{1}{2}$ hour to dig the hole, how many people are there?

12 The force of attraction, F, between two magnets varies inversely as the square of the distance, d, between them. When the magnets are 2 cm apart, the force of attraction is 18 newtons. How far apart are they if the attractive force is 2 newtons?

13 The number of tiles, n, that can be pasted using one tin of tile paste is inversely proportional to the square of the side, d, of the tile. One tin is enough for 180 tiles of side 10 cm. How many tiles of side 15 cm can be pasted using one tin?

14 The life expectancy, L, of a rat varies inversely as the square of the density, d, of poison distributed around its home.
   When the density of poison is $1\,\text{g/m}^2$ the life expectancy is 50 days.
   How long will the rat survive if the density of poison is
   a $5\,\text{g/m}^2$?   **b** $\dfrac{1}{2}\,\text{g/m}^2$?

15 When cooking snacks in a microwave oven, a cook assumes that the cooking time is inversely proportional to the power used. The five levels on his microwave have the powers shown in the table.

Level	Power used
Full	600 W
Roast	400 W
Simmer	200 W
Defrost	100 W
Warm	50 W

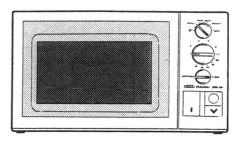

a Escargots de Bourgogne take 5 minutes on 'Simmer'. How long will they take on 'Warm'?

b Escargots a la Provençale are normally cooked on 'Roast' for 3 minutes. How long will they take on 'Full'?

16 Given $z = \dfrac{k}{x^n}$, find k and n, then copy and complete the table.

x	1	2	4	
z	100	$12\frac{1}{2}$		$\frac{1}{10}$

*17 Given $y = \dfrac{k}{\sqrt[n]{v}}$, find k and n, then copy and complete the table.

> $\sqrt[n]{v}$ means the nth root of v.

v	1	4	36	
y	12	6		$\frac{3}{25}$

5.4 Curved graphs

5.4.1 Common curves

It is helpful to know the general shape of some of the more common curves.

- **Quadratic curves** have an x^2-term as the highest power of x.

> The curve also has a line of symmetry.

For example, $y = 2x^2 - 3x + 7$ and $y = 5 + 2x - x^2$

a When the x^2-term is positive, the curve is ⌣-shaped.

b When the x^2-term is negative the curve is an inverted ⌢.

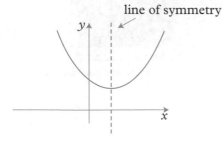

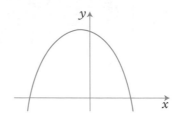

● **Cubic curves** have an x^3-term as the highest power of x.

For example, $y = x^3 + 7x - 4$ and $y = 8x - 4x^3$

a When the x^3-term is positive, the curve can be like one of the two shown below. Notice that as x gets larger, so does y.

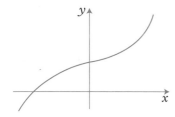

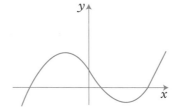

b When the x^3-term is negative, the curve can be like one of the two shown below. Notice that when x is large, y is large but negative.

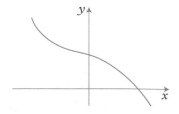

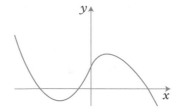

- **Reciprocal** curves have a $\frac{1}{x}$ term.

For example, $y = \frac{12}{x}$ and

$$y = \frac{6}{x} + 5$$

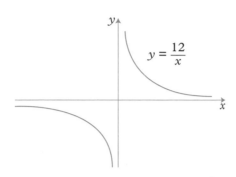

The curve has a break at $x = 0$. The x-axis and the y-axis are called **asymptotes** to the curve. The curve gets very near but never actually touches the asymptotes.

- **Exponential curves** have a term involving a^x, where a is a constant.

For example, $y = 3^x$

$$y = \left(\frac{1}{2}\right)^x$$

The x-axis is an asymptote to the curve.

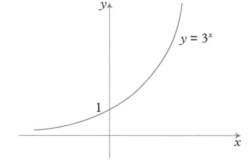

Exercise 14 Ⓜ/Ⓗ

1 What sort of curves are these? Give as much information as you can.

> For example: 'quadratic, positive x^2'

a

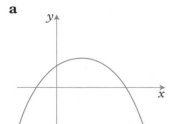

b

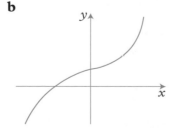

c

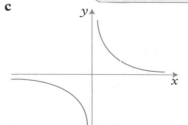

d 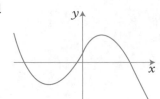   **e** 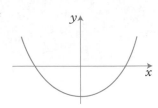   **f**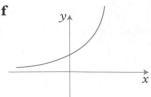

2 Draw the general shape of these curves. (Do not draw accurate graphs.)

    a $y = 3x^2 - 7x + 11$ **b** $y = 2^x$ **c** $y = \dfrac{100}{x}$

    d $y = 8x - x^2$ **e** $y = 10x^3 + 7x - 2$ ***f** $y = \dfrac{1}{x^2}$

3 Here are the equations of the six curves in question **1**, but not in the correct order.

    i $y = \dfrac{8}{x}$ **ii** $y = 2x^3 + x + 2$ **iii** $y = 5 + 3x - x^2$

    iv $y = x^2 - 6$ **v** $y = 5^x$ **vi** $y = 12 + 11x - 2x^2 - x^3$

    Write which equations fit the curves **a** to **f**.

4 Sketch the two curves given and state the number of times the curves intersect.

    a $y = x^3$,   **b** $y = x^2$,   **c** $y = x^3$,   ***d** $y = 3^x$

       $y = 10 - x$ $y = 10 - x^2$ $y = x$ $y = x^3$

5.4.2   Plotting curved graphs

EXAMPLE

Draw the graph of the function
$y = 2x^2 + x - 6$, for $-3 \leqslant x \leqslant 3$.

a   Make a table of x- and y-values.

x	−3	−2	−1	0	1	2	3
$2x^2$	18	8	2	0	2	8	18
x	−3	−2	−1	0	1	2	3
−6	−6	−6	−6	−6	−6	−6	−6
y	9	0	−5	−6	−3	4	15

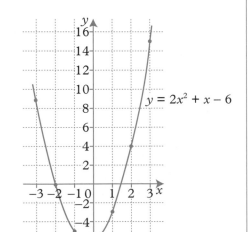

b   Draw and label axes using suitable scales.
c   Plot the points and draw a smooth curve through them with a pencil.
d   Check any points which interrupt the smoothness of the curve.
e   Label the curve with its equation.

Common errors with graphs

Avoid these mistakes. Your curve should be smooth.

A series of 'mini curves'

Flat bottom

Wrong point

Exercise 15 Ⓜ

1 a Copy and complete the table for $y = x^2 + 2x$.

x	−3	−2	−1	0	1	2	3
x^2	9	4	1	0	1	4	9
$2x$	−6	−4		0		4	
y	3	0		0			

b Draw the graph of $y = x^2 + 2x$ using a scale of 2 cm for 1 unit on the x-axis and 1 cm for 1 unit on the y-axis.

2 a Copy and complete the table for $y = x^2 - 3x$.

x	−3	−2	−1	0	1	2	3
x^2	9			0		4	
$-3x$	9			0		−6	
y	18						

b Draw the graph of $y = x^2 - 3x$ using the same scales as in question **1**.

Draw the graphs of these functions using a scale of 2 cm for 1 unit on the x-axis and 1 cm for 1 unit on the y-axis.

3 $y = x^2 + 4x$, for $-3 \leqslant x \leqslant 3$

4 $y = x^2 + 2$, for $-3 \leqslant x \leqslant 3$

5 $y = x^2 - 7$, for $-3 \leqslant x \leqslant 3$

6 $y = x^2 + x - 2$, for $-3 \leqslant x \leqslant 3$

7 $y = x^2 + 3x - 9$, for $-4 \leqslant x \leqslant 3$

8 $y = x^2 - 3x - 4$, for $-2 \leqslant x \leqslant 4$

9 $y = x^2 - 5x + 7$, for $0 \leqslant x \leqslant 6$

10 $y = 2x^2 - 6x$, for $-1 \leqslant x \leqslant 5$

> In question 10, remember: $2x^2 = 2(x^2)$.

11 $y = 2x^2 + 3x - 6$, for $-4 \leqslant x \leqslant 2$

12 $y = 3x^2 - 6x + 5$, for $-1 \leqslant x \leqslant 3$

13 $y = 2 + x - x^2$, for $-3 \leqslant x \leqslant 3$

14 $f(x) = 1 - 3x - x^2$, for $-5 \leqslant x \leqslant 2$

15 $f(x) = 3 + 3x - x^2$, for $-2 \leqslant x \leqslant 5$

16 $f(x) = 7 - 3x - 2x^2$, for $-3 \leqslant x \leqslant 3$

17 $f(x) = 6 + x - 2x^2$, for $-3 \leqslant x \leqslant 3$

18 $y = 8 + 2x - 3x^2$, for $-2 \leqslant x \leqslant 3$

19 $y = x(x-4)$, for $-1 \leqslant x \leqslant 6$

20 $y = (x + 1)(2x - 5)$, for $-3 \leqslant x \leqslant 3$.

Draw the graph of $y = \dfrac{12}{x} + x - 6$, for $1 \leqslant x \leqslant 8$.

Use the graph to find approximate values for

a the minimum value of $\dfrac{12}{x} + x - 6$

b the value of $\dfrac{12}{x} + x - 6$, when $x = 2 \cdot 25$.

· ·

Here is the table of values.

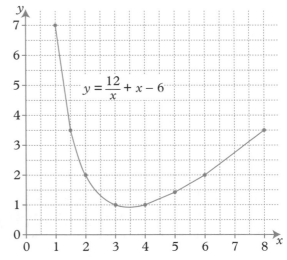

x	1	2	3	4	5	6	8
$\dfrac{12}{x}$	12	6	4	3	2·4	2	1·5
x	1	2	3	4	5	6	8
-6	−6	−6	−6	−6	−6	−6	−6
y	7	2	1	1	1·4	2	3·5

a From the graph, the minimum value of $\dfrac{12}{x} + x - 6$, (that is y) is approximately 0·9.

b At $x = 2 \cdot 25$, y is approximately 1·6.

Exercise 16 Ⓜ/Ⓗ

Draw these curves. The scales given are for one unit of x and y.

1 $y = \dfrac{12}{x}$, for $1 \leqslant x \leqslant 10$. (Scales: 1 cm for x and y)

2 $y = \dfrac{9}{x}$, for $1 \leqslant x \leqslant 10$. (Scales: 1 cm for x and y)

3 $y = \dfrac{12}{x + 1}$, for $0 \leqslant x \leqslant 8$. (Scales: 2 cm for x, 1 cm for y)

4 $y = \dfrac{8}{x - 4}$, for $-4 \leqslant x \leqslant 3 \cdot 5$. (Scales: 2 cm for x, 1 cm for y)

5 $y = \dfrac{x}{x + 4}$, for $-3 \cdot 5 \leqslant x \leqslant 4$. (Scales: 2 cm for x and y)

6 $y = \dfrac{x + 8}{x + 1}$, for $0 \leqslant x \leqslant 8$. (Scales: 2 cm for x and y)

7 $y = \dfrac{10}{x} + x$, for $1 \leqslant x \leqslant 7$. (Scales: 2 cm for x, 1 cm for y)

8 $y = 3^x$, for $-3 \leqslant x \leqslant 3$. (Scales: 2 cm for x, $\dfrac{1}{2}$ cm for y)

9 $y = \left(\dfrac{1}{2}\right)^x$, for $-4 \leqslant x \leqslant 4$. (Scales: 2 cm for x, 1 cm for y)

10 $y = 5 + 3x - x^2$, for $-2 \leqslant x \leqslant 5$. (Scales: 2 cm for x, 1 cm for y)
Find
  a the maximum value of the function $5 + 3x - x^2$
  b the two values of x for which $y = 2$.

11 $y = \dfrac{15}{x} + x - 7$, for $1 \leqslant x \leqslant 7$. (Scales: 2 cm for x and y)
From your graph find
  a the minimum value of y
  b the y-value when $x = 5 \cdot 5$.

12 $y = x^3 - 2x^2$, for $0 \leqslant x \leqslant 4$. (Scales: 2 cm for x, $\dfrac{1}{2}$ cm for y)
From your graph find
  a the y-value at $x = 2 \cdot 5$
  b the x-value at $y = 15$.

13 $y = \dfrac{1}{10}(x^3 + 2x + 20)$, for $-3 \leqslant x \leqslant 3$. (Scales: 2 cm for x and y)
From your graph find
  a the x-value where $x^3 + 2x + 20 = 0$
  b the x-value where $y = 3$.

*14 Draw the graph of
$$y = \dfrac{x}{x^2 + 1}, \text{ for } -6 \leqslant x \leqslant 6. \text{ (Scales: 1 cm for } x, \text{ 10 cm for } y)$$

*15 Draw the graph of
$$E = \dfrac{5000}{x} + 3x \text{ for } 10 \leqslant x \leqslant 80.$$

(Scales: 1 cm to 5 units for x and 1 cm to 25 units for E)
From the graph find
  a the minimum value of E,
  b the value of x corresponding to this minimum value,
  c the range of values of x for which E is less than 275.

Exercise 17 Ⓜ/Ⓗ (Mixed questions)

1 A rectangle has a perimeter of 14 cm and length x cm. Show that the width of the rectangle is $(7 - x)$ cm and hence that the area, A, of the rectangle is given by the formula, $A = x(7 - x)$.
Draw the graph, plotting x on the horizontal axis with a scale of 2 cm to 1 unit, and A on the vertical axis with a scale of 1 cm to 1 unit. Take x from 0 to 7. From the graph find
a the area of the rectangle when $x = 2 \cdot 25$ cm
b the dimensions of the rectangle when its area is 9 cm^2
c the maximum area of the rectangle
d the length and width of the rectangle corresponding to the maximum area
e what shape of rectangle has the largest area.

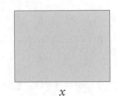

x

2 A farmer has 60 m of wire fencing which he uses to make a rectangular pen for his sheep. He uses a stone wall as one side of the pen so the wire is used for only three sides of the pen.

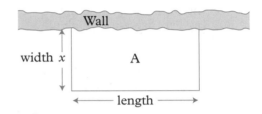

a If the width of the pen is x m, what is the length (in terms of x)?
b What is the area, A, of the pen?
c Draw a graph with area, A, on the vertical axis and the width, x, on the horizontal axis. Take values of x from 0 to 30.
d What dimensions should the pen have if the farmer wants to enclose the largest possible area?

3 A ball is thrown in the air so that t seconds after it is thrown, its height h metres above its starting point is given by the function $h = 25t - 5t^2$. Draw the graph of the function of $0 \leqslant t \leqslant 6$, plotting t on the horizontal axis with a scale of 2 cm to 1 second, and h on the vertical axis with a scale of 2 cm for 10 metres.
Use the graph to find
a the time when the ball is at its greatest height
b the greatest height reached by the ball
c the interval of time during which the ball is at a height of more than 30 m.

4 Consider the equation $y = \frac{1}{x}$.

When $x = \frac{1}{2}$, $y = \frac{1}{\frac{1}{2}} = 2$.   When $x = \frac{1}{100}$, $y = \frac{1}{\frac{1}{100}} = 100$.

As the denominator of $\frac{1}{x}$ gets smaller, the answer gets larger.

An 'infinitely small' denominator gives an 'infinitely large' answer.

You write $\frac{1}{x} \to \infty$ as $x \to 0$.

Draw the graph of $y = \frac{1}{x}$ for

$x = -4, -3, -2, -1, -0{\cdot}5, -0{\cdot}25, 0{\cdot}25, 0{\cdot}5, 1, 2, 3, 4$.
(Scales: 2 cm for x and y)

> The symbol for infinity is ∞.

5 Draw the graph of $y = x + \frac{1}{x}$ for

$x = -4, -3, -2, -1, -0{\cdot}5, -0{\cdot}25, 0{\cdot}25, 0{\cdot}5, 1, 2, 3, 4$.
(Scales: 2 cm for x and y)

6 Draw the graph of $y = \frac{2^x}{x}$, for $-4 \leqslant x \leqslant 7$, including

$x = -0{\cdot}5$, $x = 0{\cdot}5$.
(Scales: 1 cm to 1 unit for x and y)

7 At time $t = 0$, one bacterium is placed in a culture in a laboratory. The number of bacteria doubles every 10 minutes.

  a Draw a graph to show the growth of the bacteria from $t = 0$ to $t = 120$ min.
     Use a scale of 1 cm to 10 minutes across the page and 1 cm to 100 units up the page.

  b Use your graph to estimate the time taken to reach 800 bacteria.

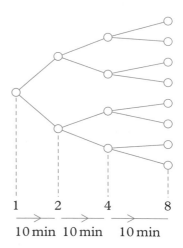

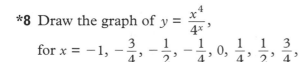

*8 Draw the graph of $y = \frac{x^4}{4^x}$,

for $x = -1, \ -\frac{3}{4}, \ -\frac{1}{2}, \ -\frac{1}{4}, \ 0, \ \frac{1}{4}, \ \frac{1}{2}, \ \frac{3}{4}$,

$1, 1{\cdot}5, 2, 2{\cdot}5, 3, 4, 5, 6, 7$.
(Scales: 2 cm to 1 unit for x, 5 cm to 1 unit for y)

  a For what values of x is the gradient of the function zero?

  b For what values of x is $y = 0{\cdot}5$?

5.5 Graphical solution of equations

With an accurately drawn graph you can find an approximate solution for a wide range of equations, many of which are impossible to solve exactly by other methods.

Draw the graph of the function $y = 2x^2 - x - 3$ for $-2 \leqslant x \leqslant 3$.
Use the graph to find approximate solutions to these equations.
a $2x^2 - x - 3 = 6$ **b** $2x^2 - x = x + 5$

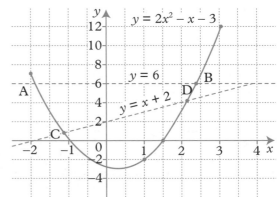

a To solve the equation $2x^2 - x - 3 = 6$, draw the line $y = 6$. At the points of intersection (A and B), y simultaneously equals both 6 and $(2x^2 - x - 3)$.
So you can write $2x^2 - x - 3 = 6$
The solutions are the x-values of the points A and B, that is $x = -1 \cdot 9$ and $x = 2 \cdot 4$ approx.

b To solve the equation $2x^2 - x = x + 5$, rearrange the equation to obtain the function $(2x^2 - x - 3)$ on the left-hand side. In this case, subtract 3 from both sides.
$2x^2 - x - 3 = x + 5 - 3$
$2x^2 - x - 3 = x + 2$
If you now draw the line $y = x + 2$, the solutions of the equation are given by the x-values of C and D, the points of intersection, that is, $x = -1 \cdot 2$ and $x = 2 \cdot 2$ approx.

Assuming that the graph of $y = x^2 - 3x + 1$ has been drawn, find the equation of the line which you should draw to solve the equation $x^2 - 4x + 3 = 0$.

Rearrange $x^2 - 4x + 3 = 0$ in order to obtain $(x^2 - 3x + 1)$ on the left-hand side.

$$x^2 - 4x + 3 = 0$$
add x $x^2 - 3x + 3 = x$
subtract 2 $x^2 - 3x + 1 = x - 2$

Remember:
'Rearrange the equation **to be solved**.'

Therefore draw the line $y = x - 2$ to solve the equation.

Exercise 18 (H)

1 In the diagram, the graphs of $y = x^2 - 2x - 3$, $y = -2$ and $y = x$ have been drawn.

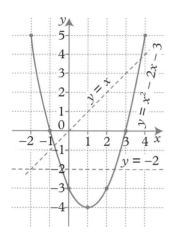

Use the graphs to find approximate solutions to these equations.

a $x^2 - 2x - 3 = -2$ **b** $x^2 - 2x - 3 = x$

c $x^2 - 2x - 3 = 0$ **d** $x^2 - 2x + 1 = 0$

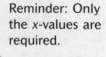

Reminder: Only the *x*-values are required.

2 The graphs of $y = x^2 - 2$, $y = 2x$ and $y = 2 - x$ are shown.

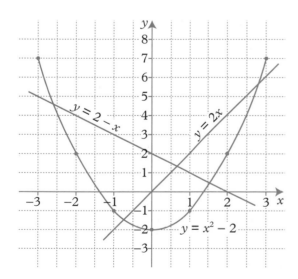

Use the graphs to solve these equations.

a $x^2 - 2 = 2 - x$

b $x^2 - 2 = 2x$

c $x^2 - 2 = 2$

In questions **3** to **6**, use a scale of 2 cm to 1 unit for x and 1 cm to 1 unit for y.

3 Draw the graphs of the functions $y = x^2 - 2x$ and $y = x + 1$ for $-1 \leqslant x \leqslant 4$. Hence find approximate solutions of the equation $x^2 - 2x = x + 1$.

4 Draw the graphs of the functions $y = x^2 - 3x + 5$ and $y = x + 3$ for $-1 \leqslant x \leqslant 5$. Hence find approximate solutions of the equation $x^2 - 3x + 5 = x + 3$.

5 Draw the graphs of the functions $y = 6x - x^2$ and $y = 2x + 1$ for $0 \leqslant x \leqslant 5$. Hence find approximate solutions of the equation $6x - x^2 = 2x + 1$.

6 a Complete the table and then draw the graph of $y = x^2 - 4x + 1$.

x	−1	0	1	2	3	4
y	6		−2		−2	1

b On the same axes, draw the graph of $y = x - 3$.
c Find the solutions of these equations.
   i $x^2 - 4x + 1 = x - 3$
   ii $x^2 - 4x + 1 = 0$ [answers to 1 dp]

In questions **7** to **9**, do **not** draw any graphs.

7 Assuming the graph of $y = x^2 - 5x$ has been drawn, find the equation of the line which you should draw to solve each of these equations.

a $x^2 - 5x = 3$ **b** $x^2 - 5x = -2$
c $x^2 - 5x = x + 4$ **d** $x^2 - 6x = 0$
e $x^2 - 5x - 6 = 0$

> You want to get $x^2 - 5x$ on the left-hand side.

8 Assuming the graph of $y = x^2 + x + 1$ has been drawn, find the equation of the line which you should draw to solve each of these equations.

a $x^2 + x + 1 = 6$ **b** $x^2 + x + 1 = 0$
c $x^2 + x - 3 = 0$ **d** $x^2 - x + 1 = 0$
e $x^2 - x - 3 = 0$

9 Assuming the graph of $y = 6x - x^2$ has been drawn, find the equation of the line which you should draw to solve each of these equations.

a $4 + 6x - x^2 = 0$ **b** $4x - x^2 = 0$
c $2 + 5x - x^2 = 0$ **d** $x^2 - 6x = 3$
e $x^2 - 6x = -2$

For questions **10** to **13**, use scales of 2 cm to 1 unit for x and 1 cm to 1 unit for y.

10 a Complete the table and then draw the graph of
$y = x^2 + 3x - 1$.

x	−5	−4	−3	−2	−1	0	1	2
y	9	3		−3	−3			9

b By drawing other graphs, solve these equations.
   i $x^2 + 3x - 1 = 0$
   ii $x^2 + 3x = 7$
   iii $x^2 + 3x - 3 = x$

11 Draw the graph of $y = x^2 - 2x + 2$ for $-2 \leqslant x \leqslant 4$. By drawing other graphs, solve these equations.
   a $x^2 - 2x + 2 = 8$ **b** $x^2 - 2x + 2 = 5 - x$
   c $x^2 - 2x - 5 = 0$

12 Draw the graph of $y = x^2 - 7x$ for $0 \leqslant x \leqslant 7$. Draw suitable straight lines to solve these equations.
   a $x^2 - 7x + 9 = 0$ **b** $x^2 - 5x + 1 = 0$

13 Draw the graph of $y = 2x^2 + 3x - 9$ for $-3 \leqslant x \leqslant 2$.
   Draw suitable straight lines to find approximate solutions of these equations.
   a $2x^2 + 3x - 4 = 0$ **b** $2x^2 + 2x - 9 = 1$

14 Draw the graph of $y = \dfrac{18}{x}$ for $1 \leqslant x \leqslant 10$, using scales of 1 cm to one unit on both axes. Use the graph to solve these equations approximately.
   a $\dfrac{18}{x} = x + 2$ **b** $\dfrac{18}{x} + x = 10$ **c** $x^2 = 18$

15 Here are five sketch graphs (see overleaf for the other two).

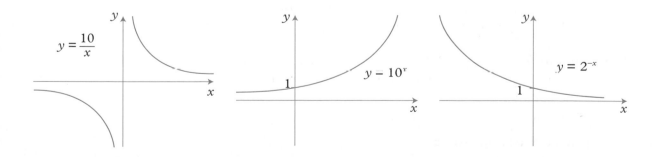

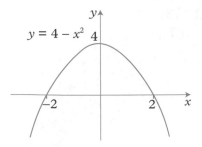 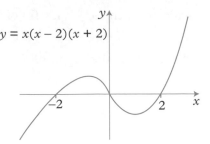

Use the graphs to make your own sketch graphs to find the
number of solutions of each of these equations.

a $\dfrac{10}{x} = 10^x$ **b** $4 - x^2 = 10^x$ **c** $x(x-2)(x+2) = \dfrac{10}{x}$

d $2^{-x} = 10^x$ **e** $4 - x^2 = 2^{-x}$ **f** $x(x-2)(x+2) = 0$

16 Draw the graph of $y = \dfrac{1}{2}x^2 - 6$ for $-4 \leqslant x \leqslant 4$, taking 2 cm to
1 unit on each axis.

  a Use your graph to solve approximately the equation
$\dfrac{1}{2}x^2 - 6 = 1$.

  b Using tables or a calculator confirm that your solutions are
approximately $\pm\sqrt{14}$ and explain why this is so.

  c Use your graph to find the two square roots of 8.

17 Draw the graph of $y = 3x^2 - x^3$ for $-2 \leqslant x \leqslant 3$. Use your graph
to find the range of values of k for which the equation
$3x^2 - x^3 = k$ has three solutions.

*18 Draw the graph of $y = 6 - 2x - \dfrac{1}{2}x^3$ for $x = \pm 2, \pm 1\frac{1}{2}, \pm 1, \pm\frac{1}{2}, 0$.
Take 4 cm to 1 unit for x and 1 cm to 1 unit for y.
Use your graph to find approximate solutions of these equations.

  a $\dfrac{1}{2}x^3 + 2x - 6 = 0$ **b** $x - \dfrac{1}{2}x^3 = 0$

Confirm that two of the solutions to the equation in part **b** are
$\pm\sqrt{2}$ and explain why this is so.

*19 Draw the graph of $y = 2^x$ for $-4 \leqslant x \leqslant 4$, taking 2 cm to one unit
for x and 1 cm to one unit for y. Find approximate solutions to
these equations.

  a $2^x = 6$ **b** $2^x = 3x$ **c** $x(2^x) = 1$

Find also the approximate value of $2^{2.5}$.

Test yourself

1 Make p the subject of the formula
$5p + 2q = 6 - q$
Simplify your answer as much as possible.

2 Make x the subject of this formula.
$4(2x - y) = 2y + 5$

(OCR, 2003)

3 Solve the inequality $7 + n > 13 - 2n$

4 a Solve the inequality $3x - 1 \leqslant 8$
 b Write down the inequality shown by the following diagram.

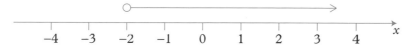

 c Write down all the integers that satisfy both inequalities
shown in parts (a) and (b).

5 Write down three incqualities which together describe the shaded
region R.

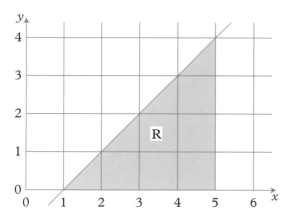

6 y is inversely proportional to x^2 and $y = 0 \cdot 8$ when $x = 2$.
 a Find an equation connecting y and x.
 b Find the value of y when $x = 3$.

(OCR, 2003)

7 y is directly proportional to the square of x.
 $y = 72$ when $x = 2$
 a Express y in terms of x.
 b Work out the value of y when $x = \frac{1}{2}$.

8 b is inversely proportional to the square of a.

When $b = 6$, $a = 2$.

Calculate the value of b when $a = 6$.

9 a Complete the table for $y = x^2 - 3x + 1$

x	−2	−1	0	1	2	3	4
y	11		1	−1		1	5

   b Draw the graph of $y = x^2 - 3x + 1$

   c Use your graph to find an estimate for the minimum value of y.

   d Use a graphical method to find estimates of the solutions to the equation $x^2 - 3x + 1 = 2x - 4$

(Edexcel, 2003)

10 a Draw the graph of $y = \dfrac{6}{x}$

     for values of x from -6 to $+6$.

   b On the same axes draw the graph of $y = x$.

   c Show why you can use these graphs to solve $x^2 = 6$.

   d Use these graphs to solve $x^2 = 6$.

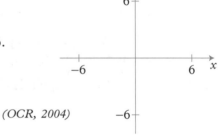

(OCR, 2004)

11 The region **R** satisfies the inequalities

$x \geqslant 2, y \geqslant 1, x + y \leqslant 6$

On a copy of the grid below, draw straight lines and use shading to show the region **R**.

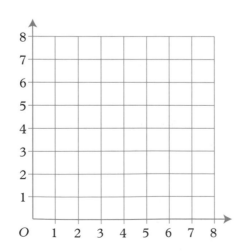

(Edexcel, 2008)

12 a Draw the graph of $y = 2^x$ for values of x from -2 to 3.

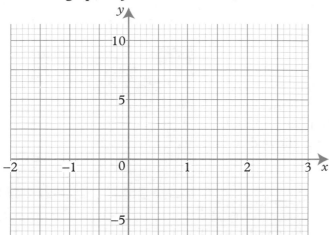

b Use your graph to solve $2^x = 6$.

(OCR, 2004)

13 a Which inequality is shown shaded on the grid?
Choose the correct answer.

$y > 2$ $\qquad\qquad$ $y \geqslant 2$

$x > 2$ $\qquad\qquad$ $x \geqslant 2$

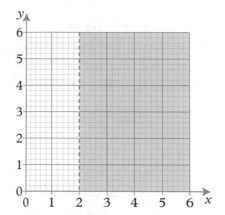

b On a copy of the grid draw lines to find the region
satisfied by the three inequalities

$$y > 2$$
$$y < x + 1$$
$$x + y < 5$$

Label the region with the letter R.

(AQA, 2007)

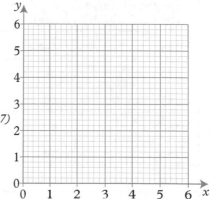

14 You are given that y is inversely proportional to x,
and that $y = 9$ when $x = 4$.

a i Find an equation connecting y and x.

 ii Find y when $x = \dfrac{1}{2}$.

b Find y when $x = y$.

(OCR, 2008)

Functional task 4

Mr Roe's horses

Mr Roe has a farm and needs to work out the cost of keeping his horses. The main costs are feed, vet's bills and keeping the horse healthy. He has volunteers to clean the stables and look after the horses in return for being allowed to ride the horses every week.

Feed

The horses are given a mixture of hay and oats. The hay is free from the farm but a 40 kg bag of oats costs £9·50. Big horses needs more feed than small horses as shown in the **feed chart**.

Horse measurement

By tradition the height of a horse is given in hands and inches. There are four inches in one hand. The formula for converting hands and inches into centimetres is:

$$c = \frac{5(4h + i)}{2}$$

Where c = number of centimetres
h = number of hands
i = number of inches

Feed chart

Height of horse (cm)	Feed per day (kg)
135	4·2
140	4·5
145	4·8
150	5·2
155	5·6
160	6·0
165	6·5
170	7·1

Mr Roe's horses

Number	Height (hands, inches)	Age (years)
1	17 hands	9
2	14 hands 2 inches	7
3	15 hands	13
4	13 hands 2 inches	6
5	16 hands 2 inches	14
6	14 hands	9
7	15 hands	15
8	13 hands 2 inches	5

Vet's bills

Each year horses need injections against Influenza and Tetanus. Illness or injury can be very expensive so Mr Roe takes out insurance to cover the cost of fees. This cost is £30 per month per horse.

Keeping the horse healthy

A farrier is needed to care for the horses's feet as the grow all the time (like human fingernails). A farrier is needed eight times a year and it costs £20 per horse per visit. Each horse also needs to be wormed seven times a year at a cost of £15 per horse per visit.

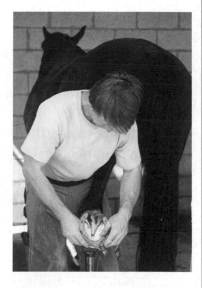

Task 1

Work out the height of each horse in centimetres.

Task 2

Work out the total cost for the eight horses, including feed, vet's bills and keeping the horse healthy.

6 Shape, space and measures 2

In this unit you will:
- learn how to draw isometric representations, plans, views and nets of 3-D objects
- learn about line and rotational symmetry and planes of symmetry
- use trigonometry in right-angled triangles
- use angles of elevation and depression and bearings
- revise reflections, rotations, translations and enlargements
- learn to use vectors
- learn how to find the sine, cosine and tangent for an angle of any size
- learn how to use the sine and cosine rules
- use circle theorems in geometrical problems.

Functional skills coverage and range:
- recognise and use 2D representations of 3D objects.

Links
Knowing the trigonometric ratios helps us understand the natural world by modelling the patterns of waves in the ocean, the movement of the tides or the motion of a pendulum in a clock. Light and sound waves also follow the sine and cosine functions.

6.1 Drawing 3-D shapes, symmetry

6.1.1 Isometric drawing

When you draw a solid on paper you are making a 2-D representation of a 3-D object.

Here are two pictures of the same cuboid, measuring $4 \times 3 \times 2$ units.

a On ordinary square grid paper

b On isometric paper [a grid of equilateral triangles]

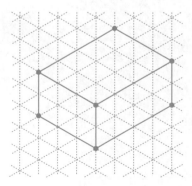

The dimensions of the cuboid cannot be taken from the first picture but they can be taken from the picture drawn on isometric paper. Instead of isometric paper you can also use 'triangular dotty' paper. Be careful to use it the right way round (as shown here).

Exercise 1 Ⓜ

In questions **1** to **3** the objects consist of 1 cm cubes joined together. Draw each object on isometric paper (or 'triangular dotty' paper). Questions **1** and **2** are already drawn on triangular dotty paper.

1 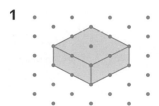 **2** 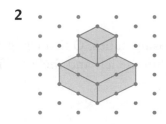 **3**

4 Here are two shapes made using four multilink cubes.

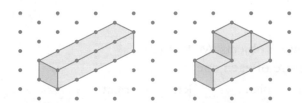

Make and then draw four more shapes, using four cubes, which are different from the two above.

5 The shape shown falls over onto its shaded face. Draw the shape after it has fallen over.

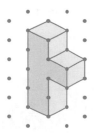

6 You need 16 small cubes.
Make these two shapes and then arrange them into a 4 × 4 × 1 cuboid by adding a third shape, which you have to find. Draw the third shape on isometric paper.
(There are two possible shapes.)

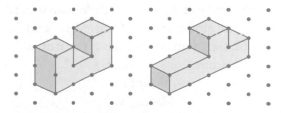

7 The diagrams show object A and three views of A, from above (the plan) and from the left and the right.

a plan

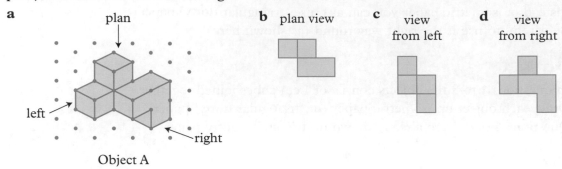

b plan view **c** view from left **d** view from right

left

right

Object A

Make objects B and C below, using cubes. On square grid paper, draw the plan view, the view from the left and the view from the right of each object.

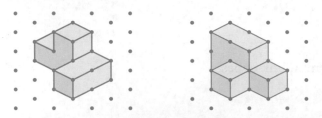

8 The diagrams show the views of three different objects. Make each of the objects, then draw an isometric picture of each one.

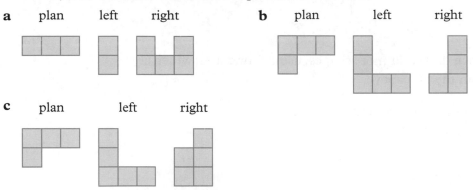

a plan left right **b** plan left right

c plan left right

6.1.2 Nets

If the cube here was made of cardboard, and you cut along some of the edges and laid it out flat, you would have the **net** of the cube.

A cube has: 8 vertices
            6 faces
            12 edges.

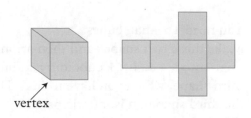

vertex

Here is the net for a square-based pyramid.

This pyramid has: 5 vertices
5 faces
8 edges.

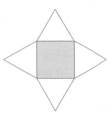

Exercise 2 Ⓜ

1 Which of these nets can be used to make a cube?

a 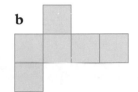　**b** 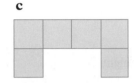　**c** 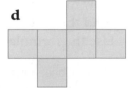　**d**

2 The numbers on opposite faces of a dice add up to 7.
Take one of the possible nets for a cube from question **1**
and show the number of dots on each face.

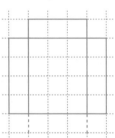

3 Here is the start of a drawing of the net of a cuboid
(a closed rectangular box) measuring 4 cm × 3 cm × 1 cm.
Copy and then complete the net.

4 This diagram needs one more square to complete the net of a cube.
Copy and cut out the shape and then draw the **four** possible
nets which would make a cube.

5 Describe the solid formed from each of these nets. State the
number of vertices and faces for each object.

a 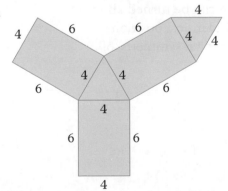　**b**

6 Sketch a possible net for each of these
   a a cuboid measuring 5 cm by 2 cm by 8 cm
   b a prism 10 cm long with a cross-section that is a right-angled triangle with sides 3 cm, 4 cm and 5 cm.

*7 The diagram shows the net of a pyramid.
The base is shaded. The lengths are in cm.
   a How many edges will the pyramid have?
   b How many vertices will it have?
   c Find the lengths a, b, c, d.
   d Use the formula

   $$V = \frac{1}{3} \text{ base area} \times \text{height}$$

   to calculate the volume of the pyramid.

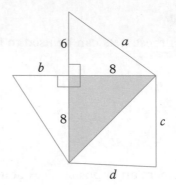

*8 This is the net of a square-based pyramid.
What are the lengths a, b, c, x, y?

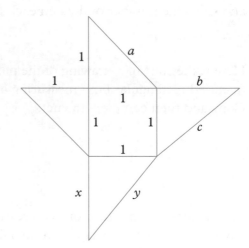

6.1.3 Symmetry

Line symmetry
The letter M has one line of symmetry, shown dotted.

Rotational symmetry
The shape can be turned about O into three identical positions.
It has rotational symmetry of order three.

Exercise 3 Ⓜ

For each shape state
a the number of lines of symmetry
b the order of rotational symmetry.

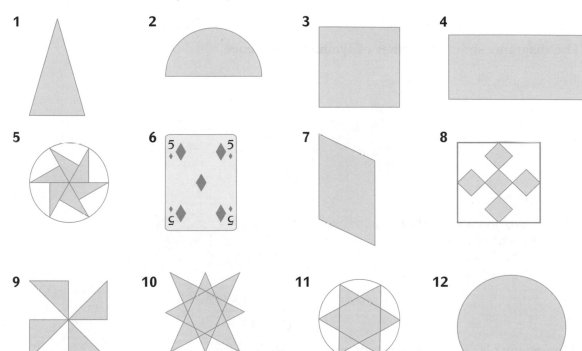

13 Here is a shape made using four squares.
  a Copy the shape and add one square so that the new shape has one line of symmetry. Do this in two different ways.
  b Copy the shape again and add one square so that the new shape has rotation symmetry of order 2.

14 a Copy this shape and shade one more triangle so that the new shape has one line of symmetry.
  b On another copy of the shape shade two more triangles so that the new shape has rotational symmetry of order 3.

15 Look at this shape.
  a Copy it and add one square so that the shape has rotational symmetry of order 2.
  b Copy the shape again and add one square so that the shape has one line of symmetry.

6.1.4 Planes of symmetry

• A plane of symmetry divides a 3-D shape into two congruent shapes. One shape must be a mirror image of the other shape.

The diagrams show two planes of symmetry of a cube.

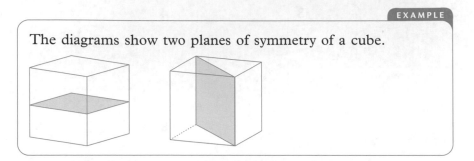

Exercise 4 Ⓜ

1 How many planes of symmetry does this cuboid have?

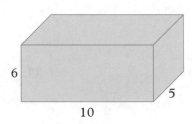

6

10

5

2 How many planes of symmetry do these shapes have?

a 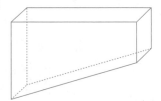   **b** 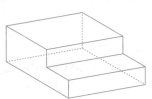   **c**

3 a Draw a diagram of a cube like the one in the example above and draw a different plane of symmetry.
  b How many planes of symmetry does a cube have?

4 Draw a pyramid with a square base so that the vertex of the pyramid is vertically above the centre of the square base.
Show any planes of symmetry by shading.

5 The diagrams show the plan view and the side view of an object.
How many planes of symmetry has this object?

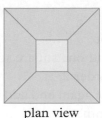

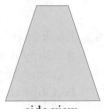

plan view   side view

6.2 Trigonometry

You use trigonometry to calculate sides and angles in triangles.
The triangle must have a right angle.

The side opposite the right angle is called
the **hypotenuse** (H). It is the longest side.

The side opposite the marked angle
is called the **opposite** (O).

The other side is called the **adjacent** (A).

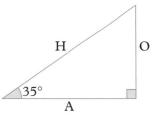

Consider two triangles, one of which is
an enlargement of the other.
It is clear that, for the angle 30°, the ratio

$$\frac{\text{opposite}}{\text{hypotenuse}} = \frac{6}{12} = \frac{2}{4} = \frac{1}{2}$$

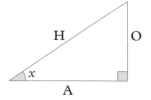

This is the same for both triangles.

6.2.1 Sine, cosine, tangent

There are three important ratios for angle x.

- $\sin x = \dfrac{O}{H}$ - $\cos x = \dfrac{A}{H}$ - $\tan x = \dfrac{O}{A}$

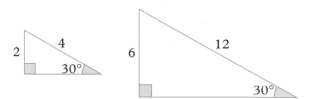

Exercise 5 Ⓜ/Ⓗ

Draw each triangle and label O, A and H in
relation to the angle marked. Write each of
these as a fraction. Question **1** is done for you.

1 $\sin w = \dfrac{O}{H} = \dfrac{3}{5}$

2 $\sin x$ **3** $\sin y$ **4** $\sin z$

5 $\cos w$ **6** $\cos x$ **7** $\cos y$

8 $\tan w$ **9** $\tan x$ **10** $\tan y$

11 $\cos z$ **12** $\tan z$

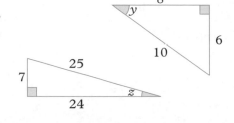

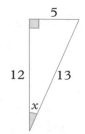

6.2.2 Finding the length of a side

EXAMPLE

a Find the length l.　　**b** Find the length x.　　**c** Find the length y.

a　$\cos 32° = \dfrac{A}{H} = \dfrac{l}{10}$

$l = 10 \times \cos 32°$

$l = 8{\cdot}48 \text{ cm (to 3 sf)}$

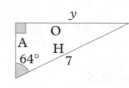

You have 'A' and 'H' so use cos.

b　$\tan 38° = \dfrac{O}{A} = \dfrac{x}{7}$

$x = 7 \times \tan 38°$

$x = 5{\cdot}47 \text{ cm (to 3 sf)}$

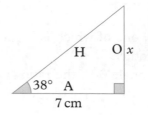

You have 'O' and 'A' so use tan.

c　$\sin 64° = \dfrac{O}{H} = \dfrac{y}{4}$

$y = 4 \times \sin 64°$

$y = 3{\cdot}60 \text{ cm (to 3 sf)}$

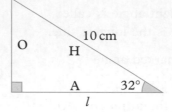

You have 'O' and 'H' so use sin.

Exercise 6 Ⓜ/Ⓗ

Find the lengths marked with letters. All lengths are in cm.
Give your answers correct to 3 sf.

Use 'sin' in questions **1**, **2**, **3**.

1

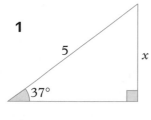

5, x, 37°

2

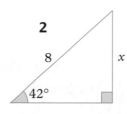

8, x, 42°

3

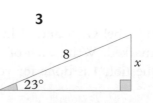

8, x, 23°

4

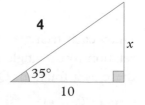

x, 35°, 10

5

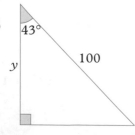

43°, y, 100

6

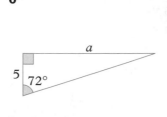

a, 5, 72°

7

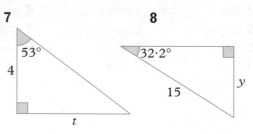

53°, 4, t

8

32·2°, 15, y

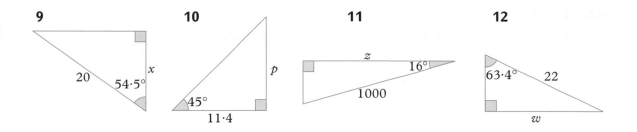

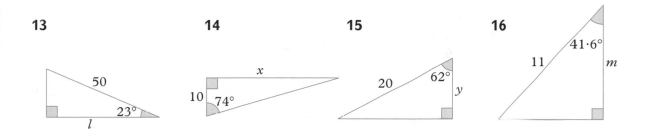

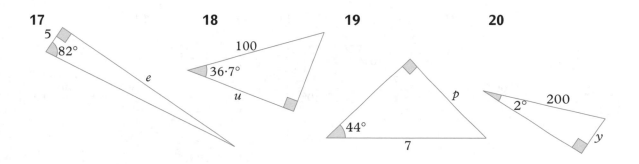

6.2.3 Finding the hypotenuse

Find the length x.

$\sin 36° = \dfrac{O}{H} = \dfrac{11}{x}$

$x \sin 36° = 11$ (Multiply by x)

$x = \dfrac{11}{\sin 36°} = 18\cdot7$ cm (to 3 sf)

> You have 'O' and 'H' so use sin.

Exercise 7 Ⓜ/Ⓗ

This exercise is more difficult. Find the lengths marked with letters.

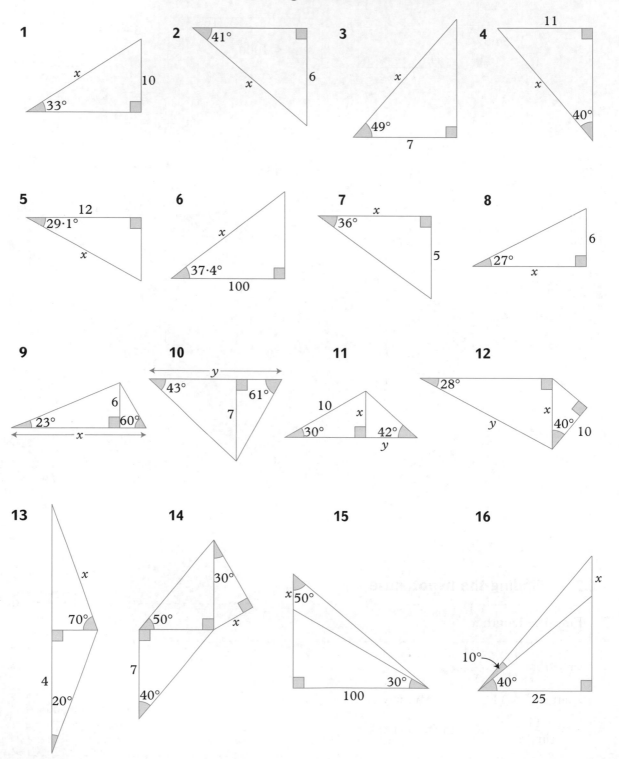

6.2.4 Finding angles

Find the angle x.

$\tan x = \dfrac{O}{A} = \dfrac{3}{8}$

$\tan x = 0.375$

$\quad x = 20.6°$ (to 1 dp)

On a calculator:

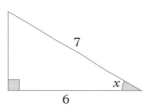

| 3 | ÷ | 8 | = | INV | tan |

You have 'O' and 'A' so use tan.

Exercise 8 Ⓜ/Ⓗ

Find the angles marked with letters. Give the answers correct to 1 dp.

1
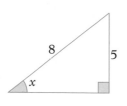
8
5
x

2
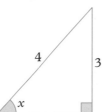
4
3
x

3
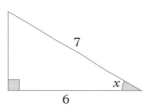
7
x
6

4
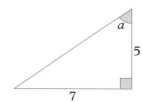
a
5
7

5

8
10
e

6
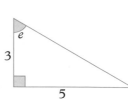
10
z
3

7
9
m
7

8
e
3
5

9
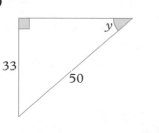
y
33
50

10
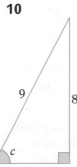
9
8
c

11
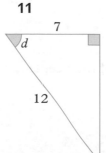
7
d
12

12
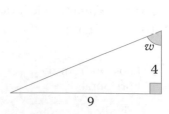
w
4
9

13

2

k

3

14

w

1000

215

15

13

4

e

16

11

6

z

17

5

3

4

x

18

y

6

4

5

19

x

7

5

6

20

11

a

7

9

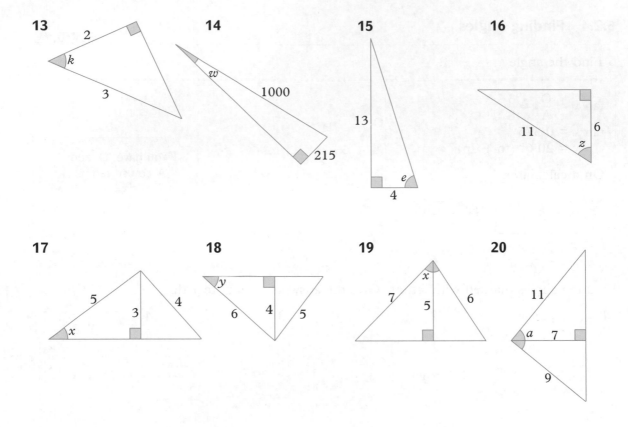

6.2.5   Angles of elevation and depression and bearings

Angle of elevation

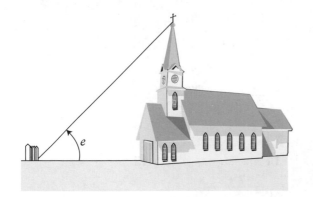

e is the angle of elevation of the steeple from the gate.

Look **up** from the horizontal.

Angle of depression

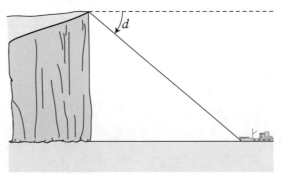

d is the angle of depression of the boat from the cliff top.

Look **down** from the horizontal.

Bearings

- Bearings are measured **clockwise** from north.

a

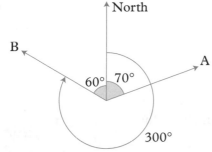

b

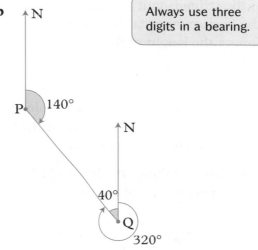

Always use three digits in a bearing.

Ship A sails on a bearing 070°.
Ship B sails on a bearing 300°.

The bearing of Q from P is 140°.
The bearing of P from Q is 320°.

Exercise 9 Ⓜ/Ⓗ

Begin each question by drawing a large clear diagram.

1 A ladder of length 4 m rests against a vertical wall so that the base of the ladder is 1·5 m from the wall.
Calculate the angle between the ladder and the ground.

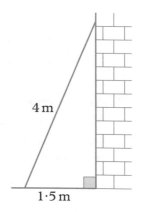

2 A ladder of length 4 m rests against a vertical wall so that the angle between the ladder and the ground is 66°. How far up the wall does the ladder reach?

3 From a distance of 20 m the angle of elevation to the top of a tower is 35°. How high is the tower?

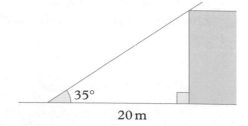

4 A point G is 40 m away from a building, which is 15 m high. What is the angle of elevation to the top of the building from G?

5 A boy is flying a kite from a string of length 60 m. If the string is taut and makes an angle of 71° with the horizontal, what is the height of the kite? Ignore the height of the boy.

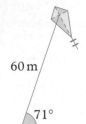

60 m

71°

6 A straight tunnel is 80 m long and slopes downwards at an angle of 11° to the horizontal. Find the vertical drop in travelling from the top to the bottom of the tunnel.

7 The diagram shows the frame of a bicycle. Find the length, *l*, of the cross bar.

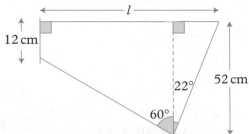

l

12 cm

22° 52 cm

60°

8 Calculate the length *x*.

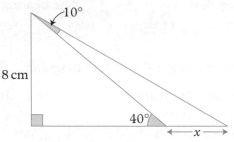

10°

8 cm

40°

x

9 AB is a chord of a circle of radius 5 cm and centre O. The perpendicular bisector of AB passes through O and also bisects the angle AOB. If ∠AOB = 100° calculate the length of the chord AB.

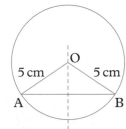

O

5 cm   5 cm

A   B

10 A ship is due south of a lighthouse L. It sails on a bearing of 055° for a distance of 80 km until it is due east of the lighthouse.
How far is it now from the lighthouse?

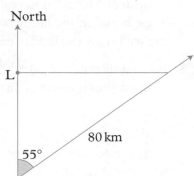

North

L

80 km

55°

11 A ship is due south of a lighthouse. It sails on a bearing of 071° for a distance of 200 km until it is due east of the lighthouse. How far is it now from the lighthouse?

12 A ship is due north of a lighthouse. It sails on a bearing of 200° at a speed of 15 km/h for five hours until it is due west of the lighthouse. How far is it now from the lighthouse?

13 From the top of a tower of height 75 m, a guard sees two prisoners, both due east of him.

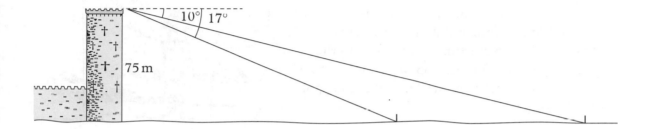

If the angles of depression of the two prisoners are 10° and 17°, calculate the distance between them.

14 From a horizontal distance of 40 m, the angle of elevation to the top of a building is 35·4°. From a point further away from the building the angle of elevation is 20·2°. What is the distance between the two points?

15 An isosceles triangle has sides of length 8 cm, 8 cm and 5 cm. Find the angle between the two equal sides.

16 The angles of an isosceles triangle are 66°, 66° and 48°. If the shortest side of the triangle is 8·4 cm, find the length of one of the two equal sides.

17 A regular pentagon is inscribed in a circle of radius 7 cm.
Find the angle a and then the length of a side of the pentagon.

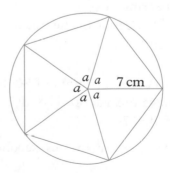

18 Find the acute angle between the diagonals of a rectangle with sides of 5 cm and 7 cm.

Exercise 10 Ⓜ/Ⓗ (Mixed questions)

1 In this diagram the chord AB is 10 cm long and the radius of the circle is 7 cm. O is the centre of the circle. Work out the angle AOB.

2 A ladder of length 5 m rests against a vertical wall. For safety reasons the angle between a ladder and the ground must be between 65° and 75°. Work out the possible safe distances from the foot of a ladder to the base of the wall.

3 An arctic explorer reached the North Pole and decided to erect his national flag and to sing his national anthem, which lasted a rather chilly four minutes. His eye was 1·6 m from the ground and the angle of elevation to the top of the 3 m flagpole was 28°. How far from the base of the flagpole did he stand?

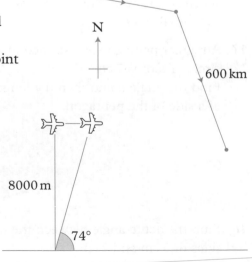

4 A ship sails 35 km on a bearing of 042°.
   a How far north has it travelled?
   b How far east has it travelled?

5 A ship sails 200 km on a bearing of 243·7°.
   a How far south has it travelled?
   b How far west has it travelled?

6 An aircraft flies 400 km from a point O on a bearing of 025° and then 700 km on a bearing of 080° to arrive at B.
   a How far north of O is B?
   b How far east of O is B?
   c Find the distance and bearing of B from O.

7 An aircraft flies 500 km on a bearing of 100° and then 600 km on a bearing of 160°.
Find the distance and bearing of the finishing point from the starting point.

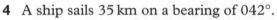

8 A plane is flying at a constant height of 8000 m. It flies vertically above me and 30 seconds later the angle of elevation is 74°.

Find the speed of the plane in metres/second.

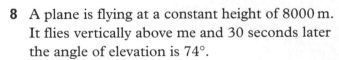

9 A hedgehog wishes to cross a road without being run over. He observes the angle of elevation of a lamp post on the other side of the road to be 27° from the edge of the road and 15° from a point 10 m back from the road.

   a How wide is the road?

   b If he can run at 1 m/s, how long will he take to cross?

   c If cars are travelling at 20 m/s, how far apart must they be if he is to survive?

10 The diagram shows a symmetrical drawbridge. When lowered, the roads AX and BY just meet in the middle. Calculate the length XY.

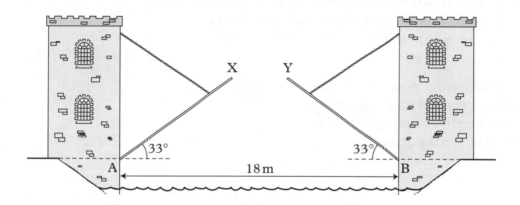

11 From a point 10 m from a vertical wall, the angles of elevation of the bottom and the top of a statue of Sir Isaac Newton, set in the wall, are 40° and 52°. Calculate the height of the statue.

12 Here are three triangles.

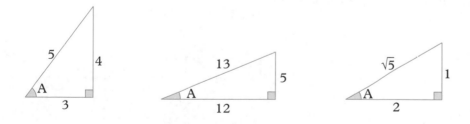

   a For each triangle write the values of sin A and cos A and work out $(\sin A)^2 + (\cos A)^2$.

   b Write what you notice.

   c Does the same result hold for a general right-angled triangle with sides a, b, c?

> A general triangle means any triangle.

13 The diagram shows part of a polygon with *n* sides and centre O.

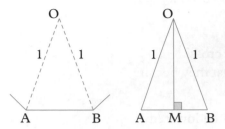

    a What is angle AOB in terms of *n*?
    b M is the midpoint of AB. What is angle MOB in terms of *n*?
    c Find the length MB and hence the length AB in terms of *n*.
    d Find an expression for the perimeter of the polygon.
    e Work out the perimeter for *n* = 100 and *n* = 1000.
      What do you notice?

*14 A rectangular paving stone 3 m by 1 m rests against a vertical wall as shown. What is the height of the highest point of the stone above the ground?

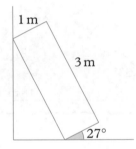

*15 A rectangular piece of paper 30 cm by 21 cm is folded so that opposite corners coincide. How long is the crease?

*16 The diagram shows the cross-section of a rectangular fish tank. When AB is inclined at 40°, the water just comes up to A.

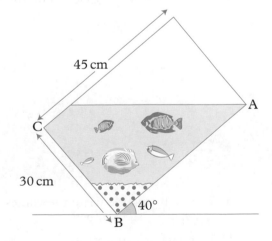

The tank is then lowered so that BC is horizontal.
What is now the depth of water in the tank?

6.2.6 Three-dimensional problems

Projections and planes

A projection is like a shadow on a surface or plane.

> ● The angle between a line and a plane is the angle between the line and its projection in the plane.

PA is a vertical pole standing at the vertex A of a horizontal rectangular field ABCD.

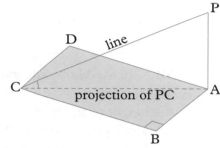

The angle between the line PC and the plane ABCD is found as follows:

a   The projection of PC in the plane ABCD is AC.
b   The angle between the line PC and the plane ABCD is the angle PCA.

Always draw a large, clear diagram. It is often helpful to redraw the particular triangle which contains the length or angle to be found.

EXAMPLE

A rectangular box with top WXYZ and base ABCD has AB = 6 cm, BC = 8 cm and WA = 3 cm.
Calculate
a   the length of AC
b   the angle between the line WC and the plane ABCD.

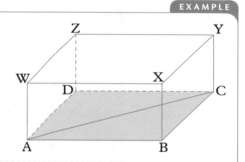

....

a   Redraw triangle ABC.
$$AC^2 = 6^2 + 8^2 = 100$$
$$AC = 10 \text{ cm}$$

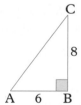

b   The projection of WC on the plane ABCD is AC.
The angle required is ∠WCA.

Redraw triangle WAC and let ∠WCA = θ.

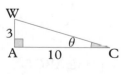

$$\tan \theta = \frac{3}{10}$$
$$\theta = 16 \cdot 7°.$$

The angle between WC and the plane ABCD is 16·7°.

Exercise 11 Ⓗ

1 In this rectangular box, find
   a AC
   b AR
   c the angle between AC and AR.

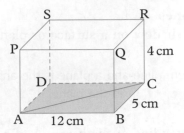

2 The vertical pole BP stands at one corner of a horizontal
   rectangular field.

   If AB = 10 m, AD = 5 m and the angle of elevation of P from
   A is 22°, calculate
   a the height of the pole
   b the angle of elevation of P from C
   c the length of a diagonal of the rectangle ABCD
   d the angle of elevation of P from D.

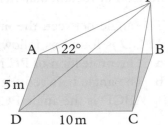

3 In this cube, find
   a BD
   b AS
   c BS
   d the angle SBD.

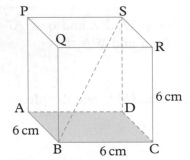

4 In this square-based pyramid, V is vertically above the
   middle of the base, AB = 10 cm and VC = 20 cm.
   Find
   a AC
   b the height of the pyramid
   c the angle between VC and the base ABCD.

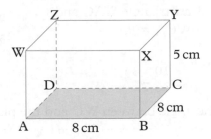

5 The diagram shows a cuboid.
   Calculate
   a the lengths of AC and AY
   b the angle between AY and the plane ABCD.

6 The diagram shows a cuboid.
Calculate
  a the lengths ZX and KX
  b the angle between NX and the plane WXYZ
  c the angle between KY and the plane KLWX.

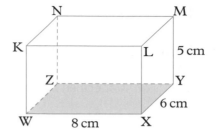

7 In this wedge, PQRS is perpendicular to ABRQ;
PQRS and ABRQ are rectangles with
AB = QR = 6 m, BR = 4 m, RS = 2 m. Find
  a BS   **b** AS
  c the angle between AS and the plane ABRQ.

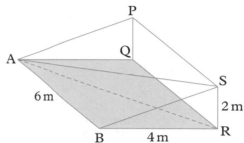

8 The pyramid VPQRS has a square base PQRS.
VP = VQ = VR = VS = 12 cm and PQ = 9 cm.
Calculate the angle between VP and the plane PQRS.

9 The pyramid VABCD has a rectangular base ABCD.
VA = VB = VC = VD = 15 cm, AB = 14 cm and BC = 8 cm.
Calculate
  a the angle between VB and the plane ABCD
  b the angle between VX and the plane ABCD where X is the
    midpoint of BC.

10 In the diagram, A, B and O are points in a horizontal plane and
P is vertically above O, where OP = h m.
A is due west of O, B is due south of O and AB = 60 m.
The angle of elevation of P from A is 25° and the angle of
elevation of P from B is 33°.
  a Find the length AO in terms of h.
  b Find the length BO in terms of h.
  c Find the value of h.

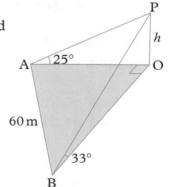

*11 The angle of elevation of the top of a tower is 38° from a point A
due south of it. The angle of elevation of the top of the tower
from another point B, due east of the tower is 29°. Find the height
of the tower if the distance AB is 50 m.

*12 An observer at the top of a tower of height 15 m sees a man due
west of him at an angle of depression 31°. He sees another man
due south at an angle of depression 17°. Find the distance
between the men.

6.3 Transformations

6.3.1 Reflection and rotation

● A **reflection** is specified by the choice of a mirror line.

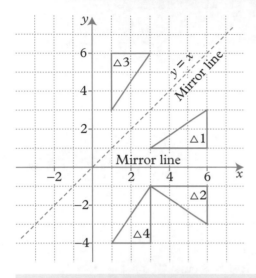

In the diagram
- **a** △2 is the image of △1 after reflection in the *x*-axis (the mirror line).
- **b** △3 is the image of △1 after reflection in the mirror line $y = x$.
- **c** △4 is the image of △2 after rotation through 90° clockwise about $(3, -1)$.
- **d** △4 is the image of △3 after rotation through 180° in either direction about $(2, 1)$.

● A **rotation** requires specification of **three** things: angle, direction and centre of rotation.
● By convention, an anticlockwise rotation is positive and a clockwise rotation is negative.

So the rotation of △2 onto △4 is −90°, centre $(3, -1)$.
It is helpful to use tracing paper to obtain the result of a rotation.

● Under reflection and rotation (and translation) the object and image are always **congruent**.

> The object and image are the same size.

Exercise 12 Ⓜ

In questions **1** and **2**, draw the object and its image after reflection in the broken line.

1

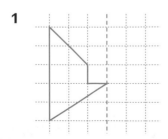

2

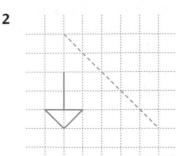

In questions **3** to **8**, draw x- and y-axes with values from -8 to $+8$.

3 a Draw the triangle DEF at $D(-6, 8)$, $E(-2, 8)$, $F(-2, 6)$. Draw the lines $x = 1$, $y = x$, $y = -x$.

  b Draw the image of $\triangle$DEF after reflection in

    i   the line $x = 1$. Label it $\triangle 1$.

    ii the line $y = x$. Label it $\triangle 2$.

    iii the line $y = -x$. Label it $\triangle 3$.

  c Write the coordinates of the image of point D in each case.

4 a Draw $\triangle 1$ at $(3, 1)$, $(7, 1)$, $(7, 3)$.

  b Reflect $\triangle 1$ in the line $y = x$ onto $\triangle 2$.

  c Reflect $\triangle 2$ in the x-axis onto $\triangle 3$.

  d Reflect $\triangle 3$ in the line $y = -x$ onto $\triangle 4$.

  e Reflect $\triangle 4$ in the line $x = 2$ onto $\triangle 5$.

  f Write the coordinates of $\triangle 5$.

5 a Draw $\triangle 1$ at $(2, 6)$, $(2, 8)$, $(6, 6)$.

  b Reflect $\triangle 1$ in the line $x + y = 6$ onto $\triangle 2$.

  c Reflect $\triangle 2$ in the line $x = 3$ onto $\triangle 3$.

  d Reflect $\triangle 3$ in the line $x + y = 6$ onto $\triangle 4$.

  e Reflect $\triangle 4$ in the line $y = x - 8$ onto $\triangle 5$.

  f Write the coordinates of $\triangle 5$.

6 a Draw a triangle PQR at $P(1, 2)$, $Q(3, 5)$, $R(6, 2)$.

  b Find the image of PQR under these rotations

    i $90°$ anticlockwise, centre $(0, 0)$; label the image $P'Q'R'$

    ii $90°$ clockwise, centre $(-2, 2)$; label the image $P''Q''R''$

    iii $180°$, centre $(1, 0)$; label the image $P\star Q\star R\star$.

  c Write the coordinates of P', P'', $P\star$.

7 a Draw $\triangle 1$ at $(1, 2)$, $(1, 6)$, $(3, 5)$.

  b Rotate $\triangle 1$ $90°$ clockwise, centre $(1, 2)$ onto $\triangle 2$.

  c Rotate $\triangle 2$ $180°$, centre $(2, -1)$ onto $\triangle 3$.

  d Rotate $\triangle 3$ $90°$ clockwise, centre $(2, 3)$ onto $\triangle 4$.

  e Write the coordinates of $\triangle 4$.

8 a Draw and label these triangles.

    $\triangle 1 : (3, 1)$, $(6, 1)$, $(6, 3)$

    $\triangle 2 : (-1, 3)$, $(-1, 6)$, $(-3, 6)$

    $\triangle 3 : (1, 1)$, $(-2, 1)$, $(-2, -1)$

    $\triangle 4 : (3, -1)$, $(3, -4)$, $(5, -4)$

    $\triangle 5 : (4, 4)$, $(1, 4)$, $(1, 2)$

  b Fully describe these rotations.

    i $\triangle 1$ onto $\triangle 2$ **ii** $\triangle 1$ onto $\triangle 3$

    iii $\triangle 1$ onto $\triangle 4$ **iv** $\triangle 1$ onto $\triangle 5$

    v $\triangle 5$ onto $\triangle 4$ **vi** $\triangle 3$ onto $\triangle 2$

6.3.2 Translation and enlargement

● A translation is specified by the choice of a column vector.

In the diagram:

a △5 is the image of △4 after translation

  with the column vector $\begin{pmatrix} 4 \\ 2 \end{pmatrix}$.

b △4 is the image of △1 after translation

  with the column vector $\begin{pmatrix} 2 \\ -3 \end{pmatrix}$.

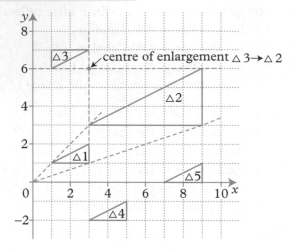

$\begin{pmatrix} 4 \\ 2 \end{pmatrix}$ means 4 units right
and 2 units up.

● An enlargement requires specification of two things: scale
factor and centre of enlargement.

c △2 is the image of △1 after enlargement by a scale factor 3 with
   centre of enlargement $(0, 0)$.

d △1 is the image of △2 after enlargement by a scale factor $\frac{1}{3}$ with
   centre of enlargement $(0, 0)$. Note the construction lines.

e △2 is the image of △3 after enlargement by a scale factor -3
   with centre of enlargement $(3, 6)$. Note that the negative scale
   factor means the object has to be inverted to form the image.

Exercise 13 Ⓜ/Ⓗ

1 Use the diagram to write the
column vector for each of
these translations

a D onto A   b B onto F
c E onto A   d A onto C
e E onto C   f C onto B
g F onto E   h B onto C.

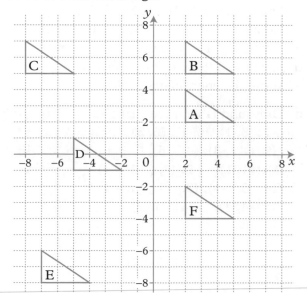

For questions **2** to **5**, copy the diagram and draw an enlargement using the centre O and the scale factor given.

2 Scale factor 2

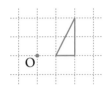

3 Scale factor 3

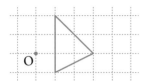

4 Scale factor 3

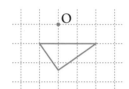

5 Scale factor −2

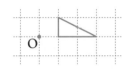

For questions **6** to **11**, draw x- and y-axes with values from 0 to 15. Enlarge the object using the centre of enlargement and scale factor given.

	Object			Centre	Scale factor
6	(2, 4)	(4, 2)	(5, 5)	(1, 2)	$+2$
7	(1, 1)	(4, 2)	(2, 3)	(1, 1)	$+3$
8	(1, 2)	(13, 2)	(1, 10)	(0, 0)	$+\dfrac{1}{2}$
9	(5, 10)	(5, 7)	(11, 7)	(2, 1)	$+\dfrac{1}{3}$
10	(1, 1)	(3, 1)	(3, 2)	(4, 3)	-2
11	(9, 2)	(14, 2)	(14, 6)	(7, 4)	$-\dfrac{1}{2}$

12 Copy the diagram, leaving space for construction lines.

    a Mark the centres for these enlargements.

        i   △1 → △2

        ii   △1 → △3

        iii △2 → △3

    b Write the scale factor for the enlargement △2 → △3.

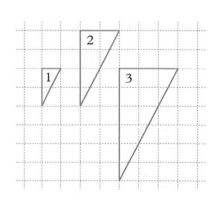

13 Copy this diagram.

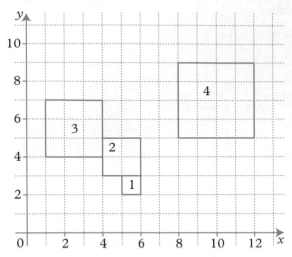

Fully describe each of these enlargements. (All scale factors are positive.)

a square 1 → square 2   **b** square 1 → square 3

c square 1 → square 4   **d** square 2 → square 4

e square 3 → square 2   **f** square 4 → square 1

g Draw the image of square 4 under enlargement with centre (10, 7) and scale factor $\frac{1}{4}$. Label the image square 5.

Questions **14** to **16** involve rotation, reflection, enlargement and translation.

14 Copy this diagram.

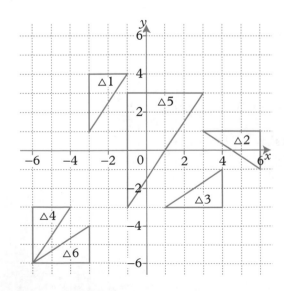

Describe fully these transformations.

a △1 → △2   **b** △1 → △3

c △4 → △1   **d** △1 → △5

e △3 → △6   **f** △6 → △4

*15 Draw x- and y-axes from -8 to $+8$. Plot and label these triangles.

$\triangle 1 : (-5, -5), (-1, -5), (-1, -3)$
$\triangle 2 : (1, 7), (1, 3), (3, 3)$
$\triangle 3 : (3, -3), (7, -3), (7, -1)$
$\triangle 4 : (-5, -5), (-5, -1), (-3, -1)$
$\triangle 5 : (1, -6), (3, -6), (3, -5)$
$\triangle 6 : (-3, 3), (-3, 7), (-5, 7)$

Fully describe these transformations.

a $\triangle 1 \rightarrow \triangle 2$ **b** $\triangle 1 \rightarrow \triangle 3$ **c** $\triangle 1 \rightarrow \triangle 4$ **d** $\triangle 1 \rightarrow \triangle 5$
e $\triangle 1 \rightarrow \triangle 6$ **f** $\triangle 5 \rightarrow \triangle 3$ **g** $\triangle 2 \rightarrow \triangle 3$

*16 Draw x- and y-axes from -8 to $+8$. Plot and label these triangles.

$\triangle 1 : (-3, -6), (-3, -2), (-5, -2)$
$\triangle 2 : (-5, -1), (-5, -7), (-8, -1)$
$\triangle 3 : (-2, -1), (2, -1), (2, 1)$
$\triangle 4 : (6, 3), (2, 3), (2, 5)$
$\triangle 5 : (8, 4), (8, 8), (6, 8)$
$\triangle 6 : (-3, 1), (-3, 3), (-4, 3)$

Fully describe these transformations.

a $\triangle 1 \rightarrow \triangle 2$ **b** $\triangle 1 \rightarrow \triangle 3$ **c** $\triangle 1 \rightarrow \triangle 4$ **d** $\triangle 1 \rightarrow \triangle 5$
e $\triangle 1 \rightarrow \triangle 6$ **f** $\triangle 3 \rightarrow \triangle 5$ **g** $\triangle 6 \rightarrow \triangle 2$

6.3.3 Successive transformations

In the diagram $\triangle 2$ is the image of $\triangle 1$ after rotation 90° clockwise about $(1, 4)$.

$\triangle 3$ is the image of $\triangle 2$ after translation $\begin{pmatrix} 3 \\ 3 \end{pmatrix}$.

So $\triangle 1$ has been moved onto $\triangle 3$ by successive transformations.

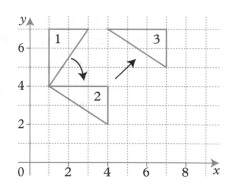

Exercise 14 Ⓜ/Ⓗ

1 Copy the diagram.
   a Rotate $\triangle 1$ 180° about $(4, 2)$.
      Label the image $\triangle 2$.
   b Reflect $\triangle 2$ in the line $y = 2$.
      Label the image $\triangle 3$.
   c Describe the **single** transformation
      which maps $\triangle 1$ onto $\triangle 3$.

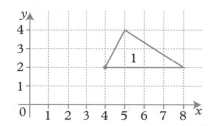

2 Copy the diagram.

  a Reflect △A in the line $y = x$.
     Label the image △B.

  b Reflect △B in the x-axis.
     Label the image △C.

  c Fully describe the single transformation
     which maps △A onto △C.

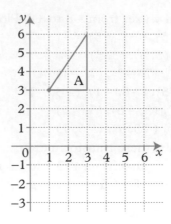

Use these transformations in questions **3** to **8**.

A denotes reflection in $x = 2$

B denotes $180°$ rotation, centre $(1, 1)$

C denotes translation $\begin{pmatrix} -6 \\ 2 \end{pmatrix}$

D denotes reflection in $y = x$

E denotes reflection in $y = 0$

F denotes translation $\begin{pmatrix} 4 \\ 3 \end{pmatrix}$

G denotes $90°$ rotation clockwise, centre $(0, 0)$

H denotes enlargement, scale factor $+\dfrac{1}{2}$, centre $(0, 0)$.

Draw x- and y-axes with values from -10 to $+10$.

3 Draw triangle LMN at L$(2, 2)$, M$(6, 2)$, N$(6, 4)$. Find the image of
   LMN under these transformations.

  a **A** followed by **C**   **b** **D** followed by **E**
  c **B** followed by **D**   **d** **E** followed by **B**.

   Write the coordinates of the image of point L in each case.

4 Draw triangle PQR at P$(2, 2)$, Q$(6, 2)$, R$(6, 4)$. Find the image of
   PQR under these transformations.

  a **F** followed by **A**   **b** **G** followed by **C**
  c **G** followed by **A**   **d** **E** followed by **H**.

   Write the coordinates of the image of point P in each case.

5 Draw triangle XYZ at X$(-2, 4)$, Y$(-2, 1)$, Z$(-4, 1)$. Find the
   image of XYZ under these combinations of transformations and
   state the equivalent single transformation in each case.

  a **E**, then **G**, then **G**   **b** **B**, then **C**
  c **A**, then **D**.

6 Draw triangle OPQ at O(0, 0), P(0, 2), Q(3, 2). Find the image of OPQ under these combinations of transformations and state the equivalent single transformation in each case.

   a **E**, then **D**   **b** **C**, then **F**

   c **C**, then **E**, then **D**   **d** **E**, then **F**, then **D**.

7 Draw triangle RST at R(−4, −1), S(−2$\frac{1}{2}$, −2), T(−4, −4). Find the image of RST under these combinations of transformations and state the equivalent single transformation in each case.

   a **G**, then **A**, then **E**   **b** **H**, then **F**

   c **F**, then **G**.

*8 The inverse of a transformation is the transformation which takes the image back to the object.

Write the inverses of the transformations **A**, **B**, … **H**.

6.4 Vectors

A vector quantity has both magnitude and direction. A translation is described by a vector, and a vector can also represent physical quantities such as velocity, force, acceleration, etc. The symbol for a vector in this book and in examination papers is a bold letter, for example **a**, **x**. In your own work show vectors by drawing a line under the letter, for example a̲. On a coordinate grid, the magnitude and direction of the vector can be shown by a column vector,

for example $\begin{pmatrix} 2 \\ 1 \end{pmatrix}$, $\begin{pmatrix} 1 \\ -3 \end{pmatrix}$

where the upper number shows the distance across the page and the lower number shows the distance up the page.

EXAMPLE

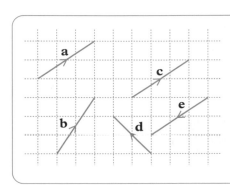

$$\mathbf{a} = \begin{pmatrix} 3 \\ 2 \end{pmatrix} \qquad \mathbf{d} = \begin{pmatrix} -2 \\ 2 \end{pmatrix}$$

$$\mathbf{b} = \begin{pmatrix} 2 \\ 3 \end{pmatrix} \qquad \mathbf{e} = \begin{pmatrix} -3 \\ -2 \end{pmatrix}$$

$$\mathbf{c} = \begin{pmatrix} 3 \\ 2 \end{pmatrix}$$

6.4.1 Equal vectors

Two vectors are equal if they have the same length **and** the same direction.

The actual position of the vector on the diagram or in space is of no consequence.

Thus in the previous example, vectors **a** and **c** are equal because they have the same magnitude and direction. Even though vector **b** also has the same length (magnitude) as **a** and **c**, it is not equal to **a** or **c** because it acts in a different direction. Likewise, vector **e** has the same length as vector **c** but acts in the reverse direction and so cannot equal **c**.
Equal vectors have identical column vectors.

6.4.2 Addition of vectors

Vectors **a** and **b** are represented by line segments. They can be added by using the 'nose-to-tail' method to give a single equivalent vector.

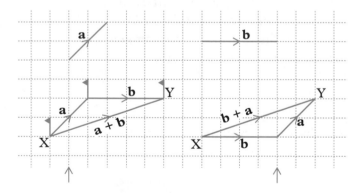

The 'tail' of vector **b** is joined to the 'nose' of vector **a**.

Alternatively the tail of **a** can be joined to the 'nose' of vector **b**.

In both cases the vector $\overrightarrow{XY}$ has the same length and direction and therefore **a + b = b + a**.

In the first diagram the flag is moved by translation **a** and then translation **b**. The translation **a + b** is the equivalent or **resultant** translation.

6.4.3 Multiplication by a scalar

A scalar quantity has magnitude but no direction (for example, mass, volume, temperature). Ordinary numbers are scalars.

When vector **x** is multiplied by 2, the result is 2**x**.

When **x** is multiplied by -3 the result is -3**x**.

EXAMPLE

The diagram shows vectors **a** and **b**. Draw a diagram to show $\overrightarrow{OP}$ and $\overrightarrow{OQ}$ such that
$\overrightarrow{OP} = 3\mathbf{a} + \mathbf{b}$ $\overrightarrow{OQ} = -2\mathbf{a} - 3\mathbf{b}$

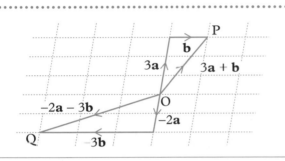

(1) A negative sign reverses the direction of the vector.
(2) The result **a** − **b** is **a** + −**b**. So, subtracting **b** is equivalent to adding the negative of **b**.

Exercise 15 (H)

In questions **1** to **15**, use this diagram to describe the vectors given in terms of **c** and **d** where $\mathbf{c} = \overrightarrow{QN}$ and $\mathbf{d} = \overrightarrow{QR}$, for example, $\overrightarrow{QS} = 2\mathbf{d}$, $\overrightarrow{TD} = \mathbf{c} + \mathbf{d}$.

```
K   J   I   H   G   F   E
L   M   N   A   B   C   D
        c
O   P   Q   R   S   T   U
        d
V   W   X   Y   Z
```

1 $\overrightarrow{AB}$	**2** $\overrightarrow{SG}$	**3** $\overrightarrow{VK}$	**4** $\overrightarrow{KH}$	**5** $\overrightarrow{OT}$
6 $\overrightarrow{WJ}$	**7** $\overrightarrow{FH}$	**8** $\overrightarrow{FT}$	**9** $\overrightarrow{KV}$	**10** $\overrightarrow{NQ}$
11 $\overrightarrow{OM}$	**12** $\overrightarrow{SD}$	**13** $\overrightarrow{PI}$	**14** $\overrightarrow{YG}$	**15** $\overrightarrow{OI}$

In questions **16** to **21**, use the same diagram to find these vectors in terms of the capital letters, starting from Q each time, for example, $3\mathbf{d} = \overrightarrow{QT}$, $\mathbf{c} + \mathbf{d} = \overrightarrow{QA}$.

16 2**c**   **17** 4**d**   **18** 2**c** + **d**
19 2**d** + **c**   **20** 3**d** + 2**c**   **21** 2**c** − **d**

In questions **22** and **23**, use this diagram. $\overrightarrow{LM}$ = **a**, $\overrightarrow{LQ}$ = **b**.

22 Write these vectors in terms of **a** and **b**.

a $\overrightarrow{GN}$ **b** $\overrightarrow{CO}$ **c** $\overrightarrow{TN}$
d $\overrightarrow{FT}$ **e** $\overrightarrow{KC}$ **f** $\overrightarrow{CJ}$

23 From your answers to question **22**, find the vector which is

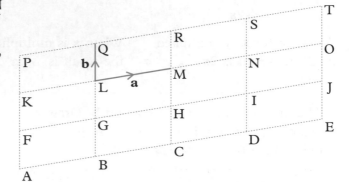

a parallel to $\overrightarrow{LR}$
b 'opposite' to $\overrightarrow{LR}$
c parallel to $\overrightarrow{CJ}$ with twice the magnitude
d parallel to the vector (**a** − **b**).

In questions **24** to **27**, write each vector in terms of **a**, **b**, or **a** and **b**.

24 a $\overrightarrow{BA}$
    b $\overrightarrow{AC}$
    c $\overrightarrow{DB}$
    d $\overrightarrow{AD}$

25 a $\overrightarrow{ZX}$
    b $\overrightarrow{YW}$
    c $\overrightarrow{XY}$
    d $\overrightarrow{XZ}$

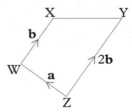

26 a $\overrightarrow{MK}$
    b $\overrightarrow{NL}$
    c $\overrightarrow{NK}$
    d $\overrightarrow{KN}$

27 a $\overrightarrow{FE}$
    b $\overrightarrow{BC}$
    c $\overrightarrow{FC}$
    d $\overrightarrow{DA}$

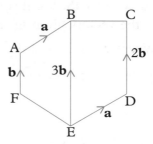

28 The points A, B and C lie on a straight line and the vector $\overrightarrow{AB}$ is **a** + 2**b**. Which of these vectors is possible for $\overrightarrow{AC}$?

   a 3**a** + 6**b**   **b** 4**a** + 4**b**   **c** **a** − 2**b**   **d** 5**a** + 10**b**

29 Find three pairs of parallel vectors from those below.

a + 3**b**	**a** − **b**	6**a** − 3**b**	2**a** + 6**b**	3**a** − 3**b**	2**a** − **b**	**a** + **b**
A	B	C	D	E	F	G

6.4.4 Vector geometry

In the diagram, OA = AP and BQ = 3OB.
N is the midpoint of PQ;
$\overrightarrow{OA} = \mathbf{a}$ and $\overrightarrow{OB} = \mathbf{b}$.

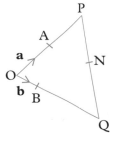

Express each of these vectors in terms of **a**, **b**, or **a** and **b**.

a $\overrightarrow{AP}$ 　　　　　　　 **b** $\overrightarrow{AB}$ 　　　　　　　 **c** $\overrightarrow{OQ}$

d $\overrightarrow{PO}$ 　　　　　　　 **e** $\overrightarrow{PQ}$ 　　　　　　　 **f** $\overrightarrow{PN}$

g $\overrightarrow{ON}$ 　　　　　　　 **h** $\overrightarrow{AN}$

....

a $\overrightarrow{AP} = \mathbf{a}$ 　　　　 **b** $\overrightarrow{AB} = -\mathbf{a} + \mathbf{b}$ 　　　　 **c** $\overrightarrow{OQ} = 4\mathbf{b}$

d $\overrightarrow{PO} = -2\mathbf{a}$ 　　　 **e** $\overrightarrow{PQ} = \overrightarrow{PO} + \overrightarrow{OQ}$ 　　　 **f** $\overrightarrow{PN} = \frac{1}{2}\overrightarrow{PQ}$
$$= -2\mathbf{a} + 4\mathbf{b} \qquad\qquad = -\mathbf{a} + 2\mathbf{b}$$

g $\overrightarrow{ON} = \overrightarrow{OP} + \overrightarrow{PN}$ 　　　 **h** $\overrightarrow{AN} = \overrightarrow{AP} + \overrightarrow{PN}$
$$= 2\mathbf{a} + (-\mathbf{a} + 2\mathbf{b}) \qquad = \mathbf{a} + (-\mathbf{a} + 2\mathbf{b})$$
$$= \mathbf{a} + 2\mathbf{b} \qquad\qquad = 2\mathbf{b}$$

Exercise 16 Ⓗ

In questions **1** to **4**, $\overrightarrow{OA} = \mathbf{a}$ and $\overrightarrow{OB} = \mathbf{b}$. Copy each diagram and
use the information given to express these vectors in terms of **a**, **b**
or **a** and **b**.

a $\overrightarrow{AP}$ 　　 **b** $\overrightarrow{AB}$ 　　 **c** $\overrightarrow{OQ}$ 　　 **d** $\overrightarrow{PO}$ 　　 **e** $\overrightarrow{PQ}$
f $\overrightarrow{PN}$ 　　 **g** $\overrightarrow{ON}$ 　　 **h** $\overrightarrow{AN}$ 　　 **i** $\overrightarrow{BP}$ 　　 **j** $\overrightarrow{QA}$

1 A, B and N are the midpoints of OP, OB
and PQ respectively.

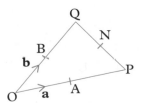

2 A and N are the midpoints of C
and PQ; BQ = 2OB.

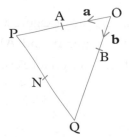

3 AP = 2OA, BQ = OB, PN = NQ.

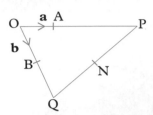

4 OA = 2AP, BQ = 3OB, PN = 2NQ.

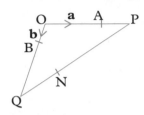

5 In △XYZ, the midpoint of YZ is M.
If $\overrightarrow{XY}$ = **s** and $\overrightarrow{ZX}$ = **t**, find $\overrightarrow{XM}$ in terms
of **s** and **t**.

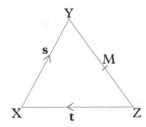

6 In △AOB, AM : MB = 2 : 1. If $\overrightarrow{OA}$ = **a**,
and $\overrightarrow{OB}$ = **b** find $\overrightarrow{OM}$ in terms of **a** and **b**.

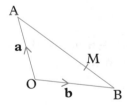

7 O is any point in the plane of the square ABCD. The vectors $\overrightarrow{OA}$,
$\overrightarrow{OB}$, and $\overrightarrow{OC}$, are **a**, **b** and **c** respectively. Find the vector $\overrightarrow{OD}$,
in terms of **a**, **b** and **c**.

8 ABCDEF is a regular hexagon with $\overrightarrow{AB}$
representing the vector **m** and $\overrightarrow{AF}$ representing
the vector **n**. Find the vector representing $\overrightarrow{AD}$.

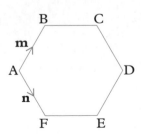

9 ABCDEF is a regular hexagon with centre O. $\overrightarrow{FA} = \mathbf{a}$ and $\overrightarrow{FB} = \mathbf{b}$.

Express these vectors in terms of **a** and/or **b**.

a $\overrightarrow{AB}$ **b** $\overrightarrow{FO}$ **c** $\overrightarrow{FC}$

d $\overrightarrow{BC}$ **e** $\overrightarrow{AO}$ **f** $\overrightarrow{FD}$

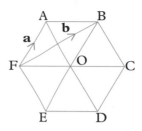

10 In the diagram, M is the midpoint of CD, BP:PM = 2:1, $\overrightarrow{AB} = \mathbf{x}$, and $\overrightarrow{AC} = \mathbf{y}$ and $\overrightarrow{AD} = \mathbf{z}$.

Express these vectors in terms of **x, y** and **z**.

a $\overrightarrow{DC}$ **b** $\overrightarrow{DM}$ **c** $\overrightarrow{AM}$

d $\overrightarrow{BM}$ **e** $\overrightarrow{BP}$ **f** $\overrightarrow{AP}$

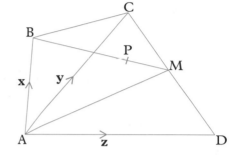

11 In the quadrilateral $\overrightarrow{OA} = 2\mathbf{a}$, $\overrightarrow{OB} = 2\mathbf{b}$, $\overrightarrow{OC} = 2\mathbf{c}$. Points P, Q, R and S are the midpoints of the sides OA, AB, BC and CO respectively.

a Express in terms of **a**, **b** and **c**

   i $\overrightarrow{AB}$

   ii $\overrightarrow{BC}$

   iii $\overrightarrow{PQ}$

   iv $\overrightarrow{QR}$

   v $\overrightarrow{PS}$.

b Describe the relationship between QR and PS.

c What sort of quadrilateral is PQRS?

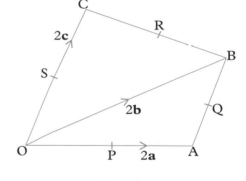

12 In the diagram, $\overrightarrow{OA} = \mathbf{a}$, $\overrightarrow{OB} = \mathbf{b}$, OC = CA, OB = BE and BD:DA = 1:2.

a Express in terms of **a** and **b**

   i $\overrightarrow{BA}$

   ii $\overrightarrow{BD}$

   iii $\overrightarrow{CD}$

   iv $\overrightarrow{CE}$.

b Explain why points C, D and E lie on a straight line.

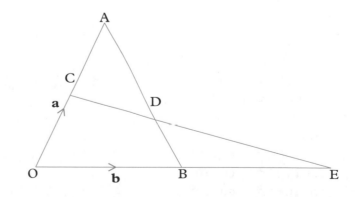

6.5 Sine, cosine, tangent for any angle

So far you have used sine, cosine and tangent only in right-angled triangles.

The circle in the diagram is of radius 1 unit with centre $(0, 0)$.
A point P with coordinates (x, y) moves round the circumference of the circle. The angle that OP makes with the positive x-axis as it turns in an anticlockwise direction is θ.

In triangle OAP, $\cos \theta = \frac{x}{1}$ and $\sin \theta = \frac{y}{1}$

The x-coordinate of P is $\cos \theta$.
The y-coordinate of P is $\sin \theta$.

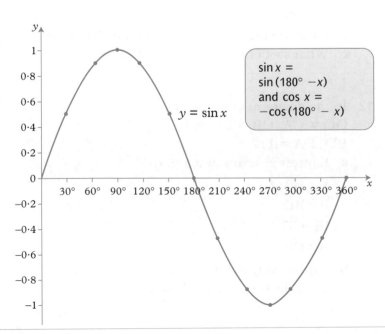

This idea is used to define the cosine and the sine of any angle, including angles greater than 90°.

Here are two angles that are greater than 90°.

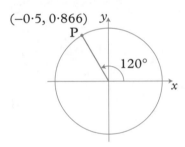

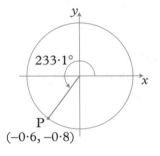

$$\cos 120° = -0.5$$
$$\sin 120° = 0.866$$

$$\cos 233.1° = -0.6$$
$$\sin 233.1° = -0.8$$

You can use a graphics calculator to show the graph of $y = \sin x$ for any range of angles. The graph shows $y = \sin x$ for x from 0° to 360°. The curve above the x-axis has symmetry about $x = 90°$ and that below the x-axis has symmetry about $x = 270°$.

$\sin x =$
$\sin (180° - x)$
and $\cos x =$
$-\cos (180° - x)$

$$\sin 150° = \sin 30°$$
$$\sin 110° = \sin 70°$$
$$\sin 163° = \sin 17°$$
and
$$\cos 150° = -\cos 30°$$
$$\cos 110° = -\cos 70°$$
$$\cos 163° = -\cos 17°$$

Exercise 17 (H)

1 a Use a calculator to find the cosines of all the angles 0°, 30°, 60°, 90°, 120°, ... 330°, 360°.

b Draw a graph of $y = \cos x$ for $0 \leqslant x \leqslant 360°$. Use a scale of 1 cm to 30° on the x-axis and 5 cm to 1 unit on the y-axis.

2 Draw the graph of $y = \sin x$, using the same angles and scales as in question **1**.

3 Find the tangent of the angles 0°, 20°, 40°, 60°, ... 320°, 340°, 360°.

a Notice that 90° and 270° are omitted. Why do you think this is?

b Draw a graph of $y = \tan x$. Use a scale of 1 cm to 20° on the x-axis and 1 cm to 1 unit on the y-axis.

c Draw a vertical dotted line at $x = 90°$ and $x = 270°$.

These lines are **asymptotes** to the curve. As the value of x approaches 90° from either side, the curve gets nearer and nearer to the asymptote but it **never** quite reaches it.

> Use a calculator to find tan 89°, tan 89·9°, tan 89·99°, tan 89·999° etc.

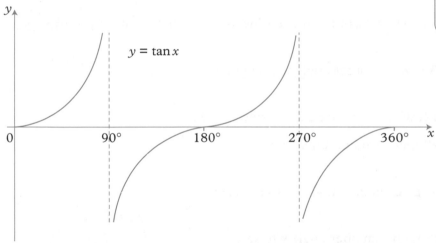

4 Use your graphs to sort these into pairs of equal value.

| cos 60° | sin 130° |

| sin 70° | cos 300° |

| sin 50° | sin 110° |

5 Sort these into pairs of equal value.

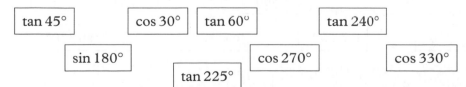

| tan 45° | cos 30° | tan 60° | tan 240° |

| sin 180° | cos 270° | cos 330° |

| tan 225° |

In questions **6** to **17** do not use a calculator.
Use the symmetry of the graphs $y = \sin x$, $y = \cos x$,
and $y = \tan x$.
Angles are from 0° to 360° and answers should be
given to the nearest degree.

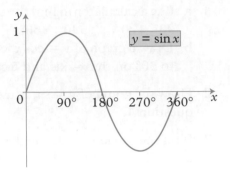

6 If $\sin 18° = 0.309$, give another angle whose sine
is 0·309.

7 If $\sin 27° = 0.454$, give another angle whose sine
is 0·454.

8 Give another angle which has the same sine as
a 40°　　**b** 70°　　**c** 130°

9 If $\cos 70° = 0.342$, give another angle whose cosine
is 0·342.

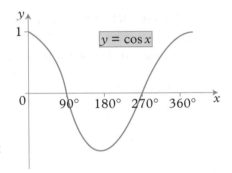

10 If $\cos 45° = 0.707$, give another angle whose cosine
is 0·707.

11 Give another angle which has the same cosine as
a 10°　　**b** 56°　　**c** 300°

12 If $\tan 40° = 0.839$, give another angle whose tangent
is 0·839.

13 If $\sin 20° = 0.342$, what other angle has a sine of 0·342?

14 If $\sin 98° = 0.990$, give another angle whose
sine is 0·990.

15 If $\tan 135° = -1$, give another angle whose tangent
is −1.

16 If $\cos 120° = -0.5$, give another angle whose
cosine is −0·5.

17 If $\tan 70° = 2.75$, give another angle whose tangent
is 2·75.

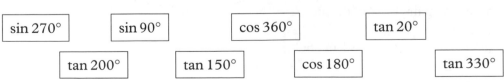

18 Sort these into pairs of equal value.

sin 270°	sin 90°	cos 360°	tan 20°

| | tan 200° | tan 150° | cos 180° | tan 330° |

Exercise 18 Ⓗ

In questions **1** to **8** give angles correct to one decimal place.

1 Find two values for x, between $0°$ and $360°$, if $\sin x = 0.848$.

2 If $\sin x = 0.35$, find two solutions for x between $0°$ and $360°$.

3 If $\cos x = 0.6$, find two solutions for x between $0°$ and $360°$.

4 If $\tan x = 1$, find two solutions for x between $0°$ and $360°$.

5 Find two solutions between $0°$ and $360°$.
  a $\sin x = 0.339$
  b $\sin x = 0.951$
  c $\sin x = \dfrac{1}{2}$
  d $\sin x = \dfrac{\sqrt{3}}{2}$

6 Find two solutions between $0°$ and $360°$.
  a $\sin x = 0.72$
  b $\cos x = 0.3$
  c $\tan x = 5$
  d $\sin x = -0.65$

7 Find **four** solutions of the equation $(\sin x)^2 = \dfrac{1}{4}$, for x between $0°$ and $360°$.

8 Find four solutions of the equation $(\tan x)^2 = 1$, for x between $0°$ and $360°$.

9 You can use the triangle in the diagram to find sin, cos and tan of $30°$ and $60°$. For example, $\sin 30° = \dfrac{1}{2}$ and $\tan 60° = \sqrt{3}$.
  Copy and complete this table.

	30°	60°	120°	150°
sin	$\dfrac{1}{2}$			
cos				
tan		$\sqrt{3}$		

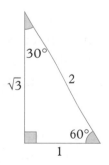

10 a Between $0°$ and $360°$, for what values of x is $\sin x > 0$?
  b Between $0°$ and $360°$, for what values of x is $\cos x < 0$?

11 a A negative angle is measured **clockwise** from the x-axis.

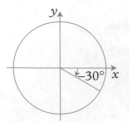

$$\sin(-30°) = -\frac{1}{2}$$

$$\cos(-30°) = \frac{\sqrt{3}}{2}$$

b An angle greater than 360° is shown in this diagram.

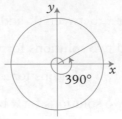

$$\sin 390° = \sin 30°$$

$$\cos 400° = \cos 40° \text{ etc.}$$

Sketch the graphs of $\sin \theta$, $\cos \theta$ and $\tan \theta$ for θ between $-360°$ and $360°$.

12 a Find three values of θ for which $\sin \theta = 1$.
   b Find four values of θ for which $\cos \theta = 0$.

13 a Write the value of $\sin (360n°)$, where n is an integer.
   b Write the value of $\cos (360n°)$, where n is an integer.

Teacher's note: The remaining questions are beyond the formal syllabus and are provided as 'extension material'.

*14 a Copy and complete.

$$\sin \theta = \frac{a}{c}, \quad \cos \theta = \boxed{-}, \quad \tan \theta = \boxed{-}$$

$$\sin \theta \div \cos \theta = \frac{a}{c} \div \boxed{-} = \boxed{-}$$

So $\dfrac{\sin \theta}{\cos \theta} = \tan \theta$

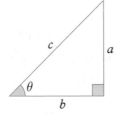

   b Check if this relationship holds for angles greater than 90°.
   Work out, with a calculator
   i $\dfrac{\sin 135°}{\cos 135°}$, $\tan 135°$
   ii $\dfrac{\sin 240°}{\cos 240°}$, $\tan 240°$
   iii $\dfrac{\sin 320°}{\cos 320°}$, $\tan 320°$

*15 Suppose the $\boxed{\cos}$ button on your calculator was broken, but the $\boxed{\sin}$ button was working. Explain how you could work out $\cos 40°$.

*16 Find all possible solutions between 0° and 360°.
   a $\sin x = \cos x$
   b $\sin 2x = \sin 60°$
   c $\tan 3x = 1$

> In part **c** there are six solutions.

*17 Draw the graph of $y = 2\sin x + 1$ for $0 \leqslant x \leqslant 180°$, taking
1 cm to 10° for x and 5 cm to 1 unit for y. Find approximate
solutions to these equations.

    a $2\sin x + 1 = 2 \cdot 3$ **b** $\dfrac{1}{(2\sin x + 1)} = 0 \cdot 5$

*18 Draw the graph of $y = 2\sin x + \cos x$ for $0 \leqslant x \leqslant 180°$, taking
1 cm to 10° for x and 5 cm to 1 unit for y.
    a   Solve these equations approximately.
        i $2\sin x + \cos x = 1 \cdot 5$
        ii $2\sin x + \cos x = 0$
    b   Estimate the maximum value of y.
    c   Find the value of x at which the maximum occurs.

*19 Draw the graph of $y = 3\cos x - 4\sin x$ for $0° \leqslant x \leqslant 220°$, taking
1 cm to 10° for x and 2 cm to 1 unit for y.
Solve these equations approximately.
    a $3\cos x - 4\sin x + 1 = 0$
    b $3\cos x = 4\sin x$

6.6   Sine and cosine rules

6.6.1   The sine rule

The sine rule enables you to calculate sides and angles in
some triangles where there is not a right angle.

In triangle ABC, use the convention that

a is the side opposite $\angle$A

b is the side opposite $\angle$B

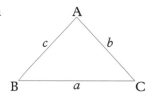

- Sine rule

    either $\dfrac{a}{\sin A} = \dfrac{b}{\sin B} = \dfrac{c}{\sin C}$ ... [1]

    or $\dfrac{\sin A}{a} = \dfrac{\sin B}{b} = \dfrac{\sin C}{c}$ ... [2]

    Use [1] when finding a **side**,
    and [2] when finding an **angle**.

You are given this
formula in the
examination.

To find a **side**,
have the **sides** on
top. To find an
angle, have the
angles on *top*.

a Find c.

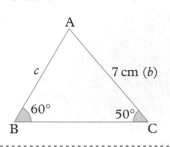

b Find d.

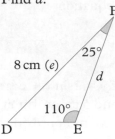

$$\frac{c}{\sin C} = \frac{b}{\sin B}$$

$$\frac{c}{\sin 50°} = \frac{7}{\sin 60°}$$

$$c = \frac{7 \times \sin 50°}{\sin 60°} = 6·19 \text{ cm (3 sf)}$$

First find $\angle D$.

$\angle D = 180° - (110° + 25°) = 45°$

$$\frac{d}{\sin D} = \frac{e}{\sin E}$$

$$\frac{d}{\sin 45°} = \frac{8}{\sin 110°}$$

$$d = \frac{8 \times \sin 45°}{\sin 110°} = 6·02 \text{ cm (3 sf)}$$

Exercise 19 Ⓗ

For questions **1** to **6**, find each side marked with a letter.

1

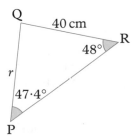

2

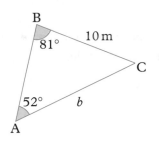

3

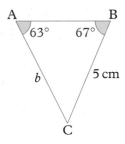

4

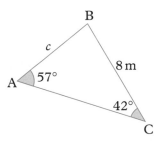

5

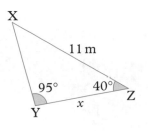

6

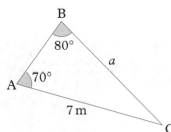

7 In △ABC, $\angle A = 61°$, $\angle B = 47°$, AC = 7·2 cm. Find BC.

8 In △XYZ, $\angle Z = 32°$, $\angle Y = 78°$, XY = 5·4 cm. Find XZ.

9 In △PQR, $\angle Q = 100°$, $\angle R = 21°$, PQ = 3·1 cm. Find PR.

10 In △LMN, $\angle L = 21°$, $\angle N = 30°$, MN = 7 cm. Find LN.

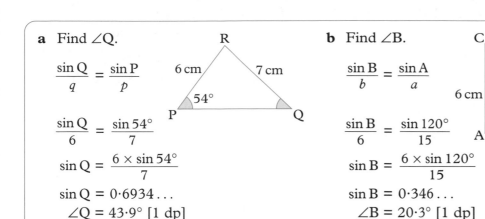

a Find ∠Q.

$$\frac{\sin Q}{q} = \frac{\sin P}{p}$$

$$\frac{\sin Q}{6} = \frac{\sin 54°}{7}$$

$$\sin Q = \frac{6 \times \sin 54°}{7}$$

$$\sin Q = 0.6934\ldots$$
$$\angle Q = 43.9° \text{ [1 dp]}$$

b Find ∠B.

$$\frac{\sin B}{b} = \frac{\sin A}{a}$$

$$\frac{\sin B}{6} = \frac{\sin 120°}{15}$$

$$\sin B = \frac{6 \times \sin 120°}{15}$$

$$\sin B = 0.346\ldots$$
$$\angle B = 20.3° \text{ [1 dp]}$$

EXAMPLE

Exercise 20 (H)

In questions **1** to **8**, find each angle marked *. All lengths are in centimetres.

1

2
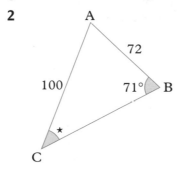

3
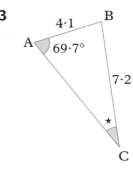

4
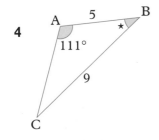

5
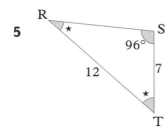

6
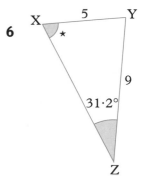

7
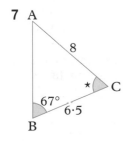

8
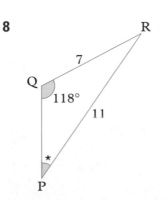

9 In △ABC, ∠A = 62°, BC = 8, AB = 7. Find ∠C.

10 In △XYZ, ∠Y = 97·3°, XZ = 22, XY = 14. Find ∠Z.

11 In △DEF, ∠D = 58°, EF = 7·2, DE = 5·4. Find ∠F.

12 In △LMN, ∠M = 127·1°, LN = 11·2, LM = 7·3. Find ∠L.

*13 The sine rule can be ambiguous.

The diagram shows two possible triangles with AB = 5·4, ∠A = 32° and BC = 3. Find the two possible values of angle C.

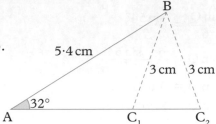

6.6.2 Cosine rule

- You use the **cosine** rule when you know either
  a two sides and the included angle or
  b all three sides.

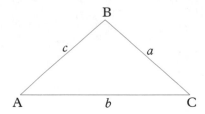

There are two forms.

1 To find the length of a side.

$$a^2 = b^2 + c^2 - (2bc \cos A)$$

or $\quad b^2 = c^2 + a^2 - (2ac \cos B)$

or $\quad c^2 = a^2 + b^2 - (2ab \cos C)$

2 To find an angle when given all three sides.

$$\cos A = \frac{b^2 + c^2 - a^2}{2bc}$$

or $\quad \cos B = \dfrac{a^2 + c^2 - b^2}{2ac}$

or $\quad \cos C = \dfrac{a^2 + b^2 - c^2}{2ab}$

> You are given a version of this formula in the examination. There is a proof of the rule in Exercise 23, question **18**, on page 335.

For an obtuse angle x, $\cos x = -\cos(180° - x)$

Examples $\cos 120° = -\cos 60°$
$\cos 142° = -\cos 38°$

a Find *b*.

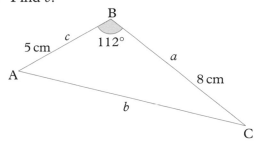

b Find angle C.

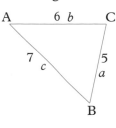

. . . .

$b^2 = a^2 + c^2 - (2ac \cos B)$

$b^2 = 8^2 + 5^2 - (2 \times 8 \times 5 \times \cos 112°)$

$b^2 = 64 + 25 - [80 \times (-0·3746)]$

$b^2 = 64 + 25 + 29·968$

(Notice the change of sign for the obtuse angle.)

$b = \sqrt{118·968} = 10·9$ cm (to 3 sf)

$\cos C = \dfrac{a^2 + b^2 - c^2}{2ab}$

$\cos C = \dfrac{5^2 + 6^2 - 7^2}{2 \times 5 \times 6} = \dfrac{12}{60} = 0·200$

$\angle C = 78·5°$ (to 1 dp)

Exercise 21 Ⓗ

Find the sides marked ⋆. All lengths are in cm.

1
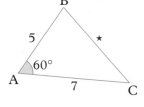

2
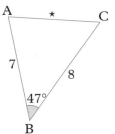

3
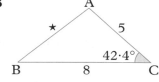

4
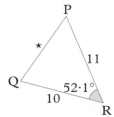

5
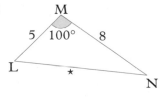

6
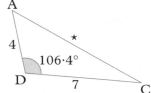

7 In △ABC, AB = 4 cm, AC = 7 cm, ∠A = 57°. Find BC.

8 In △XYZ, XY = 3 cm, ∠YZ = 3 cm, Y = 90°. Find XZ.

9 In △LMN, LM = 5·3 cm, MN = 7·9 cm, ∠M = 127°. Find LN.

10 In △PQR, ∠Q = 117°, PQ = 80 cm, QR = 100 cm. Find PR.

In questions **11** to **16**, find the angles marked ⋆.

11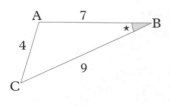

12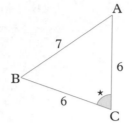

13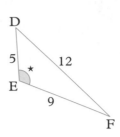

14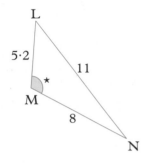

15

16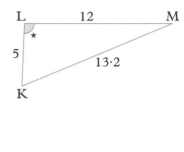

17 In △ABC, $a = 4 \cdot 3$, $b = 7 \cdot 2$, $c = 9$.
Find ∠C.

18 In △DEF, $d = 30$, $e = 50$, $f = 70$.
Find ∠E.

19 In △PQR, $p = 8$, $q = 14$, $r = 7$.
Find ∠Q.

20 In △LMN, $l = 7$, $m = 5$, $n = 4$.
Find ∠N.

6.6.3 Problems involving the sine and cosine rules

EXAMPLE

A ship sails from port P for a distance of 7 km on a bearing of 306°
and then a further 11 km on a bearing of 070° to arrive at X.
Calculate the distance from P to X.

You know two sides and the
included angle, so use
the cosine rule.

$PX^2 = 7^2 + 11^2$

$\quad - (2 \times 7 \times 11 \times \cos 56°)$

$\quad = 49 + 121 - (86 \cdot 12)$

$PX^2 = 83 \cdot 88$

$\quad PX = 9 \cdot 16$ km (to 3 sf)

The distance from P to X is 9·16 km.

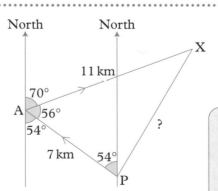

Remember:
Bearings are
measured
clockwise from
north.

Exercise 22 (H)

1 Ship B is 58 km south-east of Ship A.
Ship C is 70 km due south of Ship A.
   a How far is Ship B from Ship C?
   b What is the bearing of Ship B from
      Ship C?

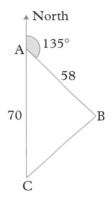

2 A destroyer D and a cruiser C leave port P at the same time.
The destroyer sails 25 km on a bearing 040° and the cruiser sails
30 km on a bearing of 320°. How far apart are the ships?

3 Two honeybees A and B leave the hive, H, at the same time.
A flies 27 m due south and B flies 9 m on a bearing of 111°.
How far apart are they?

4 Find the sides and angles marked with letters. All lengths are
in cm.

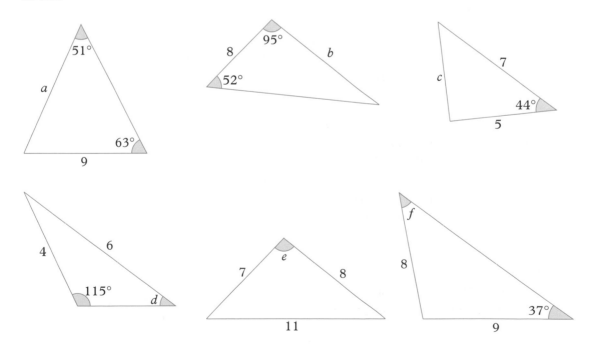

5 Find the largest angle in a triangle in which the sides are in the
ratio 5 : 6 : 8.

6 A golfer hits his ball a distance of 170 m on a hole which measures 195 m from the tee, T, to the hole. If his shot is directed 10° away from the direct line to the hole, find the distance between his ball and the hole.

7 Mr Gibson has taken up golf with alarming consequences for the local tree population. The diagram shows a typical effort, even before he has reached the 19th hole. His shot travels from the tee, hits an ash tree then an oak tree and then hits him on the head. How far is our hero from the oak tree?

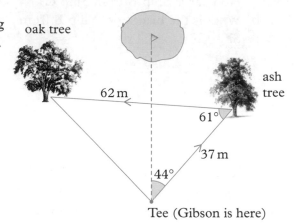

oak tree

62 m

61°

ash tree

37 m

44°

Tee (Gibson is here)

8 A rhombus has sides of length 8 cm and angles of 50° and 130°. Find the length of the longer diagonal of the rhombus.

9 From A, B lies 11 km away on a bearing of 041° and C lies 8 km away on a bearing of 341°. Find
  a the distance between B and C
  b the bearing of B from C.

10 From a lighthouse, L, an aircraft carrier, A, is 15 km away on a bearing of 112° and a submarine, S, is 26 km away on a bearing of 200°. Find
  a the distance between A and S
  b the bearing of A from S.

11 The diagram show three towns A, B, C.
  The bearing of B from A is 110°.
  The bearing of C from A is 160°.
  The bearing of C from B is 240°.
  The distance from C to B is 110 km.
  Find the distance from A to B.

12 a Find AE.
  b Find ∠EAC.
  c If the line BCD is horizontal,
    find the angle of elevation of E from A.

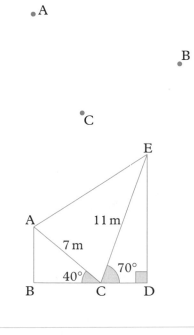

13 An aircraft flies from its base 200 km on a bearing 162°, then 350 km on a bearing 260°, and then returns directly to base. Calculate the length and bearing of the return journey.

14 Town Y is 9 km due north of town Z. Town X is 8 km from Y, 5 km from Z and somewhere to the west of the line YZ.
   a Draw triangle XYZ and find angle YZX.
   b During an earthquake, town X moves due south until it is due west of Z. Find how far it has moved.

15 The diagram shows a point, A, which lies 10 km due south of a point, B. A straight road AD is such that the bearing of D from A is 043°. P and Q are two points on this road which are both 8 km from B. Calculate the bearing of P from B.

> Find angle BPA first.

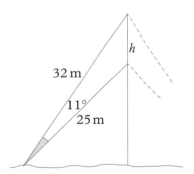

16 The diagram shows wires attached to a communications antenna. Find the length h, correct to the nearest metre.

17 In a triangle ABC, $a = \sqrt{28}$, $b = 6$ and $A = 60°$.
   a Use the cosine rule to write a quadratic equation involving c.
   b Solve the equation to find two values of c.
   c Use the sine rule to find two values of c independently. Compare your answers with those obtained in part **b**.

> For part **c**, find angle B first.

18 Here is a proof of the cosine rule.

   a Using $\triangle$ADC show that $x = b\cos C$.

   b Using Pythagoras' theorem in $\triangle$ADC, find h^2 in terms of b and x.

   c Using $\triangle$ABD find c^2 in terms of h, a and x.

   d Use your results to prove that $c^2 = a^2 + b^2 - 2ab\cos C$.

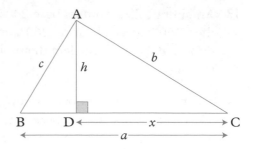

19 Calculate WX, given YZ = 15 m.

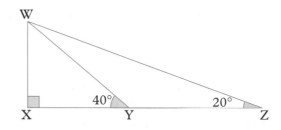

*20 Find angle DCB. Check your answer by making a scale drawing.

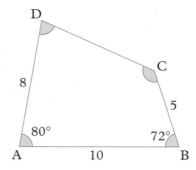

*21 Find **a** WX   **b** WZ   **c** WY.

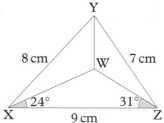

*22 The diagram shows a cube of side 10 cm from which one corner has been cut. Calculate the angle PQR.

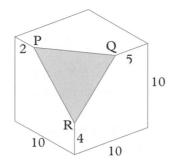

6.7 Circle theorems

6.7.1 Angles in circles

- **Theorem 1**
  The angle subtended at the centre of a circle is twice the angle subtended at the circumference.

- **Theorem 2**
  Angles subtended by an arc in the same segment of a circle are equal.

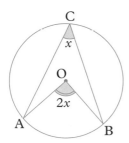

$\angle AOB = 2 \times \angle ACB$

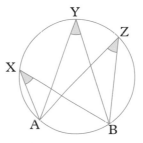

$\angle AXB = \angle AYB = \angle AZB$

A proof of Theorem 1 is given in Chapter 10, Section 10.2.

EXAMPLE

a Given $\angle ABO = 50°$, find $\angle BCA$.

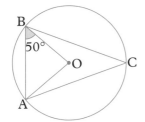

b Given $\angle BDC = 62°$ and $\angle DCA = 44°$ find $\angle BAC$ and $\angle ABD$.

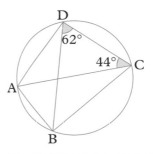

a Triangle OBA is isosceles (OA = OB).
So $\angle OAB = 50°$
and $\angle BOA = 80°$ (angle sum of a triangle)
So $\angle BCA = 40°$ (angle at the circumference)

b $\angle BDC = \angle BAC$
(both subtended by arc BC)
So $\angle BAC = 62°$
$\angle DCA = \angle ABD$
(both subtended by arc DA)
So $\angle ABD = 44°$

Exercise 23 Ⓜ/Ⓗ

Find the angles marked with letters. A line passes through the centre
only when point O is shown.

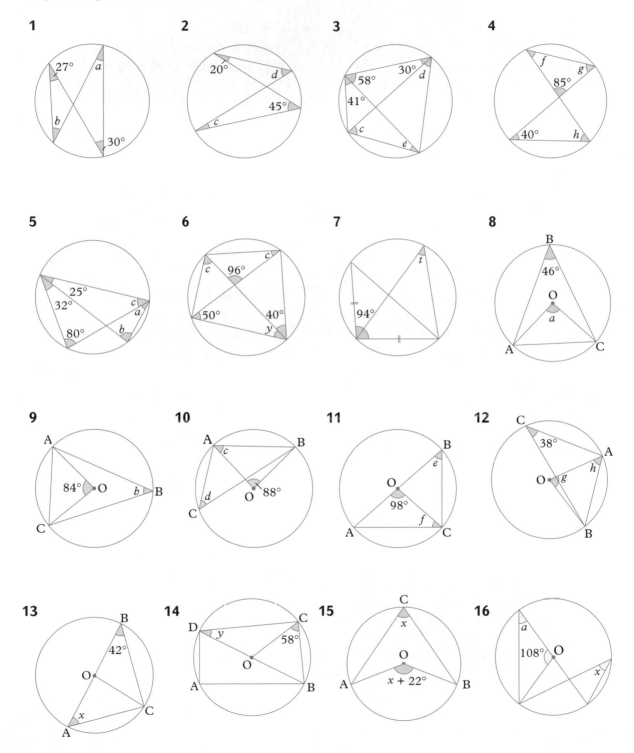

● **Theorem 3**
   The opposite angles in a cyclic quadrilateral add up to 180°
   (the angles are supplementary).

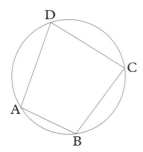

$$\angle A + \angle C = 180°$$
$$\angle B + \angle D = 180°$$

There is a proof of this result in Chapter 10, section 10.2.

EXAMPLE

Find a and x.

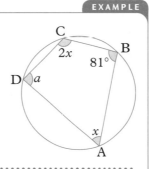

....

$a = 180° - 81°$ (opposite angles of a cyclic quadrilateral)
So $a = 99°$

$x + 2x = 180°$ (opposite angles of a cyclic quadrilateral)
      $3x = 180°$
So $x = 60°$

● **Theorem 4**
   The angle in a semicircle is a right angle.

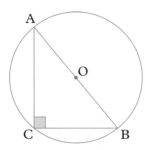

In the diagram,
AB is a diameter.
$\angle ACB = 90°$.

EXAMPLE

Find *b* given that AOB is a diameter.

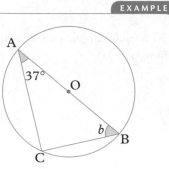

∠ACB = 90° (angle in a semicircle)
  So *b* = 180° − (90 + 37)°
      = 53°

Exercise 24 Ⓜ/Ⓗ

Find the angles marked with a letter.

1

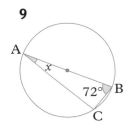

2

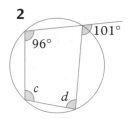

3

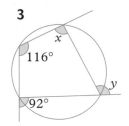

4

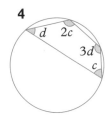

5

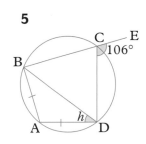

6

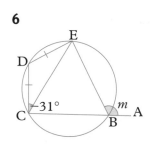

7

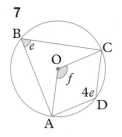

8

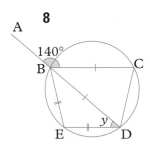

9

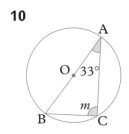

10

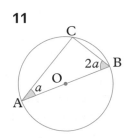

11

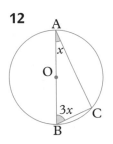

12

13

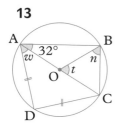

14

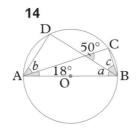

15

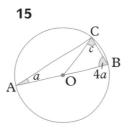

16

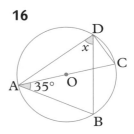

17

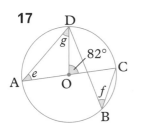

18

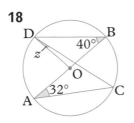

19

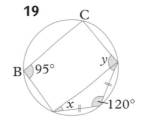

20

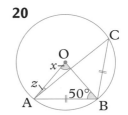

6.7.2 Tangents to circles

- **Theorem 1**
  The angle between a tangent and the radius $\angle TAO = 90°$
  drawn to the point of contact is 90°. $TA = TB$

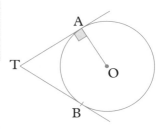

- **Theorem 2**
  From any point outside a circle just two tangents to the circle
  can be drawn and they are of equal length.

- **Theorem 3 – Alternate segment theorem**
  The angle between a tangent and a chord $\angle TAB = \angle BCA$
  through the point of contact is equal to $\angle SAC = \angle CBA$
  the angle subtended by the chord in the
  alternate segment.
  The theorem is proved in Question 16 of Exercise 26.

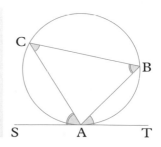

EXAMPLE

TA and TB are tangents to the circle, centre O.
Given ∠ATB = 50°, find

a ∠ABT   **b** ∠OBA   **c** ∠ACB.

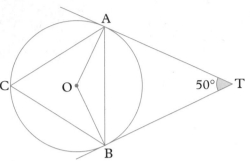

a △TBA is isosceles (TA = TB)

So ∠ABT = $\frac{1}{2}$(180° − 50°) = 65°

b ∠OBT = 90°
(tangent and radius)

So ∠OBA = 90° − 65° = 25°

c ACB = ∠ABT
(alternate segment theorem)

∠ACB = 65°

Exercise 25 Ⓜ/Ⓗ

Find the angles marked with letters.

1

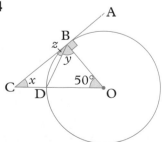

2

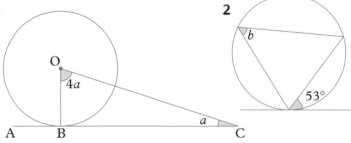

3

4

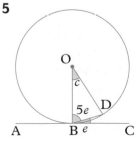

5

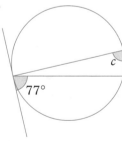

6

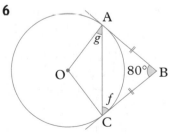

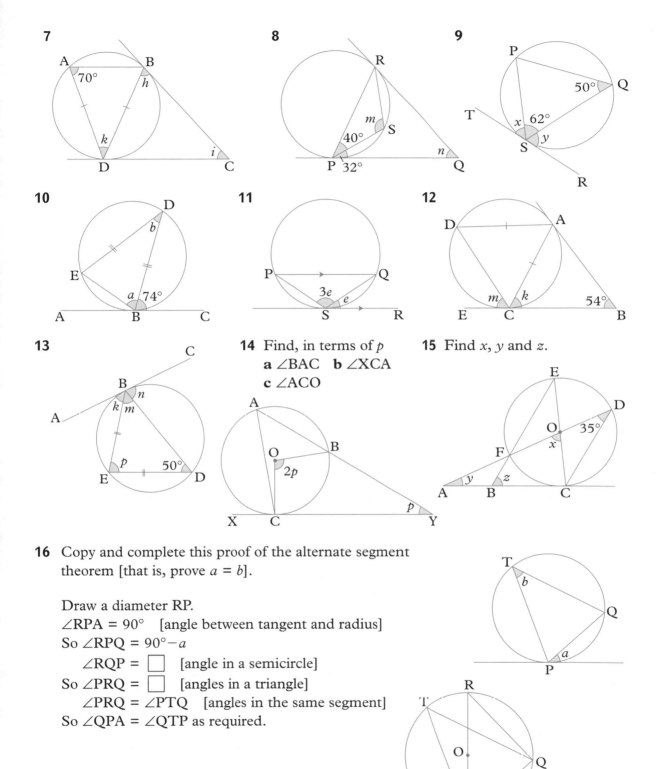

7

8

9

10

11

12

13

14 Find, in terms of p
 a ∠BAC **b** ∠XCA
 c ∠ACO

15 Find x, y and z.

16 Copy and complete this proof of the alternate segment theorem [that is, prove $a = b$].

Draw a diameter RP.
∠RPA = 90°   [angle between tangent and radius]
So ∠RPQ = 90° − a
   ∠RQP = ☐   [angle in a semicircle]
So ∠PRQ = ☐   [angles in a triangle]
   ∠PRQ = ∠PTQ   [angles in the same segment]
So ∠QPA = ∠QTP as required.

Exercise 26 (M)/(H) **(Mixed questions)**

Find the angles marked with letters. The centre of the circle is O.

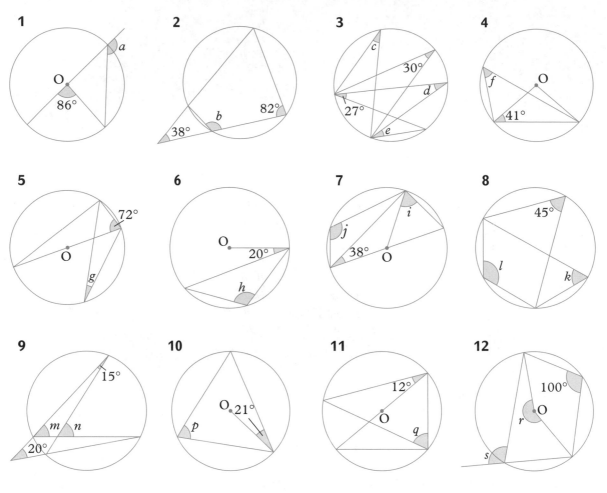

1

2

3

4

5

6

7

8

9

10

11

12

Questions **13** to **21** involve tangents to the circles.

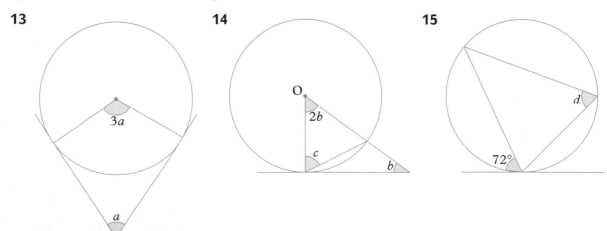

13

14

15

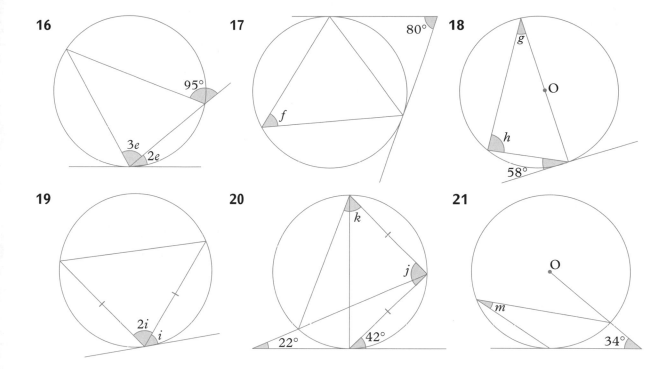

16 95° 3e 2e

17 80° f

18 g O h 58°

19 2i i

20 k j 22° 42°

21 O m 34°

Test yourself

1 Here are the front elevation, side elevation and the plan of a 3-D shape.

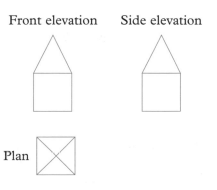

Front elevation   Side elevation

Plan

Draw a sketch of the 3-D shape.

(Edexcel, 2008)

2 a Calculate the length x.

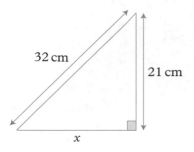

Diagram **NOT** accurately drawn

b Calculate the length y.

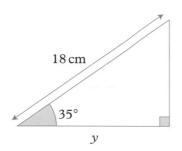

Diagram **NOT** accurately drawn

(AQA, 2007)

3 The diagram shows a cuboid with length 9 cm, width 4 cm and height 5 cm.

a Show that the length of the diagonal AB is 11·05 cm, correct to 2 decimal places.

b Calculate the angle between the diagonal AB and the base.

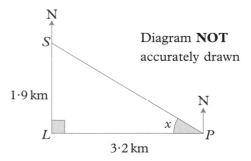

(OCR, 2004)

4 A lighthouse, L, is 3·2 km due West of a port, P. A ship, S, is 1·9 km due North of the lighthouse, L.

a Calculate the size of the angle marked x. Give your answer correct to 3 significant figures.

b Find the bearing of the port, P, from the ship, S. Give your answer correct to 3 significant figures.

(Edexcel, 2005)

5 Triangle **A** is reflected in the x-axis to give triangle **B**.
Triangle **B** is reflected in the line $x = 1$ to give triangle **C**.

Describe the **single** transformation that takes triangle **A** to
triangle **C**.

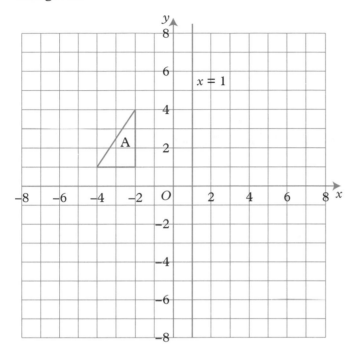

(Edexcel, 2008)

6 a Calculate h.
  b Calculate angle x.

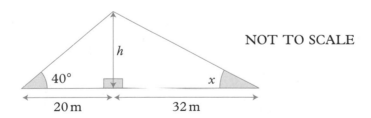

NOT TO SCALE

(OCR, 2008)

7 Shape P is enlarged with scale factor 3 and centre (2, 1) onto shape Q.
  a Describe the enlargement that would map shape Q onto shape P.
    Shape Q has an area of 36 cm^2.
  b What is the area of shape P?

(OCR, 2005)

8 Describe fully the single transformation which will map triangle **A** onto triangle **B**.

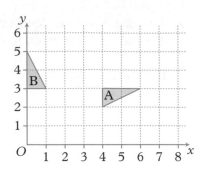

9 **a** Draw x- and y-axes with values from -5 to 5.

b Plot the points A$(0, 0)$, B$(2, 4)$, C$(4, 0)$ and draw triangle ABC.

c Draw an enlargement of triangle ABC with scale factor $1\frac{1}{2}$ and centre of enlargement $(4, 4)$.

10 ABCD is a trapezium.

$\overrightarrow{AD} = 3\overrightarrow{BC}$, $\overrightarrow{AB} = \mathbf{p}$ and $\overrightarrow{BC} = \mathbf{q}$.

Express in terms of $\mathbf{p}$ and $\mathbf{q}$
a $\overrightarrow{AD}$
b $\overrightarrow{CD}$

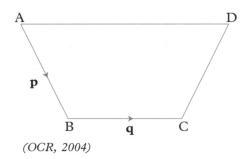

(OCR, 2004)

11 *ABCDEF* is a regular hexagon with centre *O*.
$\overrightarrow{OA} = \mathbf{a}$ and $\overrightarrow{AB} = \mathbf{b}$

a Find expressions, in terms of $\mathbf{a}$ and $\mathbf{b}$, for
i $\overrightarrow{OB}$
ii $\overrightarrow{AC}$
iii $\overrightarrow{EC}$

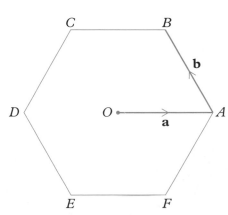

Diagram drawn accurately

b The positions of points *P* and *Q* are given by the vectors

$\overrightarrow{OP} = \mathbf{a} - \mathbf{b}$ $\qquad \overrightarrow{OQ} = \mathbf{a} + 2\mathbf{b}$

i Draw and label the positions of points *P* and *Q* on the diagram.

ii Hence, or otherwise, deduce an expression for $\overrightarrow{PQ}$.

(AQA, 2004)

12 The diagram show a square *OAPB*.
  M is the mid-point of *AP*.
  N is the mid-point of *BM*.
  AP is extended to *Q* where $AQ = 1\frac{1}{2} AP$

  $\overrightarrow{OA} = \mathbf{a}$ and $\overrightarrow{OB} = \mathbf{b}$

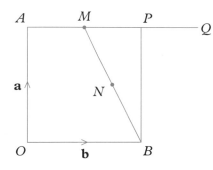

  Not drawn accurately

  Write these vectors in terms of **a** and **b**.
  Give your answers in their simplest form.

  a $\overrightarrow{OQ}$ **b** $\overrightarrow{BM}$

  c $\overrightarrow{BN}$ **d** $\overrightarrow{ON}$

  (AQA, 2007)

13 ABCD is a quadrilateral.
  BD is perpendicular to DA.
  AB = BC = 8 cm.
  Angle DAB = angle DBC = 54°.

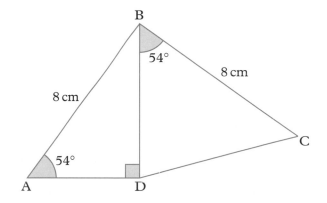

  a Show that BD = 6·47 cm, correct to 2 decimal places.
  b Calculate the area of triangle BCD.
  c Calculate CD.

  (OCR, 2004)

14 a The diagram shows the graph of $y = \cos x°$.

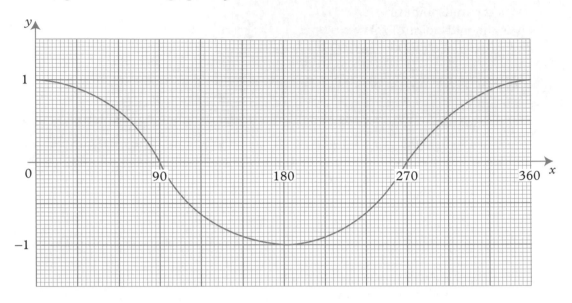

Use the graph to solve these equations for $0° \leqslant x \leqslant 360°$.

i $\cos x° = -0.4$

ii $4 \cos x° = 3$

b Sketch the graph of $y = \dfrac{1}{2} \sin x°$ for $0° \leqslant x \leqslant 360°$.

(OCR, 2004)

15 A parallelogram $ABCD$ is comprised of two right-angled triangles ACB and ACD.
AB is 90 cm and BC is 54 cm.
Calculate:

a the length of AC

b the length of BD.

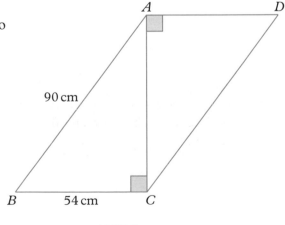

(CCEA)

16 A parallelogram has sides 5·2 cm and 9·5 cm.
The longer diagonal is 12·6 cm.
Calculate the size of an obtuse angle of the parallelogram.

(OCR, 2008)

17 The diagram shows a circle with centre O.
ABCD is a cyclic quadrilateral.
Angle AOC = 108°.
  a Prove that angle ADC is 126°.
  b AE is the tangent at A.
     Find angle CAE giving your reasons.

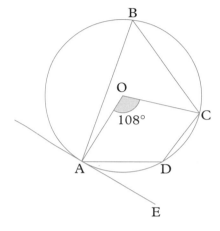

(OCR, 2004)

18 P, Q, R and S are points on the circumference of a circle.
PR is a diameter of the circle. Angle QPR = 47°.
  a Write down the value of *a*.
  b Calculate the value of *b*.

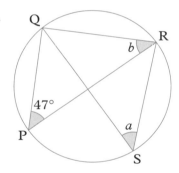

19 A O is the centre and AB
   is a diameter of the circle.
   OD is parallel to BC, and
   angle BAC = 56°.
   Calculate the size of:
   a angle ABC
   b angle AOD
   c angle ACD
   d angle OAD.

  B The tangents to the circle
   at B and C meet at T.
   Calculate the size of
   e angle BOT.

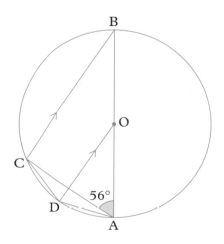

(CCEA)

20 The diagram shows two triangles PQR and PRS
with PQ = 48 cm, QR = 36 cm, ∠PQR = 19°,
∠SPR = 46° and ∠PSR = 110°.

Find the length of RS.

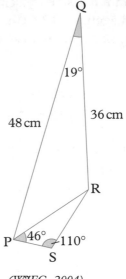

(WJEC, 2004)

7 Algebra 3

In this unit you will:
- revise the rules of indices
- learn about surds
- solve equations involving indices
- solve quadratic equations by factorising, by using the formula, and by completing the square
- use quadratic equations to solve problems
- add, subtract and simplify algebraic fractions
- transform functions.

Links

Algebra lets you describe the path of a ball as it is thrown through the air or a shape of a chain hung between two posts using quadratic and other trigonometric equations.

7.1 Indices

7.1.1 Six basic rules

- $a^n \times a^m = a^{n+m}$

  $7^2 \times 7^4 = (7 \times 7) \times (7 \times 7 \times 7 \times 7) = 7^6$

  so $\quad 7^2 \times 7^4 = 7^{2+4} = 7^6$

- $a^n \div a^m = a^{n-m}$

  $3^5 \div 3^2 = \dfrac{3 \times 3 \times 3 \times \cancel{3} \times \cancel{3}}{\cancel{3} \times \cancel{3}} = 3^3$

  so $\quad 3^5 \div 3^2 = 3^{5-2} = 3^3$

- $(a^n)^m = a^{nm}$

  $(2^2)^3 = (2 \times 2)(2 \times 2)(2 \times 2) = 2^6$

  so $\quad (2^2)^3 = 2^{2 \times 3} = 2^6$

- $a^{-n} = \dfrac{1}{a^n}$

  Consider this sequence

2^3	2^2	2^1	2^0	2^{-1}	2^{-2}
$\downarrow$	$\downarrow$	$\downarrow$	$\downarrow$	$\downarrow$	$\downarrow$
8	4	2	1	$\frac{1}{2}$	$\frac{1}{4}$

  Check: $5^3 \div 5^5 = \dfrac{\cancel{5} \times \cancel{5} \times \cancel{5}}{\cancel{5} \times \cancel{5} \times \cancel{5} \times 5 \times 5}$

  $\qquad\qquad = \dfrac{1}{5^2} = 5^{-2}$

- $a^{\frac{1}{n}}$ means 'the nth root of a'

  $3^{\frac{1}{2}} \times 3^{\frac{1}{2}} = \sqrt{3} \times \sqrt{3}$

  Using Rule 1:

  $3^{\frac{1}{2}} \times 3^{\frac{1}{2}} = 3^1$

- $a^{\frac{m}{n}}$ means 'the nth root of a raised to the power m'.

  $4^{\frac{3}{2}} = (\sqrt{4})^3 = 8$

This is the graph of $y = 2^x$.

You can use a calculator with a button marked $\boxed{x^y}$ or $\boxed{\wedge}$ to work out difficult powers. For example, for $2^{1.7}$ press $\boxed{2}$ $\boxed{\wedge}$ $\boxed{1.7}$ $\boxed{=}$

$2^{1.7} = 3.249$ (4 sf)

Some calculators have a button marked $\boxed{x^{\frac{1}{y}}}$ to work out roots of numbers. For example, the fourth root of 100 is $100^{\frac{1}{4}}$

Press $\boxed{100}$ $\boxed{x^{\frac{1}{y}}}$ $\boxed{4}$ $\boxed{=}$

$100^{\frac{1}{4}} = 3.162$ (4 sf)

You could press $\boxed{100}$ $\boxed{x^y}$ $\boxed{0.25}$ $\boxed{=}$ instead.

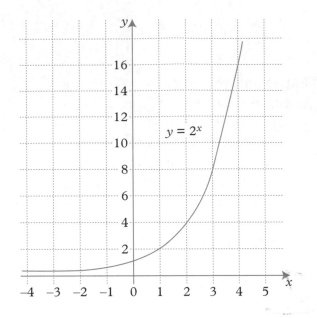

EXAMPLE

a $5 \times 5 \times 5 \times 5 = 5^4$

b $2 \times 2 \times 7 \times 7 \times 7 \times 7 = 2^2 \times 7^4$

c $6^3 \times 6^4 = 6^{3+4} = 6^7$

d $5^{10} \div 5^2 = 5^{10-2} = 5^8$

e Half of $2^8 = 2^8 \div 2^1 = 2^7$

f $18^0 = 11^0 = (-3)^0 = 1$

g $8^{-2} = \dfrac{1}{8^2} = \dfrac{1}{64}$

h $64^{\frac{1}{3}} = \sqrt[3]{64} = 4$

Exercise 1 Ⓗ

Express these in index form.

1 $3 \times 3 \times 3 \times 3$

2 $4 \times 4 \times 5 \times 5 \times 5$

3 $3 \times 7 \times 7 \times 7$

4 $2 \times 2 \times 2 \times 7$

5 $\dfrac{1}{10^3}$

6 $\dfrac{1}{5^4}$

7 $\sqrt{15}$

8 $\sqrt{1000}$

9 $a \times a \times a \times a \times a \times a \times a \times a$

10 $\sqrt[3]{17}$

11 $\dfrac{1}{11}$

12 $(\sqrt{5})^3$

In questions **13** to **32** answer 'true' or 'false.'

13 $2^3 = 8$

14 $3^2 = 6$

15 $5^3 = 125$

16 $2^{-1} = \dfrac{1}{2}$

17 $10^{-2} = \dfrac{1}{20}$

18 $3^{-3} = \dfrac{1}{9}$

19 $2^2 > 2^3$

20 $2^3 < 3^2$

21 $2^{-2} > 2^{-3}$

22 $3^{-2} < 3$

23 $1^9 = 9$

24 $-3^2 = -9$

25 $5^{-2} = \dfrac{1}{10}$ **26** $10^{-3} = \dfrac{1}{1000}$ **27** $10^{-2} > 10^{-3}$ **28** $5^{-1} = 0 \cdot 2$

29 $10^{-1} = 0 \cdot 1$ **30** $2^{-2} = 0 \cdot 25$ **31** $5^0 = 1$ **32** $16^0 = 0$

Write in a simpler form.

33 $5^2 \times 5^4$ **34** $6^3 \times 6^2$ **35** $2^3 \times 2^{10}$ **36** $3^6 \times 3^{-2}$

37 $6^7 \div 6^2$ **38** $8^5 \div 8^4$ **39** $6^2 \div 6^{-2}$ **40** Half of 2^6

41 Half of 2^{20} **42** $10^{100} \div 10^0$ **43** $5^{-2} \times 5^{-3}$ **44** $3^{-2} \times 3^{-8}$

45 $\dfrac{3^4 \times 3^5}{3^2}$ **46** $\dfrac{2^8 \times 2^4}{2^5}$ **47** $\dfrac{7^3 \times 7^3}{7^4}$ **48** $\dfrac{5^9 \times 5^{10}}{5^{20}}$

7.1.2   Fractional and negative indices

EXAMPLE

Evaluate these quantities.

a $9^{\frac{1}{2}} = \sqrt{9} = 3$

b $5^{-1} = \dfrac{1}{5}$

c $4^{-\frac{1}{2}} = \dfrac{1}{4^{\frac{1}{2}}} = \dfrac{1}{\sqrt{4}} = \dfrac{1}{2}$

d $25^{\frac{3}{2}} = (\sqrt{25})^3 = 5^3 = 125$

$a^0 = 1$ for any non-zero value of a.

e $(5^{\frac{1}{2}})^3 \times 5^{\frac{1}{2}} = 5^{\frac{3}{2}} \times 5^{\frac{1}{2}} = 5^2 = 25$

f $7^0 = 1$

Exercise 2 (H)

Evaluate these quantities.

1 $3^2 \times 3$ **2** 100^0 **3** 3^{-2}

4 $(5^{-1})^{-2}$ **5** $4^{\frac{1}{2}}$ **6** $16^{\frac{1}{2}}$

7 $81^{\frac{1}{2}}$ **8** $8^{\frac{1}{3}}$ **9** $9^{\frac{3}{2}}$

10 $27^{\frac{1}{3}}$ **11** $9^{-\frac{1}{2}}$ **12** $8^{-\frac{1}{3}}$

13 $1^{\frac{5}{2}}$ **14** $25^{-\frac{1}{2}}$ **15** $1000^{\frac{1}{3}}$

16 $2^{-2} \times 2^5$ **17** $2^4 \div 2^{-1}$ **18** $8^{\frac{2}{3}}$

19   Copy and complete the crossnumber puzzle.

Across	*Down*
1 $\sqrt{2^8}$	**1** $(1\,000\,000)^{\frac{1}{6}}$
2 $169^{\frac{1}{2}}$	**3** $3^3 + 2^2 + 1^1$
5 $7^2 \times 5^0$	**4** $10^2 - 7^0$
6 $2^4 \times 5$	**5** $20^2 + 2^1$
7 3×2^5	**6** $3^7 \div 27$
9 5^3	**8** Half of 2^7

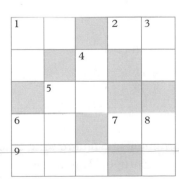

20 Find the missing index.

    a $\sqrt{80} = 80^{\square}$ **b** $\dfrac{1}{\sqrt{3}} = 3^{\square}$ **c** $(\sqrt{2})^3 = 2^{\square}$

21 Work out

    a $27^{-\frac{2}{3}}$ **b** $4^{-\frac{3}{2}}$ **c** One per cent of 100^7

Evaluate these quantities.

22 $10\,000^{\frac{1}{4}}$ **23** $100^{\frac{3}{2}}$ **24** $(100^{\frac{1}{2}})^{-3}$

25 $(9^{\frac{1}{2}})^{-2}$ **26** $(-16\cdot371)^0$ **27** $81^{\frac{1}{4}} \div 16^{\frac{1}{4}}$

28 $(5^{-4})^{\frac{1}{2}}$ **29** $1000^{-\frac{1}{3}}$ **30** $(4^{-\frac{1}{2}})^2$

31 $8^{-\frac{2}{3}}$ **32** $100^{\frac{5}{2}}$ **33** $1^{\frac{4}{3}}$

34 2^{-5} **35** $(0\cdot01)^{\frac{1}{2}}$ **36** $(0\cdot04)^{\frac{1}{2}}$

37 $(2\cdot25)^{\frac{1}{2}}$ **38** $(7\cdot63)^0$ **39** $3^5 \times 3^{-3}$

40 $\left(3\dfrac{3}{8}\right)^{\frac{1}{3}}$ **41** $\left(11\dfrac{1}{9}\right)^{-\frac{1}{2}}$ **42** $\left(\dfrac{1}{8}\right)^{-2}$

43 $\left(\dfrac{1}{1000}\right)^{\frac{2}{3}}$ **44** $\left(\dfrac{9}{25}\right)^{-\frac{1}{2}}$ **45** $(10^{-6})^{\frac{1}{3}}$

46 a Copy and complete the sentence.

    'The inverse operation of raising a positive number to power n is raising the result of this operation to power $\square$.'

  b Write two numerical examples to illustrate this result.

7.1.3 Surds

- Numbers like $\sqrt{3}$, $\sqrt{4}$, $\sqrt{11}$ are called **surds**. The following rules apply:

$$\sqrt{a} \times \sqrt{b} = \sqrt{ab} \qquad\qquad \frac{\sqrt{a}}{\sqrt{b}} = \sqrt{\frac{a}{b}}$$

EXAMPLE

  a $\sqrt{2} \times \sqrt{3} = \sqrt{6}$ **c** $\dfrac{\sqrt{27}}{\sqrt{3}} = \sqrt{\dfrac{27}{3}} = \sqrt{9} = 3$

  b $\sqrt{80} = \sqrt{16 \times 5}$

        $= \sqrt{16} \times \sqrt{5}$

        $= 4\sqrt{5}$

- It is sometimes helpful to write an expression with a denominator which is a integer. This is called rationalising the denominator.

Rationalise the denominators.

a $\dfrac{6}{\sqrt{3}}$ **b** $\dfrac{4}{\sqrt{3}-1}$

When you multiply the numerator and the denominator by the same number, the fraction is not changed.

a $\dfrac{6}{\sqrt{3}} = \dfrac{6 \times \sqrt{3}}{\sqrt{3} \times \sqrt{3}}$

$= \dfrac{6\sqrt{3}}{3}$

$= 2\sqrt{3}$

b $\dfrac{4}{\sqrt{3}-1} = \dfrac{4(\sqrt{3}+1)}{(\sqrt{3}-1)(\sqrt{3}+1)}$

$= \dfrac{4(\sqrt{3}+1)}{3-1}$

$= \dfrac{4(\sqrt{3}+1)}{2} = 2(\sqrt{3}+1)$

- A common mistake occurs with surds. $\sqrt{4}$ means 'the positive square root of 4'.

  So $\sqrt{4} = 2$ only and *not* ± 2.

  Notice that the solutions of the equation $x^2 = 4$ are $x = 2, -2$.
  You *can* write $x = \sqrt[+]{4}$ or $\sqrt{4}$.

Exercise 3 (H)

1 Sort these into four pairs of equal value.

A $\sqrt{20}$ B $\sqrt{12}$ C $\sqrt{3}$ D $2\sqrt{3}$

E 2 F $\sqrt{10} \times \sqrt{2}$ G $\dfrac{\sqrt{20}}{\sqrt{5}}$ H $\dfrac{\sqrt{15}}{\sqrt{5}}$

2 Answer 'True' or 'False'.

a $\sqrt{28} = 2\sqrt{7}$ **b** $(\sqrt{8})^2 = 8$ **c** $\sqrt{16} = \pm 4$

d $\sqrt{\dfrac{39}{3}} = \sqrt{13}$ **e** $\sqrt{4} + \sqrt{4} = \sqrt{8}$ **f** If $x^2 = 9$, $x = \pm 3$

3 Remove the brackets and simplify.

a $(1+\sqrt{2})^2$ **b** $(2-\sqrt{3})^2$ **c** $(\sqrt{2}+\sqrt{8})^2$

d $(3-\sqrt{5})^2$ **e** $(\sqrt{2}+2)^2$ **f** $(\sqrt{18}-\sqrt{2})^2$

4 Answer 'True' or 'False'.

a $\sqrt{2} + \sqrt{2} = 2\sqrt{2}$ **b** $\dfrac{\sqrt{8}}{2} = \sqrt{2}$

c $\sqrt{100} = \pm 10$ **d** $(\sqrt{2})^4 = 4$

e $\sqrt{9+16} = \sqrt{9} + \sqrt{16}$ **f** $(1+\sqrt{3})^2 = 4 + 2\sqrt{3}$

5 Without using a calculator, simplify these. Write your answers using surds where necessary.

a $\sqrt{8} \times \sqrt{2}$ **b** $\sqrt{32}$ **c** $\sqrt{300}$

d $\sqrt{18} \times 3$ **e** $\sqrt{5} + 4\sqrt{5}$ **f** $7\sqrt{3} - 2\sqrt{3}$

g $\sqrt{20} + \sqrt{45}$ **h** $\sqrt{75} - \sqrt{48}$ **i** $\dfrac{\sqrt{8}}{\sqrt{2}}$

j $\dfrac{\sqrt{27}}{\sqrt{12}}$ **k** $\dfrac{\sqrt{125}}{\sqrt{20}}$ **l** $\dfrac{\sqrt{80}}{\sqrt{45}}$

6 a Work out $\dfrac{6}{\sqrt{3}} \times \dfrac{\sqrt{3}}{\sqrt{3}}$.

b Rationalise the denominators of these fractions.

  i $\dfrac{8}{\sqrt{2}}$ **ii** $\dfrac{12}{\sqrt{3}}$ **iii** $\dfrac{12}{\sqrt{8}}$ **iv** $\dfrac{2}{\sqrt{3}}$

7 The denominator of the fraction $\dfrac{4}{(\sqrt{2} + 1)}$ can be rationalised by multiplying numerator and denominator by $(\sqrt{2} - 1)$.

a Work out $\dfrac{4}{(\sqrt{2} + 1)} \times \dfrac{(\sqrt{2} - 1)}{(\sqrt{2} - 1)}$.

b Simplify:

  i $\dfrac{4}{(\sqrt{3} + 1)}$ **ii** $\dfrac{1}{(\sqrt{5} + 2)}$ **iii** $\dfrac{10}{(\sqrt{7} - 2)}$

7.1.4 Simplifying algebraic expressions

EXAMPLE

Simplify

a $x^7 \times x^{13} = x^{7+13} = x^{20}$ **b** $x^3 \div x^7 = x^{3-7} = x^{-4} = \dfrac{1}{x^4}$

c $(x^4)^3 = x^{12}$ **d** $(3x^2)^3 = 3^3 \times (x^2)^3 = 27x^6$

e $3y^2 \times 4y^3 = 12y^5$ **f** $x^5 \div x^{-3} = x^{5-(-3)} = x^8$

g $5x^{\frac{1}{2}} \times 2x^{\frac{1}{2}} = 10x^{\frac{1}{2}+\frac{1}{2}} = 10x$ **h** $8x^{\frac{5}{2}} \div 2x^{\frac{1}{2}} = 4x^{\frac{5}{2}-\frac{1}{2}} = 4x^2$

i $(2x^{\frac{1}{2}})^4 \times x^0 = 2^4 \times x^2 \times 1 = 16x^2$

Exercise 4 (H)

Write each in a simpler form.

1 $(3^3)^2$ **2** $(5^4)^3$ **3** $(7^2)^5$ **4** $(x^2)^3$

5 $(2^{-1})^2$ **6** $(-1)^{102}$ **7** $(7^{-1})^{-2}$ **8** $(y^4)^{\frac{1}{2}}$

9 Half of 2^{22} **10** One-tenth of 10^{100}

Simplify

11 $x^3 \times x^4$ **12** $y^6 \times y^7$ **13** $z^7 \div z^3$ **14** $z^{50} \times z^{50}$

15 $m^3 \div m^2$ **16** $e^{-3} \times e^{-2}$ **17** $y^{-2} \times y^4$ **18** $w^4 \div w^{-2}$

19 $y^{\frac{1}{2}} \times y^{\frac{1}{2}}$ **20** $(x^2)^5$ **21** $x^{-2} \div x^{-2}$ **22** $w^{-3} \times w^{-2}$

23 $w^{-7} \times w^2$ **24** $x^3 \div x^{-4}$ **25** $(a^2)^4$ **26** $(k^{\frac{1}{2}})^6$

27 $e^{-4} \times e^4$ **28** $x^{-1} \times x^{30}$ **29** $(y^4)^{\frac{1}{2}}$ **30** $(x^{-3})^{-2}$

31 $z^2 \div z^{-2}$ **32** $t^{-3} \div t$ **33** $(2x^3)^2$ **34** $(4y^5)^2$

35 $2x^2 \times 3x^2$ **36** $5y^3 \times 2y^2$ **37** $5a^3 \times 3a$ **38** $(2a)^3$

39 $3x^3 \div x^3$ **40** $8y^3 \div 2y$ **41** $10y^2 \div 4y$ **42** $8a \times 4a^3$

43 $(2x)^2 \times (3x)^3$ **44** $4z^4 \times z^{-7}$ **45** $6x^{-2} \div 3x^2$ **46** $5y^3 \div 2y^{-2}$

47 $(x^2)^{\frac{1}{2}} \div (x^{\frac{1}{3}})^3$ **48** $7w^{-2} \times 3w^{-1}$ **49** $(2n)^4 \div 8n^0$ **50** $4x^{\frac{3}{2}} \div 2x^{\frac{1}{2}}$

7.1.5 Solving equations

EXAMPLE

Solve these equations.

a $2^x = 16$

b $5^y = \dfrac{1}{25}$

c $6^z = (\sqrt{6})^3$

d $3^{x-1} = 81$

· ·

Use the principle that if $a^x = a^y$ then $x = y$

a $\quad 2^x = 16 = 2^4$

$\quad$ So $x = 4$

b $\quad 5^y = \dfrac{1}{25} = 5^{-2}$

$\quad$ So $y = -2$

c $\quad 6^z = (\sqrt{6})^3 = 6^{\frac{3}{2}}$

$\quad$ So $z = \dfrac{3}{2}$

d $\quad 3^{x-1} = 81 = 3^4$

$\quad$ So $x = 5$

Exercise 5 (H)

Solve the equations for x.

1 $2^x = 8$ **2** $3^x = 81$ **3** $5^x = \dfrac{1}{5}$ **4** $10^x = \dfrac{1}{100}$

5 $3^{-x} = \dfrac{1}{27}$ **6** $4^x = 64$ **7** $6^{-x} = \dfrac{1}{6}$ **8** $100\,000^x = 10$

9 $12^x = 1$ **10** $10^x = 0{\cdot}0001$ **11** $2^x + 3^x = 13$ **12** $\left(\dfrac{1}{2}\right)^x = 32$

13 $5^{2x} = 25$ **14** $1\,000\,000^{3x} = 10$

15 Here is a sketch of $y = 3^x$.
  a Write the coordinates
    of point A.
  b Write the solution of the
    equation $3^x = 5^x$.

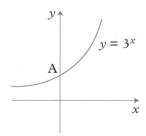

16 Find a, b and c.

   a $\dfrac{1}{8} = 2^a$ **b** $b^7 = 1$ **c** $2^{c-1} = 16$

17 Find p, q and r.

   a $x^{\frac{1}{2}} \times x^2 = x^p$ **b** 10% of $100^2 = q$ **c** $x^{-\frac{1}{4}} \times x^2 = x^r$

18 Solve these equations.

   a $10n^3 = 640$ **b** $10^n = 0{\cdot}1$ **c** $2n^3 = 0$

19 Find two solutions of the equation $x^2 = 2^x$.

20 Use a calculator to find solutions to these equations correct to three significant figures.

   a $x^x = 100$ **b** $x^x = 10\,000$

21 The last digit of 7^3 is 3 ($7^3 = 343$).

   a Copy and complete this table, which gives the last digit of 7^n.

n	1	2	3	4	5	6	7	8	9	10
Last digit of 7^n	7	9	3		7			1		

   b Write the last digit of

      i 7^{48} **ii** 7^{101} **iii** 49^{35}

22 It is given that $10^x = 3$ and $10^y = 7$. What is the value of 10^{x+y}?

23 In a laboratory there are 2 cells in a dish. The number of cells in the dish doubles every 30 minutes.

   a How many cells are in the dish after four hours?

   b After what time are there 2^{13} cells in the dish?

   c After $10\frac{1}{2}$ hours there are 2^{22} cells in the dish and an experimental fluid is added which eliminates half of the cells. How many cells are left?

> This is an example of **exponential growth**.

24 Steve's bike is ill. Its computer-controlled ignition system has a virus. The doctor has advised Steve to keep the bike warm, in which case the number of germs in the bike will decay exponentially and will be $1\,000\,000 \times 2^{-n}$ after n hours.

   a How many germs will there be after 10 hours?

   b The bike will be cured when it contains less than one germ. After how many hours will it be cured?

25 A bank pays compound interest on money invested in an account. After n years a sum of £2000 will rise to £2000 $\times$ 1·08^n. This represents exponential growth.

   a How much money is in the account after three years?

   b After how many years will the original £2000 have nearly doubled?

*26 It is given that $x + \dfrac{1}{x} = 1$, where x is not zero.

Show that $x^2 = x - 1$ and $x^3 = x^2 - x$.

Use these two expressions to show that $x^3 = -1$.

Using this value for x^3, find the value of x^6 and show that $x^7 = x$.

Hence show that $x^6 + \dfrac{1}{x^6} = 2$ and $x^7 + \dfrac{1}{x^7} = 1$.

Deduce the value of $x^{60} + \dfrac{1}{x^{60}}$ and of $x^{61} + \dfrac{1}{x^{61}}$.

7.2 Quadratic equations

7.2.1 Factorising quadratic expressions

In Chapter 3, Section 3.1, you learnt how to expand a pair of brackets like $(x + 4)(x - 3)$ to give $x^2 + x - 12$.

The reverse of this process is called factorising.

EXAMPLE

Factorise $x^2 + 6x + 8$.

....

a Find two numbers which multiply to give 8 and add up to 6.

b Put these numbers into brackets.

   So $x^2 + 6x + 8 = (x + 4)(x + 2)$

EXAMPLE

Factorise **a** $x^2 + 2x - 15$

                  b $x^2 - 6x + 8$

....

a Two numbers which multiply to give -15 and add up to $+2$ are -3 and 5.

   So $x^2 + 2x - 15 = (x - 3)(x + 5)$

b Two numbers which multiply to give $+8$ and add up to -6 are -2 and -4.

   So $x^2 - 6x + 8 = (x - 2)(x - 4)$

Exercise 6 Ⓜ/Ⓗ

Factorise these expressions.

1 $x^2 + 7x + 10$	**2** $x^2 + 7x + 12$	**3** $x^2 + 8x + 15$
4 $x^2 + 10x + 21$	**5** $x^2 + 8x + 12$	**6** $y^2 + 12y + 35$
7 $y^2 + 11y + 24$	**8** $y^2 + 10y + 25$	**9** $y^2 + 15y + 36$
10 $a^2 - 3a - 10$	**11** $a^2 - a - 12$	**12** $z^2 + z - 6$
13 $x^2 - 2x - 35$	**14** $x^2 - 5x - 24$	**15** $x^2 - 6x + 8$
16 $y^2 - 5y + 6$	**17** $x^2 - 8x + 15$	**18** $a^2 - a - 6$
19 $a^2 + 14a + 45$	**20** $b^2 - 4b - 21$	**21** $x^2 - 8x + 16$
22 $y^2 + 2y + 1$	**23** $y^2 - 3y - 28$	**24** $x^2 - x - 20$
25 $x^2 - 8x - 240$	**26** $x^2 - 26x + 165$	**27** $y^2 + 3y - 108$
28 $x^2 - 49$	**29** $x^2 - 9$	**30** $x^2 - 16$

31 The terms in the expression $2x^2 + 12x + 16$ have a common factor of 2. So $2x^2 + 12x + 16 = 2(x^2 + 6x + 8)$. Complete the factorisation.

32 Factorise

a $2x^2 + 4x - 30$	**b** $3x^2 + 21x + 30$	**c** $3x^2 + 24x + 45$
d $2n^2 - 6n - 20$	**e** $5a^2 + 5a - 30$	**f** $4x^2 - 64$

7.2.2 More difficult factorising

EXAMPLE

Factorise   **a** $3x^2 + 13x + 4$ **b** $4x^2 - 5x - 6$

....

a   **i**   Find two numbers which multiply to give (3×4), that is 12, and add up to 13. In this case the numbers are 1 and 12.

   ii   Split the '$13x$' term, $3x^2 + 1x + 12x + 4$

   iii   Factorise in pairs, $x(3x + 1) + 4(3x + 1)$

   iv   ($3x + 1$) is common, so $3x^2 + 13x + 4 = (3x + 1)(x + 4)$.

....

b   **i**   Find two numbers which multiply to give (4×-6), that is -24, and add up to -5. In this case the numbers are -8 and 3.

   ii   Split the $-5x$ term, $4x^2 - 8x + 3x - 6$

   iii   Factorise in pairs, $4x(x - 2) + 3(x - 2)$

   iv   ($x - 2$) is common, so $4x^2 - 5x - 6 = (x - 2)(4x + 3)$.

Exercise 7 Ⓜ/Ⓗ

Factorise these expressions.

1 $2x^2 + 5x + 3$ **2** $2x^2 + 7x + 3$ **3** $3x^2 + 7x + 2$ **4** $2x^2 + 11x + 12$

5 $3x^2 + 8x + 4$ **6** $2x^2 + 7x + 5$ **7** $3x^2 - 5x - 2$ **8** $2x^2 - x - 15$

9 $2x^2 + x - 21$ **10** $3x^2 - 17x - 28$ **11** $6x^2 + 7x + 2$ **12** $3x^2 - 11x + 6$

13 $3y^2 - 11y + 10$ **14** $6y^2 + 7y - 3$ **15** $10x^2 + 9x + 2$ **16** $6x^2 - 19x + 3$

17 $8x^2 - 10x - 3$ **18** $12x^2 + 23x + 10$ **19** $4y^2 - 23y + 15$ **20** $6x^2 - 27x + 30$

7.2.3 The difference of two squares

- $x^2 - y^2 = (x - y)(x + y)$

Remember this result.

EXAMPLE

Factorise
a $y^2 - 16$ **b** $4a^2 - b^2$

a $y^2 - 16 = (y - 4)(y + 4)$ **b** $4a^2 - b^2 = (2a - b)(2a + b)$

Exercise 8 Ⓜ/Ⓗ

Factorise these expressions.

1 $y^2 - a^2$ **2** $m^2 - n^2$ **3** $x^2 - t^2$

4 $y^2 - 1$ **5** $x^2 - 9$ **6** $a^2 - 25$

7 $x^2 - \frac{1}{4}$ **8** $x^2 - \frac{1}{9}$ **9** $4x^2 - y^2$

10 $a^2 - 4b^2$ **11** $25x^2 - 4y^2$ **12** $9x^2 - 16y^2$

13 $4x^2 - \frac{z^2}{100}$ **14** $x^3 - x$ **15** $a^3 - ab^2$

16 $4x^3 - x$ **17** $8x^3 - 2xy^2$ **18** $y^3 - 9y$

19 Find the exact value of $100\,003^2 - 99\,997^2$.

20 Find the exact value of $1\,500\,002^2 - 1\,499\,998^2$.

21 Rewrite 9991 as the difference of two squares. Use your answer to find the prime factors of 9991.

7.2.4   Methods of solving quadratic equations

The last three exercises have been about factorising expressions. Expressions do not have an 'equals' sign, but equations do. Quadratic equations always have an x^2-term, and often an x-term as well as a number term. They generally have two different solutions. In this section you will learn three different methods for solving quadratic equations.

Method 1 – Solution by factors

Consider the equation $a \times b = 0$, where a and b are numbers.

- The product $a \times b$ can only be zero if either a or b (or both) is equal to zero.

Can you think of other possible pairs of numbers which multiply together to give zero?

EXAMPLE

a Solve the equation $x^2 + x - 12 = 0$.   **b** Solve the equation $6x^2 + x - 2 = 0$.

a Factorising, $(x - 3)(x + 4) = 0$
either $x - 3 = 0$ or $x + 4 = 0$
so $x = 3$ or $x = -4$

b Factorising, $(2x - 1)(3x + 2) = 0$
either $2x - 1 = 0$ or $3x + 2 = 0$
so $2x = 1$ $3x = -2$
$x = \dfrac{1}{2}$ $x = -\dfrac{2}{3}$

Exercise 9 Ⓜ/Ⓗ

Solve these equations.

1 $x^2 + 7x + 12 = 0$ **2** $x^2 + 7x + 10 = 0$ **3** $x^2 + 2x - 15 = 0$
4 $x^2 + x - 6 = 0$ **5** $x^2 - 8x + 12 = 0$ **6** $x^2 + 10x + 21 = 0$
7 $x^2 - 5x + 6 = 0$ **8** $x^2 - 4x - 5 = 0$ **9** $x^2 + 5x - 14 = 0$
10 $2x^2 - 3x - 2 = 0$ **11** $3x^2 + 10x - 8 = 0$ **12** $2x^2 + 7x - 15 = 0$
13 $6x^2 - 13x + 6 = 0$ **14** $4x^2 - 29x + 7 = 0$ **15** $10x^2 - x - 3 = 0$

16 $y^2 - 15y + 56 = 0$ **17** $12y^2 - 16y + 5 = 0$ **18** $y^2 + 2y - 63 = 0$
19 $x^2 + 2x + 1 = 0$ **20** $x^2 - 6x + 9 = 0$ **21** $x^2 + 10x + 25 = 0$
22 $x^2 - 14x + 49 = 0$ **23** $6a^2 - a - 1 = 0$ **24** $4a^2 - 3a - 10 = 0$
25 $z^2 - 8z - 65 = 0$ **26** $6x^2 + 17x - 3 = 0$ **27** $10k^2 + 19k - 2 = 0$
28 $y^2 - 2y + 1 = 0$ **29** $36x^2 + x - 2 = 0$ **30** $20x^2 - 7x - 3 = 0$

*31 $x^4 - 5x^2 + 4 = 0$ ***32** $x^4 - 13x^2 + 36 = 0$
*33 $4x^4 - 17x^2 + 4 = 0$ ***34** $x^6 - 9x^3 + 8 = 0$

In questions 31–34, put $y = x^2$.

Quadratics with only two terms

a Solve the equation $x^2 - 7x = 0$. **b** Solve the equation $x^2 - 100 = 0$.

Factorising, $x(x - 7) = 0$
either $x = 0$ or $x - 7 = 0$
$x = 7$
The solutions are $x = 0$ and $x = 7$.

Rearranging, $x^2 = 100$
Take square root,
$x = 10$ or -10

Alternatively use the difference of squares, that is $(x-10)(x+10) = 0$ so either $x = 10$ or -10.

Exercise 10 Ⓜ/Ⓗ

Solve these equations.

1 $x^2 - 3x = 0$ **2** $x^2 + 7x = 0$ **3** $2x^2 - 2x = 0$ **4** $3x^2 - x = 0$

5 $x^2 - 16 = 0$ **6** $x^2 - 49 = 0$ **7** $4x^2 - 1 = 0$ **8** $9x^2 - 4 = 0$

9 $6y^2 + 9y = 0$ **10** $6a^2 - 9a = 0$ **11** $10x^2 - 55x = 0$ **12** $16x^2 - 1 = 0$

13 $y^2 - \frac{1}{4} = 0$ **14** $56x^2 - 35x = 0$ **15** $36x^2 - 3x = 0$ **16** $x^2 = 6x$

17 $x^2 = 11x$ **18** $2x^2 = 3x$ **19** $x^2 = x$ **20** $4x = x^2$

Method 2 – Solution by formula

● You can find the solutions of the quadratic equation

$ax^2 + bx + c = 0$ by using the formula $x = \dfrac{-b \pm \sqrt{(b^2 - 4ac)}}{2a}$.

If you wish to, you can derive this formula by using the method of 'completing the square' which follows on page 361.

Use this formula only after trying (and failing) to factorise.

Solve the equation $2x^2 - 3x - 4 = 0$.

In this case $a = 2$, $b = -3$, $c = -4$.

$$x = \frac{-(-3) \pm \sqrt{[(-3)^2 - (4 \times 2 \times -4)]}}{2 \times 2}$$

$$x = \frac{3 \pm \sqrt{[9 + 32]}}{4} = \frac{3 \pm \sqrt{41}}{4}$$

$$x = \frac{3 \pm 6 \cdot 403}{4}$$

either $x = \dfrac{3 + 6 \cdot 403}{4} = 2 \cdot 35$ (2 decimal places)

or $x = \dfrac{3 - 6 \cdot 403}{4} = \dfrac{-3 \cdot 403}{4} = -0 \cdot 85$ (2 decimal places).

Start by writing down the values of 'a, b and c'.

Exercise 11 Ⓗ

Solve these equations, giving answers to two decimal places where necessary.

1 $2x^2 + 11x + 5 = 0$ **2** $3x^2 + 11x + 6 = 0$ **3** $6x^2 + 7x + 2 = 0$
4 $3x^2 - 10x + 3 = 0$ **5** $5x^2 - 7x + 2 = 0$ **6** $6x^2 - 11x + 3 = 0$
7 $2x^2 + 6x + 3 = 0$ **8** $x^2 + 4x + 1 = 0$ **9** $5x^2 - 5x + 1 = 0$
10 $x^2 - 7x + 2 = 0$ **11** $2x^2 + 5x - 1 = 0$ **12** $3x^2 + x - 3 = 0$
13 $3x^2 + 8x - 6 = 0$ **14** $3x^2 - 7x - 20 = 0$ **15** $2x^2 - 7x - 15 = 0$
16 $x^2 - 3x - 2 = 0$ **17** $2x^2 + 6x - 1 = 0$ **18** $6x^2 - 11x - 7 = 0$
19 $3x^2 + 25x + 8 = 0$ **20** $3y^2 - 2y - 5 = 0$ **21** $2y^2 - 5y + 1 = 0$
22 $\frac{1}{2}y^2 + 3y + 1 = 0$ **23** $2 - x - 6x^2 = 0$ **24** $3 + 4x - 2x^2 = 0$

25 a To find the x-coordinates of A and B, solve the equation

$$x^2 - 5x + 1 = 0$$

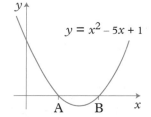

$y = x^2 - 5x + 1$

b Explain why the equation $x^2 - 5x + 10 = 0$ has no solutions.

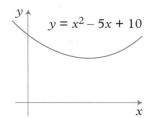

$y = x^2 - 5x + 10$

*26 Copy and complete the sentence: 'The equation $ax^2 + bx + c = 0$ has solutions if $(b^2 - 4ac)$ is _____.'

Equations leading to quadratics

The solution to a problem can involve an equation which does not at first appear to be quadratic. The terms in the equation may need to be rearranged as shown in this example.

EXAMPLE

Solve $2x(x - 1) = (x + 1)^2 - 5$

....

Multiply, out the brackets, $2x^2 - 2x = x^2 + 2x + 1 - 5$
$$2x^2 - 2x - x^2 - 2x - 1 + 5 = 0$$
$$x^2 - 4x + 4 = 0$$
Factorising, $(x - 2)(x - 2) = 0$
$$x = 2$$

In this example the quadratic has a repeated root of $x = 2$.

Exercise 12 Ⓗ

Solve these equations giving answers to two decimal places where necessary.

1 $x^2 = 6 - x$

2 $x(x + 10) = -21$

3 $3x + 2 = 2x^2$

4 $x^2 + 4 = 5x$

5 $6x(x + 1) = 5 - x$

6 $(2x)^2 = x(x - 14) - 5$

7 $(x - 3)^2 = 10$

8 $(x + 1)^2 - 10 = 2x(x - 2)$

9 $(2x - 1)^2 = (x - 1)^2 + 8$

10 $3x(x + 2) - x(x - 2) + 6 = 0$

11 $x = \dfrac{15}{x} - 22$

12 $x + 5 = \dfrac{14}{x}$

13 $4x + \dfrac{7}{x} = 29$

14 $10x = 1 + \dfrac{3}{x}$

> In questions **11–14**, multiply each term by x.

15 $2x^2 = 7x$

16 $16 = \dfrac{1}{x^2}$

17 $2x + 2 = \dfrac{7}{x} - 1$

18 $\dfrac{2}{x} + \dfrac{2}{x + 1} = 3$

19 $\dfrac{3}{x - 1} + \dfrac{3}{x + 1} = 4$

20 $\dfrac{2}{x - 2} + \dfrac{4}{x + 1} = 3$

21 Hassan says you can make it easier to solve some equations by dividing through by a common factor.

So, by dividing through by 2, the equation $2x^2 + 10x - 48 = 0$

can be written as $x^2 + 5x - 24 = 0$

Is Hassans's method OK?

Check by solving the two equations above. Are the solutions the same?

22 Solve these equations.

a $4x^2 - 10x - 6 = 0$

b $3x^2 - 6x - 72 = 0$

*23 One of the solutions published by Cardan in 1545 for the solution of cubic equations is given below. For an equation in the form $x^3 + px = q$

$$x = \sqrt[3]{\left[\sqrt{\left(\dfrac{p}{3}\right)^3 + \left(\dfrac{q}{2}\right)^2} + \dfrac{q}{2}\right]} - \sqrt[3]{\left[\sqrt{\left(\dfrac{p}{3}\right)^3 + \left(\dfrac{q}{2}\right)^2} - \dfrac{q}{2}\right]}$$

Use the formula to solve these equations, giving answers to 4 significant figures where necessary.

a $x^3 + 7x = -8$

b $x^3 + 6x = 4$

c $x^3 + 3x = 2$

d $x^3 + 9x - 2 = 0$

Method 3 – Solution by completing the square

Look at the function $f(x) = x^2 + 6x$

Completing the square, this becomes $f(x) = (x + 3)^2 - 9$

This is done as follows.

i 3 is half of 6 and gives $6x$ when $(x + 3)^2$ is expanded.

ii If you add 3 to the square term, you need to subtract 9 from the expression to cancel the $+9$ obtained.

Here are some more examples.

a $x^2 - 12x = (x - 6)^2 - 36$

b $x^2 + 3x = \left(x + \dfrac{3}{2}\right)^2 - \dfrac{9}{4}$

c $\quad x^2 + 6x + 1 = (x + 3)^2 - 9 + 1$

$\qquad\qquad\qquad = (x + 3)^2 - 8$

d $x^2 - 10x - 17 = (x - 5)^2 - 25 - 17$

$\qquad\qquad\qquad = (x - 5)^2 - 42$

e $\quad 2x^2 - 8x + 3 = 2\left[x^2 - 4x + \dfrac{3}{2}\right]$

$\qquad\qquad\qquad = 2\left[(x - 2)^2 - 4 + \dfrac{3}{2}\right]$

$\qquad\qquad\qquad = 2\left[(x - 2)^2 - \dfrac{5}{2}\right]$

> Check that the two sides of the equations are equal.

EXAMPLE

Solve the quadratic equation $x^2 - 6x + 7 = 0$ by completing the square.

....

$x^2 - 6x + 7 = (x - 3)^2 - 9 + 7$

$(x - 3)^2 - 9 + 7 = 0$

$(x - 3)^2 \qquad\quad = 2$

$\therefore x - 3 = +\sqrt{2} \quad$ or $\quad -\sqrt{2}$

$\qquad x = 3 + \sqrt{2} \quad$ or $\quad 3 - \sqrt{2}$

So, $\quad x = 4{\cdot}41 \quad$ or $\quad 1{\cdot}59$ to 2 dp

---

EXAMPLE

Given $f(x) = x^2 - 8x + 18$, show that $f(x) \geqslant 2$ for all values of x.

....

Completing the square, $f(x) = (x - 4)^2 - 16 + 18$

$\qquad\qquad\qquad\qquad f(x) = (x - 4)^2 + 2$

Now $(x - 4)^2$ is always greater than or equal to zero because it is 'something squared.'

$$\therefore \quad f(x) \geqslant 2$$

Exercise 13 Ⓗ

In questions **1** to **10**, complete the square for each expression by writing each one in the form $(x + a)^2 + b$ where a and b can be positive or negative.

1 $x^2 + 8x$ **2** $x^2 - 12x$ **3** $x^2 + x$

4 $x^2 + 4x + 1$ **5** $x^2 - 6x + 9$ **6** $x^2 + 2x - 15$

7 $x^2 + 16x + 5$ **8** $x^2 - 10x$ **9** $x^2 + 3x$

10 Solve the equations by completing the square.

    a $x^2 - 8x + 12 = 0$ **b** $x^2 + 10x + 21 = 0$ **c** $x^2 - 4x - 5 = 0$

11 Solve these equations by completing the square.

    a $x^2 + 4x - 3 = 0$ **b** $x^2 - 3x - 2 = 0$ **c** $x^2 + 12x = 1$

12 Try to solve the equation $x^2 + 6x + 10 = 0$, by completing the square. Explain why you can find no solutions.

13 Given $f(x) = x^2 + 6x + 12$, show that $f(x) \geqslant 3$ for all values of x.

14 Given $g(x) = x^2 - 7x + \dfrac{1}{4}$, show that the least possible value of $g(x)$ is -12.

*15 To complete the square with the expression $f(x) = 2x^2 - 12x + 7$, begin by taking out the factor 2 (that is, the coefficient of x^2).

$$2x^2 - 12x + 7 = 2\left(x^2 - 6x + \frac{7}{2}\right)$$

Continue by completing the square inside the brackets.

$$f(x) = 2\left[(x - 3)^2 - 9 + \frac{7}{2}\right] = 2\left[(x - 3)^2 - \frac{11}{2}\right]$$

Complete the square for each expression.

    a $2x^2 + 4x + 1$ **b** $3x^2 - 6x + 2$ **c** $2x^2 + x + 2$

*16 If $f(x) = x^2 + 4x + 7$ find

    a the smallest possible value of $f(x)$

    b the value of x for which this smallest value occurs

    c the greatest possible value of $\dfrac{1}{(x^2 + 4x + 7)}$.

*17 Given $y = x^2 - x + 1$, find

    a the lowest value of y

    b the value of x for which this lowest value occurs

    c the greatest possible value of $\dfrac{1}{(x^2 - x + 1)}$.

*18 Simplify $(x + 4)(x + 2) - 2(x + 2)$, and explain why the expression can never be negative, whatever the value of x.

7.2.5   Using quadratic equations to solve problems

EXAMPLE

The perimeter of a rectangle is 42 cm. If the diagonal is 15 cm, find the width of the rectangle.

....

Let the width of the rectangle be x cm.
Since the perimeter is 42 cm, the sum of the length and the width is 21 cm.
So length of rectangle = $(21 - x)$ cm
By Pythagoras' theorem

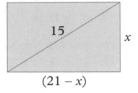

$$x^2 + (21 - x)^2 = 15^2$$
$$x^2 + (21 - x)(21 - x) = 15^2$$
$$x^2 + 441 - 42x + x^2 = 225$$
$$2x^2 - 42x + 216 = 0$$
$$x^2 - 21x + 108 = 0$$
$$(x - 12)(x - 9) = 0$$
$$x = 12$$
$$\text{or} \quad x = 9$$

Note that the dimensions of the rectangle are 9 cm and 12 cm, whichever value of x is taken.

So the width of the rectangle is 9 cm.

Exercise 14 Ⓗ

Solve by forming a quadratic equation.

1 Two numbers, which differ by 3, have a product of 88. Find them.

> Call the numbers x and $x + 3$.

2 The product of two consecutive odd numbers is 143. Find the numbers.

> If the first odd number is x, what is the next odd number?

3 The height of a photo exceeds the width by 7 cm. If the area is 60 cm^2, find the height of the photo.

4 The length of a rectangle exceeds the width by 2 cm. If the diagonal is 10 cm long, find the width of the rectangle.

5 The area of the rectangle exceeds the area of the square by $24\,\text{m}^2$. Find x.

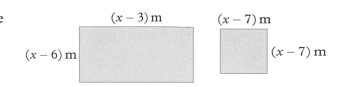

6 Three consecutive integers are written as x, $x + 1$, $x + 2$.
The square of the largest number is 45 less than the sum of the squares of the other numbers. Find the three numbers.

7 $(x - 1)$, x and $(x + 1)$ represent three positive integers.
The product of the three numbers is five times their sum.
a Write an equation in x.
b Show that your equation simplifies to $x^3 - 16x = 0$.
c Factorise $x^3 - 16x$ completely.
d Hence find the three positive integers.

8 An aircraft flies a certain distance on a bearing of $045°$ and then twice the distance on a bearing of $135°$. Its distance from the starting point is then $350\,\text{km}$. Find the length of the first part of the journey.

9 The area of rectangle A is twice the area of rectangle B.

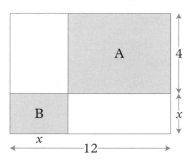

Find x.

10 The perimeter of a rectangle is $68\,\text{cm}$. If the diagonal is $26\,\text{cm}$, find the dimensions of the rectangle.

11 A stone is thrown in the air. After t seconds its height, h, above sea level is given by the formula $h = 80 + 3t - 5t^2$. Find the value of t when the stone falls into the sea.

12 The total surface area of a cylinder, A, is given by the formula

$$A = 2\pi r^2 + 2\pi rh.$$

Given that $A = 200\,\text{cm}^2$ and $h = 10\,\text{cm}$, find the value of r, correct to 1 dp.

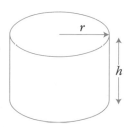

13 A rectangular pond, 6 m × 4 m, is surrounded by a uniform path of width x. The area of the path is equal to the area of the pond. Find x.

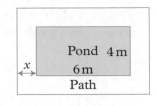

14 The perimeters of a square and a rectangle are equal.
The length of the rectangle is 11 cm and the area of the square is 4 cm^2 more than the area of the rectangle. Find the side of the square.

15 The sequence 3, 8, 15, 24, ... can be written (1×3), (2×4), (3×5), (4×6)...
a Write an expression for the nth term of the sequence.
One term in the sequence is 255.
b Form an equation and hence find what number term it is.

16 In Figure 1, ABCD is a rectangle with AB = 12 cm and BC = 7 cm. AK = BL = CM = DN = x cm. If the area of KLMN is 54 cm^2 find x.

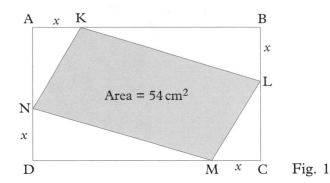

Fig. 1

17 In Figure 1, AB = 14 cm BC = 11 cm and AK = BL = CM = DN = x cm. If the area of KLMN is now 97 cm^2, find x.

18 When each edge of a cube is decreased by 1 cm, its volume is decreased by 91 cm^3. Find the length of a side of the original cube.

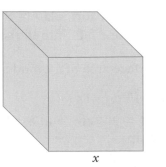

19 A cyclist travels 40 km at a speed x km/h. Find the time taken in terms of x. Find the time taken when his speed is reduced by 2 km/h. If the difference between the times is 1 hour, find the original speed, x km/h.

20 An increase of speed of 4 km/h on a journey of 32 km reduces the time taken by 4 hours. Find the original speed.

21 A train normally travels 60 miles at a certain speed. One day, due to bad weather, the train's speed is reduced by 10 mph so that the journey takes 3 hours longer. Find the normal speed.

22 A number exceeds four times its reciprocal by 3. Find the number.

23 Two numbers differ by 3. The sum of their reciprocals is $\dfrac{7}{10}$; find the numbers.

24 The numerator of a fraction is 1 less than the denominator. When both numerator and denominator are increased by 2, the fraction is increased by $\dfrac{1}{12}$. Find the original fraction.

25 Paper often comes in standard sizes A0, A1, A2, A3 etc. The large sheet shown is size A0; A0 can be cut in half to give two sheets of A1; A1 can be cut in half to give two sheets of A2 and so on.

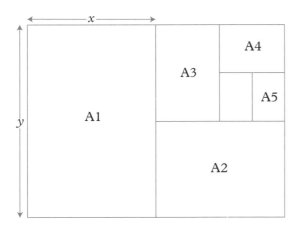

Call the sides of A1 x and y, as shown.
a What are the sides of A0 in terms of x and y?
b All the sizes of paper are similar. Form an equation involving x and y.
c Size A0 has an area of 1 m^2. Calculate the values of x and y correct to the nearest mm.
d Hence calculate the long side of a sheet of A4. Check your answer by measuring a sheet of A4.

7.3 Algebraic fractions

Simplify these fractions.

a **i** $\dfrac{3a}{5a^2} = \dfrac{3 \times \cancel{a}}{5 \times a \times \cancel{a}}$

$\qquad = \dfrac{3}{5a}$

ii $\dfrac{3y + y^2}{6y} = \dfrac{y(3 + y)}{6y}$

$\qquad = \dfrac{3 + y}{6}$

iii $\dfrac{x^2 - x - 2}{x^2 + 4x + 3}$

$\qquad = \dfrac{(x - 2)(x + 1)}{(x + 1)(x + 3)}$

$\qquad = \dfrac{(x - 2)}{(x + 3)}$

b Write as a single fraction:

i $\dfrac{4\cancel{x}}{3} \times \dfrac{2}{x^2} = \dfrac{8}{3x}$

ii $\dfrac{5}{n} \div \dfrac{3}{n} = \dfrac{5}{\cancel{n}} \times \dfrac{\cancel{n}}{3} = \dfrac{5}{3}$

Exercise 15 (H)

1 Simplify as far as possible.

a $\dfrac{25}{35}$
   b $\dfrac{5y^2}{y}$
   c $\dfrac{y}{2y}$
   d $\dfrac{8x^2}{2x^2}$

e $\dfrac{2x}{4y}$
   f $\dfrac{6y}{3y}$
   g $\dfrac{5ab}{10b}$
   h $\dfrac{8ab^2}{12ab}$

2 Sort these into four pairs of equivalent expressions.

A $\dfrac{x^2}{2x}$
   B $\dfrac{x(x + 1)}{x^2}$
   C $\dfrac{x^2 + x}{x^2 - x}$
   D $\dfrac{3x + 6}{3x}$

E $\dfrac{x + 1}{x}$
   F $\dfrac{x + 2}{x}$
   G $\dfrac{x}{2}$
   H $\dfrac{x + 1}{x - 1}$

3 Write as a single fraction.

a $\dfrac{3x}{2} \times \dfrac{2a}{3x}$
   b $\dfrac{5mn}{3} \times \dfrac{2}{n}$
   c $\dfrac{3y^2}{3} \times \dfrac{2x}{9y}$
   d $\dfrac{2}{q} \div \dfrac{a}{2}$

e $\dfrac{4x}{3} \div \dfrac{x}{2}$
   f $\dfrac{x}{5} \times \dfrac{y^2}{x^2}$
   g $\dfrac{a^2}{5} \div \dfrac{a}{10}$
   h $\dfrac{x^2}{x^2 + 2x} \div \dfrac{x}{x + 2}$

4 Simplify as far as possible.

a $\dfrac{7a^2b}{35ab^2}$
   b $\dfrac{(2a)^2}{4a}$
   c $\dfrac{7yx}{8xy}$
   d $\dfrac{3x}{4x - x^2}$

e $\dfrac{5x + 2x^2}{3x}$
   f $\dfrac{9x + 3}{3x}$
   g $\dfrac{4a + 5a^2}{5a}$
   h $\dfrac{5ab}{15a + 10a^2}$

5 Copy and complete.

a $\dfrac{x^2}{3} \times \dfrac{\square}{x} = 2x$
   b $\dfrac{8}{x} \div \dfrac{2}{x} = \square$
   c $\dfrac{a}{3} + \dfrac{a}{3} = \dfrac{\square}{3}$

6 Sort these into four pairs of equivalent expressions.

A $\dfrac{x^2}{3x}$ B $\dfrac{x}{2} \times \dfrac{x}{2}$ C $\dfrac{12x + 6}{6}$ D $\dfrac{x}{5} - \dfrac{2}{5}$

E $\dfrac{2x^2 + x}{x}$ F $\dfrac{x(x + 1)}{3x + 3}$ G $\dfrac{x - 2}{5}$ H $\dfrac{ax^2}{4a}$

7 Write in a simpler form.

a $\dfrac{5x + 10y}{15xy}$ **b** $\dfrac{18a - 3ab}{6a^2}$ **c** $\dfrac{4ab + 8a^2}{2ab}$ **d** $\dfrac{(2x)^2 - 8x}{4x}$

8 Simplify as far as possible.

a $\dfrac{x^2 + 2x}{x^2 - 3x}$ **b** $\dfrac{x^2 - 3x}{x^2 - 2x - 3}$ **c** $\dfrac{x^2 + 4x}{2x^2 - 10x}$

d $\dfrac{x^2 + 6x + 5}{x^2 - x - 2}$ **e** $\dfrac{x^2 - 4x - 21}{x^2 - 5x - 14}$ **f** $\dfrac{x^2 + 7x + 10}{x^2 - 4}$

7.4   Transformation of curves

7.4.1   Four transformations

The notation f(x) means 'function of x'. A function of x is an expression which (usually) varies, depending on the value of x. Examples of functions are:

f(x) = $x^2 + 3$;   f(x) = $\dfrac{1}{x} + 7$;   f(x) = sin x.

Imagine a box which performs the function f on any input.

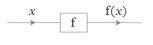

If f(x) = $x^2 + 7x + 2$,   f(3) = $3^2 + 7 \times 3 + 2 = 32$

> Here are **four** ways in which any function f(x) can be transformed.
> **i** f(x) + a **ii** f($x - a$)   **iii** af(x)   **iv** f(ax)

i $y = \mathbf{f}(x) + a$

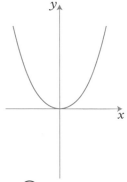 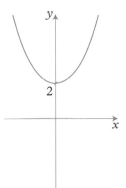 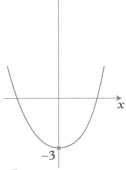

① $y = \mathrm{f}(x)$ ② $y = \mathrm{f}(x) + 2$ ③ $y = \mathrm{f}(x) - 3$

- Can you describe the transformation from ① to ②? From ① to ③?

ii $y = f(x - a)$

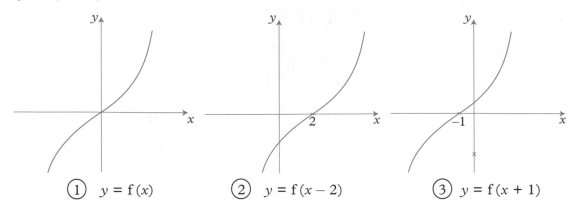

① $y = f(x)$ ② $y = f(x - 2)$ ③ $y = f(x + 1)$

● Can you describe the transformation from ① to ②? From ① to ③?

iii $y = af(x)$

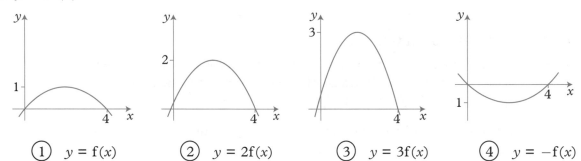

① $y = f(x)$ ② $y = 2f(x)$ ③ $y = 3f(x)$ ④ $y = -f(x)$

● Describe the transformations: from ① to ②; from ① to ③; from ① to ④

iv $y = f(ax)$

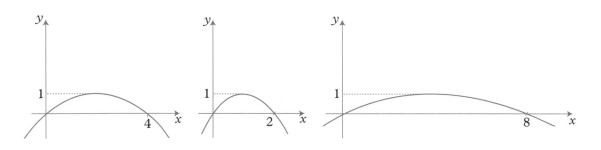

① $y = f(x)$ ② $y = f(2x)$ ③ $y = f\left(\frac{1}{2}x\right)$

● Can you describe the transformation from ① to ②?
● Can you describe the transformation from ① to ③?

This is the most difficult one:

From ① to ② the transformation is a stretch parallel to the x-axis by a scale factor $\frac{1}{2}$.

From ② to ① the transformation is a stretch parallel to the x-axis by a scale factor 2.

Similarly f(3x) would be a stretch with scale factor $\frac{1}{3}$ and $f\left(\frac{1}{5}x\right)$ would be a stretch with scale factor 5.

7.4.2 Summary of rules for curve transformations

i $y = f(x) + a$ Translation by a units parallel to the y-axis.

ii $y = f(x - a)$ Translation by a units parallel to the x-axis. (note the negative sign).

iii $y = af(x)$ Stretch parallel to the y-axis by a scale factor a.

iv $y = f(ax)$ Stretch parallel to the x-axis by a scale factor $\frac{1}{a}$ (note the inverse of a).

> Notice that $y = -f(x)$ is a reflection in the x-axis and that $y = f(-x)$ is a reflection in the y-axis.

EXAMPLE

From the sketch of $y = f(x)$, draw
a $y = f(x + 3)$,
b $y = f(3x)$

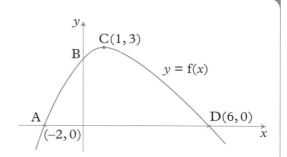

....

a f(x + 3) is a translation of −3 units parallel to the x-axis.

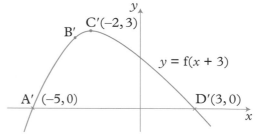

b f(3x) is a stretch parallel to the x-axis by a scale factor $\frac{1}{3}$. Notice that B remains in the same place and that the y-coordinate of C remains unchanged.

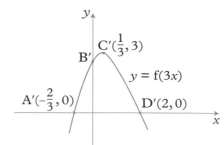

Exercise 16 (H)

(ICT) A computer or calculator which sketches curves can be used effectively in this exercise, although it is not essential.

1 a Draw an accurate graph of $y = x^2$ for values of x from -3 to 3.
Label the graph $y = f(x)$.

b On the same axes draw an accurate graph of $y = x^2 + 4$ and label the graph $y = f(x) + 4$.

c On the same axes draw the graph of $y = (x - 1)^2$ and label the graph $y = f(x - 1)$.

Scales: x from -3 to $+3$, 2 cm = 1 unit
y from   0 to 14, 1 cm = 1 unit.

2 This is the sketch graph of $y = f(x)$.
a Sketch the graph of $y = f(x) + 3$.
b Sketch the graph of $y = f(x + 1)$.
c Sketch the graph of $y = -f(x)$.

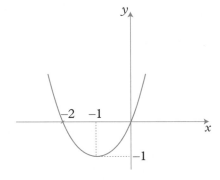

3 This is the sketch graph of $y = f(x)$.
a Sketch the graph of $y = f(x) - 2$.
b Sketch the graph of $y = f(x - 7)$.
Give the new coordinates of the point A on the two sketches.

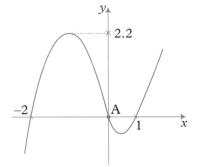

4 This is the sketch of $y = f(x)$ which passes through A, B, C.
Sketch these curves, giving the new coordinates of A, B, C in each case.

a $y = -f(x)$
b $y = f(x - 2)$
c $y = f(2x)$

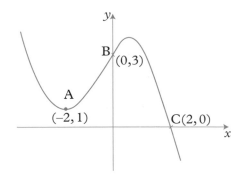

In questions **5** to **16** each graph shows a different function f(x). On
square grid paper draw a sketch to show the given transformation.
The scales are 1 square = 1 unit on both axes.

5

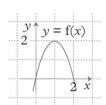

Sketch $y = f(x) + 3$.

6

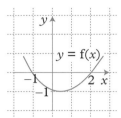

Sketch $y = f(x - 1)$.

7

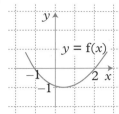

Sketch $y = 2f(x)$.

8

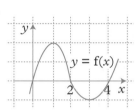

Sketch $y = f(x + 2)$.

9

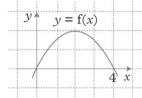

Sketch $y = f(2x)$.

10

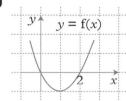

Sketch $y = f\left(\dfrac{x}{2}\right)$.

11

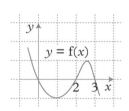

Sketch $y = 3f(x)$.

12

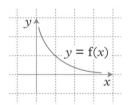

Sketch $y = f(x) - 2$.

13

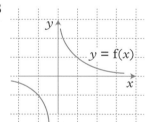

Sketch $y = f(x - 2)$.

14

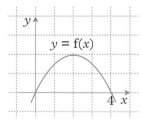

Sketch $y = f(4x)$.

15

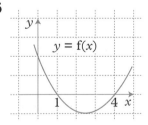

Sketch $y = 2f(x)$.

16

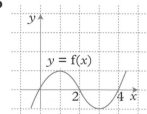

Sketch $y = -f(x)$.

Exercise 17 (H)

1 On the same axes, sketch and label the graphs of
   a $y = x^2$ **b** $y = x^2 - 4$ **c** $y = x^2 + 2$

2 On the same axes, sketch and label the graphs of
   a $y = x^2$ **b** $y = (x - 3)^2$ **c** $y = (x + 2)^2$

3 On the same axes, sketch and label the graphs of
   a $y = x^2$ **b** $y = 4x^2$ **c** $y = \frac{1}{2}x^2$

4 On the same axes, sketch and label the graphs of
   a $y = x^3$ **b** $y = (x - 1)^3$ **c** $y = (x - 1)^3 + 4$

5 On the same axes, sketch and label the graphs of
   a $y = x(x - 1)$ **b** $y = 2x(2x - 1)$

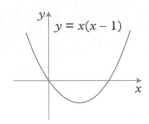

6 On the same axes, sketch and label
   a $y = \frac{1}{x}$ **b** $y = \frac{1}{x - 2}$

7 On square grid paper copy the sketch of
   $y = \cos x$.
   Using different colours, sketch
   a $y = \cos(x + 90°)$
   b $y = 2\cos x$
   c $y = \cos 2x$

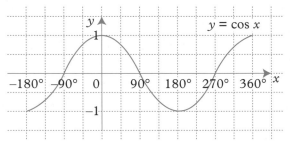

8 On square grid paper copy the sketch graphs of $y = \sin x$ and $y = \tan x$.

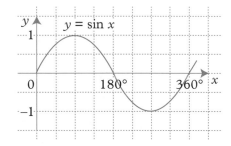 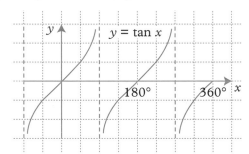

Using different colours, sketch the graphs of
   a $y = \sin\left(\frac{x}{2}\right)$ **b** $y = \tan(x - 90°)$ **c** $y = 2\sin x$ **d** $y = \tan 2x$

9 **a** Draw an accurate graph of $y = (x + 2)^2$ for values of x from -5
   to 1. Label the graph $y = f(x)$.

   b Draw an accurate graph of $y = \left(\frac{1}{2}x + 2\right)^2$ for values of x from -9 to 1.

   Label the graph $y = f\left(\frac{1}{2}x\right)$.

   Scales: x from -10 to 1, 1 cm = 1 unit
            y from   0 to 10, 1 cm = 1 unit.

   Describe the transformation from $y = (x + 2)^2$ onto $y = \left(\frac{1}{2}x + 2\right)^2$.

10 It is possible to perform two or more successive transformations on the same curve. This is a sketch of $y = f(x)$.

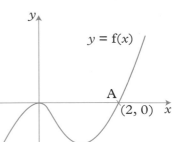

   a Sketch $y = f(x + 1) + 5$.
   b Sketch $y = f(x - 3) - 4$.

Show the new coordinates of the point A on each sketch.

*11 Find the equation of the curve obtained when the graph of $y = x^2 + 3x$ is
   a translated 5 units in the direction ↑
   b translated 2 units in the direction →
   c reflected in the x-axis.

*12 $f(x) = x^2$ and $g(x) = x^2 - 4x + 7$.
   a If $g(x) = f(x - a) + b$, find the values of a and b.
   b Hence sketch the graphs of $y = f(x)$ and $y = g(x)$ showing the transformation from f to g.

*13 Here are the sketches of $y = f(x)$ and $y = g(x)$.

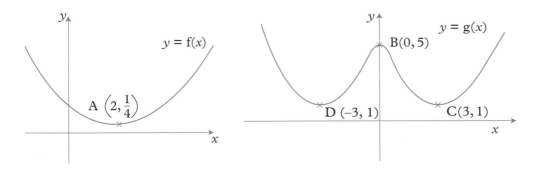

   a Sketch $y = \dfrac{1}{f(x)}$, showing the new coordinates of A.

   b Sketch $y = \dfrac{1}{g(x)}$, showing the new coordinates of B, C, D.

Test yourself

1 a i Simplify.
$$\sqrt{20} \times \sqrt{5}$$

   ii Rationalise the denominator and simplify.
$$\frac{20}{\sqrt{5}}$$

   b Change $0 \cdot 40\dot{3}$ to a fraction.

(OCR, 2008)

2 a Solve $\frac{x}{4} + 1 = 6$

   b Solve $\frac{4}{y + 1} = 3$

   c Factorise fully $6ab^2 - 2ab$

   d Factorise fully $3x^2 + 5x - 12$

(AQA, 2007)

3 Rationalise the denominator of $\frac{2 + \sqrt{3}}{\sqrt{3}}$
Simplify your answer fully.

(AQA, 2004)

4 This sketch graph shows a curve with equation $y = pq^x$.
The curve passes through the points $(1, 5)$ and $(4, 320)$.
Calculate the value of p and the value of q.

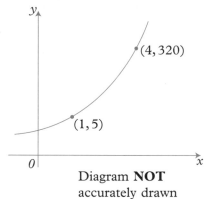

Diagram **NOT** accurately drawn

(Edexcel, 2005)

5 The diagram shows the graph of an equation
of the form $y = x^2 + bx + c$

Find the values of b and c.

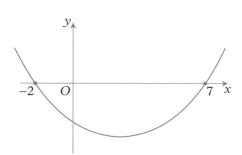

6 The diagram below shows a 6-sided shape.
All the corners are right angles.
All the measurements are given in centimetres.

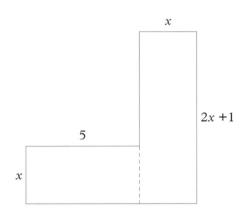

Diagram **NOT**
accurately drawn

The area of the shape is 95 cm^2.
a Show that $2x^2 + 6x - 95 = 0$
b Solve the equation
$2x^2 + 6x - 95 = 0$
Give your solutions correct to 3 significant figures.

(Edexcel, 2008)

7 a Factorise $2n^2 - n - 3$.
b Hence, or otherwise, write 187 as the product of two prime factors.

8 A rectangle has length $(x + 5)$ cm
and width $(x - 1)$ cm.
A corner is removed from the
rectangle as shown.
a Show that the shaded area
is given by $x^2 + 4x - 11$.
b The shaded area is 59 cm^2.
   i Show that $x^2 + 4x - 70 = 0$.
  ii Calculate the value of x.

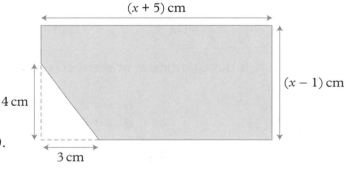

(OCR, 2005)

9 Here is a sketch of $y = 16 - (x - 2)^2$
The curve meets the x-axis at A, the y-axis at B
and has a maximum value at C.
Find the coordinates of
a B   **b** C   **c** A.

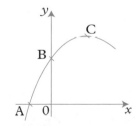

10 a Simplify.

i $\dfrac{7(x-3)}{(x-3)^2}$ **ii** $\left(\dfrac{3x^2}{y^3}\right)^4$ **iii** $\dfrac{4x-x^2}{16-x^2}$

(OCR, 2003)

11 The diagram shows a sketch of part of the curve $y = \cos x°$.

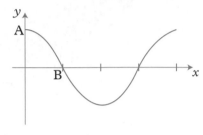

  a Write down the coordinates of
    i   point A
    ii point B
  b Sketch the graph of $y = \cos 2x°$.

12 This is a sketch of the curve with equation $y = f(x)$.
It passes through the origin O and has one vertex at P(2, 3)

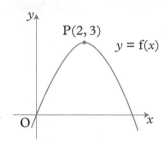

  Write down the coordinates of the vertex of the curve with equation
  a $y = f(-x)$ **b** $y = f(x + 1)$
  c $y = f(2x)$ **d** $y = f(x) + 3$

13 The diagram shows a sketch of $y = f(x)$.

  Sketch the graphs
  a $y = f(x) + 2$,
  b $y = -f(x)$

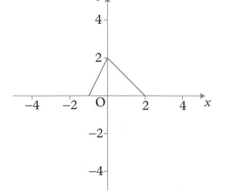

(OCR, 2005)

14 a Show that $(\sqrt{32} + \sqrt{2})^2 = 50$

b The diagram shows a triangle *ABC* of area 30 cm².
The length of *AB* is $4\sqrt{3}$ cm.

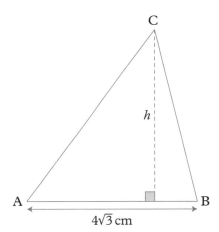

NOT accurately drawn

$4\sqrt{3}$ cm

Calculate the perpendicular height, *h*, of the triangle.
Write your answer in the form $n\sqrt{3}$ where *n* is an integer.

(AQA, 2007)

8 Handling data

In this unit you will:
- find the mean, median, mode and range of a set of data
- draw and use frequency tables and grouped frequency tables
- draw and interpret bar charts, pie charts and frequency polygons
- draw and interpret stem-and-leaf diagrams
- learn about the handling data cycle
- learn about sampling methods and bias
- draw and interpret cumulative frequency diagrams, box plots and histograms.

Functional skills coverage and range:
- collect and represent discrete and continuous data, using ICT where appropriate
- use and interpret statistical measures, tables and diagrams for discrete and continuous data
- use statistical methods to investigate situations.

Links
Statisticians use some complicated methods for ensuring that their data is correct and not biased! This is important for data gathered from medical trials, for example, to make sure that any effects seen come from the drug or operation itself not from other causes such as patients' other medical conditions.

8.1 Averages and range

8.1.1 Mean, median and mode

A single number called an average can be used to represent the whole range of a set of data, whether the data are discrete (for example, exam marks) or continuous (for example, heights).

The median
The data are arranged in order from the smallest to the largest; the middle number is then selected. This is really the central number of the range and is called the median.
If there are two 'middle' numbers, the median is in the middle of these two numbers.

The mean
All the data are added up and the total divided by the number of items. This is called the mean and is equivalent to sharing out all the data evenly.

The mode

The item which occurs most frequently in a frequency table is selected. This is the most popular value and is called the mode (from the French 'a la mode' meaning 'fashionable').

Each average has its purpose and sometimes one is preferable to the other.

The median is fairly easy to find and has an advantage in being hardly affected by untypical values such as very large or very small values that occur at the ends of a distribution.
Look at these examination marks.

20, 21, 21, 22, 23, 23, 25, 27, 27, 27, 29, 98, 98
↑

Clearly the **median** value is 25.
The **mean** of the data is 35·5. It is easier to use in further work such as standard deviation, but clearly it does not give a true picture of the centre of distribution of the data.
The **mode** of these data is 27. It is easy to find and it eliminates some of the effects of extreme values. However it does have disadvantages, particularly in data which have two 'most popular' values, and it is not widely used.

8.1.2 Range

In addition to knowing the centre of a distribution, it is useful to know the range or spread of the data.

range = (largest value) − (smallest value)

For the examination marks, range = 98 − 20 = 78.

EXAMPLE

Find the median, the mean, the mode and the range of this set of 10 numbers: 5, 4, 10, 3, 3, 4, 7, 4, 6, 5.

....

a Arrange the numbers in order of size to find the median.
3, 3, 4, 4, 4, 5, 5, 6, 7, 10
↑

The median is the 'average' of 4 and 5.
So median = 4·5

b mean = $\dfrac{(5 + 4 + 10 + 3 + 3 + 4 + 7 + 4 + 6 + 5)}{10}$

= $\dfrac{51}{10}$ = 5·1

c mode = 4 because there are more 4s than any other number.
d range = 10 − 3 = 7

Exercise 1 (M)

1 Find the median, the mean, the mode and the range of these sets of numbers
   a 3, 12, 4, 6, 8, 5, 4
   b 7, 21, 2, 17, 3, 13, 7, 4, 9, 7, 9
   c 12, 1, 10, 1, 9, 3, 4, 9, 7, 9
   d 8, 0, 3, 3, 1, 7, 4, 1, 4, 4.

2 Copy and complete.
   a The mean of 3, 5, 6 and ☐ is 6.
   b The mean of 7, 8, ☐ and 4 is 8.

3 The frequency table shows the test results for a class of 30 students.

Mark	3	4	5	6	7	8
Frequency	2	5	4	7	6	6

What was the modal mark?

4 **a** Calculate the mean of 2, 5, 7, 7, 4.
   b **Hence** find the mean of 32, 35, 37, 37, 34.

5 The total mass of five men is 380 kg. Calculate the mean mass of the men.

6 The temperature in a greenhouse was measured at midnight every day for a week. The results (in °C) were
     1   −2   4   0   5   2   −1
What was the range of the temperatures?

7 The table shows the age, height and weight of seven children

	Mike	Steve	Dora	Sam	Pat	Rayan	Gary
Age (years)	16	15	17	15	16	15	16
Height (cm)	169	180	170	175	172	163	164
Weight (kg)	50	50	52	44	51	41	48

   a What was the median age?
   b What was the median height?
   c What was the median weight?

8 The range for the eight numbers shown is 40.
Find the **two** possible values of the missing number.

    13   5   27

        19   ?

    42   11   33

9 The mean weight of ten people in a lift is 70 kg. The weight limit for the lift is 1000 kg. Roughly how many more people can get into the lift?

10 There were ten cowboys in a saloon. The mean age of the men was 25 and the range of their ages was 6. Write each statement below and then write next to it whether it is true, possible or false.
   a The youngest man was 18 years old.
   b All the men were at least 20 years old.
   c The oldest person was 4 years older than the youngest.
   d Every man was between 20 and 26 years old.

11 These are the salaries of five employees in a small business.
   Mr A : £22 500   Mr B : £17 900   Mr C : £21 400
   Mr D : £22 500   Mr E : £85 300.
   a Find the mean, median and mode of their salaries.
   b Which does **not** give a fair average? Explain why in one sentence.

12 A farmer has 32 cattle to sell. Their weights in kg are

81	81	82	82	83	84	84	85
85	86	86	87	87	88	89	91
91	92	93	94	96	150	152	153
154	320	370	375	376	380	381	390

[Total weight = 5028 kg]

On the telephone to a potential buyer, the farmer describes the cattle and says the average weight is 'over 157 kg'.
   a Find the mean weight and the median weight.
   b Which average has the farmer used to describe his animals? Does this average describe the cattle fairly?

13 A gardening magazine sells seedlings of a plant through the post and claims that the average height of the plants after one year's growth will be 85 cm. A sample of 24 of the plants were measured after one year with these results (in cm)

6	7	7	9	34	56	85	89
89	90	90	91	91	92	93	93
93	94	95	95	96	97	97	99

(The sum of the heights is 1788 cm.)

   a Find the mean and the median height of the sample.
   b Is the magazine's claim about average height justified?

14 The mean weight of five men is 76 kg. The weights of four of the men are 72 kg, 74 kg, 75 kg and 81 kg. What is the weight of the fifth man?

15 The mean length of six rods is 44·2 cm. The mean length of five of them is 46 cm. How long is the sixth rod?

16 **a** The mean of 3, 7, 8, 10 and x is 6. Find x.
   b The mean of 3, 3, 7, 8, 10, x and x is 7. Find x.

17 The mean height of 12 men is 1·70 m, and the mean height of 8 women is 1·60 m. Find
   a the total height of the 12 men
   b the total height of the 8 women
   c the mean height of the 20 men and women.

18 The total weight of 6 rugby players is 540 kg and the mean weight of 14 ballet dancers is 40 kg. Find the mean weight of the group of 20 rugby players and ballet dancers.

19 Write five numbers so that

the mean is 6
the median is 5
the mode is 4.

20 Find five numbers so that the mean, median, mode and range are all 4.

21 The numbers 3, 5, 7, 8 and N are arranged in ascending order. If the mean of the numbers is equal to the median, find N.

22 The mean of five numbers is 11.
   The numbers are in the ratio $1:2:3:4:5$.
   Find the smallest number.

*23 The median of five consecutive integers is N.
   a Find the mean of the five numbers.
   b Find the mean and the median of the squares of the integers.
   c Find the difference between the two values in part **b**.

8.1.3 Calculating the mean from a frequency table

The frequency table shows the weights of the eggs bought in a supermarket. Find the mean, median and modal weight.

Weight	58 g	59 g	60 g	61 g	62 g	63 g
Frequency	3	7	11	9	8	2

a Mean weight of eggs

$$= \frac{(58 \times 3) + (59 \times 7) + (60 \times 11) + (61 \times 9) + (62 \times 8) + (63 \times 2)}{(3 + 7 + 11 + 9 + 8 + 2)}$$

$$= \frac{2418}{40} = 60.45 \text{ g}$$

b There are 40 eggs so the median weight is the weight between the 20th and 21st numbers. By inspection, both the 20th and 21st weights are 60 g.

So median weight = 60 g

c The modal weight = 60 g   [60 g has the highest frequency]

> 60 g has the highest frequency.

Exercise 2 Ⓜ ⭐

1 The frequency table shows the weights of the 40 apples sold in a shop.

Weight	70 g	80 g	90 g	100 g	110 g	120 g
Frequency	2	7	9	11	8	3

Calculate the mean weight of the apples.

2 The frequency table shows the price of a packet of butter in 30 different shops.

Price	49p	50p	51p	52p	53p	54p
Frequency	2	3	5	10	6	4

Calculate the mean price of a packet of butter.

3 A box contains 50 nails of different lengths as shown in the frequency table.

Length of nail	2 cm	3 cm	4 cm	5 cm	6 cm	7 cm
Frequency	4	7	9	12	10	8

Calculate the mean length of the nails.

4 The tables give the distribution of marks obtained by two classes in a test. For each table, find the mean, median and mode.

a

Mark	0	1	2	3	4	5	6
Frequency	3	5	8	9	5	7	3

b

Mark	15	16	17	18	19	20
Frequency	1	3	7	1	5	3

5 A teacher conducted a mental arithmetic test for 26 students and the marks out of 10 are shown in the table.

Mark	3	4	5	6	7	8	9	10
Frequency	6	3	1	2	0	5	5	4

a Find the mean, median and mode.
b The teacher congratulated the class saying that 'over three-quarters were above average'. Which average justifies this statement?

6 The table shows the number of goals scored in a series of football matches.

Number of goals	1	2	3
Number of matches	8	8	x

a If the mean number of goals is 2·04, find x.
b If the modal number of goals is 3, find the smallest possible value of x.
c If the median number of goals is 2, find the largest possible value of x.

7 The table shows the results of a survey on the number of occupants per car.

Number of occupants	1	2	3	4
Number of cars	7	11	7	x

   a If the mean number of occupants is $2\frac{1}{3}$, find x.

   b If the mode is 2, find the largest possible value of x.

   c If the median is 2, find the largest possible value of x.

8 The marks obtained by the members of a class are summarised in the table.

Mark	x	y	z
Frequency	a	b	c

Calculate the mean mark in terms of a, b, c, x, y and z.

8.1.4 Data in groups

EXAMPLE

The results of 51 students in a test are given in the frequency table.
Find the **a** mean   **b** median   **c** mode.

Mark	30–39	40–49	50–59	60–69
Frequency	7	14	21	9

In order to find the mean you approximate by saying each interval is represented by its midpoint. For the 30–39 interval you say there are 7 marks of 34·5 [that is $(30 + 39) \div 2 = 34\cdot5$].

a Mean $= \dfrac{(34\cdot5 \times 7) + (44\cdot5 \times 14) + (54\cdot5 \times 21) + (64\cdot5 \times 9)}{(7 + 14 + 21 + 9)}$

        $= 50\cdot7745098$

        $= 51$ (2 sf)

b The median is the 26th mark, which is in the interval 50–59.
   You cannot find the exact median.

c The **modal group** is 50–59. You cannot find an exact mode.

Don't forget the mean is only an **estimate** because you do not have the raw data and you have made an assumption with the midpoint of each interval.

Later you will find out how to get an estimate of the median by drawing a cumulative frequency curve.

Exercise 3 Ⓜ ⭐

1 The table gives the number of words in each sentence of a page in a book.

   a Copy and complete the table.

   b Work out an estimate for the mean number of words in a sentence.

Number of words	Frequency f	Midpoint x	fx
1–5	6	3	18
6–10	5	8	40
11–15	4		
16–20	2		
21–25	3		
Totals	20	—	

2 The results of 24 students in a test are given in the table.

Mark	40–54	55–69	70–84	85–99
Frequency	5	8	7	4

Find the midpoint of each group of marks and calculate an estimate of the mean mark.

3 The table shows the number of letters delivered to the 26 houses in a street.

Calculate an estimate of the mean number of letters delivered per house.

Number of letters delivered	Number of houses (frequency)
0–2	10
3–4	8
5–7	5
8–12	3

4 The histogram shows the heights of the 60 athletes in the 2006 British athletics team.

   a Calculate an estimate for the mean height of the 60 athletes.

   b Explain why your answer is an **estimate** for the mean height.

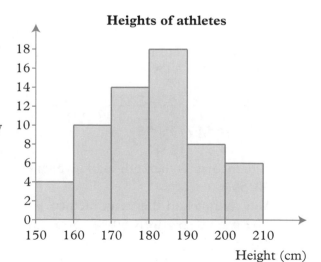

Heights of athletes

8.2 Data presentation

The results of a statistical investigation are known as data.

Raw data are data in the form that they were collected, for example, the number of peas in 40 pods.

```
5  3  6  5  4  6  6  7  4  6
4  7  7  3  7  4  7  5  7  5
6  7  6  7  5  6  6  7  6  6
5  3  6  4  6  5  7  3  6  4
```

Data in this form are difficult to interpret.
Data can be presented in a variety of different ways that make it easier to interpret.

8.2.1 Frequency tables

This presentation is in the form of a tally chart. The tally provides a numerical value for the frequency of occurrence.
From the example above, there are 4 pea pods that contain 3 peas, etc.

Number of peas	Tally	Frequency (f)
3	||||	4
4	⊬⊬ |	6
5	⊬⊬ ||	7
6	⊬⊬ ⊬⊬ |||	13
7	⊬⊬ ⊬⊬	10

8.2.2 Display charts

You can use a pie chart, a bar chart or a frequency polygon to display data. This pie chart illustrates the data about the number of peas in a pod.

Number of peas	Angle on pie chart
3	$\frac{4}{40} \times 360 = 36°$
4	$\frac{6}{40} \times 360 = 54°$
5	$\frac{7}{40} \times 360 = 63°$
6	$\frac{13}{40} \times 360 = 117°$
7	$\frac{10}{40} \times 360 = 90°$

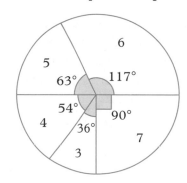

Here is the same data displayed in a bar chart and a frequency polygon.

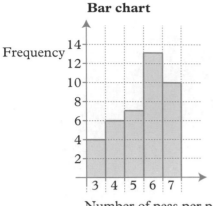

Bar chart

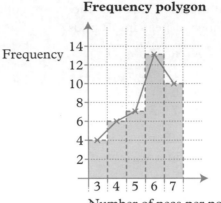

Frequency polygon

Note that in the bar chart and the frequency polygon, the vertical axis is always used to show frequency. It is the frequency that most clearly shows the mode, that is, that more pods contained 6 peas than any other number of peas.

In the frequency polygon, the boxes of the bar chart are replaced by a line joining the tops of their midpoints.

The example above relates to **discrete** data, that is individually distinct number quantities. The next section is about **continuous** data, which are derived from measurements, for example, height, weight, age, time.

> You collect discrete data by counting. You collect continuous data by measuring.

8.2.3   Continuous data

Rounded data

You round each measurement and record the frequency of this rounded quantity as for discrete data. For example, the lengths of 36 pea pods can be rounded to the nearest mm and then recorded as raw data — a pea pod measuring 59·2 mm is recorded as one that has a length of 59 mm.

Rounded data – lengths of 36 pea pods

```
52  80  65  82  77  60  72  83  63
78  84  75  53  73  70  86  55  88
85  59  76  86  73  89  91  76  92
66  93  84  62  79  90  73  68  71
```

Grouped data

You group the measurements into classes with defined class boundaries. In this **grouped frequency table**, the data on lengths of pea pods are grouped into the classes shown in the left-hand column.

Length (*l* mm)	Tally	Frequency
50 < *l* ≤ 60	IIII	4
60 < *l* ≤ 70	IHI I	6
70 < *l* ≤ 80	IHI IHI II	12
80 < *l* ≤ 90	IHI IHI	10
90 < *l* ≤ 100	IIII	4

So, the pea pod measuring 59·2 mm is one of 4 recorded in the class 50 < *l* ≤ 60.

Here are the grouped data displayed in a bar chart and a frequency polygon.

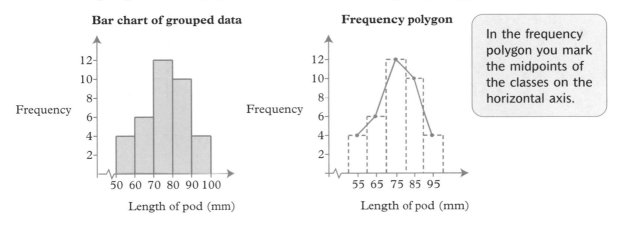

In the frequency polygon you mark the midpoints of the classes on the horizontal axis.

In some cases where there are lots of small classes, the midpoints of the frequency polygon are joined with a curve. The diagram is then called a **frequency curve**.

Exercise 4 Ⓜ ⭐Ⓕ

1 Karine and Jackie intend to go skiing in February. They have information about the expected snowfall in February for two possible places.

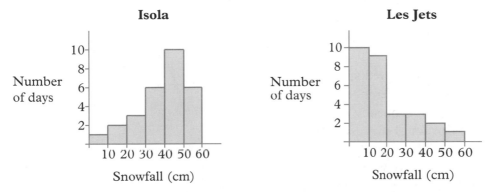

Decide where you think they should go. It doesn't matter where you decide, but you **must** say why, using the charts above to help you explain.

2 The chart shows information about people who use the Internet regularly

a About what percentage of boys in the 5–9 age group used the Internet regularly?

b In what age groups did more women than men use the Internet?

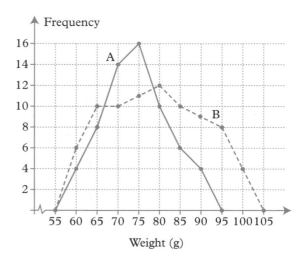

3 In a survey, Jake counted the number of people in 100 cars passing a set of traffic lights. Here are the results.

Number of people in car	0	1	2	3	4	5	6
Frequency	0	10	35	25	20	10	0

a Draw a bar chart to illustrate the data.

b On the same graph draw the frequency polygon. Here the bar chart has been started. For frequency, use a scale of 1 cm for 5 units.

4 The diagram shows two frequency polygons giving the distribution of the weights of players in two different sports A and B.

a How many people played sport A?

b Comment on two differences between the two frequency polygons.

c Either for A or for B suggest a sport where you would expect the frequency polygon of weights to have this shape. Explain in one sentence why you have chosen that sport.

5 A scientist at an agricultural college is studying the effect of a new fertiliser for raspberries. She measures the heights of the plants and also the total weight of fruit collected. She does this for two sets of plants: one with the new fertiliser and one without it. Here are the frequency polygons:

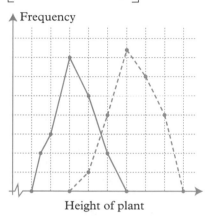

a What effect did the fertiliser have on the heights of the plants?
b What effect was there on the weights of fruit collected?

6 The diagram illustrates the production of apples in two countries.
In what way could the pictorial display be regarded as misleading?

UK
470 thousand
tonnes

France
950 thousand
tonnes

7 The graph shows the performance of a company in the year in which a new manager was appointed. In what way is the graph misleading?

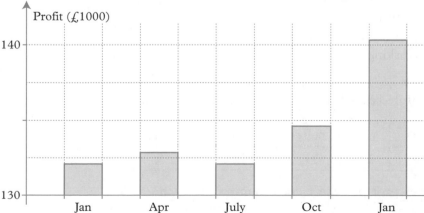

8 In one week Tom spends £120 on various items as shown in the pie chart.

  a What fraction of his money did Tom spend on
  i food
  ii fares
  iii rent
  iv savings
  v entertainment
  vi clothes?

  b How much of the £120 did Tom spend on
  i food
  ii fares
  iii rent
  iv savings?

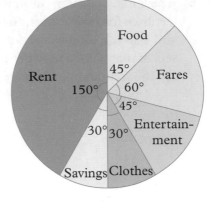

9 The total cost of a holiday was £900.
The pie chart shows how this cost was made up.
  a How much was spent on food?
  b How much was spent on travel?
  c How much was spent on the hotel?
  d How much was spent on other items?

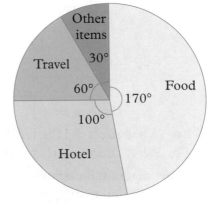

10 The pie chart shows how Sandra spends her time in a maths lesson which lasts 60 minutes.
  a How much time does Sandra spend
  i getting ready to work
  ii talking
  iii sharpening a pencil?
  b Sandra spends 3 minutes working. What is the angle on the pie chart for the time spent working?

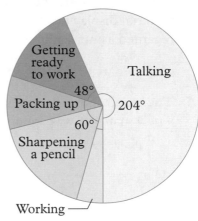

11 A quantity of scrambled eggs is made using these ingredients.

Ingredient	eggs	milk	butter	cheese	salt/pepper
Mass	450 g	20 g	39 g	90 g	1 g

Calculate the angles on a pie chart corresponding to each ingredient.

12 A firm making artificial sand sold its products in four countries.

    5% were sold in Spain

   15% were sold in France

   15% were sold in Germany

   65% were sold in the UK.

What angles would you use on a pie chart representing this information?

13 The pie chart illustrates the sales of various brands of petrol.

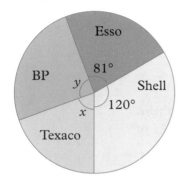

  a What percentage of sales does Esso have?

  b If Texaco accounts for $12\frac{1}{2}\%$ of total sales, calculate the angles x and y.

14 The cooking times for meals L, M and N are in the ratio $3:7:x$. On a pie chart, the angle corresponding to L is $60°$. Find x.

8.2.4 Stem-and-leaf diagrams

You can display data in groups in a stem-and-leaf diagram.

Here are the marks of 20 girls in a science test.

54	42	61	47	24	43	55	62	30	27
28	43	54	46	25	32	49	73	50	45

Put the marks into groups 20–29, 30–39, ... 70–79.
Choose the tens digit as the 'stem' and the units as the 'leaf'.

The first four marks are shown [54, 42, 61, 47]

Stem (tens)	Leaf (units)
2	
3	
4	2 7
5	4
6	1
7	

It is helpful to have a key.

↓

Key 5|4 means 54

The complete diagram is below … and then with the leaves in
numerical order:

Stem	Leaf
2	4 7 8 5
3	0 2
4	2 7 3 3 6 9 5
5	4 5 4 0
6	1 2
7	3

Stem	Leaf
2	4 5 7 8
3	0 2
4	2 3 3 5 6 7 9
5	0 4 4 5
6	1 2
7	3

The diagram shows the shape of the distribution. It is also easy to
find the mode, the median and the range.

Back-to-back stem plots

Two sets of data can be compared using a **back-to-back stem plot**.
The two sets of leaves share the same stem. Here are the marks of
20 boys who took the same science test as the girls.

33	55	63	74	20	35	40	67	21	38
51	64	57	48	46	67	44	59	75	56

Here are the boys' marks entered onto
a back-to-back stem plot.

Boys	Stem	Girls
1 0	2	4 5 7 8
8 5 3	3	0 2
8 6 4 0	4	2 3 3 5 6 7 9
9 7 6 5 1	5	0 4 4 5
7 7 4 3	6	1 2
5 4	7	3

Here is
the key.

You can see that the boys achieved higher
marks than the girls in this test.

Key (boys)
1 | 5 means 51

Key (girls)
5 | 4 means 54

Exercise 5 Ⓜ ⭐

1 The diagram shows the marks of 24 students in a test.

41	23	35	15	40	39	47	29
52	54	45	27	28	36	48	51
59	65	42	32	46	53	66	38

Stem	Leaf
1	
2	3
3	5
4	1
5	
6	

Key
2 | 3 means 23

a Copy and complete the stem-and-leaf diagram.
The first three entries are shown.
b Write the range of the marks.

2 Draw a stem-and-leaf diagram for each set of data.

a

24	52	31	55	40	37	58	61	25	46
44	67	68	75	73	28	20	59	65	39

b

30	41	53	22	72	54	35	47
44	67	46	38	59	29	47	28

Stem	Leaf
2	
3	
4	
5	
6	
7	

3 Here is the stem-and-leaf diagram showing the masses, in kg, of some people in a lift.
  a Write the range of the masses
  b How many people were in the lift?
  c What is the median mass?

Key
3 | 2 means 32 kg

Stem (tens)	Leaf (units)
3	2 5
4	1 1 3 7 8
5	0 2 5 8
6	4 8
7	1
8	2

4 In this question the stem shows the units digit and the leaf shows the first digit after the decimal point.
Draw the stem and leaf diagram using these data.

2·4	3·1	5·2	4·7	1·4	6·2	4·5	3·3
4·0	6·3	3·7	6·7	4·6	4·9	5·1	5·5
1·8	3·8	4·5	2·4	5·8	3·3	4·6	2·8

Key
3 | 7 means 3·7

Stem	Leaf
1	
2	
3	
4	
5	
6	

  a What is the median?
  b Write the range.

5 Here is a back-to-back stem plot showing the pulse rates of several people.

Men			Women
5 1	4		
7 4 2	5		3
8 2 0	6		2 1
5 2	7		4 4 5 8 9
2 6	8		2 5 7
4	9		2 8

Key (men)
1 | 4 means 41

Key (women)
5 | 3 means 53

  a How many men were tested?
  b What was the median pulse rate for the women?
  c Write a sentence to describe the main features of the data.

8.3 Comparing sets of data

To compare two sets of data we need to write at least two things:
- Compare an average (mean, median or mode)
- Compare the range of each set of data (this measures how spread out the data is).

The weights of the players in two basketball teams are (in kg):

Team A: 58, 60, 60, 68, 81, 83, 94
Team B: 62, 66, 66, 72, 74, 76, 79, 81

Team A: median weight = 68 kg; range = 94 − 58 = 36 kg
Team B: median weight = 73 kg; range = 81 − 62 = 19 kg

Compare the weight of the players in Team A and Team B.
....

The median weight for Team A is lower than that for Team B and the range for Team A is much greater than that for Team B. The weights of the players in Team A are more spread out.

Exercise 6 Ⓜ ⭐

1 The six members of the Pearce family weigh 39 kg, 48 kg, 52 kg, 56 kg, 58 kg and 59 kg.
The five members of the Taylor family weigh 35 kg, 41 kg, 63 kg, 64 kg and 75 kg.
  a For each family work out the median weight and the range of the weights.
  b Write a sentence, similar to the one in the example, to compare the weights of the people in the two families.

2 Two marks for two classes in a test are shown in the stem-and-leaf diagrams.
  a Find the median and range for each class.
  b Write a sentence to compare the marks for the two classes.

Class 10M

5	6 7
6	0 1
7	3 4 6
8	2 7

Class 10S

4	1 2 4
5	5 8 8
6	2
7	1 7
8	0 8

Key 5 | 6 = 56

8.4 Scatter diagrams

Sometimes it is interesting to discover if there is a relationship
(or **correlation**) between two sets of data.
Examples:

- Do tall people weigh more than short people?
- If you spend longer revising for a test, will you get a higher mark?
- Do tall parents have tall children?
- If there is more rain, will there be less sunshine?
- Does the number of Olympic gold medals won by British athletes
  affect the rate of inflation?

If there is a relationship, it will be easy to spot if your data are
plotted on a scatter diagram – that is a graph in which one set
of data is plotted on the horizontal axis and the other on the
vertical axis.

EXAMPLE

Each month the average outdoors temperature was recorded
together with the number of therms of gas used to heat
a house. The scatter diagram shows the results.

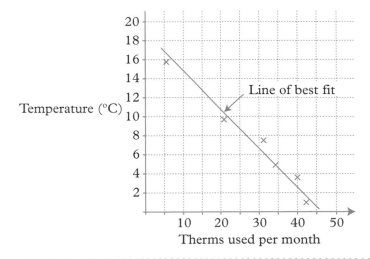

Clearly there is a high degree of **correlation** between these
two figures. British Gas do in fact use weather forecasts as
their main short-term predictor of future gas consumption
over the whole country. A **line of best fit** has been drawn 'by
eye'. You can estimate that if the outdoor temperature is 12 °C
then about 17 therms of gas will be used.

Note: You can only predict within the range of values given.
If you extend the line for temperatures below zero the line of best fit predicts that about 60 therms will be used when the temperature is $-4\,°C$. But $-4\,°C$ is well outside the range of the values plotted so the prediction is not valid. [Perhaps at $-4\,°C$ a lot of people might stay in bed and the gas consumption would not increase by much. The point is that you don't know!]

a The line in the example has a negative gradient and you say that there is **negative correlation**.

b If the line of best fit has a positive gradient you say that there is **positive correlation**.

c Some data when plotted on a scatter diagram does not appear to fit any line at all. In this case there is no correlation.

Note that when a scatter diagram shows no correlation it does **not** necessarily imply that there is 'no relationship' between the variables. It shows that there is no **linear** relationship. There might, for example, be an inverse or an exponential relationship.

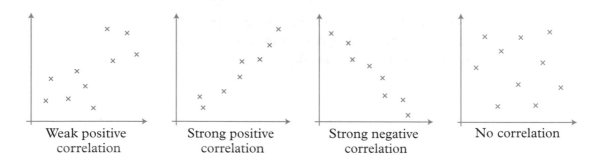

Weak positive correlation   Strong positive correlation   Strong negative correlation   No correlation

Exercise 7

1 For this question you need to make some measurements of people in your class.

a Measure everyone's height and armspan to the nearest cm.

Plot the measurements on a scatter diagram. Is there any correlation?

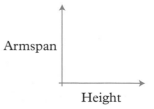

b Now measure the circumference of everyone's head just above the eyes. Plot head circumference and height on a scatter graph. Is there any correlation?

c Decide as a class which other measurements [for example, pulse rate] you can (fairly easily) take and plot these to see if any correlation exists.

d Which pair of measurements gave the best correlation?

2 Plot the points given on a scatter diagram, with t across the page and z up the page. Draw axes with values from 0 to 20. Describe the correlation, if any, between the values of t and z. [That is, 'strong positive', 'weak negative' etc.]

a

t	8	17	5	13	19	7	20	5	11	14
z	9	16	7	13	18	10	19	8	11	15

b

t	4	9	13	16	17	6	7	18	10
z	5	3	11	18	6	11	18	12	16

c

t	12	2	17	8	3	20	9	5	14	19
z	6	13	8	15	18	2	12	9	12	6

3 Describe the correlation, if any, in each scatter diagram.

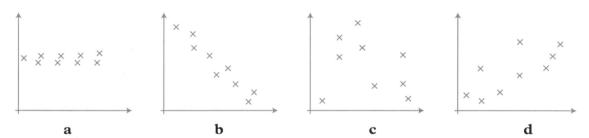

| **a** | **b** | **c** | **d** |

4 Plot the points in the table on a scatter diagram, with s across the page and h up the page. Draw axes with values from 0 to 20.

s	3	13	20	1	9	15	10	17
h	6	13	20	6	12	16	12	17

  a Draw a line of best fit.
  b What value would you expect for h when s is 6?

5 The table shows the marks of seven students in the two papers of a physics examination.

Paper 1	20	32	40	60	71	80	91
Paper 2	15	25	40	50	64	75	84

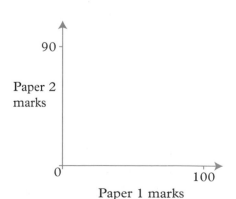

  a Plot the marks on a scatter diagram, using a scale of 1 cm to 10 marks, and draw a line of best fit.
  b Rana scored a mark of 50 on Paper 1. What would you expect her to get on Paper 2?

6 The table shows
   i the engine size in litres of various cars
   ii the distance travelled in km on one litre of petrol.

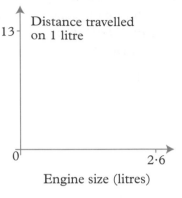

Engine	0·8	1·6	2·6	1·0	2·1	1·3	1·8
Distance	13·0	10·2	5·4	12·0	7·8	11·2	8·5

a Plot the figures on a scatter diagram using a scale of 5 cm to 1 litre across the page and 1 cm to 1 km up the page. Draw a line of best fit.
b A car has a 2·3 litre engine. How far would you expect it to go on one litre of petrol?

7 The data shows the latitude of ten cities in the northern hemisphere and the average high temperatures.

City	Latitude (degrees)	Mean high temperature (°F)
Bogota	5	66
Casablanca	34	72
Dublin	53	56
Hong Kong	22	77
Istanbul	41	64
St Petersburg	60	46
Manila	15	89
Mumbai	19	87
Oslo	60	50
Paris	49	59

a Draw a scatter diagram and draw a line of best fit. Plot latitude across the page with a scale of 2 cm to 10°. Plot temperature up the page from 40 °F to 90 °F with a scale of 2 cm to 10 °F.
b Which city lies well off the line?
   Do you know what factor might cause this apparent discrepancy?
c The latitude of Shanghai is 31°N. What do you think its mean high temperature is?

8 What sort of pattern would you expect if you took readings of these variables and drew a scatter diagram?
   a cars on roads; accident rate
   b sales of perfume; advertising costs
   c birth rate; rate of inflation
   d outside temperature; sales of ice cream
   e height of adults; age of same adults

8.5 The handling data cycle

1 Specifying the problem and planning	Start with a problem or a question you wish to answer, for example • Do most children go to the secondary school of their choice? • Do adults watch more television than children?
2 Collecting data	• You have to decide what data (information) you need to collect to answer the question. • You may decide to collect the data yourself. When you collect data by a survey or by an experiment the data are called **primary data**. • If you look up published data in a book or on the Internet they are called **secondary data**.
3 Processing and presenting the data	• You usually have to **process** the data by, for example, calculating averages or frequencies so that you can **represent** the data in a pictorial form such as a pie chart or a frequency diagram.
4 Interpreting and discussing results	• Finally you need to **interpret** the results and draw conclusions so that you can answer the original question. • You need to look for patterns in the data and for any possible exceptions. • Sometimes the results you obtain suggest that you need to modify the way you did the survey. You may need to go again through stages **1**, **2**, **3** and **4**.

8.5.1 Collecting data

Simple data collection sheet

• A **tally chart** is often a good way to get started. Here is a tally chart for the responses to a question about the amount of homework set to students in a school.

Response	Tally	Frequency
Not enough	⊥⊦⊣ ∣∣∣∣	9
About right	⊥⊦⊣ ⊥⊦⊣ ⊥⊦⊣ ⊥⊦⊣ ∣∣	22
Too much	⊥⊦⊣ ⊥⊦⊣ ∣∣∣	13
Don't know	∣∣∣	3

> You could represent these data in a bar chart or pie chart.

8.5.2 Questionnaires

Most surveys are conducted using questionnaires. It is important to design the questionnaire well so that:

a people will cooperate and will answer the questions honestly
b the questions are not biased
c the answers to the questions can be analysed and presented for ease of understanding.

Checklist

A Provide an introduction to the sheet so that the person you are asking knows the purpose of the questionnaire.

> 'Proposed new traffic lights'

B Make the questions easy to understand and specific to answer.
Do **not** ask vague questions like this ⟶

> Did you see much of the Olympics on TV?

The answers could be:

>   'Yes, a lot'
>   'Not much'
>   'Only the best bits'
>   'Once or twice a day'

You would find it hard to analyse those answers.

A **better** question is:

> 'How much of the Olympic coverage did you watch?' Tick one box
>
> None at all   ☐
> Up to 1 hour per day   ☐
> 1 to 2 hours per day   ☐
> More than 2 hours per day ☐

C Make sure that the questions are not **leading** questions. It is human nature not to contradict the questioner. Remember that the survey is to find out opinions of other people, not to support your own.
Do **not** ask
>   'Do you agree that the BBC has the best sports coverage?'
A better question is

> 'Which of these has the best sports coverage?'
>
> BBC   ITV   Channel 4   Satellite TV
> ☐   ☐   ☐   ☐

You might ask for one tick or possibly numbers 1, 2, 3, 4 to show an order of preference.

D If you are going to ask sensitive questions (about age or income, for example), design the question with care so as not to offend or embarrass.

Do **not** ask:

'How old are you?'

or 'Give your date of birth'

A better question is:

Tick one box for your age group.

15–17	18–20	21–30	31–50
☐	☐	☐	☐

E Do not ask more questions than necessary and put the easy questions first.

Exercise 8 Ⓜ ⭐

In questions **1** to **7** explain why the question is not suitable for a questionnaire. Write an improved question in each case.

Write some questions with 'yes/no' answers and some questions which involve multiple responses.

Remember to word your questions simply.

1 Which sort of holiday do you like best?

2 What do you think of the new head teacher?

3 How long do you watch television each day?

2–3 hrs	3–4 hrs	5–6 hrs

4 How much would you pay to use the new car park?

☐ less than £1   ☐ more than £2·50

5 Do you agree that English and Maths are the most important subjects at school?

6 Do you or your parents often hire DVDs from a shop?

7 Do you agree that we get too much homework?

8 Some students designed a questionnaire to find out peoples' views about television. Comment on their questionnaire and write an improved version.

Name _____   Sex   M/F

Age   _____

- How much television do you watch?

☐   ☐   ☐

not much   quite a lot   a lot

- What is your favourite programme on TV?

- Do you agree that there should be more stations like MTV?

☐ agree   ☐ disagree

- Do you like nature programmes?

☐   ☐   ☐   ☐

No   Not really   Sometimes   I love them

---

9 Write a suitable question to find out what **type** of TV programme people of your age watch most. For example: comedy, romance, sport etc.

10 Here is another style of question which can be useful.

Which of these statements best describes your attitude to using a computer.

A I like using them for all sorts of things.
B I use them when I have to.
C I hate them.

Please ring   A   B   C

Design a questionnaire to test the truth of these statements.
a Most people choose to shop in a supermarket where it is easy to park a car.
b Children of school age watch more television than their parents.
c Boys prefer action films and girls prefer some sort of story.
d Most people who smoke have made at least one serious attempt to give it up.

8.5.3 Two-way tables

A school collects data about the reading ability of six-year-olds. Children are given a short passage and they 'pass' if they can read over three-quarters of the words. The results are given in this two-way table.

	Boys	Girls	Total
Can read	464	682	
Cannot read	317	388	
Total			

Useful information can be found from the table. Begin by working out the totals for each row and column.

Find also the 'grand total' by adding together the totals in **either** the rows **or** columns.

	Boys	Girls	Total
Can read	464	682	1146
Cannot read	317	388	705
Total	781	1070	1851

↑
'Grand total'

EXAMPLE

Answer these questions using the above two-way table.
a What percentage of the boys can read?
b Similarly, what is the percentage of girls who can read?

....

a Out of 781 boys, 464 can read.
So percentage of boys who can read $= \frac{464}{781} \times 100\%$

$= 59 \cdot 4\%$ (1 dp)

b Out of 1070 girls, 682 can read.
So percentage of girls who can read $= \frac{682}{1070} \times 100\%$

$= 63 \cdot 7\%$ (1 dp)

So a slightly higher percentage of the girls can read.

Exercise 9 Ⓜ

Give percentages correct to one decimal place.

1 Here are ten shapes.

Copy and complete this two-way table to show the different shapes.

	Shaded	Unshaded
Squares		
Triangles		

2 The students in a class were asked to name their favourite sport.
Here are the results.

Boy	Football	Boy	Swimming	Boy	Football
Girl	Hockey	Boy	Football	Girl	Football
Girl	Football	Girl	Hockey	Girl	Swimming
Boy	Hockey	Girl	Swimming	Boy	Football
Boy	Football	Girl	Hockey	Boy	Hockey

  a Record the results in a two-way table.
  b How many boys were in the class?
  c What percentage of the boys chose hockey?

3 This incomplete two-way table shows details
of seven year-olds who can/cannot ride
a bicycle without stabilisers.

	Girls	Boys	Total
Can cycle	95		215
Cannot cycle		82	
Total			476

  a Copy the table and work out the
    missing numbers.
  b What percentage of the girls can cycle?
  c What percentage of the boys can cycle?

4 Mrs Kotecha collected these data to help assess the risk
of various drivers who apply for car insurance with her company.
She needs to know if men drivers are more or less likely
than women drivers to be involved in motor accidents.

	Men	Women	Total
Had accident	75	88	
Had no accident	507	820	
Total			

  a Copy the table and work out the totals for the rows
    and columns.
  b What percentage of the men had accidents?
  c What percentage of the women had accidents?
  d What conclusions, if any, can you draw?

8.6 Sampling and bias

8.6.1 Surveys

Statisticians frequently carry out surveys to investigate a characteristic of a population. A population is the set of all possible items to be observed, not necessarily people.

Surveys are done for many reasons.

- Newspapers seek to know voting intentions before an election.
- Advertisers try to find out which features of a product appeal most to the public.
- Research workers testing new drugs need to know their effectiveness and their possible side effects.
- Government agencies conduct tests for the public's benefit such as the percentage of first class letters arriving the next day or the percentage of '999' calls answered within 5 minutes.

Census

A census is a survey in which data are collected from every member of the population of a country. A census is done in Great Britain every ten years; 1971, 1981, 1991, 2001, ... Data about age, educational qualifications, race, distance travelled to work, etc., is collected from every person in the country on the given 'Census Day'. Census forms take several years to prepare and many years for thorough analysis of the data collected.

> Because no person must be omitted, a census is a huge and very expensive task.

Samples

If a full census is not possible or practical, then it is necessary to take a sample from the population. Two reasons are:

- it will be much cheaper and quicker than a full census
- the object being tested may be destroyed in the test, like testing the average lifetime of a new light bulb or the new design of a car tyre.

The choice of a sample prompts two questions.

1 Does the sample truly represent the whole population?
2 How large should the sample be?

After taking a sample, it is assumed that the result for the sample reflects the whole population. For example, if 20% of a sample of 1000 people say they will vote in a particular way, then it is assumed that 20% of the whole electorate would do so likewise, subject of course to some error.

8.6.2 Simple random sampling

This is a method of sampling in which every item of the population has an equal chance of being selected.

Selection techniques

Out of a school of 758 students, ten are to be selected to take part in an educational experiment in which homework is banned. How would you select these students?
First give each student a three-digit number from 001 to 758.

<u>Method 1</u>

Each person's number is written on a piece of card, the cards are placed in a hat and mixed up. Ten cards are then drawn.

<u>Method 2</u>

Use a random number table like this one.

This table is published by kind permission of the Department of Statistics, University College, London

Random number table

11 74	26 93	81 44	33 93	08 72	32 79	73 31	18 22	64 70	68 50
43 36	12 88	59 11	01 64	56 23	93 00	90 04	99 43	64 07	40 36
93 80	62 04	78 38	26 80	44 91	55 75	11 89	32 58	47 55	25 71
49 54	01 31	81 08	42 98	41 87	69 53	82 96	61 77	73 80	95 27
36 76	87 26	33 37	94 82	15 69	41 95	96 86	70 45	27 48	38 80
07 09	25 23	92 24	62 71	26 07	06 55	84 53	44 67	33 84	53 20
43 31	00 10	81 44	86 38	03 07	52 55	51 61	48 89	74 29	46 47
61 57	00 63	60 06	17 36	37 75	63 14	89 51	23 35	01 74	69 93
31 35	28 37	99 10	77 91	89 41	31 57	97 64	48 62	58 48	69 19
57 04	88 65	26 27	79 59	36 82	90 52	95 65	46 35	06 53	22 54
09 24	34 42	00 68	72 10	71 37	30 72	97 57	56 09	29 82	76 50
97 95	53 50	18 40	89 48	83 29	52 23	08 25	21 22	53 26	15 87
93 73	25 95	70 43	78 19	88 85	56 67	16 68	26 95	99 64	45 69
72 62	11 12	25 00	92 26	82 64	35 66	65 94	34 71	68 75	18 67
61 02	07 44	18 45	37 12	07 94	95 91	73 78	66 99	53 61	93 78
97 83	98 54	74 33	05 59	17 18	45 47	35 41	44 22	03 42	30 00
89 16	09 71	92 22	23 29	06 37	35 05	54 54	89 88	43 81	63 61
25 96	68 82	20 62	87 17	92 65	02 82	35 28	62 84	91 95	48 83
81 44	33 17	19 05	04 95	48 06	74 69	00 75	67 65	01 71	65 45
11 32	25 49	31 42	36 23	43 86	08 62	49 76	67 42	24 52	32 45

Start anywhere in the table and go either up, down, left, or right to read numbers in groups of three digits. Starting on the bottom right-hand corner going up and then down, would give the numbers

553 108 ~~797~~ 049 370 071 605 372 ~~824~~ ~~915~~ 586 670 684

Why were 797, 824, 915 rejected?

So students with the above 10 numbers will take part in the experiment.

Method 3   Random number generator
Use a calculator or a computer.
On the CASIO calculator, random numbers are generated by pressing
RAN # after using SHIFT to get this function. The numbers produced are three-digit numbers between .001 and .999. Ignore the decimal point.
Although these are not true random numbers due to their method of generation, they are adequate for school use.

Random two-digit numbers are obtained simply by omitting the last digit.

Find 6 two-digit numbers between 01 and 60.

Ignoring the decimal points, a calculator gave these numbers.
~~819~~ 453 480 ~~718~~ ~~891~~ 326 050 217 ~~990~~ 437

So use: 45 48 32 05 21 43

8.6.3 Stratified random sampling

Some populations separate naturally into a number of sub-groups or strata. Providing that the strata are quite distinct and that every member of the population belongs to one and only one stratum, then the population can be sampled using the method of stratified random sampling.

s.   pl.
datum – data
stratum – strata

Opinion pollsters use this method when conducting polls on voting intentions. Their claim is that results are accurate to about 3 per cent.

Method

- Separate the population into suitable strata. Find what proportion of the population is in each stratum.
- Select a sample from each stratum proportional to the stratum size, by simple random sampling.

In Year 10 of a school, the 180 students are split into three groups for games; 90 play football, 55 play hockey and 35 play badminton.
Use a stratified random sample of 30 students to estimate the mean weight of the 180 students.

....

The sample size from each of the three groups (strata) must be proportional to the stratum size. So the 30 students are made up by selecting

from the football group, $\dfrac{90}{180} \times 30 = 15$

from the hockey group, $\dfrac{55}{180} \times 30 = 9 \cdot 16$ (say 9)

from the badminton group, $\dfrac{35}{180} \times 30 = 5 \cdot 83$ (say 6)

The three sample means were calculated as

football $53 \cdot 2$ kg   hockey $49 \cdot 4$ kg   badminton $46 \cdot 4$ kg

So the mean for the
population of 180 students $= \dfrac{53 \cdot 2 + 49 \cdot 4 + 46 \cdot 4}{3}$

$= 49 \cdot 7$ kg (3 sf)

Exercise 10 (H)

1 The table shows the number of students in each year group at a new school.

Year 7	162
Year 8	161
Year 9	157
Year 10	63
Year 11	58

Naseem takes a stratified sample of 60 students from this school. Work out how many students from each year group should be in Naseem's sample.

2 The table gives the number of employees in five firms.

Firm	Employees
A	398
B	1011
C	409
D	207
E	1985

A stratified sample of 100 people is taken.
How many people should be selected from each firm?

8.6.4 Bias

It is **not** possible to choose a random sample using personal
judgement. Random samples may give results containing errors
but these can be predicted and allowed for. The type of error
introduced with judgmental sampling is unpredictable and thus
corrections for it cannot be made. This type of unpredictable error
is called **bias**.

Bias can come from a variety of sources including non-random
sampling. For example, for data collected by questionnaire,
non-response from subjects can introduce bias into the results.
Qualities which these subjects have in common will not be
represented in the collected data. For data collected by a street
survey, the time of day and location may well mean that there
is a sector or sectors of the population who are not
questioned.

Perhaps the most famous mistake in opinion sampling occurred in
America in 1936. A magazine carried out an enormous poll in which
over two million people were asked to state their preference for
President, either the Democrat (F D Roosevelt) or the Republican
(A E Landon). The sample was taken by selecting the two million
people by simple random sampling from the telephone directories.
The poll showed a large majority would vote for the Republican,
but in the election the Democrat, Roosevelt, won. In 1936
poorer people did not usually have a phone so the sample that
was taken was not representative of the whole population of
voters.

A common cause of bias occurs when the questions asked in the survey are not clear or are leading questions. This was considered earlier in the section on questionnaires.

Exercise 11 (H)

In questions **1** to **4** decide whether the method of choosing people to answer questions is satisfactory or not. Consider whether or not the sample suggested might be **biased** in some way. Where necessary suggest a better way of obtaining a sample.

1 A teacher, with responsibility for school meals, wants to hear students' opinions on the meals currently provided. She waits next to the dinner queue and questions the first 50 students as they pass.

2 An opinion pollster wants to canvass opinion about our European neighbours. He questions drivers as they are waiting to board their ferry at Dover.

3 A journalist wants to know the views of local people about a new one-way system in the town centre. She takes the electoral roll for the town and selects a random sample of 200 people.

4 A pollster working for the BBC wants to know how many people are watching a new series which is being shown. She questions 200 people as they are leaving a supermarket between 10:00 and 12:00 one Thursday.

8.7 GCSE statistics project

8.7.1 Your own work

● Almost certainly the best idea for a statistics project will be an idea which **you** think of because **you** want to know the answer. You have to decide what data to collect and how you are going to collect them. You need to make a plan and then design an experiment or survey. You must decide what primary or secondary data will be suitable.

- At this point stop to consider two points.
  a Will the survey help to answer the question?
  b How many people do I need for the survey?

- Having collected the data you need to **process** them and then represent them, usually in pictorial form. You may need to calculate averages or percentages and then draw pie charts, scatter graphs, stem-and-leaf diagrams and so on.

  Do not be afraid to use colours.

> Both of these questions can be answered by first doing a **pilot survey**, that is, a quick mini-survey of 5–10 people. Some of the questions may need changing straightaway.

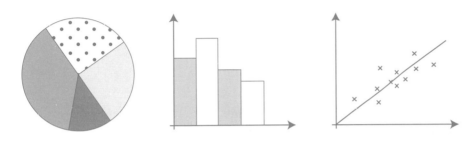

- Finally you interpret the results of your survey, making sure that you refer back to the initial question or problem. Write a **conclusion** in which you summarise the main findings of your work. Make sure that your conclusions are justified by the evidence in your report. You might suggest ways in which your work could be **extended** if more time was available.

- Remember: The best project will be on a topic which **you** find interesting. If you cannot think of a suitable idea you can use one from the list below which some of the author's students have enjoyed investigating.

  a Most people choose to shop in a supermarket where it is easy to park a car.
  b Children of school age watch more television than their parents.
  c Boys prefer action films and girls prefer some sort of story.
  d Most people who smoke have made at least one serious attempt to give it up.
  e Young people are more superstitious than old people.
  f Given a free choice, most girls would hardly ever choose to wear a dress in preference to something else.

g More babies are born in the winter than in the summer.
h Most cars these days use unleaded petrol.
i The school day should start at 0800 and end at 1400.
j Older people sleep less than younger people.

Below is a project about people in a factory which provides practice in doing a GCSE statistics project.

8.7.2 Statistics project practice

Exercise 12

A sample of 50 employees from a factory were asked to anonymously complete a questionnaire which asked for their gender, their age and how much they were currently earning.

The results are summarised in the three tables on page 418.
Table 1 has been sorted by gender (female and male).
Table 2 has been sorted by age.
And **Table 3** has been sorted by wage.

1 a Use **Table 1** to draw a pie chart showing the number of employees from the sample in each of these four categories

> females under 30
> males under 30
> females 30 or over
> males 30 and over

Summarise what the pie chart shows.

> There are 360° for 50 employees so for 1 employee the angle on the pie chart is 7·2°.

b If the sample is representative of the whole factory population, how many people in each of the categories above would there be if the factory employed 1500 people?

2 a On two **separate** diagrams draw scatter graphs of wage against age for employees under 30 and employees over 30 (put age on the horizontal axis). Use **Table 2** to help you with this.
b Comment on any correlation.

3 With the help of **Table 3**, copy and complete this frequency chart for wages. Note that a wage of exactly £20 000 should be put in the 20 000–25 000 group.

Wage, £	Mid-value, x	Frequency, f	fx
15 000–20 000	17 500	4	70 000
20 000–25 000	22 500		
25 000–30 000	27 500		
30 000–35 000	32 500		
35 000–40 000	37 500	2	75 000
40 000–45 000	42 500		
125 000–130 000	127 500		
Total		50	A

b Calculate the mean wage from the frequency chart.

c Calculate the mean wage from the raw data (the 50 individual wages in the table).

d Which of these two values for the mean is more reliable? Why is that?

> Mean = A ÷ 50.

4 Use the frequency chart from question **3** to help you draw a bar chart for these data.

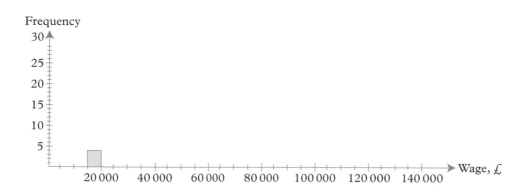

5 a Calculate the median wage with the help of **Table 3**.

b What is the range of the wages?

c Which do you think is a better estimate of the average wage for the factory, the mean or the median?

d Explain your answer to part **c**.

Table 1

	Gender	Age	Wage, £
1	F	17	18500
2	F	18	19000
3	F	18	19500
4	F	20	21000
5	F	22	22000
6	F	23	21500
7	F	26	23000
8	F	28	23500
9	F	30	24000
10	F	31	25500
11	F	31	25000
12	F	33	27500
13	F	37	25000
14	F	42	29000
15	F	46	27000
16	F	48	29500
17	F	52	28500
18	F	53	28000
19	M	17	18000
20	M	18	20000
21	M	19	21000
22	M	19	20500
23	M	22	22000
24	M	24	23500
25	M	26	24000
26	M	27	25000
27	M	29	25000
28	M	30	27500
29	M	33	28000
30	M	35	28000
31	M	36	29500
32	M	37	42000
33	M	39	28000
34	M	40	27500
35	M	41	125000
36	M	43	28500
37	M	46	40000
38	M	46	27000
39	M	47	37000
40	M	48	27500
41	M	49	28000
42	M	50	27000
43	M	50	29000
44	M	52	28000
45	M	54	27500
46	M	58	36000
47	M	57	28000
48	M	59	29000
49	M	60	28500
50	M	62	29000

Table 2

	Gender	Age	Wage, £
1	M	17	18000
2	F	17	18500
3	F	18	19000
4	F	18	19500
5	M	18	20000
6	M	19	20500
7	M	19	21000
8	F	20	21000
9	F	22	22000
10	M	22	22000
11	F	23	21500
12	M	24	23500
13	F	26	23000
14	M	26	24000
15	M	27	25000
16	F	28	23500
17	M	29	25000
18	F	30	24000
19	M	30	27500
20	F	31	25000
21	F	31	25500
22	F	33	27500
23	M	33	28000
24	M	35	28000
25	M	36	29500
26	F	37	25000
27	M	37	42000
28	M	39	28000
29	M	40	27500
30	M	41	125000
31	F	42	29000
32	M	43	28500
33	F	46	27000
34	M	46	27000
35	M	46	40000
36	M	47	37000
37	M	48	27500
38	F	48	29500
39	M	49	28000
40	M	50	27000
41	M	50	29000
42	M	52	28000
43	F	52	28500
44	F	53	28000
45	M	54	27500
46	M	57	28000
47	M	58	36000
48	M	59	29000
49	M	60	28500
50	M	62	29000

Table 3

	Gender	Age	Wage, £
1	M	17	18000
2	F	17	18500
3	F	18	19000
4	F	18	19500
5	M	18	20000
6	M	19	20500
7	M	19	21000
8	F	20	21000
9	F	23	21500
10	F	22	22000
11	M	22	22000
12	F	26	23000
13	M	24	23500
14	F	28	23500
15	M	26	24000
16	F	30	24000
17	M	27	25000
18	M	29	25000
19	F	31	25000
20	F	37	25000
21	F	31	25500
22	F	46	27000
23	M	46	27000
24	M	50	27000
25	M	30	27500
26	F	33	27500
27	M	40	27500
28	M	48	27500
29	M	54	27500
30	M	33	28000
31	M	35	28000
32	M	39	28000
33	M	49	28000
34	M	52	28000
35	F	53	28000
36	M	57	28000
37	M	43	28500
38	F	52	28500
39	M	60	28500
40	F	42	29000
41	M	50	29000
42	M	59	29000
43	M	62	29000
44	M	36	29500
45	F	48	29500
46	M	58	36000
47	M	47	37000
48	M	46	40000
49	M	37	42000
50	M	41	125000

8.8 Cumulative frequency

8.8.1 Quartiles, interquartile range

The range is a simple measure of spread but one extreme (very high or very low) value can have a big effect.
The **interquartile range** is a better measure of spread.

> **EXAMPLE**
>
> Find the quartiles and the interquartile range for these numbers.
>
> 12 6 4 9 8 4 9 8 5 9 8 10
>
> ....
>
> Put the numbers in order. 4 4 5 | 6 8 8 | 8 9 9 | 9 10 12
>
> lower quartile   median   upper quartile
> is 5·5   is 8   is 9
>
> Interquartile range = upper quartile – lower quartile
>   = 9 – 5·5 = 3·5

8.8.2 Box plot

A **box plot** or **box-and-whisker diagram** shows the spread of a set of data. It shows the quartiles (Q_1 and Q_3) and the median.
The 'whiskers' extend to the lowest and the highest values and show the range.

Here is a box plot. Note that they can run horizontally or vertically.

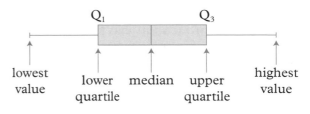

Exercise 13 Ⓜ/Ⓗ

For each set of data, work out
a the lower and upper quartile
b the interquartile range.

1 1 1 4 4 5 8 8 8 9 10 11 11

2 5 2 6 4 1 9 3 2 8 4 1 0

3 7 9 11 15 18 19 23 27

4 0 0 1 1 2 3 3 4 4 4 6 6 6 7 7

5 Here are three box plots. Estimate **a** the median
                                         b the interquartile range
                                         c the range.

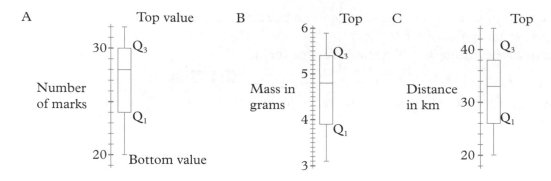

8.8.3 Cumulative frequency

● The total of the frequencies up to a particular value is called the **cumulative frequency**.

Data from a frequency table can be used to calculate cumulative frequencies. These new values, when plotted and joined, form a cumulative frequency curve, sometimes called an S-shaped curve.

It is a simple matter to find the median from the halfway point of a cumulative frequency curve.
Other points of location can also be found from this curve. The cumulative frequency axis can be divided into 100 parts.

● The **upper quartile** is at the 75% point
● The **lower quartile** is at the 25% point

The quartiles are particularly useful in finding the central 50% of the range of the distribution; this is known as the interquartile range.

● **Interquartile range** = (upper quartile) − (lower quartile)

The interquartile range is an important measure of spread in that it shows how widely the data are spread.
Half the distribution is in the interquartile range. If the interquartile range is small, then the middle half of the distribution is bunched together.

EXAMPLE

In a survey, 200 people were asked to state their weekly earnings. The results were plotted on a cumulative frequency curve.

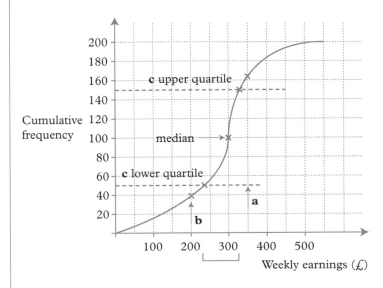

a How many people earned up to £350 a week?

b How many people earned more than £200 a week?

c Find the interquartile range.

....

a From the curve, about 165 people earned up to £350 per week.

b About 40 people earned up to £200 per week.
There are 200 people in the survey, so 160 people earned more than £200 per week.

c Lower quartile (50 people) = £235
Upper quartile (150 people) = £325
∴   Interquartile range = 325 − 235
                              = £90

A pet shop owner likes to weigh all his mice every week as a check on their state of health. The table shows the weights of the 80 mice.

Weight w (g)	Frequency	Cumulative frequency	Weight represented by cumulative frequency
$0 < w \leqslant 10$	3	3	$\leqslant 10$ g
$10 < w \leqslant 20$	5	8	$\leqslant 20$ g
$20 < w \leqslant 30$	5	13	$\leqslant 30$ g
$30 < w \leqslant 40$	9	22	$\leqslant 40$ g
$40 < w \leqslant 50$	11	33	$\leqslant 50$ g
$50 < w \leqslant 60$	15	48	$\leqslant 60$ g
$60 < w \leqslant 70$	14	62	$\leqslant 70$ g
$70 < w \leqslant 80$	8	70	$\leqslant 80$ g
$80 < w \leqslant 90$	6	76	$\leqslant 90$ g
$90 < w \leqslant 100$	4	80	$\leqslant 100$ g

The table also shows the cumulative frequency.

Plot a cumulative frequency curve and hence estimate
a the median   **b** the interquartile range.

From the cumulative frequency curve,

median = 55 g
lower quartile = 36 g
upper quartile = 68 g
interquartile range = (68 − 36) g
= 32 g

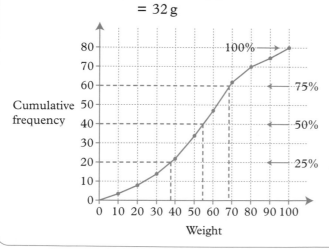

The points on the graph are plotted at the upper limit of each group of weights.

Exercise 14 (M)/(H)

1 The graph shows the cumulative frequency curve for the marks
of 40 students in an examination.

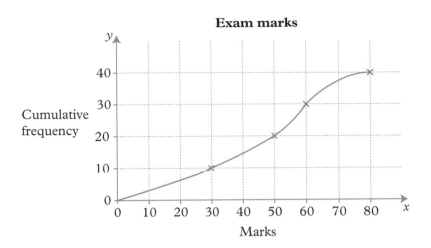

From the graph estimate
  a the median mark
  b the mark at the lower quartile and at the upper quartile
  c the interquartile range
  d the pass mark if three-quarters of the students passed.

2 The graph shows the cumulative frequency curve for the marks of
60 students in an examination.

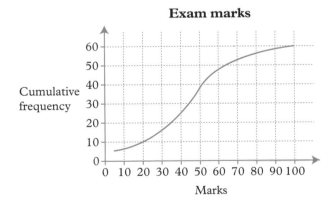

From the graph estimate
  a the median mark
  b the mark at the lower quartile and at the upper quartile
  c the interquartile range
  d the pass mark if two-thirds of the students passed.

3 The lifetimes of 500 electric light bulbs were measured in a laboratory. The results are shown in the cumulative frequency diagram.

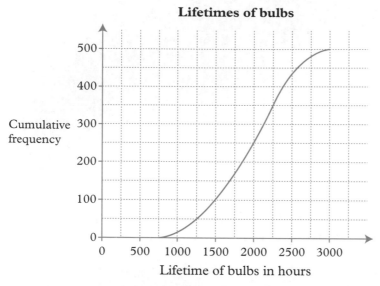

Lifetimes of bulbs

a How many bulbs had a lifetime of 1500 hours or less?
b How many bulbs had a lifetime of between 2000 and 3000 hours?
c After how many hours were 70% of the bulbs dead?
d What was the shortest lifetime of a bulb?

4 A photographer measures all the snakes required for a scene in a film involving a snake pit.
a Draw a cumulative frequency curve for the results.

Length l (cm)	Frequency	Cumulative frequency	Upper limit
$0 < l \leqslant 10$	0	0	$\leqslant 10$
$10 < l \leqslant 20$	2	2	$\leqslant 20$
$20 < l \leqslant 30$	4	6	$\leqslant 30$
$30 < l \leqslant 40$	10	16	$\leqslant 40$
$40 < l \leqslant 50$	17		⋮
$50 < l \leqslant 60$	11		⋮
$60 < l \leqslant 70$	3		⋮
$70 < l \leqslant 80$	3		⋮

Use a scale of 2 cm for 10 units across the page for the lengths and 2 cm for 10 units up the page for the cumulative frequency. Remember to plot points at the **upper** end of the classes (10, 20, 30 etc).
b Find **i** the median   **ii** the interquartile range.

5 As part of a medical inspection, a nurse measures the heights of 48 students in a school.

a Draw a cumulative frequency curve for the results.

Height (cm)	Frequency	Cumulative frequency
$140 < h \leqslant 145$	2	2 [$\leqslant 145$ cm]
$145 < h \leqslant 150$	4	6 [$\leqslant 150$ cm]
$150 < h \leqslant 155$	8	14 [$\leqslant 155$ cm]
$155 < h \leqslant 160$	9	
$160 < h \leqslant 165$	12	
$165 < h \leqslant 170$	7	
$170 < h \leqslant 175$	4	
$175 < h \leqslant 180$	2	

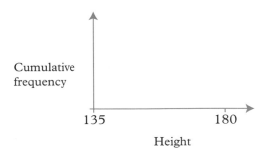

Use a scale of 2 cm for 5 units across the page and 2 cm for 10 units up the page.

b Find **i** the median **ii** the interquartile range.

6 Hugo and Boris are brilliant darts players. They recorded their scores over 60 throws. Here are Hugo's scores.

Score (x)	$30 < x \leqslant 60$	$60 < x \leqslant 90$	$90 < x \leqslant 120$	$120 < x \leqslant 150$	$150 < x \leqslant 180$
Frequency	10	4	13	23	10

a Draw a cumulative frequency curve.
Use a scale of 2 cm for 20 points across the page and 2 cm for 10 throws up the page.

b For Hugo, find **i** his median score
ii the interquartile range of his scores.

For his 60 throws, Boris had a median score of 105 and an interquartile range of 20.

c Which of the players is more consistent? Give a reason.

7 The life of a Dodgy photocopying machine is tested by timing how long it works before breaking down. The table shows the results for 50 machines.

Time, t (hours)	$2 < t \leqslant 4$	$4 < t \leqslant 6$	$6 < t \leqslant 8$	$8 < t \leqslant 10$	$10 < t \leqslant 12$
Frequency	9	6	15	15	5

a Draw a cumulative frequency graph.
[t across the page, 1 cm = 1 hour; C.F. up the page]

b Find **i** the median life of a machine
ii the interquartile range.
c The makers claim that their machines will work for at least 10 hours.
What percentage of the machines do not match this description?

8 In an international competition 60 students from Britain and
France did the same science test.

Marks	Britain frequency	France frequency	Britain cum. freq.	France cum. freq.
1–5	1	2	1 [≤5·5]	2 [≤5·5]
6–10	2	5	3 [≤10·5]	7 [≤10·5]
11–15	4	11	7 [≤15·5]	
16–20	8	16		
21–25	16	10		
26–30	19	8		
31–35	10	8		

> The upper class boundaries for the marks are 5·5, 10·5, 15·5 etc.

The cumulative frequency graph should be plotted for values
≤5·5, ≤10·5, ≤15·5 and so on.

a Using the same axes, draw the cumulative frequency curves for
the British and French results.
Use a scale of 2 cm for 5 marks across the page
and 2 cm for 10 people up the page.
b Find the median mark for each country.
c Find the interquartile range for the British results.
d Describe in one sentence the main difference between the
two sets of results.

8.9 Histograms

8.9.1 Drawing a histogram

In a histogram, the **area** of each bar shows the frequency of the data.
Histograms resemble bar charts but are not to be confused with
them; in bar charts the height of each bar shows the frequency.
Histograms often have bars of varying widths. Because the area
of the bar represents frequency, the height must be adjusted to
correspond with the width of the bar. The vertical axis is not labelled
frequency but **frequency density**.

● **Frequency density** = $\dfrac{\text{frequency}}{\text{class width}}$

You can use histograms to represent both discrete data and continuous data, but their main purpose is for use with continuous data.

Draw a histogram from the table for the distribution of ages of passengers travelling on a flight to New York.

Ages	Frequency
$0 \leqslant x < 20$	28
$20 \leqslant x < 40$	36
$40 \leqslant x < 50$	20
$50 \leqslant x < 70$	30
$70 \leqslant x < 100$	18

Note that the data have been collected into class intervals of different widths.

To draw the histogram, the heights of the bars must be adjusted by calculating frequency density.

Ages	Frequency	Frequency density (f.d.)
$0 \leqslant x < 20$	28	$28 \div 20 = 1 \cdot 4$
$20 \leqslant x < 40$	36	$36 \div 20 = 1 \cdot 8$
$40 \leqslant x < 50$	20	$20 \div 10 = 2$
$50 \leqslant x < 70$	30	$30 \div 20 = 1 \cdot 5$
$70 \leqslant x < 100$	18	$18 \div 30 = 0 \cdot 6$

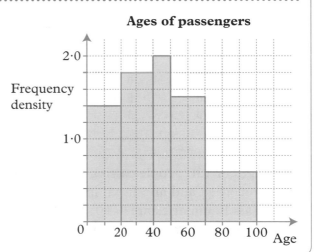

Ages of passengers

Exercise 15 (H)

1 Kylie measured the lengths of 20 copper nails. She recorded the results in a frequency table. Calculate the frequency densities and draw the histogram as started below.

Length *l* (in mm)	Frequency	Frequency density (f.d.)
$0 \leqslant l < 20$	5	$5 \div 20 = 0 \cdot 25$
$20 \leqslant l < 25$	5	
$25 \leqslant l < 30$	7	
$30 \leqslant l < 40$	3	

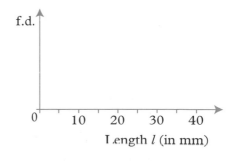

2 The frequency table shows the volumes of 55 containers.
Calculate the frequency densities and draw the histogram.

Volume (mm³)	Frequency	Frequency density
$0 < V \leqslant 5$	5	$5 \div 5 = 1$
$5 < V \leqslant 10$	3	$3 \div$
$10 < V \leqslant 20$	12	
$20 < V \leqslant 30$	17	
$30 < V \leqslant 40$	13	
$40 < V \leqslant 60$	5	

3 Thirty students in a class are weighed on the first day of term.
Draw a histogram to represent these data.
Note that the weights do not start at zero. This can be shown on the graph by a broken axis.

Weight w (kg)	Frequency
$30 < w \leqslant 40$	5
$40 < w \leqslant 45$	7
$45 < w \leqslant 50$	10
$50 < w \leqslant 55$	5
$55 < w \leqslant 70$	3

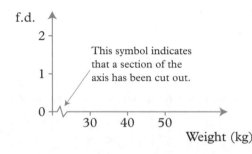

f.d.

This symbol indicates that a section of the axis has been cut out.

Weight (kg)

4 The ages of 120 people passing through a turnstile were recorded and are shown in the frequency table.
The class boundaries are 0, 10, 15, 20, 30, 40.
Draw the histogram for the data.

Age A (yrs)	Frequency
$0 < A \leqslant 10$	18
$10 < A \leqslant 15$	46
$15 < A \leqslant 20$	35
$20 < A \leqslant 30$	13
$30 < A \leqslant 40$	8

5 The table shows the daily profit made at a beach resort sailing club

Draw a histogram for the data.

Profit P ($)	Frequency
$0 < P \leqslant 40$	16
$40 < P \leqslant 80$	48
$80 < P \leqslant 100$	30
$100 < P \leqslant 120$	40
$120 < P \leqslant 160$	36
$160 < P \leqslant 240$	16

6 Another common notation is used here for the masses of plums picked in an orchard.

Mass (g)	20–	30–	40–	60–	80–
Frequency	11	18	7	5	0

The initial 20– means $20\,\text{g} \leqslant \text{mass} < 30\,\text{g}$.
Draw a histogram with class boundaries at 20, 30, 40, 60, 80.

8.9.2 Finding frequencies from histograms

You can draw up a frequency table from the data in a histogram. Then you can find the actual size of the sample tested as well as the frequencies in each class.

The histogram shows children's heights. Draw up the relevant frequency table and find the size of the sample which was tested.

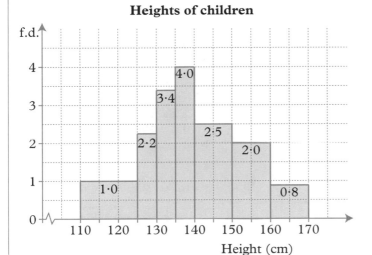

Use   frequency density = $\dfrac{\text{frequency}}{\text{class width}}$

So   frequency = (frequency density) × (class width)

Height x (cm)	f.d.	Frequency
110 < x ≤ 125	1·0	1·0 × 15 = 15
125 < x ≤ 130	2·2	2·2 × 5 = 11
130 < x ≤ 135	3·4	3·4 × 5 = 17
135 < x ≤ 140	4·0	4·0 × 5 = 20
140 < x ≤ 150	2·5	2·5 × 10 = 25
150 < x ≤ 160	2·0	2·0 × 10 = 20
160 < x ≤ 170	0·8	0·8 × 10 = 8
		Total frequency = 116

So there were 116 children in the sample measured.

Remember: the area of each bar represents frequency.

Exercise 16 Ⓗ

1 From the histogram find
  a how many of the lengths are in these
    intervals
    i 40–60 cm   **ii** 30–40 cm
  b the total frequency.

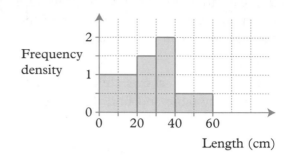

2 For this histogram find the total
frequency.

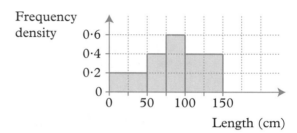

3 The histogram shows the ages of the
trees in a small wood. How many
trees were in the wood altogether?

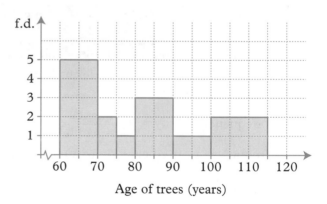

4 One day a farmer weighs all the hens' eggs
which he collects. The results are shown in
this histogram.
There were 8 eggs in the class 50–60 g.
Work out the numbers on the frequency
density axis and hence find the total number
of eggs collected.

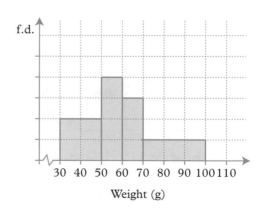

5 A different way of giving frequencies is to draw a key to show what a certain area represents. The histogram below shows the size of the area which represents 4 people.

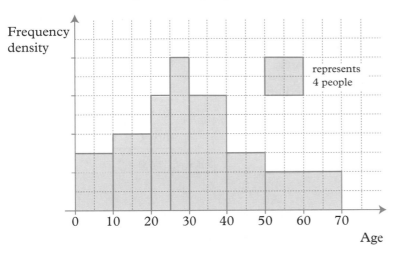

a How many people are aged between 0 and 10?
b How many people are aged between 10 and 20?
c How many people are represented in the whole histogram?

6 A farmer takes a sample of 92 trees from a field and measures their heights.

Height (cm)	Frequency
100–120	16
120–140	24
140–150	18
150–160	10
160–200	24

He draws a bar 4 cm high to represent the trees of height 100–120 cm.
How high will the bar be representing
a 120–140 cm
b 140–150 cm
c 160–200 cm?

Test yourself

1 Emma repairs bicycles.
She keeps records of the cost of the repairs.

The table gives information about the costs of all repairs which she carried out in one week.

Find the class interval in which the median lies.

Cost (£C)	Frequency
$0 < C \leqslant 10$	3
$10 < C \leqslant 20$	7
$20 < C \leqslant 30$	6
$30 < C \leqslant 40$	8
$40 < C \leqslant 50$	9

(Edexcel, 2005)

2 Jane records the times taken by 30 pupils to complete a number puzzle.

  a Calculate an estimate of the mean time taken to complete the puzzle.
  b Which time interval contains the median time taken to complete the puzzle?

Time, t (minutes)	Number of pupils
$2 < t \leqslant 4$	3
$4 < t \leqslant 6$	6
$6 < t \leqslant 8$	7
$8 < t \leqslant 10$	8
$10 < t \leqslant 12$	5
$12 < t \leqslant 14$	1

(AQA, 2003)

3 A quiz has five questions.
The table shows the number of correct answers given by the people who took the quiz.
  a Calculate the mean number of correct answers. You **must** show your working.
  b A mark of 4 is given for every correct answer. A mark of -1 is given for every blank or incorrect answer.
  Find the mean mark.

Number of correct answers	Number of people
0	6
1	10
2	13
3	21
4	49
5	1

(AQA, 2007)

4 Jason collected some information about the heights of 19 plants.

This information is shown in the stem and leaf diagram.

```
1 | 1  2  3  3
2 | 3  3  5  9  9
3 | 0  2  2  6  6  7        Key 4|8 means 48 mm
4 | 1  1  4  8
```

Find the median.

(Edexcel, 2008)

5 Naomi wants to find out how often adults go to the cinema.

She uses this question on a questionnaire.

"How many times do you go to the cinema?"

☐ Not very often   ☐ Sometimes   ☐ A lot

a Write down **two** things wrong with this question.
b Design a better question for her questionnaire to find out how often adults go to the cinema.
You should include some response boxes.

(Edexcel, 2008)

6 Chandni wants to survey pupils in her school about their reading habits.
a Write a question that would help Chandni to investigate how often pupils in her school read for pleasure.
Include a response section.

b There are 1000 pupils in Chandni's school.

Chandni samples 50 pupils at random and asks them to complete her survey.
She finds that 16 of the pupils in the sample read comics.
Estimate the number of pupils in the school who read comics.

(AQA, 2003)

7 The table shows the number of students in each year group at a school.

Year group	7	8	9	10	11
Number of students	190	145	145	140	130

Jenny is carrying out a survey for her GCSE Mathematics project. She uses a stratified sample of 60 students according to year group. Calculate the number of year 11 students that should be in her sample.

(Edexcel, 2005)

8 The government wants to survey students studying science at University about their views on becoming teachers.
They decide to survey science students at Surrey University.
237 students do science at Surrey University.

The cumulative percentage table of students doing each science is

Geology	Physics	Chemistry	Biology

0%   18%   42%   66%   100%
Cumulative percentage

a The government decide to do a 10% stratified sample. Write down the numbers from each category that they should sample.

Geology	Physics	Chemistry	Biology

b Give one other factor they should take into account when selecting the sample to ensure an unbiased sample.

(AQA, 2003)

9 A theatre has 500 seats.

The theatre manager recorded the audience size at each performance of a pantomime.

The results are summarised in the table below.

Audience size (A)	$0 < A \leqslant 100$	$100 < A \leqslant 200$	$200 < A \leqslant 300$	$300 < A \leqslant 400$	$400 < A \leqslant 500$
Frequency	2	18	11	14	35

The cumulative frequency table for the audience size is given below.

Audience size (A)	$A \leqslant 100$	$A \leqslant 200$	$A \leqslant 300$	$A \leqslant 400$	$A \leqslant 500$
Cumulative frequency	2	20	31	45	80

a Explain how the cumulative frequency values have been calculated.

b The cumulative frequency graph for the audience size is shown below.

  i Use the graph to find an estimate of the median audience size.

  ii Use the graph to estimate on how many occasions the theatre was **over** half full.

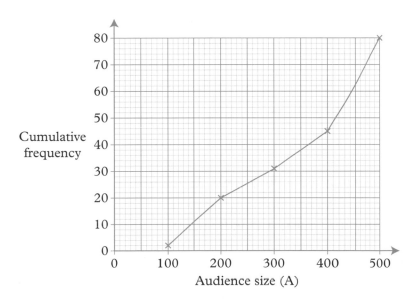

(OCR, 2008)

10 The table shows information about the number of hours that 120 children used a computer last week.

Number of hours (h)	Frequency
$0 < h \leqslant 2$	10
$2 < h \leqslant 4$	15
$4 < h \leqslant 6$	30
$6 < h \leqslant 8$	35
$8 < h \leqslant 10$	25
$10 < h \leqslant 12$	5

a Work out an estimate for the mean number of hours that the children used a computer.
Give your answer correct to two decimal places.

b Copy and complete the cumulative frequency table.

Number of hours (h)	Frequency
$0 < h \leqslant 2$	10
$0 < h \leqslant 4$	
$0 < h \leqslant 6$	
$0 < h \leqslant 8$	
$0 < h \leqslant 10$	
$0 < h \leqslant 12$	

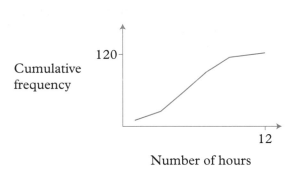

c Draw a cumulative frequency graph for your table.

d Use your graph to find an estimate for the number of children who used a computer for **less** than 7 hours last week.

(Edexcel, 2005)

11 56 boys and 52 girls took an English test.
The box plots show the distributions of their marks.

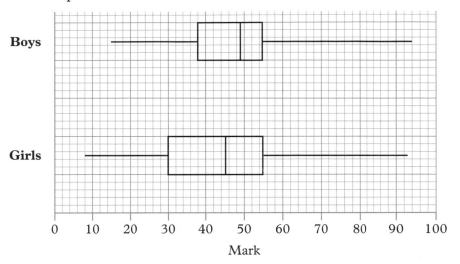

Give **two** differences between the boys' marks and the girls' marks.

(AQA, 2004)

12 The histogram shows the time, in seconds, it took for 60 girls to swim one length of a pool.

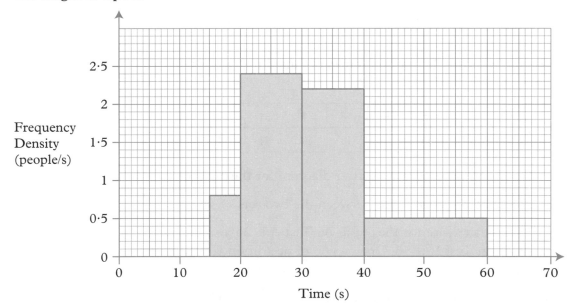

How many of the girls took 30 seconds or longer to swim one length?

(OCR, 2004)

13 A shop selling CD players collected data on the lengths of time
that the players worked for before developing a fault.

Time (*t* years)	$0 < t \leqslant 1$	$1 < t \leqslant 2$	$2 < t \leqslant 5$	$5 < t \leqslant 10$
Frequency	18	52	21	15

Complete the histogram to represent these data.

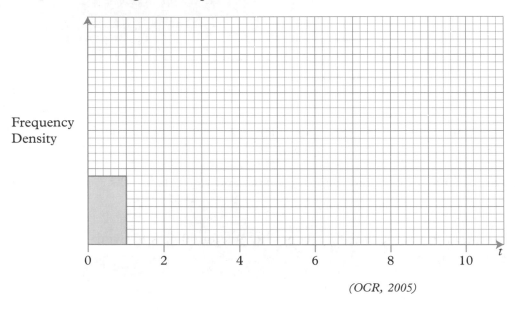

(OCR, 2005)

14 The table shows the pairs of scores obtained by 8 pupils on two
types of tests.

Test A	21	6	43	48	8	31	29	14
Test B	58	94	28	18	84	41	54	71

a On graph paper draw a scatter diagram for these results.

b Describe the correlation between the two sets of scores.

c The mean score for the pupils on Test A is 25 and the mean
score on Test B is 56. Draw a line of best fit on your scatter
diagram.

d Another pupil sat Test A and was given a score of 18, but was
absent for Test B. Use your line of best fit to estimate the score
on Test B for this pupil.

(WJEC, 2003)

9 Probability

In this unit you will:
- estimate and calculate probabilities
- learn about the probabilities of combined, exclusive and independent events
- draw and use tree diagrams to find probabilities
- learn about conditional probabilities.

Functional skills coverage and range:
- use probability to assess the likelihood of an outcome
- understand and use equivalences between fractions, decimals and percentages.

Links
Estimating probabilities is the basis for many financial industries including insurance, banking and mortgage lending. It is also useful for working out risks such as the risk of rain if you want to go out for the day.

9.1 Relative frequency

If you wanted to work out the probability of a drawing pin landing point up you could conduct an experiment in which you dropped the pin many times.

- If the pin lands 'point up' on x occasions out of a total number of N trials, the **relative frequency** of landing 'point up' is $\frac{x}{N}$.

When an experiment is repeated many times you can use the relative frequency as an estimate of the probability of the event occurring.

The graph shows how the ratio $\frac{x}{N}$ settles down to a consistent value as N becomes large (say 1000 or 10 000 trials). So the relative frequency is a more reliable guide to the probability of the event occurring when the number of trials is large.

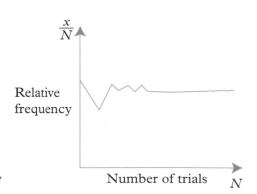

The probability of an event is a measure of how likely it is to occur. This probability can be any number between 0 and 1 (inclusive).

There are four different ways of estimating probabilities.

Method A   Use symmetry (the theoretical probability).
- The probability of rolling a 3 on a fair dice is $\frac{1}{6}$.

    This is because all the scores, 1, 2, 3, 4, 5, 6 are equally likely.

● Similarly the probability of getting a head when tossing a fair coin is $\frac{1}{2}$.

Method B Conduct an experiment or survey to collect data.

● The dice shown has five faces, not all the same shape.
  To estimate the probability of the dice landing with a '1' showing, you could conduct an experiment to see what happened in, say, 500 trials.
● You might want to know the probability that the next car going past the school gates will be driven by a woman.
  You could conduct a survey in which the drivers of cars are recorded over a period of time.

Method C Look at past data.

● If you wanted to estimate the probability of your plane crashing as it lands at Heathrow airport you could look at accident records at Heathrow over the last five years or so.

Method D Make a subjective estimate.

You have to use this method when the event is not repeatable. It is not really a 'method' in the same sense as are methods A, B, C.

● You might want to estimate the probability of England beating France in a soccer match next week. You could look at past results but these could be of little value for all sorts of reasons. You might consult 'experts' but even they are notoriously inaccurate in their predictions.

Exercise 1 Ⓜ

In questions **1** to **10** state which method A, B, C or D you would use to estimate the probability of the event given.

1 The probability that a person chosen at random from a class will be left-handed.

2 The probability that there will be snow in the ski resort to which a school party is going in February next year.

3 The probability of drawing an 'ace' from a pack of playing cards.

4 The probability that you hole a six-foot putt when playing golf.

5 The probability that the world record for running 1500 m will be under 3 min 20 seconds by the year 2020.

6 The probability of winning the National Lottery.

7 The probability of rolling a 3 using a dice which is suspected of being biased.

8 The probability that a person selected at random would vote 'Labour' in a general election tomorrow.

9 The probability that a train will arrive within ten minutes of its scheduled arrival time.

10 The probability that the current Wimbledon Ladies Champion will successfully defend her title next year.

11 Alex rolls a dice and records the number of ones that he gets.

Number of rolls	10	25	50	100	300	1000	2000
Number of ones	1	3	5	12	54	160	338

Relative frequency of getting a one

Number of rolls

Plot a graph of relative frequency of getting one against the number of rolls.

12 Conduct an experiment where you cannot predict the result. You could roll a dice with a piece of 'Blu-tack' stuck to it. Or make a spinner where the axis is not quite in the centre. Or drop a drawing pin.
Conduct the experiment many times and work out the relative frequency of a 'success' after every 10 or 20 trials. Plot a relative frequency graph to see if the results settle down to a consistent value.

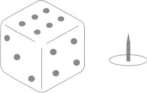

13 The spinner has an equal chance of giving any digit from 0 to 9.
Four friends did an experiment where they spun the pointer a different number of times and recorded the number of zeros they got. Here are their results.

	Number of spins	Number of zeros	Relative frequency
Steve	10	2	0·2
Mike	150	14	0·093
Nick	200	41	0·205
Jason	1000	104	0·104

One of the four recorded his results incorrectly. Say who you think this was and explain why.

9.2 Working out probabilities

9.2.1 Calculating probabilities using symmetry

You can calculate the probability of an event occurring using symmetry.

For example, when you toss a coin you have an equal chance of getting a 'head' or a 'tail'. So the probability of spinning a 'head' is a half.

You write 'p (spinning a head) = $\frac{1}{2}$'.

> You could also write this as 0·5 or 50%.

EXAMPLE

A single card is drawn from a pack of 52 playing cards. Find the probability of these results
a the card is a queen
b the card is a club
c the card is the jack of hearts.

....

There are 52 equally likely outcomes of the 'trial' (drawing a card).

a p (queen) = $\frac{4}{52}$ = $\frac{1}{13}$

b p (club) = $\frac{13}{52}$ = $\frac{1}{4}$

c p (jack of hearts) = $\frac{1}{52}$

13 Spades
13 Hearts
13 Diamonds
13 Clubs

Total 52

Exercise 2 Ⓜ

1 If one card is picked at random from a pack of 52 playing cards, what is the probability that it is
   a a king
   b the ace of clubs
   c a heart?

> Note: You will not be tested on your knowledge of playing cards in your examination.

2 Nine counters numbered 1, 2, 3, 4, 5, 6, 7, 8, 9 are placed in a bag. One is taken out at random. What is the probability that it is
   a a '5'   **b** divisible by 3
   c less than 5   **d** divisible by 4?

3 A bag contains 5 green balls, 2 red balls and 4 yellow balls.
One ball is taken out at random. What is the probability
that it is
  a green
  b red
  c yellow?

4 A cash bag contains two 20p coins, four 10p coins, five 5p coins,
three 2p coins and three 1p coins. Find the probability that
one coin selected at random is
  a a 10p coin
  b a 2p coin
  c a silver coin.

5 One card is selected at random from those shown.
Find the probability of selecting
  a a heart,
  b an ace,
  c the 10 of clubs,
  d a spade,
  e a heart or a diamond.

6 A pack of playing cards is well shuffled and a card is drawn.
Find the probability that the card is
  a a jack
  b a queen or a jack
  c the ten of hearts
  d a club higher than the 9 (count the ace as high).

7 One ball is selected at random from those
in the diagram. Find the probability of
selecting
  a a white ball
  b a yellow or a black ball
  c a ball which is not red.

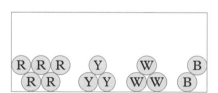

R = red
Y = yellow
W = white
B = black

8 Jade has three queens and one king. She
shuffles the cards and takes one without
looking. Jade asks two of her friends about
the probability of getting a king.

Kim says

'It is $\frac{1}{3}$ because there are 3 queens and 1 king.'

Megan says

'It is $\frac{1}{4}$ because there are 4 cards and only 1 king.'

Which of her friends is right?

9 The numbers of matches in ten boxes are

48, 46, 45, 49, 44, 46, 47, 48, 45, 46

One box is selected at random. Find the probability of the box containing

a 49 matches

b 46 matches

c more than 47 matches.

10 A bag contains 9 balls, all of which are blue or white. Jane selects a ball and then replaces it. She repeats this several times.
Here are her results (B = blue, W = white):

B W B W B B B W B B W B B W B

B B W W B B B B W B W B B W B

How many balls of each colour do you think there were in the bag?

11 There are eight balls in a bag. Asif takes a ball from the bag, notes its colour and then returns it to the bag. He does this 25 times.
Here are the results.

a What is the smallest number of red balls there could be in the bag?

b Asif says 'There cannot be any blue balls in the bag because there are no blues in my table'. Explain why Asif is wrong.

Red	4
White	11
Black	10

12 Cards with numbers 1, 2, 3, 4, 5, 6, 7, 8 are shuffled and then placed face down in a line. The cards are then turned over one at a time from the left. In this case the first card is a '3'.

Find the probability that the next card will be

a the 6

b an even number

c higher than 1.

13 Refer back to the scenario in question 12. Suppose the second card is a 7.

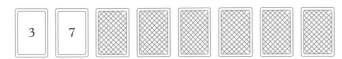

Find the probability that the next card will be

a the 5

b an odd number

c higher than 6.

14 a A bag contains 5 red balls, 6 green balls and 2 black balls. Find the probability of selecting

    i a red ball   **ii** a green ball.

b One black ball is removed from the bag. Find the new probability of selecting

    i a red ball   **ii** a black ball.

15 A pack of playing cards is split so that all the picture cards (kings, queens, jacks) are in Pile A and all the other cards are in Pile B.

pile A   pile B

Find the probability of selecting

a the queen of clubs from pile A

b the seven of spades from pile B

c any heart from pile B.

16 One ball is selected at random from a bag containing 12 balls, of which x are white.

a What is the probability of selecting a white ball?

When a further 6 white balls are added to the bag the probability of selecting a white ball is doubled.

b Find x.

17 A large firm employs 3750 people.

One person is chosen at random.

What is the probability that that person's birthday is on a Monday in the year 2008?

Write your answer as a decimal to two d.p.

18 The numbers on a set of 28 dominoes are:

6	6	6	6	6	6	6		5	5	5
6	5	4	3	2	1	0		5	4	3

5	5	5		4	4	4	4	4		3	3
2	1	0		4	3	2	1	0		3	2

3	3		2	2	2		1	1		0
1	0		2	1	0		1	0		0

  a What is the probability of drawing a domino from a full set with
    i at least one six on it?
    ii at least one four on it?
    iii at least one two on it?
  b What is the probability of drawing a 'double' from a full set?
  c If I draw a double five which I do not return to the set, what is the probability of drawing another domino with a five on it?

9.2.2 Expectation

A fair dice is rolled 240 times. How many times would you expect to roll a number greater than 4?

You can roll a 5 or a 6 out of the six equally likely outcomes.

$$\therefore p \text{ (number greater than 4)} = \frac{2}{6}$$

$$= \frac{1}{3}$$

Expected number of successes = (probability of a success) × (number of trials)

Expected number of scores greater than

$$4 = \frac{1}{3} \times 240$$

$$= 80$$

If you actually rolled 100 numbers greater than 4, would the dice be fair?

Exercise 3 (M) ⭐

1 Abdul rolls a fair dice 300 times. How many times
would you expect him to roll
  a an even number
  b a 'six'?

2 a This spinner has four equal sectors.
  How many 3s would you expect in 100 spins?
  b If you had 50 3s in 100 spins, could you say the
  spinner was fair?

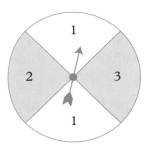

3 About one in eight of the population is left-handed. How many
left-handed people would you expect to find in a firm employing
400 people?

4 A bag contains a large number of marbles of which one in five
is red. If you randomly select one marble on 200 occasions how
many times would you expect to select a red marble?

5 This spinner is used for a simple game. Anna
pays 10p and then spins the pointer. She wins
the amount indicated.
  a What is the probability that Anna wins nothing?
  b If the game is played by 200 people how
  many times would someone win 50p?
  c 10 people play the game and 2 win 50p.
  Is the game fair?

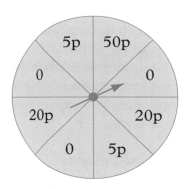

6 These number cards are shuffled and put into a pile.

John picks one card at random and does not replace it. He then
picks a second card.
  a If the first card was the '11', find the probability that John selects
  an even number with the second draw.
  b If the first card was the '8', find the probability that he selects a
  number higher than 9 with the second draw.

7 When the ball is passed to Vinnie, the probability that he kicks it is 0·2 and the probability that he heads it is 0·1. Otherwise he will miss the ball completely, fall over and claim a foul.
In one game his ever-optimistic team mates pass the ball to Vinnie 150 times.
   a How often would you expect him to head the ball?
   b How often would you expect him to miss the ball?

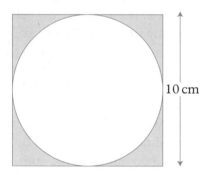

8 A small pack of 20 cards consists of the ace, king, queen, jack and ten of all four suits. Kate takes a card and then replaces it. This procedure is repeated 100 times.
How many times would you expect Kate to select
   a an ace
   b the queen of spades
   c a red card
   d any king or queen?

*9 Tarquin chooses a point at random from inside the square.
   a What is the probability that he chooses a point which lies outside the circle?
   Give your answer as a decimal to 4 sf.
   b In a computer simulation 5000 points are chosen using a random number generator. How many points would you expect to be chosen which lie outside the circle?

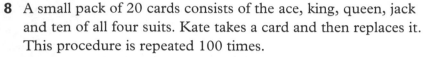

10 cm

9.2.3 Combined events

When a 10p coin and a 50p coin are tossed together two events are occurring
● tossing the 10p coin
● tossing the 50p coin.

You can list all the possible outcomes in a table.

10p	50p
head	head
head	tail
tail	head
tail	tail

Note that a **sample space** is a list of all the possible outcomes and a **sample space diagram** is a table or diagram which shows all the possible outcomes.

EXAMPLE

a Rana throws a red dice and a black dice together.
Show all the possible outcomes.

....

You could list them in pairs with the black
dice first

(1, 1), (1, 2), (1, 2), ... (1, 6)
(2, 1), (2, 2), ...
and so on.

In this example it is easier to see all the possible
outcomes if you show them on a grid.
The × shows 2 on the black dice and 6 on the red.
The ○ shows 5 on the black dice and 3 on the red.

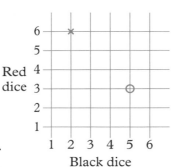

b Rob throws two dice. Find the probability that Rob gets a total of
four on the two dice.

....

There are 36 equally likely outcomes when two dice are thrown.
Rob can obtain a total of 4 in three different ways: (1, 3) (2, 2) or (3, 1)

So the probability that Rob gets a total of 4 = $\dfrac{3}{36}$ = $\dfrac{1}{12}$.

Exercise 4 Ⓜ

1 Daren throws three coins (10p, 20p, 50p)

a List all the possible ways in which they
could land.

10p	20p	50p
H	H	H
H	H	T
⋮	⋮	⋮

b What is the probability of getting three heads?

2 List all the possible outcomes when four coins are tossed together.
How many outcomes are there altogether?

3 A black dice and a white dice are thrown together.
   a Draw a grid to show all the possible outcomes.
   b How many ways can you get a total of nine on the two dice?
   c What is the probability that you get a total of nine?

4 Simon spins these two spinners together.

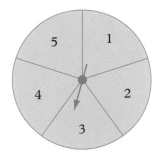

A	B
1	1
1	2
1	3
2	1
⋮	⋮

   Copy and complete the table to list all the possible outcomes. Find
   the probability of obtaining
   a total of 4   **b** the same number on each spinner.

5 Four friends, Wayne, Xavier, Yves and Zara, each write their name
   on a card and the four cards are placed in a hat. Wayne chooses
   two cards to decide who does the maths homework that night.
   List all the possible combinations.
   What is the probability that Xavier and Yves have to do
   the homework?

6 Alec spins the spinner and throws the dice at the same time.

   a Draw a grid to show all the possible outcomes.
   b A 'win' occurs when the number on the spinner is greater than or
       equal to the number on the dice. Find the probability of a 'win'.

7 Jane throws a red dice and a blue dice at the same time. Show all
   the possible outcomes in a systematic way. Find the probability
   that Jane obtains
   a a total of 10   **b** a total of 12
   c a total less than 6   **d** the same number on both dice.

8 Ed throws a dice; when he has recorded the result, he throws the dice a second time. Display all the possible outcomes of the two throws.
Find the probability that Ed gets

  a a total of 4 from the two throws

  b a total of 8 from the two throws

  c a total between 5 and 9 inclusive from the two throws

  d a number on the second throw which is double the number on the first throw

  e a number on the second throw which is four times the number on the first throw.

9 Florence picks two cards at random from the 3, 4, 5 and 6 of hearts.
Find the probability that the total of the two cards is more than 9.

10 a How many possible outcomes are there when six coins are tossed together?

  b What is the probability of tossing six heads?

11 Shirin and Dipika are playing a game in which three coins are tossed. Shirin wins if there are no heads or one head. Dipika wins if there are either two or three heads. Is the game fair to both players?

12 Students X, Y and Z play a game in which four coins are tossed.

X wins if there is 0 or 1 head.
Y wins if there are 2 heads.
Z wins if there are 3 or 4 heads.

Is the game fair to all three players?

13 Four cards numbered 2, 4, 5 and 7 are mixed up and placed face down.

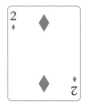   

In a game you pay 10p to select two cards. You win 25p if the total of the two cards is nine.

How much would you expect to win or lose if you played the game 12 times?

*14 Two dice and two coins are thrown at the same time.
Find the probability of obtaining
a two heads and a total of 12 on the dice.
b a head, a tail and a total of 9 on the dice.

Exercise 5 A practical exercise Ⓜ/Ⓗ

ICT The ⟨RAN #⟩ button on a calculator generates random numbers between 0·000 and 0·999. It can be used to simulate tossing three coins.

You could say any **odd** digit is a **tail** and any **even** digit is a **head**.

So the number 0·568 represents THH
and 0·605 represents HHT

Use the ⟨RAN #⟩ button to simulate the tossing of three coins.

'Toss' the three coins 32 times and work out the relative frequencies of

a three heads and **b** two heads and a tail.

Compare your results with the values that you would expect to get theoretically.

> You could use a random number generator on a computer for this exercise.

H H T

2 A calculator can be used to simulate throwing imaginary dice with 10 faces numbered 0, 1, 2, 3, 4, 5, 6, 7, 8, 9.

The ⟨RAN #⟩ button gives a random 3-digit number between 0·000 and 0·999. You can use the first two digits after the point to represent the numbers on two ten-sided dice.

So ⟨0.763⟩ means 7 on one dice and 6 on the second (Ignore the '3') and ⟨0.031⟩ means 0 on one dice and 3 on the second (Ignore the '1').

You add the numbers on the two 'dice' to give the total.

So for $\boxed{0.763}$ the total is 13 and for $\boxed{0.031}$ the total is 3.

Press the $\boxed{\text{RAN}\#}$ button lots of times and record the totals in a tally chart.

Total	0	1	2	$\cdots$	18
Tally		||	||||	$\cdots$	

This grid can be used to predict the result. There are 100 equally likely outcomes like 3, 7 or 4, 1. The ten results shown on the grid each give a total of 9 which is the most likely score.

So in 100 throws you could expect to get a total of 9 on ten occasions.

Plot your results on a frequency graph. Compare your results with the values you would expect.

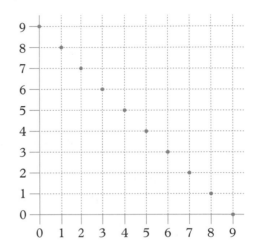

3 At Gibson Academy there are six forms (A, B, C, D, E, F) in Year 10 and the Mathematics Department has been asked to work out a schedule so that, over 5 weeks, each form can play each of the others in a basketball competition.

The games are all played at lunch time on Wednesdays and three games have to be played at the same time.

For example in Week 1 you could have

> Form A *v* Form B
> Form C *v* Form D
> Form E *v* Form F

Work out a schedule for the remaining four weeks so that each team plays each of the others. Check carefully that each team plays every other team just once.

9.3 Probabilities of exclusive events

9.3.1 Exclusive events

- Events are **mutually exclusive** if they cannot occur at the same time.

Examples
- Selecting an ace ⎫ from a
  Selecting a ten ⎭ pack of cards

- Tossing a 'head'
  Tossing a 'tail'

- The total sum of the probabilities of mutually exclusive events is 1.
- The probability of something happening is 1 minus the probability of it not happening.

EXAMPLE

A bag contains 3 black balls, 4 blue balls and 1 white ball. Find the probability of selecting
a a white ball
b a ball which is not white
c a black ball
d a ball which is not black.

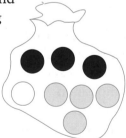

There are 8 balls in the bag.

a p (white ball) $= \dfrac{1}{8}$

b p (not white ball) $= 1 - \dfrac{1}{8} = \dfrac{7}{8}$

c p (black ball) $= \dfrac{3}{8}$

d p (not black ball) $= 1 - \dfrac{3}{8} = \dfrac{5}{8}$

9.3.2 The 'OR' rule: the addition rule

- For exclusive events A and B
  $p(A \text{ or } B) = p(A) + p(B)$

Many questions involving exclusive events can be done **without** using the addition rule. You can decide for yourself which method is easier. See the example below.

EXAMPLE

A card is selected at random from a pack of 52 playing cards. What is the probability of selecting any king or queen?

a Count the number of ways in which a king **or** queen can occur. That is 8 ways.

p (selecting a king or a queen) $= \dfrac{8}{52} = \dfrac{2}{13}$

b Since 'selecting a king' and 'selecting a queen' are exclusive events, you could use the addition law.

p (selecting a king) $= \dfrac{4}{52}$

p (selecting a queen) $= \dfrac{4}{52}$

So, as before, p (selecting a king or a queen)

$= \dfrac{4}{52} + \dfrac{4}{52} = \dfrac{8}{52} = \dfrac{2}{13}.$

Non-exclusive events

If the events are not exclusive, the addition rule **cannot** be used.

Here is a spinner with numbers and colours. The events 'spinning a red' and 'spinning a 1' are **not** exclusive because a red and a 1 can occur at the same time.

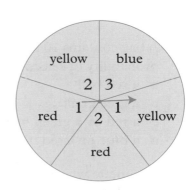

Exercise 6 (H)

1 A bag contains a large number of balls including some red balls.
   The probability of selecting a red ball is $\frac{1}{5}$. What is the probability
   of selecting a ball which is not red?

2 A card is selected from a pack of 52.
   Find the probability of selecting:
   a a king
   b a card which is not a king
   c any picture card (king, queen or jack)
   d a card which is not a picture card.

3 On a roulette wheel the probability of getting '21' is $\frac{1}{36}$. What is
   the probability of not getting '21'?

4 A motorist does a survey at some traffic lights on his way to
   work every day. He finds that the probability that the lights
   are red when he arrives is 0·24. What is the probability that
   the lights are not red?

5 Government birth statistics show that the probability of a baby
   being a boy is 0·506.
   What is the probability of a baby being a girl?

6 The spinner has 8 equal sectors.
   Find the probability of
   a spinning a 5
   b not spinning a 5
   c spinning a 2
   d not spinning a 2
   e spinning a 7
   f not spinning a 7.

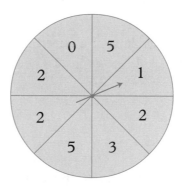

7 A bag contains 10 red balls, 5 blue balls and 7 green balls. Find the
   probability of selecting at random:
   a a red ball
   b a green ball
   c a red **or** a green ball
   d a blue **or** a red ball.

8 A roulette wheel has the numbers 1 to 36 once only. What is the
   probability of spinning either a 10 or a 20?

9 A bag contains a large number of balls coloured red, white, black or green. The table shows the probabilities of selecting each colour.

Colour	red	white	black	green
Probability	0·3	0·1		0·3

Find the probability of selecting a ball
a which is black
b which is not white.

10 A fair dice is rolled. What is the probability of rolling either a 1 or a 6?

11 From a pack of cards I have already selected the king, queen, jack and ten of diamonds.
What is the probability that on my next draw I will select either the ace or the nine of diamonds?

12 In a survey the number of people in cars is recorded. When a car passes the school gates this table shows the probability of it having 1, 2, 3, . . . occupants.

Number of people	1	2	3	4	more than 4
Probability	0·42	0·23		0·09	0·02

a Find the probability that the next car past the school gates contains **i** 3 people **ii** less than 4 people.
b One day 2500 cars passed the gates. How many of the cars would you expect to have 2 people inside?

13 Thirty students were asked to state the activities they enjoyed from swimming (S), tennis (T) and hockey (H). The numbers in each set are shown.
One student is randomly selected.
a Which of these pairs of events are exclusive?
  i 'selecting a student from S', 'selecting a student from H'
  ii 'selecting a student from S', 'selecting a student from T'.
b What is the probability of selecting a student who enjoyed either hockey or tennis?

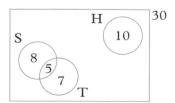

14 This spinner has four equal sectors.

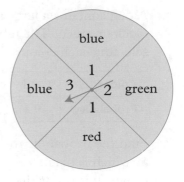

    a Which of these pairs of events are exclusive?
       i 'spinning 1', 'spinning green'
      ii 'spinning 3', 'spinning 2'
     iii 'spinning blue', 'spinning 1'.
    b What is the probability of spinning either
       a 1 or a green?

15 Here is a spinner with unequal sectors.
The tables show the probability of getting each colour
and number

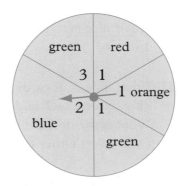

red	0·1
blue	0·3
orange	0·2
green	0·4

1	0·5
2	0·3
3	0·2

    a What is the probability of spinning either 1 or 2?
    b What is the probability of spinning either blue or green?
    c Why is the probability of spinning either a 1 or a green **not**
       0·5 + 0·4?

16 According to a weather forecaster in Norway, 'The probability of
snow on Saturday is 0·6 and the probability of snow on Sunday
is 0·5 so it is certain to snow over the weekend.' Explain why this
statement is wrong.

9.4 Probabilities of independent events, tree diagrams

9.4.1 Independent events

- Two events are independent if the occurrence of one event is
  unaffected by the occurrence of the other.

So, obtaining a head on one coin and a tail on another coin when the
coins are tossed at the same time are independent.
Similarly 'tossing a coin and getting a head' and 'rolling a dice and
getting a 2' are independent events.

9.4.2   The 'AND' rule: the multiplication rule

● For independent events A and B
  p (A and B) = p (A) × p (B)

The multiplication rule only works for **independent** events.

Two coins are tossed and the possible outcomes are listed.

  HH   HT   TH   TT

The probability of tossing two heads is $\frac{1}{4}$.

By the multiplication rule for independent events

p (two heads) = p (head on first coin) × p (head on second coin)

So p(two heads) = $\frac{1}{2} \times \frac{1}{2} = \frac{1}{4}$ as before.

EXAMPLE

a A fair coin is tossed and a fair dice is rolled.
   Find the probability of obtaining a 'head'
   and a 'six'.

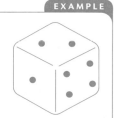

···

The two events are independent.

p(head **and** six) = p(head) × p(six) = $\frac{1}{2} \times \frac{1}{6} = \frac{1}{12}$

b When the two spinners are spun, what is the probability of getting a B
   on the first and a 3 on the second?

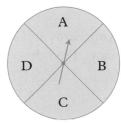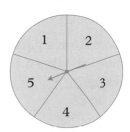

···

The events 'B on the first spinner' and '3 on the second spinner'
are independent.

∴ p(spinning B and 3) = p(B) × p(3) = $\frac{1}{4} \times \frac{1}{5} = \frac{1}{20}$

Exercise 7 (H)

1 A card is drawn from a pack of playing cards and a dice
   is thrown. Events A and B are
   A: 'a jack is drawn from the pack'
   B: 'a one is thrown on the dice'.
   a Write the values of p (A), p (B).
   b Write the value of p (A and B).

2 A coin is tossed and a dice is thrown. Write the probability of
   obtaining
   a a 'head' on the coin
   b an odd number on the dice
   c a 'head' on the coin and an odd number on the dice.

3 Box A contains 3 blue balls and 3 white balls.
   Box B contains 1 blue ball and 4 white balls.

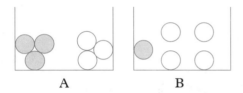

   One ball is randomly selected from Box A and one from Box B.
   What is the probability that both balls are blue?

4 In an experiment, Ann draws a card from a pack of playing
   cards and throws a dice.
   Find the probability of Ann obtaining
   a a card which is an ace and a six on the dice
   b the king of clubs and an even number on the dice
   c a heart and a 'one' on the dice.

5 Joe takes a card at random from a pack of playing cards and
   replaces it. After shuffling, he takes a second card. Find the
   probability of Joe obtaining
   a two cards which are clubs
   b two kings
   c two picture cards.

6 Maria takes a ball at random from a bag containing 3 red balls,
   4 black balls and 5 green balls. She replaces the first ball and
   takes a second.
   Find the probability of Maria obtaining
   a two red balls   **b** two green balls.

p (two red balls) =
p (red ball 1st
draw and red ball
2nd draw)

7 The letters of the word 'INDEPENDENT' are written on individual cards and the cards are put into a box. A card is selected and then replaced and then a second card is selected. Find the probability of obtaining
a the letter 'P' twice
b the letter 'E' twice.

8 A fruit machine has three independent reels and pays out a jackpot of £100 when three apples are obtained. Each reel has 15 pictures. The first reel has 3 apples, the second has 4 apples and the third has 2 apples. Find the probability of winning the jackpot.

9 Three coins are tossed and two dice are thrown at the same time.
Find the probability of obtaining
a three heads and a total of 12 on the dice
b three tails and a total of 9 on the dice.

10 A coin is biased so that it shows 'heads' with a probability of $\frac{2}{3}$.
The same coin is thrown three times. Find the probability of obtaining
a two tails on the first two throws
b a head, a tail and a head (in that order).

11 A fair dice and a biased dice are thrown together. The probabilities of throwing the numbers 1 to 6 are shown for the biased dice.

Fair

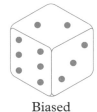

Biased

$$p(6) = \frac{1}{4}; \quad p(1) = \frac{1}{12}$$
$$p(2) = p(3) = p(4) = p(5) = \frac{1}{6}$$

Find the probability of obtaining a total of 12 on the two dice.

12 Philip and his sister toss a coin to decide who does the washing up.
If the result is heads Philip does it. If the result is tails
his sister does it. What is the probability that Philip does the
washing up every day for a week (7 days)?

13 Here is the answer sheet for five questions in
a multiple choice test.
What is the probability of getting all five
correct by guessing?

Answer sheet			
1. (A)	(B)	(C)	
2. (A)	(B)		
3. (A)	(B)	(C)	
4. (A)	(B)	(C)	(D)
5. (A)	(B)		

14 This spinner has six equal sectors.

Find the probability of getting
a 20 Qs in 20 trials
b no Qs in n trials
c at least one Q in n trials.

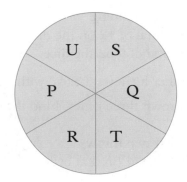

9.4.3 Tree diagrams

EXAMPLE

A bag contains 5 red balls and 3 green balls.
A ball is drawn at random and then replaced.
Another ball is drawn.

What is the probability that both balls are green?

∙∙

The branch marked ∗ involves the selection
of a green ball twice.
You find the probability of this event by
simply multiplying the fractions on the
two branches.

∴ p (two green balls) $= \dfrac{3}{8} \times \dfrac{3}{8} = \dfrac{9}{64}$

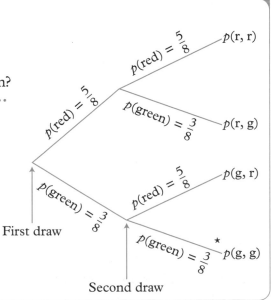

A bag contains 5 red balls and 3 green balls.
A ball is selected at random and **not** replaced.
A second ball is then selected.
Find the probability of selecting
a two green balls
b one red ball and one green ball.

· ·

Notice that since the first ball is not
replaced there are only 7 balls in the
bag for the second selection.

a p (two green balls) $= \dfrac{3}{8} \times \dfrac{2}{7}$

$\qquad = \dfrac{3}{28}$

b p (one red, one green) $= \left(\dfrac{5}{8} \times \dfrac{3}{7}\right) + \left(\dfrac{3}{8} \times \dfrac{5}{7}\right)$

$\qquad = \dfrac{15}{28}$

$p(\text{red}) = \dfrac{5}{8}$

$p(\text{green}) = \dfrac{3}{8}$

First draw

$p(\text{red}) = \dfrac{4}{7}$ $\qquad p(\text{r, r}) = \dfrac{5}{8} \times \dfrac{4}{7}$

$p(\text{green}) = \dfrac{3}{7}$ $\qquad p(\text{r, g}) = \dfrac{5}{8} \times \dfrac{3}{7}$

$p(\text{red}) = \dfrac{5}{7}$ $\qquad p(\text{g, r}) = \dfrac{3}{8} \times \dfrac{5}{7}$

$p(\text{green}) = \dfrac{2}{7}$ $\qquad p(\text{g, g}) = \dfrac{3}{8} \times \dfrac{2}{7}$

Second draw

Notice that you can add here because the events 'red then green' and 'green then
red' are exclusive.
As a check, all the fractions at the ends of the branches should add up to 1.

So $\left(\dfrac{5}{8} \times \dfrac{4}{7}\right) + \left(\dfrac{5}{8} \times \dfrac{3}{7}\right) + \left(\dfrac{3}{8} \times \dfrac{5}{7}\right) + \left(\dfrac{3}{8} \times \dfrac{2}{7}\right) = \dfrac{20}{56} + \dfrac{15}{56} + \dfrac{15}{56} + \dfrac{6}{56}$

$\qquad\qquad\qquad\qquad\qquad\qquad\qquad\qquad\qquad\qquad\qquad\quad = 1$

Exercise 8 Ⓜ/Ⓗ

1 A bag contains 10 discs; 7 are black and 3 white.
A disc is selected, and then replaced. A second
disc is selected. Copy and complete the tree
diagram showing all the probabilities and outcomes.

Find the probability that
a both discs are black
b both discs are white.

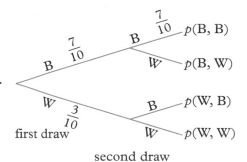

2 A bag contains 5 red balls and 3 green balls.
A ball is drawn and then replaced before a ball is drawn again.
Draw a tree diagram to show all the possible outcomes.
Find the probability that
a two green balls are drawn
b the first ball is red and the second is green.

3 A bag contains 5 black balls, 3 blue balls
and 2 white balls. A ball is drawn and
then replaced. A second ball is drawn.

Find the probability of drawing
a two black balls
b one blue ball and one white ball (in any order)
c two white balls.

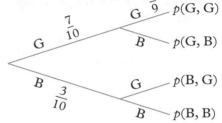

4 A bag contains 7 green discs and 3 blue discs. A disc
is drawn and **not** replaced. A second disc is drawn.
Copy and complete the tree diagram.

Find the probability that
a both discs are green
b both discs are blue.

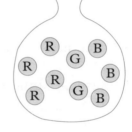

5 A bag contains 4 red balls, 2 green balls
and 3 blue balls. A ball is drawn and not
replaced. A second ball is drawn.

Find the probability of drawing
a two blue balls
b two red balls
c one red ball and one blue ball (in any order)
d one green ball and one red ball (in any order).

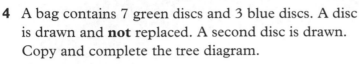

6 A six-sided dice is thrown three times.
Complete the tree diagram, showing at
each branch the two events:
'three' and 'not three' (written $\overline{3}$).

Find the probability of throwing a total of
a three threes
b no threes
c one three
d at least one three (use part **b**).

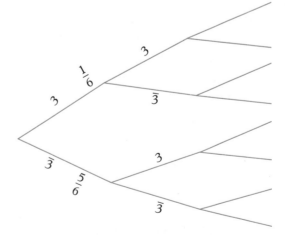

7 A card is drawn at random from a pack of 52 playing cards. The card
is replaced and a second card is drawn. This card is replaced and a
third card is drawn. What is the probability of drawing
a three hearts   **b** at least two hearts   **c** exactly one heart?

8 A bag contains 6 red marbles and 4 yellow marbles. A marble is drawn at random and **not** replaced. Two further draws are made, again without replacement. Find the probability of drawing
 a three red marbles
 b three yellow marbles
 c no red marbles
 d at least one red marble (use part **c**).

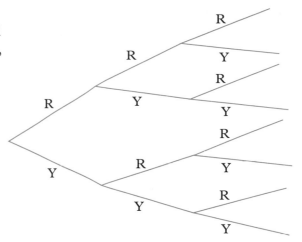

9 When a cutting is taken from a geranium the probability that it grows is $\frac{3}{4}$. Two cuttings are taken. What is the probability that

 a both cuttings grow
 b neither of them grows?

10 A dice has its six faces marked 0, 1, 1, 1, 6, 6. Two of these dice are thrown together and the total score is recorded.
 Draw a tree diagram.

 a How many different totals are possible?
 b What is the probability of obtaining a total of 7?

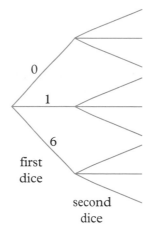

11 Louise loves the phone! When the phone rings $\frac{2}{3}$ of the calls are for Louise, $\frac{1}{4}$ are for her brother, Herbert, and the rest are for me.
 On Christmas Day I answered the phone twice.
 Find the probability that
 a both calls were for Louise
 b one call was for me and one was for Herbert.

12 Bag A contains 3 blue balls and 1 white ball.
Bag B contains 2 blue balls and 3 white balls.

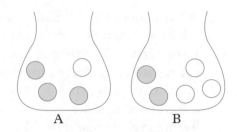

A ball is chosen at random from each bag in turn.
Find the probability of taking
  a a white ball from each bag
  b two balls of the same colour.

13 A teacher decides to award exam grades A, B or C by a new,
fairer method. Out of 20 students, three are to receive As, five Bs
and the rest Cs. She writes the grades on 20 pieces of
paper and invites the students to draw their exam result,
going through the class in alphabetical order.
Find the probability that

  a the first three students all get grade A
  b the first three students all get grade B
  c the first four students all get grade B.
(Do not cancel down the fractions.)

There are 10 boys and 12 girls in a class.
Two students are chosen at random.
What is the probability that one boy and one girl are chosen in
either order?

Here two students are chosen. It is like choosing
one student without replacement and then
choosing a second.
Here is the tree diagram.

$$p\,(b, g) + p\,(g, b) = \left(\frac{10}{22} \times \frac{12}{21}\right) + \left(\frac{12}{22} \times \frac{10}{21}\right)$$

$$= \frac{40}{77}$$

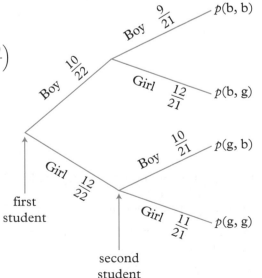

Exercise 9 (H)

1 There are 1000 components in a box of which 10 are known to be defective. Two components are selected at random. What is the probability that
  a both are defective **b** neither is defective
  c just one is defective?
  (Do **not** simplify your answers.)

2 There are 10 boys and 15 girls in a class. Two students are chosen at random. What is the probability that
  a both are boys **b** both are girls
  c one is a boy and one is a girl?

3 Two similar spinners are spun together and the scores are added.
  a What is the most likely total score on the two spinners?
  b What is the probability of obtaining this score on three successive spins of the two spinners?

4 A bag contains 3 red, 4 white and 5 green balls. Three balls are selected without replacement.

  Find the probability that the three balls chosen are
  a all red **b** all green **c** one of each colour.
  d If the selection of the three balls was carried out 1100 times, how often would you expect to choose three red balls?

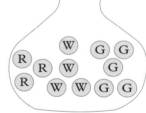

5 The diagram represents 15 students in a class and the diagram shows the sets G, S and F.
  G represents those who are girls
  S represents those who are swimmers
  F represents those who believe in Father Christmas.

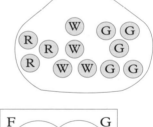

  A student is chosen at random. Find the probability that the student
  a can swim **b** is a girl swimmer
  c is a boy swimmer who believes in Father Christmas.

  Two students are chosen at random. Find the probability that
  d both are boys **e** neither can swim.

6 There are 500 ball bearings in a box of which 100 are known to be undersize. Three ball bearings are selected at random. What is the probability that
  a all three are undersize **b** none is undersize?
  Give your answers as decimals correct to three significant figures.

> ● Remember: When a question says '2 balls are drawn' or '3 people are chosen', you draw a tree diagram **without** replacement.

7 There are 9 boys and 15 girls in a class. Three students are chosen at random. What is the probability that
   a all three are boys   **b** all three are girls
   c one is a boy and two are girls?
  Give your answers as fractions.

8 A box contains x milk chocolates and y plain chocolates. Two chocolates are selected at random. Find, in terms of x and y, the probability of choosing
   a a milk chocolate on the first choice
   b two milk chocolates
   c one of each sort
   d two plain chocolates.

9 A pack of z cards contains x 'winning cards'. Two cards are selected at random. Find, in terms of x and z, the probability of choosing
   a a 'winning' card on the first choice
   b two 'winning cards' in the two selections
   c exactly one 'winning' card in the pair.

10 When a golfer plays any hole, he will take 3, 4, 5, 6, or 7 strokes with probabilities of $\frac{1}{10}, \frac{1}{5}, \frac{2}{5}, \frac{1}{5}$, and $\frac{1}{10}$ respectively.
  He never takes more than 7 strokes.
  Find the probability of these events:
   a scoring 4 on each of the first three holes
   b scoring 3, 4 and 5 (in that order) on the first three holes
   c scoring a total of 28 for the first four holes
   d scoring a total of 10 for the first three holes.

11 Assume that births are equally likely on each of the seven days of the week. Two people are selected at random. Find the probability that
   a both were born on a Sunday
   b both were born on the same day of the week.

12 a A playing card is drawn from a standard pack of 52 and then replaced. A second card is drawn. What is the probability that the second card is the same as the first?
   b A card is drawn and **not** replaced. A second card is drawn. What is the probability that the second card is **not** the same as the first?

9.5 Conditional probability

The performance of a climber is affected by the weather. When it rains, the probability that he will fall on a mountain is $\frac{1}{10}$, whereas in dry weather the probability that he will fall is $\frac{1}{50}$. In the climbing seasons, the probability of rain on any day is $\frac{1}{4}$.

What is the probability that he falls on the mountain on a day chosen at random?

Draw a tree diagram.

p (he falls) $= p(r, f) + p(\bar{r}, f)$

$$= \left(\frac{1}{4} \times \frac{1}{10}\right) + \left(\frac{3}{4} \times \frac{1}{50}\right) = \frac{1}{25}$$

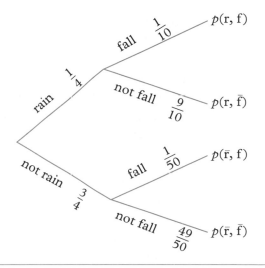

Exercise 10 Ⓗ

1 Bag A contains 3 white balls and 3 blue balls. Bag B contains 1 white ball and 3 blue balls.

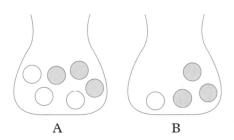

A B

A ball is taken at random from Bag A and placed in Bag B. A ball is then chosen from Bag B. What is the probability that the ball taken from B is white?

2 On a Monday or a Thursday Mr Gibson paints a 'masterpiece' with a probability of $\frac{1}{5}$. On any other day, the probability of producing a 'masterpiece' is $\frac{1}{100}$. In common with other great painters, Gibson never knows what day it is. Find the probability that on one day chosen at random, he will in fact paint a masterpiece.

3 If a hedgehog crosses a certain road before 7:00 a.m., the probability of being run over is $\frac{1}{10}$. After 7:00 a.m., the corresponding probability is $\frac{3}{4}$. The probability of the hedgehog waking up early enough to cross before 7:00 a.m. is $\frac{4}{5}$.

What is the probability of these events:
a the hedgehog waking up too late to reach the road before 7:00 a.m.
b the hedgehog waking up early and crossing the road in safety
c the hedgehog waking up late and crossing the road in safety
d the hedgehog waking up early and being run over
e the hedgehog crossing the road in safety?

4 In Stockholm the probability of snow falling on a day in January is 0·2. If it snows on one day, the probability of snow on the following day increases to 0·35.
What is the probability that for two consecutive days
a it will snow on both days
b it will not snow on either day
c it will snow on just one of the two days?

5 Two boxes are shown containing red and white balls. One ball is selected from Box A and placed in Box B.
A ball is then selected from Box B and placed in Box A. What is the probability that Box A now contains 4 red balls?

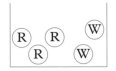

Box A

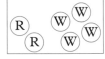

Box B

6 Michelle has been revising for a multiple choice exam in history but she has only had time to learn 70% of the facts being tested. If there is a question on any of the facts she has revised she will get that question right. Otherwise she will simply guess one of the five possible answers.
a A question is chosen randomly from the paper. What is the probability that she will get it right?
b If the paper has 50 questions what mark do you expect her to get?

7 Surgeons can operate to cure Pythagoratosis but the success rate at the first attempt is only 65%. If the first operation fails the operation can be repeated but this time the success rate is only 20%. After a second failure there is so little chance of success that surgeons will not operate again.

There is an outbreak of the disease at Boston College and 38 students contract the disease.
How many of these students do you expect can be saved after both operations?

Test yourself

1 Here is a 5-sided spinner.

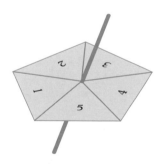

The sides of the spinner are labelled 1, 2, 3, 4 and 5
The spinner is biased.
The probability that the spinner will land on each of the numbers 1, 2, 3 and 4 is given in the table.

Number	1	2	3	4	5
Probability	0·15	0·05	0·2	0·25	x

Work out the value of x.

(Edexcel, 2008)

2 Tony throws a biased dice 100 times.
The table shows his results
He throws the dice once more.
a Find an estimate for the probability that he will get a 6.
Emma has a biased coin.
The probability that the biased coin will land on a head is 0·7
Emma is going to throw the coin 250 times.
b Work out an estimate for the number of times the coin
will land on a head.

Score	Frequency
1	12
2	13
3	17
4	10
5	18
6	30

(Edexcel, 2004)

3 The probability that a boy is left-handed is 0·2
The probability that a girl is left-handed is 0·3
A school has 480 boys and 520 girls.
a Estimate the number of left-handed students in the school.
b A student is picked at random from the whole school.
Estimate the probability that the student is left-handed.

(AQA, 2007)

4 The diagram shows a fair octagonal spinner.

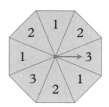

a Dave spins the spinner 20 times.
The results are shown in this table.

Spin	1	2	3	4	5	6	7	8	9	10	11	12	13	14	15	16	17	18	19	20
Result	1	3	1	2	3	2	1	2	1	2	3	3	2	1	2	2	1	2	3	1

   i What is the relative frequency of the spinner landing on 1?
   ii Steve also spins the spinner 20 times.
     Explain why Steve may not get the same results as Dave.
b How many times would you expect a result of 3 if you spin the
spinner 1000 times?

(AQA, 2007)

5 The probability that a biased dice will land on a two is 0·12.
The dice is thrown 300 times.
  a Estimate how many times the dice will land on a two.
The probability that the same dice will land on a five is 0·35.
  b Work out the probability that on its next throw it will land on
    either a two or a five.

6 Here is a 4-sided spinner.
The sides of the spinner are labelled 1, 2, 3 and 4.
The spinner is biased.
The probability that the spinner will land on each of the
numbers 2 and 3 is given in the table.
The probability that the spinner will land on 1 is equal to the
probability that it will land on 4.

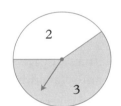

Number	1	2	3	4
Probability	x	0·3	0·2	x

  a Work out the value of x.
Sarah is going to spin the spinner 200 times.
  b Work out an estimate for the number of times it will land on 2.

(Edexcel, 2005)

7 The diagram shows a spinner.
When the arrow is spun the probability of scoring 2 is 0·3
The arrow is spun twice and the scores are added.

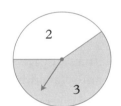

  a Copy and complete the tree diagram.

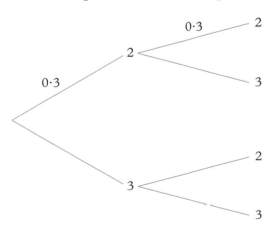

  b What is the probability that the total score is 4?

(AQA, 2003)

8 Tom and Sam each take a driving test.
  The probability that Tom will pass the driving test is 0·8
  The probability that Sam will pass the driving test is 0·6
  a Copy and complete the probability tree diagram.

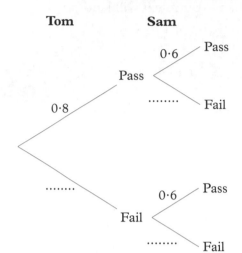

Tom   **Sam**

Pass ⟨ 0·6 Pass
       .... Fail

0·8

....

0·6 Pass
Fail ⟨
       .... Fail

  b Work out the probability that both Tom and Sam will pass the
    driving test.
  c Work out the probability that only one of them will pass the
    driving test.

    (Edexcel, 2008)

9 The letters of the word 'PROCESSES' are written on individual
  cards and placed in a bag.

P R O C E S S E S

  A letter is selected at random from the bag and then replaced.
  a What is the probability that the letter is
    i   E
    ii a vowel (the vowels are A, E, I, O, U)
    iii T
    iv not S?
  The experiment is repeated 360 times.
  b How many times would you expect the letter S to be chosen?

    (CCEA)

10 The diagram shows a fair spinner with numbers as shown.

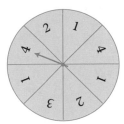

Work out the probability of getting exactly one 4 in two spins.

(OCR, 2009)

10 Using and applying mathematics

> **In this unit you will:**
> - practise problem-solving tasks
> - practise proving/disproving conjectures
> - solve mathematical puzzles.

10.1 Problem-solving tasks

There are a large number of possible starting points for problem-solving tasks here so it may be possible to allow students to choose tasks which appeal to them. On other occasions the same tasks may be set to a whole class.

Here are a few guidelines for students:
a If the set problem is too complicated try an easier case.
b Draw your own diagrams.
c Make tables of your results and be systematic.
d Look for patterns.
e Is there a rule or formula to describe the results?
f Can you **predict** further results?
g Can you **prove** any rules which you may find?
h Where possible extend the task further by asking questions like 'what happens if . . .'

10.1.1 Opposite corners

Here the numbers are arranged in 9 columns.

1	2	3	4	5	6	7	8	9
10	11	12	13	14	15	16	17	18
19	20	21	22	23	24	25	26	27
28	29	30	31	32	33	34	35	36
37	38	39	40	41	42	43	44	45
46	47	48	49	50	51	52	53	54
55	56	57	58	59	60	61	62	63
64	65	66	67	68	69	70	71	72
73	74	75	76	77	78	79	80	81
82	83	84	85	86	87	88	89	90

In the 2×2 square . . .

6	7
15	16

$6 \times 16 = 96$
$7 \times 15 = 105$

. . . the difference between them is 9.

In the 3×3 square . . .

22	23	24
31	32	33
40	41	42

$22 \times 42 = 924$
$24 \times 40 = 960$

. . . the difference between them is 36.

Investigate to see if you can find any rules or patterns
connecting the size of square chosen and the difference.
If you find a rule, use it to **predict** the difference for larger squares.
Test your rule by looking at squares like 8 × 8 or 9 × 9.

Can you **generalise** the rule?

[What is the difference for a square of size $n \times n$?]

Can you **prove** the rule?

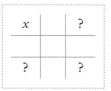

In a 3 × 3 square...

What happens if the numbers are arranged in six columns or
seven columns?

1	2	3	4	5	6
7	8	9	10	11	12
13	14	15	16	17	18
19					

1	2	3	4	5	6	7
8	9	10	11	12	13	14
15	16	17	18	19	20	21
22						

10.1.2 Hiring a car

You are going to hire a car for one week (7 days).
Which of the firms below should you choose?

Gibson car hire	Snowdon rent-a-car	Hav-a-car
£170 per week unlimited mileage	£10 per day 6·5p per mile	£60 per week 500 miles without charge 22p per mile over 500 miles.

Work out as detailed an answer as possible.

10.1.3 Half-time score

The final score in a football match was 3–2. How many different
scores were possible at half-time?

Investigate for other final scores where the difference between the teams is
always one goal. [1–0, 5–4, etc.]. Is there a pattern or rule which would tell
you the number of possible half-time scores in a game which finished 58–57?
Suppose the game ends in a draw. Find a rule which would tell you
the number of possible half-time scores if the final score was 63–63.
Investigate for other final scores [3–0, 5–1, 4–2, etc.].
Find a rule which gives the number of different half-time scores for
any final score (say $a–b$).

10.1.4 Squares inside squares

Here is a square drawn inside a square. Is there a connection between the **area** of the inside square and the numbers $\begin{pmatrix} 2 \\ 3 \end{pmatrix}$ which describe the angle through which the square is rotated?

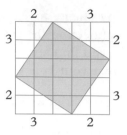

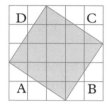

Method

You calculate the area of the inside square by subtracting the areas of triangles A, B, C and D from the area of the outside square.

area of outside square = $5 \times 5 = 25$ squares

area of each triangle = $\frac{1}{2} \times 2 \times 3 = 3$ squares

so, area of inside square = $25 - (4 \times 3)$

$= 13$ squares

Your task is to investigate the areas of different squares drawn inside other squares in a similar way.

Method

A Start with simple cases.

    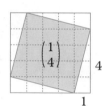

Find the areas of the inside squares where the top number is 1. Can you find a connection or rule which you can use to find the area of the inside square **without** all the working with subtracting areas of triangles?

B Look at more difficult cases.

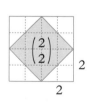   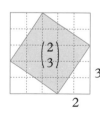   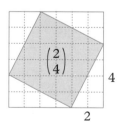

Again try to find a rule which you can use to find the area of the inside square directly.

C Look at **any** size square.

Try to find a rule and, if possible, use algebra to show why it **always** works.

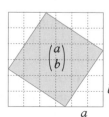

10.1.5 Maximum box

a You have a square sheet of card 24 cm by 24 cm.
You can make a box (without a lid) by cutting squares from the corners and folding up the sides.
What size corners should you cut out so that the volume of the box is as large as possible?
Try different sizes for the corners and record the results in a table.

Length of the side of the corner square (cm)	Dimensions of the open box (cm)	Volume of the box (cm³)
1	22 × 22 × 1	484
2		
–		
–		

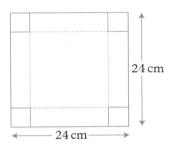

24 cm

24 cm

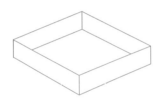

Now consider boxes made from different sized cards:
15 cm × 15 cm and 20 cm by 20 cm.
What size corners should you cut out this time so that the volume of the box is as large as possible?
Is there a connection between the size of the corners cut out and the size of the square card?

b Investigate the situation when the card is not square. Take rectangular cards where the length is twice the width (20 × 10, 12 × 6, 18 × 9 etc.)
Again, for the maximum volume is there a connection between the size of the corners cut out and the size of the original card?

10.1.6 What shape tin?

You need a cylindrical tin which will contain a volume of 600 cm³ of drink.

What shape should you make the tin so that you use the minimum amount of metal?
In other words, for a volume of 600 cm³, what is the smallest possible surface area?

r	h	A
2	?	?
3	?	?
⋮		

Make a table.

What shape tin should you design to contain a volume of 1000 cm³?

10.1.7 Painting cubes

The large cube in the diagram consists of 27 unit cubes.
All six faces of the large cube are painted blue.

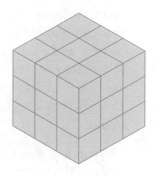

- How many unit cubes have 3 blue faces?
- How many unit cubes have 2 blue faces?
- How many unit cubes have 1 blue face?
- How many unit cubes have 0 blue faces?

Answer the four questions for a cube which is $n \times n \times n$.

10.1.8 Discs

a You have five blue discs and five white discs which are arranged in a line as shown.

The aim is to get all the blue discs to the right-hand end and all the white discs to the left-hand end.

The only move allowed is to interchange two neighbouring discs.

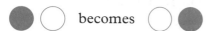 becomes

How many moves does it take?
How many moves would it take if you had fifty blue discs and fifty white discs arranged alternately?

b Suppose the discs are arranged in pairs

 ... etc.

How many moves would it take if you had fifty blue discs and fifty white discs arranged like this?

c Now suppose you have three colours black, white and blue arranged alternately.

 ... etc.

You want to get all the black discs to the right, the blue discs to the left and the white discs in the middle.
How many moves would it take if you have 30 discs of each colour?

In both cases work with a smaller number of discs until you can see a pattern.

10.1.9 Diagonals

In a 4×7 rectangle, the diagonal passes through 10 squares.

Draw rectangles of your own choice and count the number of squares through which the diagonal passes. A rectangle is 640×250. How many squares will the diagonal pass through?

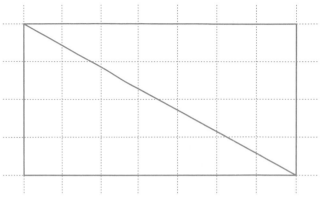

10.1.10 Chess board

Start with a small board, just 4×4.
How many squares are there? [It is not just 16!]
How many squares are there on an 8×8 chess board?
How many squares are there on an $n \times n$ chess board?

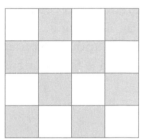

10.1.11 Find the connection

Work through the flow diagram several times, using a calculator.

What do you notice?
Try different numbers for x (suggestions: 11, 5, 8, 27)
What do you notice?

What happens if you take the square root three times?

Suppose in the flow diagram you change 'Multiply by x' to 'Divide by x'. What happens now?

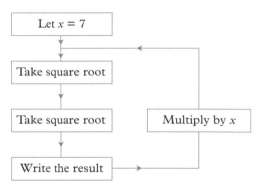

Suppose in the flow diagram you change 'Multiply by x' to 'Multiply by x^2'. What happens now?

You can vary the number of times you take the square root and whether you 'Multiply by x', 'Multiply by x^2', 'Divide by x' etc.

10.1.12 Spotted shapes

For this investigation you need square dotted paper. If you do not have any, you can make your own using a felt tip pen and squared paper.

The rectangle in Diagram 1 has 10 dots on the perimeter ($p = 10$) and 2 dots inside the shape ($i = 2$).
The area of the shape is 6 square units ($A = 6$).

Diagram 1

The triangle in Diagram 2 has 9 dots on the perimeter ($p = 9$) and 4 dots inside the shape ($i = 4$).
The area of the triangle is $7\frac{1}{2}$ square units $\left(A = 7\frac{1}{2}\right)$.

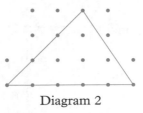

Diagram 2

Draw more shapes of your own design and record the values for p, i and A in a table. Make some of your shapes more difficult like the one in Diagram 3.

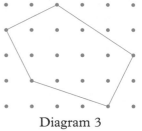

Diagram 3

This is quite difficult.
Be systematic, recording the values for i, p, A in a table.
Start by drawing several shapes with $i = 0$. Find a connection between p and A. Now increase i by one at a time.
Try to find a formula connecting p, i and A.

10.2   Proof and conjectures

10.2.1   Conjectures

A conjecture (or hypothesis) is simply a statement which may or may not be true. In mathematics people are constantly looking for rules or formulae to make calculations easier. Suppose you carry out an investigation and after looking at the results you spot what you think might be a rule.
You could make the conjecture that:

> 'x^2 is always greater than x'

or that 'the angle in a semicircle is always 90°'
or that 'the expression $x^2 + x + 41$ gives a prime number for all integer values of x'

There are three possibilities for any conjecture. It may be
a true   **b** false   **c** not proven.

(a) True
To show that a conjecture is true it is not enough to simply find lots of examples where it is true. You have to **prove** it for **all** values. Proof comes later in this section.

(b) False
To show that a conjecture is false it is only necessary to find one **counter-example**.

EXAMPLE

Consider the conjecture 'x^2 is always greater than x'.

....

It is true that $2^2 > 2$; $3 \cdot 1^2 > 3 \cdot 1$; $(-5)^2 > 5$

But $\left(\frac{1}{2}\right)^2$ is **not** greater than $\frac{1}{2}$.

This is a counter-example so the conjecture is false.
You have **disproved** the conjecture.

EXAMPLE

Consider the conjecture:
'the expression $x^2 + x + 41$ gives a prime number for all integer values of x'.

If you try the first few values of x, you get these results.

x	1	2	3	4	5	6
$x^2 + x + 41$	43	47	53	61	71	83

The expression does give prime numbers for these values of x and indeed, it does so for all values of x up to 39.
But when $x = 40$, $x^2 + x + 41 = 1681$
$$= 41 \times 41$$

So you do not always obtain a prime number.
The conjecture is, therefore, false.

Notice that it takes only **one** counter-example to disprove a conjecture.

(c) Not proven
Suppose you have a conjecture for which you cannot find a counter-example but which you also cannot prove. In this case the conjecture is **not proven**.

Until recently, the most famous example of a conjecture not proven was 'Fermat's Last Theorem' which stated that there are no whole numbers a, b, c for which $a^3 + b^3 = c^3$ [or further, that $a^n + b^n = c^n$ $(n > 2)$].

For 358 years no mathematician could prove the conjecture but no one could find a counter-example. In 1994 Andrew Wiles, a British mathematician, finally proved the theorem after devoting himself to it for seven years. The proof is extremely long (100 pages) and it is said that only a handful of people in the world understand it.

Exercise 1 (H)

In questions **1** to **12**, consider the conjecture given. Some are true and some are false. Write a counter-example where the conjecture is false. If you cannot find a counter-example, state that the conjecture is 'not proven' as you are **not** asked to prove it.

1 $(n + 1)^2 = n^2 + 1$ for all values of n.

2 $\frac{1}{x}$ is always less than x.

3 $\sqrt{n + 1}$ is always smaller than $\sqrt{n} + 1$.

4 For all values of n, $2n$ is greater than $n - 2$.

5 For any set of numbers, the median is always smaller than the mean.

6 $\sqrt{x}$ is always less than x.

7 The diagonals of a parallelogram never intersect at right angles.

8 The product of two irrational numbers is always another irrational number.

9 If n is even $5n^2 + n + 1$ is odd.

10 All numbers in the sequence 31, 331, 3331, 33 331, 333 331, 3 333 331, 33 333 331, 333 333 331, ... are prime.

> Divide by 17.

11 $n^2 + n > 0$ for all values of n.

12 The expression $a^n - a$ is always a multiple of n.

13 Here is a circle with 2 points on its circumference and 2 regions.

This circle has 3 points ($n = 3$) and 4 regions ($r = 4$).

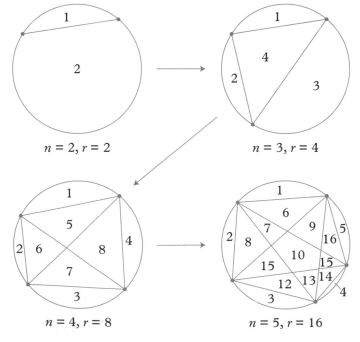

$n = 2, r = 2$

$n = 3, r = 4$

$n = 4, r = 8$

$n = 5, r = 16$

Each time, every point on the circumference is joined to all the others.

a Do you think a conjecture is justified at this stage?
Consider:
'The circle with 6 points on its circumference will have 32 regions'.
Or again:
'With n points on the circle the number of regions, r, is given by $r = 2^{n-1}$'.

Draw the next two circles in the sequence with $n = 6$ and $n = 7$.
Be careful to avoid this sort of thing where three lines pass through a single point.

Under a magnifying glass you might see this: and a region might be lost.
What can you say about your conjecture at this stage?

b When you have the results for $n = 6$ and $n = 7$ you might get somewhere (although it is quite difficult) by looking at **differences** in the results.

10.2.2 Logical argument

● Here is a statement
Triangle PQR is equilateral $\Rightarrow$ angle QRP $= 60°$.
The statement can be read:
If triangle PQR is equilateral, then angle QRP equals $60°$.
or
Triangle PQR is equilateral implies that angle QRP equals $60°$.

> $\Rightarrow$ means 'implies'.

● Here is another statement.
$$a > b \Rightarrow a - b > 0$$
This can be read:
If a is greater than b, then $a - b$ is greater than zero
or
a is greater than b implies that $a - b$ is greater than zero.

Note $x = 4 \Rightarrow x^2 = 16$ is true
but $x^2 = 16 \Rightarrow x = 4$ is not true because x could be -4.

Exercise 2 (H)

[Teacher note: See also Revision Exercise 14 which has a similar theme.]

1 Write these as sentences.

a Today is Monday $\Rightarrow$ tomorrow is Tuesday.

b It is raining $\Rightarrow$ there are clouds in the sky.

c Abraham Lincoln was born in 1809 $\Rightarrow$ Abraham Lincoln is dead.

2 State whether these statements are true or false.

a $x - 4 = 3 \Rightarrow x = 7$

b n is even $\Rightarrow n^2$ is even

c $a = b \Rightarrow a^2 = b^2$

d $x > 2 \Rightarrow x = 3$

e $a + b$ is odd $\Rightarrow ab$ is even (a and b are whole numbers).

f $x^2 = 4 \Rightarrow x = 2$

g $x = 45° \Rightarrow \tan x = 1$

h $5^x = 1 \Rightarrow x = 0$

i $pq = 0 \Rightarrow p = 0$

j a is odd $\Rightarrow a$ can be written in the form $2k + 1$, where k is an integer.

3 Think of pairs of statements like those in question **1**. Link the statements with the $\Rightarrow$ sign.

10.2.3 Proof

- In the field of science a theory, like Newton's theory of gravitation, can never be proved. It can only be considered highly likely using all the evidence available at the time. The history of science contains many examples of theories which were accepted at the time but were later shown to be untrue when more accurate observation was possible.

- A mathematical proof is far more powerful. Once a theorem is proved mathematically it will **always** be true. Pythagoras proved his famous theorem over 2500 years ago and when he died he knew it would never be disproved.
  A proof starts with simple facts which are accepted. The proof then argues logically to the result which is required.

- It is most important to realise that a result **cannot** be proved simply by finding thousands or even millions of results which support it.

  In 1738 Euler made the conjecture that there were no solutions to the equation $a^4 + b^4 + c^4 = d^4$.

  For 250 years this result remained unproved but it was then shown that $2\,682\,440^4 + 15\,365\,639^4 + 18\,796\,760^4 = 20\,615\,673^4$, thereby showing that Euler's conjecture was false.

Algebraic proof

EXAMPLE

Prove that, if a and b are odd numbers, then $a + b$ is even.

....

Proof

If a is odd, there is remainder 1 when a is divided by 2.
So, a may be written in the form $(2m + 1)$ where m is a whole number.
Similarly b may be written in the form $(2n + 1)$.

So, $a + b = 2m + 1 + 2n + 1$
$\qquad\quad = 2(m + n + 1)$

So, $a + b$ is even, as required.

EXAMPLE

Prove that the answer to every line of the pattern below is 8.

$3 \times 5 - 1 \times 7$
$4 \times 6 - 2 \times 8$
$5 \times 7 - 3 \times 9$
$\qquad \vdots \qquad\quad \vdots$

....

Proof

The nth line of the pattern is $(n + 2)(n + 4) - n(n + 6)$
$\qquad\qquad\qquad\qquad\qquad = n^2 + 6n + 8 - (n^2 + 6n)$
$\qquad\qquad\qquad\qquad\qquad = 8$

The result is proved.

The next example is much harder but very interesting.

Prove that the product of 4 consecutive numbers is always one less than a square number.

for example, $1 \times 2 \times 3 \times 4 = 24$ or $3 \times 4 \times 5 \times 6 = 360$
$$= 25 - 1 \qquad\qquad\qquad = 19^2 - 1$$

. . . .

Proof

Let the first number be n.
Then the next three consecutive numbers are $(n + 1)$, $(n + 2)$, $(n + 3)$.

Find the product
$$n(n + 1)\,(n + 2)\,(n + 3)$$
$$= (n^2 + n)\,(n^2 + 5n + 6)$$
$$= n^4 + 6n^3 + 11n^2 + 6n$$

Write the product as $(n^4 + 6n^3 + 11n^2 + 6n + 1) - 1$.
Note that this does not alter the value of the product.

The expression in the brackets factorises as
$(n^2 + 3n + 1)\,(n^2 + 3n + 1)$
So, the product is $(n^2 + 3n + 1)^2 - 1$

Since $(n^2 + 3n + 1)^2$ is a square number, the result is proved.

Exercise 3 Ⓗ

1 If a is odd and b is even, prove that ab is even.

2 If a and b are both odd, prove that ab is odd.

3 Prove that the sum of two odd numbers is even.

4 Prove that the square of an even number is divisible by 4.

5 Prove that the answer to every line of the pattern below is 3.

$$2 \times 4 - 1 \times 5 =$$
$$3 \times 5 - 2 \times 6 =$$
$$4 \times 6 - 3 \times 7 =$$
$$\vdots \qquad\quad \vdots$$

See the first example on the previous page.

6 In any three consecutive numbers prove that the product of the first and the third number is one less than the square of the middle number.

7 Prove that the product of four consecutive numbers is divisible by 4.

8 Prove that the sum of the squares of any two consecutive integers is an odd number.

9 Prove that the sum of the squares of five consecutive numbers is divisible by 5.

For example, $2^2 + 3^2 + 4^2 + 5^2 + 6^2 = 90 = 5 \times 18$

Begin by writing the middle number as n, so the other numbers are $n - 2, n - 1, n + 1, n + 2$.

10 Find the sum of the squares of three consecutive numbers and then subtract 2. Prove that the result is always 3 times a square number.

11 A proof of Pythagoras' theorem.

Four right-angled triangles with sides a, b and c are drawn around the tilted square as shown.

Area of large square $= (a + b)^2$ ①

Also area of large square = area of 4 triangles + area of tilted square   ②

Equate ① and ② and hence complete the proof of Pythagoras' theorem.

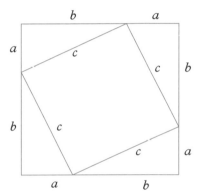

12 Here is the 'proof' that $1 = 2$.
Let $a = b$

$\Rightarrow ab = b^2$ [multiply by b]

$\Rightarrow ab - a^2 = b^2 - a^2$ [subtract a^2]

$\Rightarrow a(b - a) = (b + a)(b - a)$ [factorise]

$\Rightarrow a = b + a$ [divide by $(b - a)$]

$\Rightarrow a = a + a$ [from top line]

$\Rightarrow 1 = 2$

Which step in the argument is not allowed?

Geometric proof

Over 2000 years ago Greek mathematicians were very keen on proofs. Euclid wrote a best seller called 'The Elements' in which he first of all stated some very basic assumptions called **axioms**. He then proceeded to prove one result after another, basing each new result on a previous result which had already been proved. Examples of his axioms are 'parallel lines do not meet in either direction' and 'only one straight line can be drawn between two points'.

When you are writing a geometric proof of your own you do not need to go right back to Euclid's axioms. What you should do is state clearly after each line what facts you have used. For example 'angles in a triangle add up to 180°' or 'opposite angles are equal.'

EXAMPLE

Prove that the angle at the centre of a circle is twice the angle at the circumference.

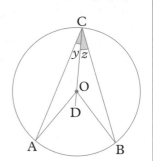

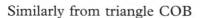

Proof
Draw the straight line COD.
Let $\angle ACO = y$ and $\angle BCO = z$.

In triangle AOC

	$AO = OC$	(radii)
So	$\angle OCA = \angle OAC$	(isosceles triangle)
and	$\angle COA = 180° - 2y$	(angle sum of triangle)
So	$\angle AOD = 2y$	(angles on a straight line)

Similarly from triangle COB

	$\angle DOB = 2z$
Now	$\angle ACB = y + z$
and	$\angle AOB = 2y + 2z$
So	$\angle AOB = 2 \times \angle ACB$ as required.

EXAMPLE

Prove that opposite angles in a cyclic quadrilateral add up to 180°.

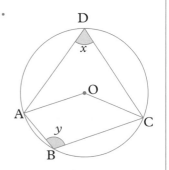

Proof
Draw radii OA and OC.
Let $\angle ADC = x$ and $\angle ABC = y$.

$\angle AOC$ obtuse $= 2x$ (angle at the centre)
$\angle AOC$ reflex $= 2y$ (angle at the centre)
$2x + 2y = 360°$ (angles at a point)
So $\quad x + y = 180°$ as required.

Exercise 4 (H)

1 Triangle ABC is isosceles.

Prove that ∠CBD = 2 × ∠CAB.

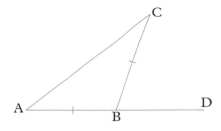

2 Triangle PQR is isosceles and ∠PSQ = 90°.

Prove that PS bisects ∠QPR.

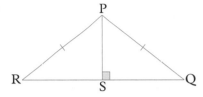

3 Two chords of a circle AB and CD intersect at X.

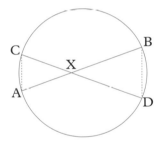

Use similar triangles to prove that AX × BX = CX × DX.
This result is called the Intersecting Chords Theorem.

4 Line ATB touches a circle at T and TC is a diameter. AC and BC cut the circle at D and E respectively. Prove that the quadrilateral ADEB is cyclic.

5 Prove that the angle in a semicircle is a right angle.

6 TC is a tangent to a circle at C and BA produced meets this tangent at T.

> 'Produced' means extended.

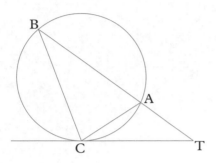

Show that triangles TCA and TBC are similar and hence prove that $TC^2 = TA \times TB$.

7 Given that BOC is a diameter and that $\angle ADC = 90°$, prove that AC bisects $\angle BCD$.

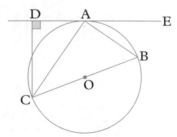

10.3 Puzzles and games

10.3.1 Crossnumbers without clues

Here are five crossnumber puzzles with a difference.
There are no clues, only answers, and it is your task to find where the answers go.

a Copy out the crossnumber pattern or ask your teacher for a photocopy.

b Fit all the given numbers into the correct spaces.
Tick off the numbers from the lists as you write them in the square.

1

2 digits	3 digits	4 digits	5 digits	6 digits
14	111	3824	24 715	198 264
25	127	4210	54 073	338 472
26	131	8072	71 436	387 566
30	249	8916	72 180	414 725
42	276	9603	82 125	
52	328			
53	571			*7 digits*
57	609			8 592 070
61	653			
99	906			
	918			

2

2 digits	3 digits	4 digits	5 digits	6 digits
19	106	1506	21 362	134 953
26	156	1624	21 862	727 542
41	180	2007	57 320	
63	215	2637	83 642	
71	234	4214		
76	263	4734		
	385	5216		
	427	5841		
	725	9131		
	872	9217		

The next three are more difficult but they are possible! Don't give up.

3

2 digits	3 digits	4 digits	5 digits	6 digits
26	306	3654	38 975	582 778
28	457	3735	49 561	585 778
32	504	3751	56 073	728 468
47	827	3755	56 315	
49	917	3819	56 435	
52	951	6426	57 435	*7 digits*
70		7214	58 535	8 677 056
74		7315	58 835	
		7618	66 430	
		7643	77 435	
		9847	77 543	

4

2 digits	3 digits	4 digits	5 digits	6 digits
11	121	2104	14 700	216 841
17	147	2356	24 567	588 369
18	170	2456	25 921	846 789
19	174	3714	26 759	861 277
23	204	4711	30 388	876 452
31	247	5548	50 968	
37	287	5678	51 789	
58	324	6231	78 967	
61	431	6789	98 438	
62	450	7630		
62	612	9012		**7 digits**
70	678	9921		6 645 678
74	772			
81	774			
85	789			
94	870			
99				

*5

2 digits		3 digits	4 digits	5 digits	6 digits
12	47	129	2096	12 641	324 029
14	48	143	3966	23 449	559 641
16	54	298	5019	33 111	956 782
18	56	325	5665	33 210	
20	63	331	6462	34 509	
21	67	341	7809	40 551	
23	81	443	8019	41 503	
26	90	831	8652	44 333	**7 digits**
27	91	923		69 786	1 788 932
32	93			88 058	5 749 306
38	98			88 961	
39	99			90 963	
46				94 461	
				99 654	

10.3.2 Estimating game

This is a game for two players. On square grid paper draw an answer grid with the numbers shown.

Answer grid

891	7047	546	2262	8526	429
2548	231	1479	357	850	7938
663	1078	2058	1014	1666	3822
1300	1950	819	187	1050	3393
4350	286	3159	442	2106	550
1701	4050	1377	4900	1827	957

The players now take turns to choose two numbers from the question grid and multiply them on a calculator. They cross out the answers on the answer grid. (Each player uses a different colour.)

The game continues until all the numbers in the answer grid have been crossed out. The object is to get four answers in a line (horizontally, vertically or diagonally). The winner is the player with most lines of four. A line of **five** counts as **two** lines of four. A line of **six** counts as **three** lines of four.

Question grid

11	26	81
17	39	87
21	50	98

10.3.3 The chess board problem

a On the 4×4 square, four objects have been placed, subject to the restriction that nowhere are there two objects on the same row, column or diagonal.

Subject to the same restrictions:
i   find a solution for a 5×5 square, using five objects
ii find a solution for a 6×6 square, using six objects
iii find a solution for a 7×7 square, using seven objects
iv find a solution for a 8×8 square, using eight objects.

It is called the chess board problem because the objects could be 'Queens' which can move any number of squares in any direction.

b Suppose you remove the restriction that no two Queens can be on the same row, column or diagonal. Is it possible to attack every square on an 8×8 chess board with less than eight Queens? Try the same problem with other pieces like knights or bishops.

10.3.4 Creating numbers

Using only the numbers 1, 2, 3 and 4 once each and the operations $+, -, \times, \div, !$ create every number from 1 to 100.
You can use the numbers as powers and you must use all of the numbers 1, 2, 3 and 4.

Examples $1 = (4 - 3) \div (2 - 1)$ $20 = 4^2 + 3 + 1$
$68 = 34 \times 2 \times 1$ $100 = (4! + 1)(3! - 2!)$

4! is pronounced 'four factorial' and means $4 \times 3 \times 2 \times 1$ (that is, 24) similarly $3! = 3 \times 2 \times 1 = 6$ and $5! = 5 \times 4 \times 3 \times 2 \times 1 = 120$

10.3.5   Designing square patterns

The object is to design square patterns of different sizes. The patterns are all to be made from smaller tiles all of which are themselves square.

Designs for a 4 × 4 square:

a This design consists of four tiles each 2 × 2. The pattern is rather dull.

b Suppose that the design must contain at least one 1 × 1 square.

This design is more interesting and consists of seven tiles.

1 Try the 5 × 5 square. Design a pattern which divides the 5 × 5 square into eight smaller squares.

2 Try the 6 × 6 square. Here you must include at least one 1 × 1 square. Design a pattern which divides the 6 × 6 square into nine smaller squares. Colour in the final design to make it look interesting.

3 The 7 × 7 square is more difficult. With no restrictions, design a pattern which divides the 7 × 7 square into nine smaller squares.

4 Design a pattern which divides an 8 × 8 square into ten smaller squares. You must not use a 4 × 4 square.

5 Design a pattern which divides a 9 × 9 square into ten smaller squares. You can use only one 3 × 3 square.

6 Design a pattern which divides a 10 × 10 square into eleven smaller squares. You must include a 3 × 3 square.

7 Design a pattern which divides an 11 × 11 square into eleven smaller squares. You must include a 6 × 6 square.

Test yourself

1 Show that the sum of **any** three consecutive integers is always a multiple of 3.

(AQA, 2003)

2 Tom is investigating the equation $y = x^2 - x + 5$

He starts to complete a table of values of y for some integer values of x.

x			−2	−1	0	1	2	3			
y			11	7	5	5	7	11			

Tom says, "When x is an integer, y is **always** a prime number".

Find a counter-example to show that Tom is wrong.

Explain your answer.

(AQA, 2004)

3 a Solve $x^2 + x + 11 = 14$

Give your solutions correct to 3 significant figures.

$y = x^2 + x + 11$

The value of y is a prime number when $x = 0, 1, 2$ and 3
The following statement is **not** true.
'$y = x^2 + x + 11$ is **always** a prime number when x is an integer'

b Show that the statement is not true.

(Edexcel, 2004)

4 a Write down an expression, in terms of n, for the nth multiple of 5.

b Hence or otherwise
  i prove that the sum of two consecutive multiples of 5 is always an odd number.
  ii prove that the product of two consecutive multiples of 5 is always an even number.

(Edexcel, 2004)

5 a n is a positive integer.
    i Explain why $n(n + 1)$ must be an even number.
    ii Explain why $2n + 1$ must be an odd number.
  b Expand and simplify $(2n + 1)^2$
  c Prove that the square of any odd number is always 1 more than a multiple of 8.

(AQA, 2004)

6 Prove that,

$$(n + 1)^2 - (n - 1)^2$$

is a multiple of 4, for all positive integer values of n.

(Edexcel, 2003)

7 ABC is an isosceles triangle.

M is the midpoint of AC.

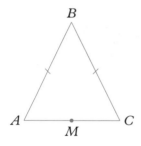

Prove that triangles ABM and CBM are congruent.

(AQA, 2007)

11 Revision

11.1 Revision exercises (non-calculator)

Exercises 1, 2, 3 contain questions on material for GCSE grades D and C. The questions in exercises 4 to 9 are on material for grades C to A*.

Revision exercise 1

1 Copy this bill and complete it by filling in the four blank spaces.

8 rolls of wallpaper at £3·20 each = £☐

3 tins of paint at £☐ each = £20·10

☐ brushes at £2·40 each = £ 9·60

Total = £☐

2 Write each sequence and find the next two numbers.
   a 2, 9, 16, 23
   b 20, 18, 16, 14
   c −5, −2, 1, 4
   d 128, 64, 32, 16
   e 8, 11, 15, 20

3 A man buys 500 pencils at 2·4 pence each. What change does he receive from £20?

4 Every day at school Stephen buys a roll for 14p, crisps for 11p and a drink for 21p. How much does he spend in pounds in the whole school year of 200 days?

5 An athlete runs 25 laps of a track in 30 minutes 10 seconds.
   a How many seconds does he take to run 25 laps?

   b How long does he take to run one lap, if he runs the 25 laps at a constant speed?

6 A pile of 250 tiles is 2 m thick. What is the thickness of one tile in cm?

7 Work out
   a 20% of £65
   b 37% of £400
   c 8·5% of £2000.

8 In a test, the marks of nine students were 7, 5, 2, 7, 4, 9, 7, 6, 6. Find
   a the mean mark
   b the median mark
   c the modal mark.

9 Work out
   a −6 − 5   b −7 + 30
   c −13 + 3   d −4 × 5
   e −3 × (−2)   f −4 + (−10)

10 Given $a = 3$, $b = -2$ and $c = 5$, work out
   a $b + c$ b $a - b$
   c ab d $a + bc$

11 Solve these equations.
   a $x - 6 = 3$ b $x + 9 = 20$
   c $x - 5 = -2$ d $3x + 1 = 22$

12 Which of these nets can be used to make a cube?

a   b

c

 Cut out the nets on squared paper.

Revision exercise 2

1 Here is a sequence.

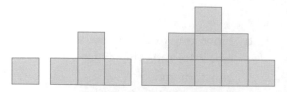

  a Draw the next diagram in this sequence.

  b Write the number of squares in each diagram.

  c Describe in words the sequence you obtain in part **b**.

  d How many squares will there be in the diagram which has 13 squares on the base?

2 Solve these equations.

  a $3x - 1 = 20$ **b** $4x + 3 = 4$

  c $5x - 7 = -3$

3 A bag contains 3 red balls and 5 white balls. Find the probability of selecting

  a a red ball

  b a white ball.

4 A box contains 2 yellow discs, 4 blue discs and 5 green discs. Find the probability of selecting

  a a yellow disc

  b a green disc

  c a blue or a green disc.

5 Point B is on a bearing 120° from point A. The distance from A to B is 110 km.

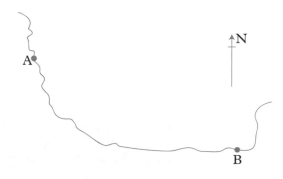

  a Draw an accurate diagram showing the positions of A and B. Use a scale of 1 cm to 10 km.

  b Ship S is on a bearing 072° from A. Ship S is on a bearing 325° from B. Show S on your diagram and state the distance from S to B.

6 Look at this diagram.

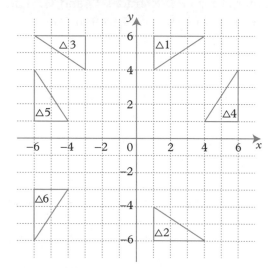

Describe fully these transformations.

  a $\triangle 1 \rightarrow \triangle 2$ **b** $\triangle 1 \rightarrow \triangle 3$

  c $\triangle 1 \rightarrow \triangle 4$ **d** $\triangle 5 \rightarrow \triangle 1$

  e $\triangle 5 \rightarrow \triangle 6$ **f** $\triangle 4 \rightarrow \triangle 6$

7 Plot and label these triangles.

  $\triangle 1$: $(-3, -6)$, $(-3, -2)$, $(-5, -2)$

  $\triangle 2$: $(-5, -1)$, $(-5, -7)$, $(-8, -1)$

  $\triangle 3$: $(-2, -1)$, $(2, -1)$, $(2, 1)$

  $\triangle 4$: $(6, 3)$, $(2, 3)$, $(2, 5)$

  $\triangle 5$: $(8, 4)$, $(8, 8)$, $(6, 8)$

  $\triangle 6$: $(-3, 1)$, $(-3, 3)$, $(-4, 3)$

Describe fully these transformations.

  a $\triangle 1 \rightarrow \triangle 2$

  b $\triangle 1 \rightarrow \triangle 3$

  c $\triangle 1 \rightarrow \triangle 4$

  d $\triangle 1 \rightarrow \triangle 5$

  e $\triangle 1 \rightarrow \triangle 6$

  f $\triangle 3 \rightarrow \triangle 5$

  g $\triangle 6 \rightarrow \triangle 2$

8 Find the shaded area. All lengths are in cm.

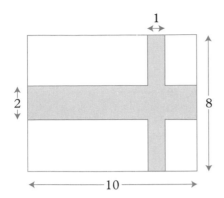

Revision exercise 3

1 Write each statement correctly.
 a $t + t + t = t^3$
 b $a^2 \times a^2 = 2a^2$
 c $2n \times n = 2n^2$

2 Look at this number pattern.
 $(2 \times 1) - 1 = 2 - 1$
 $(3 \times 3) - 2 = 8 - 1$
 $(4 \times 5) - 3 = 18 - 1$
 $(5 \times 7) - 4 = 32 - 1$
 $(6 \times a) - 5 = b - 1$
 a What number does the letter a stand for?
 b What number does the letter b stand for?
 c Write the next line in the pattern.

3 a Plot and label these triangles
 $\triangle1$: $(-3, 4)$, $(-3, 8)$, $(-1, 8)$
 $\triangle5$: $(-8, -2)$, $(-8, -6)$, $(-6, -2)$
 b Draw $\triangle2$, $\triangle3$, $\triangle4$, $\triangle6$ and $\triangle7$ as follows:
 i $\triangle1 \to \triangle2$: translation $\begin{pmatrix} 9 \\ -4 \end{pmatrix}$
 ii $\triangle2 \to \triangle3$: translation $\begin{pmatrix} -4 \\ -8 \end{pmatrix}$
 iii $\triangle3 \to \triangle4$: reflection in the line $y = x$

iv $\triangle5 \to \triangle6$: rotation 90° anticlockwise, centre $(-4, -1)$
 v $\triangle6 \to \triangle7$: rotation 180°, centre $(0, -1)$.
 c Write the coordinates of the 'pointed ends' of $\triangle2$, $\triangle3$, $\triangle4$, $\triangle6$, and $\triangle7$.

4 The faces of a round and a square clock have exactly the same area. If the round clock has a radius of 10 cm, how wide is the square clock?

5 The tables show the rail fares for adults and part of a British Rail timetable for trains between Cambridge and Bury St. Edmunds.

Fares for **one** adult

Cambridge

£1·00	Dullingham			
£1·20	40p	Newmarket		
£1·30	£1·00	60p	Kennett	
£2·00	£1·30	£1·20	80p	Bury St. Edmunds

Train times

Cambridge	11:20
Dullingham	11:37
Newmarket	11:43
Kennett	11:52
Bury St. Edmunds	12:06

 a How much would it cost for four adults to travel from Dullingham to Bury St. Edmunds?
 b How long does this journey take?

6 The diagram shows a regular octagon with centre O.
Find angles a and b.

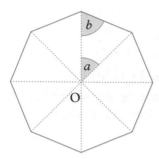

7 In December 2006, a factory employed 220 men, each man being paid £200 per week.
  a Calculate the total weekly wage bill for the factory.
  b In January 2007, the work force of 220 was reduced by 10 per cent.
  Find the number of men employed at the factory after the reduction.
  c Also in January 2007, the weekly wage of £200 was increased by 10 per cent. Find the new weekly wage.
  d Calculate the total weekly wage bill for the factory in January 2007.

8 $1 + 3 = 2^2$ $1 + 3 + 5 = 3^2$
  a $1 + 3 + 5 + 7 = x^2$
  Calculate x.
  b $1 + 3 + 5 + \ldots + n = 100$
  Calculate n.

9 This graph shows a car journey from Gateshead to Middlesbrough and back again.

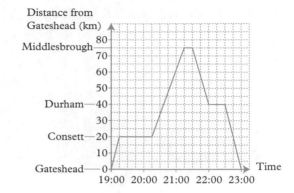

  a Where is the car
    i at 19:15
    ii at 22:15
    iii at 22:45?

  b How far is it
    i from Consett to Middlesbrough
    ii from Durham to Gateshead?

  c At what speed does the car travel
    i from Gateshead to Consett
    ii from Consett to Middlesbrough
    iii from Middlesbrough to Durham
    iv from Durham to Gateshead?

  d For how long is the car stationary during the journey?

10 A photo 21 cm by 12 cm is enlarged as shown.

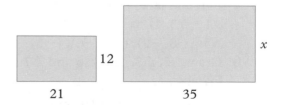

  a What is the scale factor of the enlargement?
  b Work out the length x.

11 The distance-time graphs for several objects are shown. Decide which line represents each of these.
- hovercraft from Dover
- car ferry from Dover
- cross-channel swimmer from Dover
- marker buoy outside harbour
- train from Dover
- car ferry from Calais

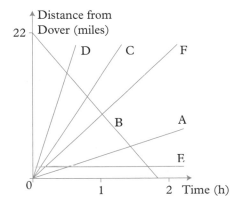

12 a The mean mass of 10 boys in a class is 56 kg. Calculate their total mass.
  b Another boy, whose mass is 67 kg, joins the group. Calculate the mean mass of the 11 boys.

Revision exercise 4

1 The equations of these lines are $y = 3x$, $y = 6$, $y = 10 - x$ and $y = \frac{1}{2}x - 3$.

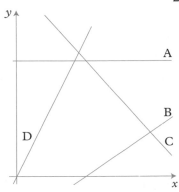

Find the equation corresponding to each line.

2 Find x.
  a $2^x = 32$ **b** $\frac{4}{x} = 7$

3 Find the median of 6, 3, 8, 2 and 9.

4 a A lies on a bearing of $040°$ from B. Calculate the bearing of B from A.
  b The bearing of X from Y is $115°$. Calculate the bearing of Y from X.

5 Given $a = 3$, $b = 4$ and $c = -2$, evaluate
  a $2a^2 - b$ **b** $a(b - c)$ **c** $2b^2 - c^2$

6 Work out $6578 \div 26$.

7 Increase £5000 by 5%.

8 Convert $\frac{3}{8}$ to a decimal.

9 Work out
  a $\frac{3}{5} \times \frac{3}{4}$ **b** $\frac{7}{12} - \frac{1}{8}$

10 $a = \frac{1}{2}$, $b = \frac{1}{4}$. Which one of the following has the greatest value?
  i ab **ii** $a + b$ **iii** $\frac{a}{b}$
  iv $\frac{b}{a}$ **v** $(ab)^2$

11 a On a map, the distance between two points is 16 cm. Calculate the scale of the map if the actual distance between the points is 8 km.
  b On another map, two points appear 1·5 cm apart and are in fact 60 km apart. Calculate the scale of the map.

12 Given that $s - 3t = rt$, express
  a s in terms of r and t
  b r in terms of s and t

13 Work out an estimate of the value of $\frac{61·51 \times 4·93}{31·33}$ correct to one significant figure.

14 The shaded region A is formed by the lines $y = 2$, $y = 3x$ and $x + y = 6$. Write the three inequalities which define A.

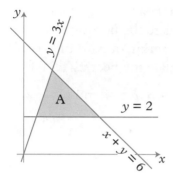

Revision exercise 5

1 Phil buys a camera for £300 and then sells it for £315. Work out his percentage profit.

2 The sides of a square are increased by 10%. By what percentage is the area increased?

3 Work out x, if $4^x = 16^3$.

4 Twenty-seven small wooden cubes fit exactly inside a cubical box without a lid. How many of the cubes are touching the sides or the bottom of the box?

5 Given that $x = 4$, $y = 3$, $z = -2$, evaluate

   a $2x(y + z)$ **b** $(xy)^2 - z^2$

   c $x^2 + y^2 + z^2$ **d** $(x + y)(x - z)$

   e $\sqrt{x(1 - 4z)}$ **f** $\dfrac{xy}{z}$

6 When two dice are thrown simultaneously, what is the probability of obtaining the same number on both dice?

7 A target consists of concentric circles of radii 3 cm and 9 cm.

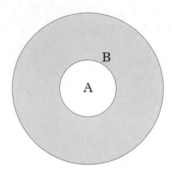

   a Find the area of A, in terms of π.

   b Find the ratio $\dfrac{\text{area of B}}{\text{area of A}}$.

8 A motorist travelled 200 miles in five hours. Her average speed for the first 100 miles was 50 mph. What was her average speed for the second 100 miles?

9 The perimeter of this rectangle is 40 cm. Find x.

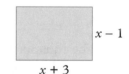

10 The table shows the number of students in a class who scored marks 3 to 8 in a test.

Marks	3	4	5	6	7	8
Number of students	2	3	6	4	3	2

Find

   a the mean mark

   b the modal mark

   c the median mark.

11 In the diagram, triangles ABC and EBD are similar but DE is **not** parallel to AC. Given that AD = 5 cm, DB = 3 cm and BE = 4 cm, calculate the length of BC.

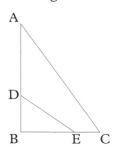

12 Nadia said: 'I thought of a number, multiplied it by 6, then added 15. My answer was less than 200'.

   a Write Nadia's statement in symbols, using x as the starting number.

   b Nadia actually thought of a prime number. What was the largest prime number she could have thought of?

Revision exercise 6

1 a A piece of meat, initially weighing 2·4 kg, is cooked and subsequently weighs 1·9 kg. What is the percentage loss in weight?

   b An article is sold at a 6% loss for £225·60. What was the cost price?

2 In the diagram, the equations of the lines are $2y = x - 8$, $2y + x = 8$, $4y = 3x - 16$ and $4y + 3x = 16$.

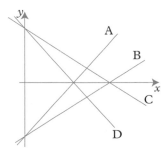

Find the equation corresponding to each line.

3 a Calculate the speed (in metres per second) of a slug which moves a distance of 30 cm in 1 minute.

   b Calculate the time taken for a bullet to travel 8 km at a speed of 5000 m/s.

   c Calculate the distance flown, in a time of four hours, by a pigeon which flies at a speed of 12 m/s.

4 Solve the simultaneous equations
$$7c + 3d = 29$$
$$5c - 4d = 33$$

5 Describe the single transformation which maps

   a $\triangle$ABC onto $\triangle$DEF

   b $\triangle$ABC onto $\triangle$PQR

   c $\triangle$ABC onto $\triangle$XYZ

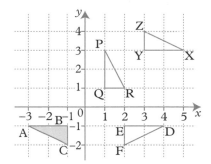

6 Sketch the curve $y = \sin x$ for x from 0° to 360°. If $\sin 28° = 0·469$, give another angle whose sine is 0·469.

7 Sketch the curve $y = \tan x$ for x from 0° to 360°. Find two solutions of the equation $\tan x = 1$.

8 Given that $y = \dfrac{k}{k + w}$

   a find the value of y when $k = \dfrac{1}{2}$ and $w = \dfrac{1}{3}$

   b express w in terms of y and k.

9 It is given that $y = \frac{k}{x}$ and that
$1 \leqslant x \leqslant 10$.

  a If the smallest possible value of y is 5, find the value of the constant k.

  b Find the largest possible value of y.

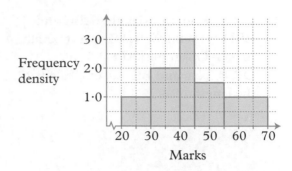

Frequency density

Marks

a If the pass mark was 45, how many students passed the test?

b How many took the test altogether?

Revision exercise 7

1 What values of x satisfy the inequality $15 - 4x > 12$?

2 Work out $30\,000 \times 5$ million and write the answer in standard form.

3 Factorise **a** $x^2 + 8x + 15$
              b $x^2 + x - 6$
              c $5x^2 - 30x$

4 In the diagram the area of the smaller square is $10\,\text{cm}^2$. Find the area of the larger square.

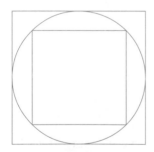

5 A bag contains 8 balls of which 2 are red and 6 are white. Ahmed selects a ball and does not replace it. He then takes a second ball. Find the probability of Ahmed obtaining
  a two red balls   **b** two white balls
  c one ball of each colour.

6 Several students took a science test and the results are shown in the histogram.

7 Draw a histogram for these data giving the ages of people at a disco.

Ages	Frequency	Frequency density
$14 \leqslant x < 16$	10	
$16 \leqslant x < 17$	18	
$17 \leqslant x < 18$	26	
$18 \leqslant x < 21$	30	
$21 \leqslant x < 26$	40	

8 The shaded region B is formed by the lines $x = 0$, $y = x - 2$ and $x + y = 7$.

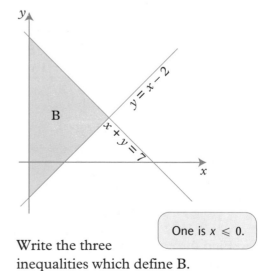

One is $x \leqslant 0$.

Write the three inequalities which define B.

9 Estimate the answer correct to one significant figure. Do not use a calculator.

   a $(612 \times 52) \div 49{\cdot}2$

   b $(11{\cdot}7 + 997{\cdot}1) \times 9{\cdot}2$

   c $\sqrt{\left(\dfrac{91{\cdot}3}{10{\cdot}1}\right)}$ **d** $\pi\sqrt{5{\cdot}2^2 + 18{\cdot}2^2}$

10 In the quadrilateral PQRS, PQ = QS = QR, PS is parallel to QR and ∠QRS = 70°.

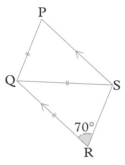

   Calculate

   a ∠RQS   **b** ∠PQS.

11 The triangle PQR is rolled clockwise along the line AB.

   Draw the locus of P as the triangle rotates around Q and then R.

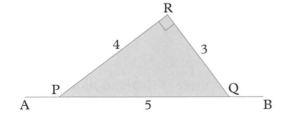

Revision exercise 8

1 Solve these equations by factorising.

   a $x^2 = 2x + 15$

   b $x^2 + 12 = 8x$

2 One solution of the equation $2x^2 - 7x + k = 0$ is $x = -\dfrac{1}{2}$. Find the value of k.

3 Two girls walk at the same speed from A to B. Aruni takes the large semicircle and Deepa takes the three small semicircles. Who arrives at B first?

4 The radii of two spheres are in the ratio 2 : 5. The volume of the smaller sphere is 16 cm³. Calculate the volume of the larger sphere.

5 The surface areas of two similar jugs are 50 cm² and 450 cm² respectively.

   a If the height of the larger jug is 10 cm, find the height of the smaller jug.

   b If the volume of the smaller jug is 60 cm³, find the volume of the larger jug.

6 Find the angles marked with letters. (O is the centre of the circle.)

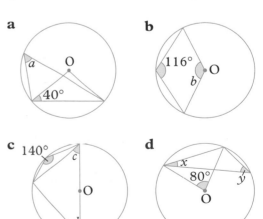

7 Estimate the probability that the next President of Russia was born in April.

8 The probability that it will be wet today is $\frac{1}{6}$. If it is dry today, the probability that it will be wet tomorrow is $\frac{1}{8}$. What is the probability that both today and tomorrow will be dry?

9 Two dice are thrown. What is the probability that the **product** of the numbers on top is
 a 12 **b** 4 **c** 11?

10 In a mixed school there are twice as many boys as girls and ten times as many girls as teachers. Using the letters b, g, t to represent the number of boys, girls and teachers, find an expression for the total number of boys, girls and teachers. Give your answer in terms of b only.

Revision exercise 9

1 **Sketch** the following curves for positive and negative values of x.
 a $y = x^2 + 3$
 b $y = \frac{1}{x}$
 c $y = x^3$

2 A car is an enlargement of a model, the scale factor being 10.
 a If the windscreen of the model has an area of 100 cm², find the area of the windscreen on the actual car (answer in m²).
 b If the capacity of the boot of the car is 1 m³, find the capacity of the boot on the model (answer in cm³).

3 Solve these equations.
 a $4(y + 1) = \dfrac{3}{1 - y}$
 b $4(2x - 1) - 3(1 - x) = 0$

4 A coin is tossed four times. What is the probability of obtaining at least three 'heads'?

5 ABCD is a parallelogram and AE bisects angle A. Prove that DE = BC.

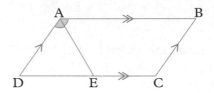

6 Without using a calculator, work out
 a $9^{-\frac{1}{2}} + \left(\dfrac{1}{8}\right)^{\frac{1}{3}} + (-3)^0$
 b $(1000)^{-\frac{1}{3}} - (0{\cdot}1)^2$

7 a Given that $x - z = 5y$, express z in terms of x and y.
 b Given that $mk + 3m = 11$, express m in terms of k.
 c For the formula $T = C\sqrt{z}$, express z in terms of T and C.

8 Here is the graph of $y = f(x)$.

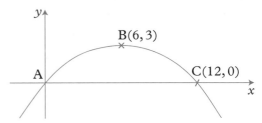

Draw three sketches to show
 a $y = f(x) - 3$
 b $y = f(x - 3)$
 c $y = f(3x)$
Give the new coordinates of A, B and C on each sketch.

9 The sides of a right-angled triangle have lengths $(x - 3)$ cm, $(x + 11)$ cm and $2x$ cm, where $2x$ is the hypotenuse. Find x.

10 In the parallelogram OABC, M is the midpoint of AB and N is the midpoint of BC.

If $\overrightarrow{OA}$ = **a** and $\overrightarrow{OC}$ = **c**, express in terms of **a** and **c**

a $\overrightarrow{CA}$ **b** $\overrightarrow{ON}$ **c** $\overrightarrow{NM}$

Describe the relationship between CA and NM.

11 Solve these equations.

a $\frac{x+3}{x} = 2$

b $x^2 = 5x$

c $(7^x)^2 = 1$

12 Draw the graph of $y = \frac{5}{x} + 2x - 3$, for $\frac{1}{2} \leqslant x \leqslant 7$, taking 2 cm to one unit for x and 1 cm to one unit for y. Use the graph to find approximate solutions to the equation $\frac{5}{x} + 2x = 9$.

11.2 Revision exercises (calculator allowed)

Revision exercise 10

1 Calculate the area of each shape.

8 cm

←6·4 cm→

2 A motorist travelled 800 miles during May, when the cost of petrol was 90 pence per litre. In June the cost of petrol increased by 10% and he reduced his mileage for the month by 5%.

a What was the cost, in pence per litre, of petrol in June?

b How many miles did he travel in June?

3 Copy the diagrams and then calculate x, correct to 3 sf.

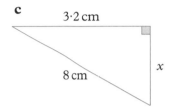

a

4 cm

x

6 cm

b

6 cm

x

11 cm

c

3·2 cm

8 cm

x

4 Solve the equation $3x^2 + 1 = 12$, giving your answers correct to 1 dp.

5 Work out on a calculator, correct to 4 sf

a $3{\cdot}61 - (1{\cdot}6 \times 0{\cdot}951)$

b $\frac{(4{\cdot}65 + 1{\cdot}09)}{(3{\cdot}6 - 1{\cdot}714)}$

6 The area of a square is 58 cm^2. How long is each side of the square?

7 Calculate the volume of each shape.

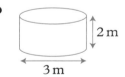

a

2 m

3·2 m

5·4 m

b

2 m

3 m

8 Work out, using a calculator.

a $(-3{\cdot}2) \times 4{\cdot}65$ **b** $(-8{\cdot}3)^2 - 10{\cdot}45$

c $\frac{1}{5} + \frac{3}{7}$ **d** $2\frac{1}{4} \div \frac{3}{5}$

9 Dave buys a car for £600 and later sells it for £850. Calculate his profit as a percentage of the cost price.

10 A metal ingot is in the form of a solid cylinder of length 7 cm and radius 3 cm.

a Calculate the volume, in cm^3, of the ingot.

The ingot is to be melted down and used to make cylindrical coins of thickness 0·3 cm and radius 1·2 cm.

b Calculate the volume, in cm³, of each coin.

c Calculate the number of coins which can be made from the ingot, assuming that there is no wastage of metal.

Revision exercise 11

1 The pump shows the price of petrol in a garage.

One day I buy £20 worth of petrol. How many litres do I buy?

2 Given that OA = 10 cm and ∠AOB = 70° (where O is the centre of the circle), calculate

a the arc length AB

b the area of minor sector AOB.

3 Throughout his life, Mr Cram's heart has beat at an average rate of 72 beats per minute. Mr Cram is sixty years old. How many times has his heart beat during his life? Give the answer in standard form correct to two significant figures.

4 The edges of a cube are all increased by 10%. What is the percentage increase in the volume?

5 A cylinder of radius 8 cm has a volume of 2 litres. Calculate the cylinder height.

6 The dimensions of the rectangle are correct to the nearest cm.

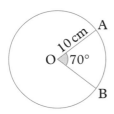

Give the maximum and minimum values for the area of the rectangle consistent with these data.

7 The fraction $\frac{139}{99}$ gives an approximate value for $\sqrt{2}$. What is the percentage error in using this fraction? Give your answer correct to 2 sf. ['Percentage error' means the error as a percentage of the exact value.]

8 Evaluate these using a calculator (answers to 4 sf).

a $\dfrac{0·74}{0·81 \times 1·631}$ **b** $\sqrt{\left(\dfrac{9·61}{8·34 - 7·41}\right)}$

c $\left(\dfrac{0·741}{0·8364}\right)^4$ **d** $\dfrac{8·4 - 7·642}{3·333 - 1·735}$

9 The mean of four numbers is 21.

a Calculate the sum of the four numbers. Six other numbers have a mean of 18.

b Calculate the mean of the ten numbers.

10 Given BD = 1 m, calculate the length AC.

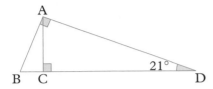

11 Use the method of trial and improvement to find a solution of the equation $x^5 = x^3 + 1$, giving your answer correct to 2 decimal places.

12 Given that x is an acute angle and that $3 \tan x - 2 = 4 \cos 35 \cdot 3°$ calculate
   a $\tan x$
   b the value of x in degrees correct to 1 dp.

Revision exercise 12

1 A supermarket sells its 'own-label' raspberry jam in two sizes.

Which jar represents the better value for money? You are given that 1 kg = 2·20 lb.

2 Sketch the curve $y = \cos x$ for x from 0° to 360°.
   a If $\cos 70° = 0·342$, find another angle whose cosine is 0·342.
   b Find two values of x if $\cos x = 0·5$.

3 Evaluate these and give the answers to 3 significant figures.
   a $\sqrt[3]{(9·61 \times 0·0041)}$
   b $\left(\dfrac{1}{9·5} - \dfrac{1}{11·2} \right)^3$

c $\dfrac{15·6 \times 0·714}{0·0143 \times 12}$ **d** $\sqrt[4]{\left(\dfrac{1}{5 \times 10^3} \right)}$

4 Calculate the side or angle marked with a letter.

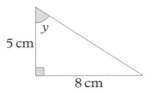

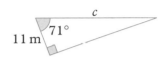

5 A cube of side 10 cm is melted down and made into five identical spheres. Calculate the radius of each sphere.

6 In figure 1 a circle of radius 4 cm is inscribed in a square. In figure 2 a square is inscribed in a circle of radius 4 cm.

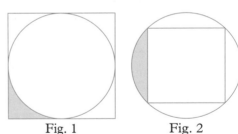

Fig. 1   Fig. 2

Calculate the shaded area in each diagram.

7 The dimensions of the cylinder are accurate to the nearest mm.

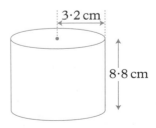

Work out the minimum possible volume of the cylinder. Give your answer to 3 sf.

8 The figure shows a cube of side 10 cm.

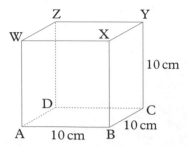

Calculate
a the length of AC
b the angle YAC.

9 A bag contains x green discs and 5 blue discs. A disc is selected and replaced. A second disc is drawn. Find, in terms of x, the probability of selecting
a a green disc on the first draw
b a green disc on the first and second draws.

10 A copper pipe has external diameter 18 mm and thickness 2 mm. The density of copper is 9 g/cm^3 and the price of copper is £150 per tonne. What is the cost of the copper in a length of 5 m of this pipe?

11 The mean height of 10 boys is 1·60 m and the mean height of 15 girls is 1·52 m. Find the mean height of the 25 boys and girls.

12 Find x.

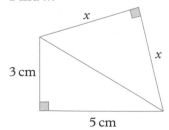

13 If the values of p are $-2 \leqslant p \leqslant 4$, find
a the smallest possible value of p^2
b the largest possible value of p^2
c the smallest possible value of p^3.

Revision exercise 13

1 The mass of the planet Jupiter is about 350 times the mass of the Earth. The mass of the Earth is approximately $6·03 \times 10^{21}$ tonnes. Give an estimate correct to 2 significant figures for the mass of Jupiter.

2 Work out the difference between one ton and one tonne.

1 tonne = 1000 kg
1 ton = 2240 lb
1 lb = 454 g

Give your answer to the nearest kg.

3 A regular octagon of side length 20 cm is to be cut out of a square card.

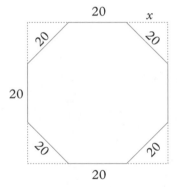

a Find the length x and hence find the size of the smallest square card from which this octagon can be cut.
b Calculate the area of the octagon, correct to 3 sf.

4 Solve these equations, correct to 2 dp.
a $3x^2 - 5x - 11 = 0$
b $(x - 2)^2 = 2x$

5 A formula for z is $z = ut - x^2$. The values of u, t and x are 5·2, 8·8 and 6·3 respectively, all correct to one decimal place. Work out the minimum possible value of z consistent with these data.

6 Calculate the length of AB.

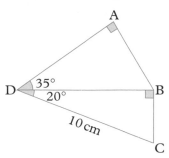

7 Find the mean and median of these numbers.

5, 17, 23, 11, 4, 8, 15, 32, 9, 12

8 Two lighthouses A and B are 25 km apart and A is due west of B. A submarine S is on a bearing of 137° from A and on a bearing of 170° from B. Find the distance of S from A and the distance of S from B.

9 The diagram shows a rectangular block. AY = 12 cm, AB = 8 cm, BC = 6 cm.

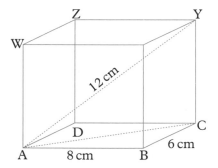

Calculate
a the length YC
b the angle ∠YAZ.

10 The square has sides of length 3 cm and the arcs have centres at the corners. Find the shaded area.

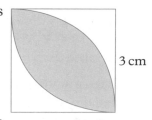

11 Given that cos ∠ACB = 0·6, AC = 4 cm, BC = 5 cm and CD = 7 cm, find the length of AB and AD.

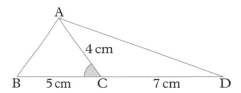

12 A sphere of radius 5 cm is melted down and made into a solid cube. Find the length of a side of the cube.

13 If AB = AC, find angle *x*, to the nearest degree.

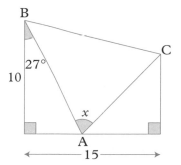

Revision exercise 14

Which of the following statements are true?

1 All prime numbers are odd numbers.

2 Every positive integer greater than 10 has an even number of factors.

3 If $x^2 = x$, then *x* must be the number 1.

4 The translation $\begin{pmatrix} 0 \\ 0 \end{pmatrix}$ is the only transformation that leaves **any** shape completely unchanged.

5 All non-negative numbers are positive.

6 If you add a given number to both the numerator and denominator of a fraction, then the new fraction is equivalent to the original fraction.

7 If both the numerator and denominator of a fraction are squared, then the new fraction is equivalent to the original fraction.

8 If you multiply both the numerator and denominator of a fraction by a given non-zero number, then the new fraction is equivalent to the original fraction.

9 $a \times (b \times c) = (a \times b) \times c$

10 $a \div (b \div c) = (a \div b) \div c$

11 All mathematical curves cross the x-axis or the y-axis or both.

12 If a quadrilateral has exactly 2 lines of symmetry, then it must be a rectangle.

13 Except for 1, no cube number is also a square number.

14 x^2 is never equal to $5x + 14$.

15 If n is a positive integer, then $n^2 + n + 5$ is a prime number.

16 If m and n are positive integers, then $6m + 4n + 13$ is an odd number.

17 No square number differs from a cube number by exactly 2.

18 A polygon having all its sides equal is a **regular** polygon.

19 If $a^2 = 7^2$, then a must be 7.

20 $\dfrac{a + b}{c} = \dfrac{a}{c} + \dfrac{b}{c}$

21 $\dfrac{a}{b + c} = \dfrac{a}{b} + \dfrac{a}{c}$

22 An enlargement always changes the area of a shape, unless the scale factor of the enlargement is 1.

23 $a \times (b + c) = ab + ac$

24 $a \times (b \times c) = ab \times ac$

25 $(a + b)^3 = a^3 + b^3$

26 $\sqrt{x + y} = \sqrt{x} + \sqrt{y}$

27 x^2 is never less than x.

28 $(a + b)(c + d) = ac + bd$

29 2^x is always positive.

30

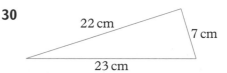

This triangle (not drawn to scale) is a **right-angled** triangle.

31 Suppose that you accurately draw the straight line $y = x$ on a set of axes. Then if you take a protractor and measure the acute angle between the line and the x-axis, it will be 45°.

32 If you have an unlimited supply of 5p, 7p and 11p stamps, then you can make up any amount above 13p from just these stamps.

33 If n is a positive integer greater than 1, then $2^n < n^3$.

34 There is exactly one point which lies on **both** of the straight lines $y = \dfrac{1}{2}x + 5$ and $x - 2y = 3$.

35 If n is a positive integer, then $n^4 - 10n^3 + 35n^2 - 48n + 24$ is equal to $2n$.

36 If the number x is multiplied by 78·39, and the result is then divided by 78·39, then the final answer is x.

37 If the number x is increased by 8·3%, and the result is then decreased by 8·3%, then the final answer is x.

38 Written as a fraction, π is $\frac{22}{7}$.

39 If triangle ABC is isosceles, then $\angle ABC = \angle ACB$.

40 If the product of two numbers is 8, then one of the numbers must be 8.

41 If the product of two numbers is 0, then one of the numbers must be 0.

42 A cuboid has 6 faces, 12 edges, and 8 vertices.

43 A pyramid has 5 faces, 8 edges, and 5 vertices.

44 The number 133! (that is, the number obtained by working out the product $1 \times 2 \times 3 \times 4 \times 5 \times \ldots \times 132 \times 133$) ends in exactly 26 noughts.

Answers

1 Number 1

page 1 **Exercise 1** Ⓜ

1 20 **2** 400 **3** 80   **4** 6   **5** 6000 **6** 20 000 **7** 5 000 000 **8** 800 000 **9** 200
10 70 **11** 10 **12** 800 **13 a** 720 **b** 5206 **c** 16 430 **d** 500 000 **e** 300 090 **f** 8500
14 a 8753 **b** 3578   **15 a** 75 423 **b** 23 574   **16 a** 257 **b** 3221 **c** 704
17 a 1392 **b** 26 611 **c** 257 900   **18 a** 0 **b** 52 000

page 3 **Exercise 2** Ⓜ

1 True **2** False **3** True **4** True **5** True **6** False **7** True **8** True
9 $50 + 7 + \frac{2}{10}$ **10 a** 235·1 **b** 67·23   **c** 98·32 **d** 3·167 **11** 0·2, 0·31, 0·41
12 0·58, 0·702, 0·75   **13** 0·41, 0·43, 0·432   **14** 0·6, 0·609, 0·61 **15** 0·04, 0·15, 0·2, 0·35
16 0·18, 0·81, 1·18, 1·8 **17** 0·061, 0·07, 0·1, 0·7 **18** 0·009, 0·025, 0·03, 0·2
19 0·01, 0·05, 0·1, 0·11, 0·2, 0·205, 0·25 CARWASH   **20 a** 32·51 **b** 0·853   **c** 1·16
21 a 5·69 **b** 0·552 **c** 1·30   **22 a** £3·50 **b** £0·15 **c** £0·03 **d** £0·10 **e** £12·60 **f** £0·08
23 a True **b** False **c** True **d** True

page 4 **Exercise 3** Ⓜ

1 3497 **2** 2435 **3** 785   **4** 91 745   **5** 212 **6** 41   **7** 859 **8** 208
9 270 **10** 5000 **11** 365   **12** 856   **13** 2528 **14** 64 568 **15** 85 **16** 324
17 639 **18** 325 **19** 52 r 1 *or* $52\frac{1}{7}$ **20** 52   **21** 2018 **22** 4569 **23** 7   **24** 1080
25 1492 **26** 524 **27** 5800   **28** 188   **29** 1641 **30** 365   **31** 254 **32** 21 200

page 4 **Exercise 4** Ⓜ

1 a
```
    2  8  5
 +  5  1  4
 ─────────
    7  9  9
```
b
```
    6  3  7
 +  2  5  2
 ─────────
    8  8  9
```
c
```
    6  3  5
 +  3  4  4
 ─────────
    9  7  9
```

2 a
```
    3  5  6
 +  5  2  6
 ─────────
    8  8  2
```
b
```
    2  2  4
 +  5  3  7
 ─────────
    7  6  1
```
c
```
    3  8  8
 +  4  2  5
 ─────────
    8  1  3
```

3 a
```
       4  8
 ×        3
 ─────────
    1  4  4
```
b
```
       3  3
 ×        7
 ─────────
    2  3  1
```
c
```
    3  2  1
 ×        5
 ─────────
 1  6  0  5
```

4 a $\boxed{1}\ \boxed{5}\ \boxed{0} \div 3 = 50$ **b** $\boxed{1}\ \boxed{5} \times 4 = 60$
  c $9 \times \boxed{9} = 81$ **d** $\boxed{1}\ \boxed{1}\ \boxed{5}\ \boxed{2} \div 6 = 192$

5 a
```
    4 ⬚4⬚ 5
  + 2 8 ⬚5⬚
    ⬚7⬚ 3 0
```
b
```
    4 ⬚2⬚ 7
  + ⬚1⬚ 7 ⬚7⬚
    6 0 4
```
c
```
    ⬚5⬚ 3 ⬚5⬚
  + 2 ⬚6⬚ 4
    7 9 9
```

6 a ⬚3⬚ ⬚5⬚ × 7 = 245   **b** ⬚5⬚ ⬚8⬚ × 10 = 580

   c 32 ÷ ⬚4⬚ = 8   **d** ⬚9⬚ ⬚5⬚ ⬚0⬚ ÷ 5 = 190

page 5 Exercise 5 Ⓜ

1 10·14 **2** 20·94 **3** 26·71 **4** 216·95 **5** 9·6 **6** 23·1 **7** 9·14 **8** 17·32
9 0·062 **10** 1·11 **11** 4·36 **12** 2·41 **13** 1·36 **14** 6·23 **15** 2·46 **16** 8·4
17 2·8 **18** 10·3 **19** 0·18 **20** 4·01 **21** 6·66 **22** 41·11 **23** 3·6 **24** 6·44
25 105·2 cm **26** £8·96

page 6 Exercise 6 Ⓜ

1 0·06 **2** 0·15 **3** 0·12 **4** 0·006 **5** 1·8 **6** 3·5 **7** 1·8 **8** 0·8
9 0·36 **10** 0·014 **11** 1·26 **12** 2·35 **13** 8·52 **14** 3·12 **15** 0·126 **16** 127·2
17 0·17 **18** 0·327 **19** 0·126 **20** 0·34 **21** 0·055 **22** 0·52 **23** 1·3 **24** 0·001

page 6 Exercise 7 Ⓜ

1 2·1 **2** 3·1 **3** 4·36 **4** 4 **5** 4 **6** 2·5 **7** 16 **8** 200
9 70 **10** 0·92 **11** 30·5 **12** 6·2 **13** 12·5 **14** 122 **15** 212 **16** 56
17 60 **18** 1500 **19** 0·3 **20** 0·7 **21** 0·5 **22** 3·04 **23** 5·62 **24** 0·78
25 0·14 **26** 3·75 **27** 0·075 **28** 0·15 **29** 1·22 **30** 163·8 **31** 1·75 **32** 18·8

page 7 Exercise 8 Ⓜ

1 7·91 **2** 22·22 **3** 7·372 **4** 0·066 **5** 466·2 **6** 1·22 **7** 1·67 **8** 1·61
9 16·63 **10** 24·1 **11** 26·7 **12** 3·86 **13** 0·001 **14** 1·56 **15** 0·0288 **16** 2·176
17 0·02 **18** 0·0001 **19** 355 **20** 20 **21** £12·80 **22** 34 000 **23** 18 **24** 0·569
25 2·335 **26** 1620 **27** 8·8 **28** 1200 **29** 0·00175 **30** 13·2 **31** 200 **32** 0·804
33 0·08 **34** 0·077 **35** 0·0009 **36** 0·01 **37 a** 20 **b** 184 **c** 0·63 **d** 540 **e** 3

38 a

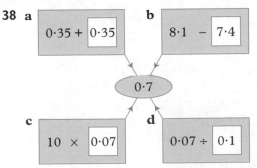

39 a True **b** True **c** True **d** False **e** False **f** True **g** True **h** False **i** True **j** True
40 a > **b** > **c** < **d** < **e** > **f** >

page 9 Exercise 9 Ⓜ

1 805 **2** 459 **3** 650 **4** 1333 **5** 2745 **6** 1248 **7** 4522 **8** 30 368
9 28 224 **10** 8568 **11** 46 800 **12** 66 281 **13** 57 602 **14** 89 516 **15** 97 525

page 10 Exercise 10 Ⓜ

1 32 **2** 25 **3** 18 **4** 13 **5** 35 **6** 22 r 2 *or* $22\frac{1}{13}$ **7** 23 r 24 *or* $23\frac{12}{17}$

8 18 r 10 *or* $18\frac{10}{41}$ **9** 27 r 18 *or* $27\frac{3}{4}$ **10** 13 r 31 *or* $13\frac{31}{52}$ **11** 35 r 6 *or* $35\frac{2}{9}$

12 25 r 28 *or* $25\frac{28}{31}$ **13** 64 r 37 *or* $64\frac{37}{64}$ **14** 151 r 17 *or* $151\frac{17}{18}$ **15** 2961 r 15 *or* $2961\frac{15}{19}$

page 10 Exercise 11 Ⓜ

1 £47·04 **2** 46 **3** 7592 **4** 15 with 20p change **5** 8 **6** £14 million **7** £85
8 £80·64 **9** £21 600 **10** 54 **11 a** 6·72 **b** 2·795 **c** 3·976 **12 a** 4·9 **b** 3·5 **c** 2·0

13 a

End of year	1	2	3	4	5
No. of gerbils	40	120	360	1080	3240

b 608

page 12 Exercise 12 Ⓜ

Part A

¹5	2	²5		³3	5		⁴2
2	⁵2	4	1	7			1
⁶1	7	2		0		⁷6	4
8		⁸1	⁹7	0		8	
	¹⁰1	0	1		¹¹5	4	9
¹²8	7		¹³5	¹⁴5	5		
7		¹⁵4		¹⁶3	0	¹⁷4	¹⁸6
¹⁹7	2	9		5		²⁰6	6

Part B

¹5	8	²4		³3	5		⁴3
8		⁵3	4	2	7		6
⁶3	9	9		8		⁷3	4
6		⁸1	⁹8	3		1	
	¹⁰7	6	4		¹¹9	5	0
¹²8	7		¹³5	¹⁴6	7		
5		¹⁵9		¹⁶6	8	¹⁷7	¹⁸5
¹⁹8	2	4		7		²⁰3	6

Part C

¹1	1	²5		³3	9		⁴1
8		⁵1	4	4	0		2
⁶6	5	0		7		⁷6	5
0		⁸5	⁹7	0		4	
	¹⁰2	1	9		¹¹6	6	6
¹²1	4		¹³3	¹⁴4	1		
2		¹⁵4		¹⁶9	0	¹⁷5	¹⁸0
¹⁹7	0	5		0		²⁰8	1

Part D

¹1	7	²5		³2	4		⁴0
2		⁵9	9	6	0		0
⁶9	9	9		9		⁷3	7
5		⁸9	⁹0	5		8	
	¹⁰1	6	0		¹¹8	0	7
¹²3	6		¹³5	¹⁴7	2		
0		¹⁵2		¹⁶2	3	¹⁷4	¹⁸5
¹⁹5	0	1		9		²⁰2	2

page 15 **Exercise 13** Ⓜ

1 a 1, 2, 3, 6   **b** 1, 3, 5, 15   **c** 1, 2, 3, 6, 9, 18   **d** 1, 3, 7, 21   **e** 1, 2, 4, 5, 8, 10, 20, 40
2 2, 3, 5, 7, 11, 13, 17, 19   **3** Multiple answers. Two examples are 2 + 3 = 5 and 2 + 5 = 7
4 3, 11, 19, 23, 29, 31, 37, 47, 59, 61, 67, 73
5 a $2 \times 2 \times 2 \times 3 \times 5 \times 5$ **b** $3 \times 3 \times 7 \times 11$ **c** $2 \times 2 \times 2 \times 2 \times 2 \times 7 \times 11$
   d $2 \times 3 \times 3 \times 3 \times 5 \times 13$ **e** $2 \times 2 \times 2 \times 2 \times 2 \times 5 \times 5 \times 5$ **f** $2 \times 2 \times 5 \times 7 \times 7 \times 23$

6

	Prime number	Multiple of 3	Factor of 16
Number greater than 5	7	9	8
Odd number	5	3	1
Even number	2	6	4

7 a 7   **b** 50   **c** 1   **d** 5   **8 a** 1, 4, 9, 16, 25, 36, 49, 64, 81, 100, 121, 144, 169, 196, 225
9 1, 8, 27, 64, 125   **10 a** 25   **b** 27   **c** 125   **11** 13

12 (6) (30) (19)

13 $13 + 15 + 17 + 19 = 64 = 4^3$
   $21 + 23 + 25 + 27 + 29 = 125 = 5^3$
   $31 + 33 + 35 + 37 + 39 + 41 = 216 = 6^3$

14 No, because for example 2 + 3 is an odd number.
15 a No   **b** Yes   **c** No   **d** Yes   **e** Yes   **16** 8   **17** 2, 5, 7, 43, 61, 89
18 $7 = 3 + 2^2$ $9 = 5 + 2^2$ $11 = 7 + 2^2$ $13 = 5 + 2^3$ $15 = 7 + 2^3$ $17 = 13 + 2^2$
   $19 = 11 + 2^3$ $21 = 13 + 2^3$ $23 = 19 + 2^2$ $25 = 17 + 2^3$ $27 = 23 + 2^2$
19 Yes it is.   **20** No, for example when $n = 4$ and $a = 2$, $a^n - a = 2^4 - 2 = 16 - 2 = 14$

page 17 **Exercise 14** Ⓜ

 1 a 2, 4, 6, 8, 10, 12   **b** 5, 10, 15, 20, 25, 30   **c** 10   **2 a** 4, 8, 12, 16   **b** 12, 24, 36, 48   **c** 12
 3 a 18   **b** 24   **c** 70   **d** 12   **e** 10   **f** 63   **4** 12   **5** 6   **6 a** 6   **b** 11   **c** 9   **d** 6   **e** 12   **f** 10
 7 a 6   **b** 40   **c** Multiple answers e.g. 22, 33   **d** 5 and 2 or 10 and 1   **8** 15   **9** 21
10 a $2^4 \times 7 \times 3^2$ $2^3 \times 3 \times 5 \times 7$ **b** 168   **c** 7
11 a $2^3 \times 3^2 \times 5^2 \times 11$ $2 \times 3^2 \times 5 \times 11 \times 13$ **b** 990   **c** 22

page 19 **Exercise 15** Ⓜ

1 i ii A = +15,   B = ×2,
   C = −22,   D = ÷4

2 i ii A = square,   B = ×3,
   C = −10,   D = ÷2

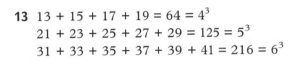

a 7 → A → 22 → B → 44 → C → 22 → D → 5·5 →
b 10 → A → 25 → B → 50 → C → 28 → D → 7 →
c 16 → A → 31 → B → 62 → C → 40 → D → 20 →
d $\frac{1}{2}$ → A → $15\frac{1}{2}$ → B → 31 → C → 9 → D → $2\frac{1}{4}$ →

a 4 → A → 16 → B → 48 → C → 38 → D → 19 →
b 5 → A → 25 → B → 75 → C → 65 → D → $32\frac{1}{2}$ →
c 6 → A → 36 → B → 108 → C → 98 → D → 49 →
d 8 → A → 64 → B → 192 → C → 182 → D → 91 →

3 A = ×4,   B = square root,   C = −10,   D = ×(−2)
4 A = reciprocal,   B = +1,   C = square,   D = ÷3
5 A = +3,   B = cube,   C = ÷(−2),   D = +100

page 20 **Exercise 16** Ⓜ

1 **a** 4   **b** 6   **c** 6   **d** 15   **e** 21   **f** 20   **g** 18   **h** 30

2 **a** $\frac{3}{4}$ **b** $\frac{1}{4}$ **c** $\frac{2}{5}$ **d** $\frac{1}{4}$ **e** $\frac{3}{5}$ **f** $\frac{8}{9}$ **g** $\frac{4}{5}$ **h** $\frac{1}{3}$ **i** $\frac{3}{5}$ **j** $\frac{7}{8}$

3 **a** $\frac{2}{4}$ **b** $\frac{5}{9}$ **c** $\frac{2}{97}$ **4** $\frac{3}{12}=\frac{1}{4}$, $\frac{6}{15}=\frac{2}{5}$, $\frac{10}{45}=\frac{2}{9}$, $\frac{5}{6}=\frac{30}{36}$, $\frac{3}{7}=\frac{12}{28}$

5 **a** $\frac{2}{3}=\frac{8}{12}$, $\frac{1}{2}=\frac{6}{12}$, $\frac{1}{4}=\frac{3}{12}$ **b** $\frac{1}{4}$, $\frac{1}{2}$, $\frac{2}{3}$ **6** **a** $\frac{5}{6}=\frac{10}{12}$, $\frac{2}{3}=\frac{8}{12}$, $\frac{1}{4}=\frac{3}{12}$ **b** $\frac{1}{4}$, $\frac{2}{3}$, $\frac{5}{6}$

7 $\frac{1}{2}$, $\frac{3}{5}$, $\frac{7}{10}$ **8** $\frac{7}{12}$, $\frac{3}{4}$, $\frac{5}{6}$ **9** $\frac{3}{8}$, $\frac{1}{2}$, $\frac{3}{4}$ **10** $\frac{2}{15}$, $\frac{1}{3}$, $\frac{7}{15}$ **11** **a** $\frac{2}{3}$ **b** $\frac{2}{5}$ **c** $\frac{5}{8}$ **d** $\frac{3}{4}$

12 **a** $\frac{7}{4}=1\frac{3}{4}$ **b i** $2\frac{2}{3}$ **ii** $2\frac{2}{5}$ **iii** $2\frac{1}{4}$ **iv** $1\frac{1}{6}$ **v** $7\frac{1}{2}$ **13** **a** $\frac{7}{5}$ **b** $\frac{5}{2}$ **c** $\frac{15}{4}$ **d** $\frac{13}{3}$ **e** $\frac{15}{7}$

page 22 **Exercise 17** Ⓜ

1 **a** $\frac{3}{8}$ **b** $\frac{9}{10}$ **c** $\frac{11}{15}$ **2** **a** $\frac{5}{6}$ **b** $\frac{5}{12}$ **c** $\frac{7}{12}$ **d** $\frac{9}{20}$ **e** $1\frac{11}{20}$ **f** $1\frac{1}{2}$ **g** $1\frac{7}{12}$ **h** $3\frac{7}{12}$

3 **a** $\frac{1}{6}$ **b** $\frac{5}{12}$ **c** $\frac{7}{15}$ **d** $\frac{1}{10}$ **e** $1\frac{1}{12}$ **f** $\frac{1}{6}$

4

+	$\frac{1}{4}$	$\frac{3}{8}$	$\frac{1}{3}$
$\frac{1}{2}$	$\frac{3}{4}$	$\frac{7}{8}$	$\frac{5}{6}$
$\frac{1}{8}$	$\frac{3}{8}$	$\frac{1}{2}$	$\frac{11}{14}$
$\frac{1}{5}$	$\frac{9}{20}$	$\frac{23}{40}$	$\frac{8}{15}$

5 **a** $\frac{1}{9}$ **b** $\frac{1}{5}$ **c** $\frac{5}{8}$ **d** $\frac{9}{100}$ **e** 7   **f** 14   **g** £22   **h** £20

6 **a** $\frac{8}{15}$ **b** $\frac{5}{42}$ **c** $\frac{15}{26}$ **d** $1\frac{1}{6}$ **e** $\frac{5}{8}$ **f** $9\frac{1}{6}$ **g** $\frac{4}{15}$ **h** $\frac{3}{14}$

page 24 **Exercise 18** Ⓜ

1 **a** $\frac{4}{1}\times\frac{3}{2}=6$ **b** $\frac{2}{3}\times\frac{5}{3}=1\frac{1}{9}$ **2** $\frac{8}{9}$ **b** 6   **c** $\frac{4}{5}$ **d** $\frac{3}{16}$ **e** $5\frac{1}{3}$ **f** 16   **g** 1   **h** 70

3 **a** $\frac{5}{12}$ **b** $4\frac{1}{2}$ **c** $1\frac{2}{3}$ **d** $\frac{5}{16}$ **e** $1\frac{7}{8}$ **f** $2\frac{5}{8}$ **g** $1\frac{9}{26}$ **h** 8

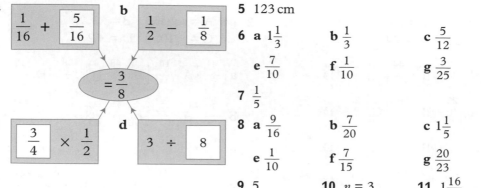

4 **a**   **b**   **5** 123 cm

6 **a** $1\frac{1}{3}$ **b** $\frac{1}{3}$ **c** $\frac{5}{12}$ **d** $1\frac{2}{3}$

e $\frac{7}{10}$ **f** $\frac{1}{10}$ **g** $\frac{3}{25}$ **h** $4\frac{2}{3}$

7 $\frac{1}{5}$

c   **d**   **8** **a** $\frac{9}{16}$ **b** $\frac{7}{20}$ **c** $1\frac{1}{5}$ **d** $1\frac{1}{6}$

e $\frac{1}{10}$ **f** $\frac{7}{15}$ **g** $\frac{20}{23}$ **h** $1\frac{3}{5}$

9 5   **10** $n=3$ **11** $1\frac{16}{17}$ **12** 9

13 a $a = 1, b = 2$ **b** $a = 1, b = 5$ **c** $a = 1, b = 13$ *or* $a = 5, b = 7$

14

×	$\frac{2}{5}$	$\frac{3}{4}$	$\frac{2}{3}$
$\frac{1}{3}$	$\frac{2}{15}$	$\frac{1}{4}$	$\frac{2}{9}$
$\frac{1}{2}$	$\frac{1}{5}$	$\frac{3}{8}$	$\frac{1}{3}$
$\frac{1}{4}$	$\frac{1}{10}$	$\frac{3}{16}$	$\frac{1}{6}$

15 144 ml   **16 a** line $3 = \frac{15}{16}$; line $4 = \frac{31}{32}$ **b** $1 - \frac{1}{2^{n+1}}$

page 27 Exercise 19 Ⓜ/Ⓗ

1 C   **2** A   **3** C   **4** C   **5** A   **6** B   **7** C   **8** A   **9** A   **10** C   **11** C   **12** B   **13** C
14 B   **15** A   **16** C   **17** A   **18** C   **19** B   **20** C   **21** B   **22** A   **23** B   **24** B   **25** A

page 28 Exercise 20 Ⓜ/Ⓗ

1 £8000   **2** £6   **3** Approx. 20 minutes   **4** C   **5** B   **6** A   **7** A
8 a 89·89   **b** 4·2   **c** 358·4   **d** 58·8   **e** 0·3   **f** 2·62
9 a 4·5   **b** 462   **c** 946·4   **d** 77·8   **e** 0·2   **f** 21   **10** £5200
11 20   **12** Yes, it is   **13** Yes   **14** Cost $\approx 200 \times £2 \div 40 = £10$ so he is not correct.   **15** 50

page 30 Exercise 21 Ⓜ

1 a 5600   **b** 600   **c** 28 200   **d** 100   **e** 11 400   **f** 200
2 a 18   **b** 225   **c** 4
3 a 0·7   **b** 8·81   **c** 0·726   **d** 1·18   **e** 0·9   **f** 8·22
   g 0·075   **h** 11·726   **i** 20·2   **j** 6·67   **k** 0·3   **l** 0·0725
4 a 2·7   **b** 189   **c** 3   **d** 0·36   **e** 1·7   **f** 0·0416
   g 0·04   **h** 800   **i** 9300   **j** 8·1   **k** 6·10   **l** 8

page 32 Exercise 22 Ⓜ

1 a 3:2   **b** 3:5   **c** 1:4   **d** 5:2   **e** 3:4   **f** 8:5   **g** 3:2:4   **h** 3:2:5
2 a 1:6   **b** 1:50   **c** 1:1·6   **d** 1:0·75   **3 a** 2·4:1   **b** 2·5:1   **c** 0·8:1   **4** £15:£25
5 £36:£84   **6** £70:£28   **7** £2·10:£6·30   **8** 15 kg:75 kg:90 kg
9 46 mins:69 mins:69 mins   **10** £39   **11** 5:3   **12 a** 1:3   **b** 2:7   **c** 3:5
13 £200   **14** 3:7   **15** $\frac{1}{7}x$ **16** 6   **17** £120   **18** 300 g   **19** 625

page 33 Exercise 23 Ⓜ

1 12·3 km   **2** 4·71 km   **3** 50 cm   **4** 64 cm   **5** 5·25 cm
6 40 m × 30 m, 12 cm², 1200 m²   **7** 1 m², 6 m²   **8** 0·32 km² (or 320 000 m²)
9 a i 14 m   **ii** 6 m   **iii** 4 m   **b** 8 m   **c** 14 m   **d** 2 cm   **e** 12 m   **f** 42 m²

page 35 Exercise 24 Ⓜ

1 £1·10   **2** 6 hours   **3** 6 days   **4** $2\frac{1}{2}$ litres   **5** 60 km   **6** 119 g   **7** 260 mins
8 $2\frac{1}{4}$ weeks   **9** 6 men   **10 a** 12 pens   **b** 2100

page 36 **Exercise 25** (M)/(L)

1 −4	**2** −12	**3** −11	**4** −3	**5** −5	**6** 4	**7** −5	**8** −8	**9** 19
10 −17	**11** −4	**12** −5	**13** −11	**14** 6	**15** −4	**16** 6	**17** 0	**18** −18
19 −3	**20** −11	**21** −12	**22** 4	**23** 4	**24** 0	**25** −8	**26** −3	**27** 3
28 −12	**29** 18	**30** −5	**31** −66	**32** 98				

page 37 **Exercise 26** (M)/(L)

1 −8	**2** −7	**3** 1	**4** 1	**5** 9	**6** 11	**7** −8	**8** 42	**9** 4	**10** 15
11 −7	**12** −9	**13** −1	**14** −7	**15** 0	**16** 11	**17** −14	**18** 0	**19** 17	**20** 3
21 −1	**22** −3	**23** 12	**24** −9	**25** 3	**26** 0	**27** 8	**28** 2		

29 a

+	−5	1	6	−2
3	−2	4	9	1
−2	−7	−1	4	−4
6	1	7	12	4
−10	−15	−9	−4	−12

b

+	−3	2	−4	7
5	2	7	1	12
−2	−5	0	−6	5
10	7	12	6	17
−6	−9	−4	−10	1

page 38 **Exercise 27** (M)/(L)

1 −6	**2** −4	**3** −15	**4** 9	**5** −8	**6** −15	**7** −24	**8** 6
9 12	**10** −18	**11** −21	**12** 25	**13** −60	**14** 21	**15** 48	**16** −16
17 −42	**18** 20	**19** −42	**20** −66	**21** −4	**22** −3	**23** 3	**24** −5
25 4	**26** −4	**27** −4	**28** −1	**29** −2	**30** 4	**31** −16	**32** −2
33 −4	**34** 5	**35** −10	**36** 11	**37** 16	**38** −2	**39** −4	**40** −5
41 64	**42** −27	**43** −600	**44** 40	**45** 2	**46** 36	**47** −2	**48** −8

49 a

×	4	−3	0	−2
−5	−20	15	0	10
2	8	−6	0	−4
10	40	−30	0	−20
−1	−4	3	0	2

b

×	−2	5	−1	−6
3	−6	15	−3	−18
−3	6	−15	3	18
7	−14	35	−7	−42
2	−4	10	−2	−12

Test 1

1 −16	**2** 64	**3** −15	**4** −2	**5** 15	**6** 18	**7** 3	**8** −6	**9** 11	**10** −48
11 −7	**12** 9	**13** 6	**14** −18	**15** −10	**16** 8	**17** −6	**18** −30	**19** 4	**20** −1

Test 2

1 −16	**2** 6	**3** −13	**4** 42	**5** −4	**6** −4	**7** −12	**8** −20	**9** 6	**10** 0
11 36	**12** −10	**13** −7	**14** 10	**15** 6	**16** −18	**17** −9	**18** 15	**19** 1	**20** 0

Test 3

1 100   **2** −20   **3** −8   **4** −7   **5** −4   **6** 10   **7** 9   **8** −10   **9** 7   **10** 35
11 −20   **12** −6   **13** −10   **14** −7   **15** −19   **16** −1   **17** −5   **18** −13   **19** 0   **20** 8

Test 4

1 5   **2** 2   **3** −20   **4** 4   **5** $-\dfrac{1}{2}$ **6** −5   **7** 1   **8** 2   **9** 6   **10** 12
11 0   **12** −8   **13** 1   **14** −1000   **15** −1

page 39 **Exercise 28** Ⓜ

1 6   **2** 17   **3** $6^2 + 7^2 + 42^2 = 43^2$, $10^2 + 11^2 + 110^2 = 111^2$ **4** 400   **5** 200 g
6 128 cm²   **7** $\dfrac{1}{66}$ **8** 5

9 a
$$
\begin{array}{r}
5\;\boxed{4} \\
9 \times \\ \hline
\boxed{4}\;\boxed{8}\;6 \\
\end{array}
$$

b
$$
\begin{array}{r}
\boxed{5}\;7 \\
\boxed{8} \times \\ \hline
4\;\boxed{5}\;6 \\
\end{array}
$$

c
$$
\begin{array}{r}
5\;\boxed{2} \\
\boxed{2} \times \\ \hline
1\;\boxed{0}\;4 \\
\end{array}
$$

10 a $\dfrac{1}{3}$ **b** $a = 2$, $b = 8$ or vice versa

page 40 **Exercise 29** Ⓜ

1

×	$\dfrac{2}{3}$	$\dfrac{3}{4}$	$\dfrac{1}{5}$
$\dfrac{1}{2}$	$\dfrac{1}{3}$	$\dfrac{3}{8}$	$\dfrac{1}{10}$
$\dfrac{1}{4}$	$\dfrac{1}{6}$	$\dfrac{3}{16}$	$\dfrac{1}{20}$
$\dfrac{2}{5}$	$\dfrac{4}{15}$	$\dfrac{3}{10}$	$\dfrac{2}{25}$

2 a $66\,667^2 = 4\,444\,488\,889$ **b** $44\,444\,448\,888\,889$
  c 66 666 667
3 a $a = 100$, $b = 1$ is one solution.
  b $a = 100$, $b = 2$ is one solution.
4 $47 \rightarrow$ prime, $26 \rightarrow 2 \times 13$, $40 \rightarrow 2 \times 2 \times 2 \times 5$,
  $25 \rightarrow 5 \times 5$, $63 \rightarrow 3 \times 3 \times 7$, $71 \rightarrow$ prime
5 $5 \times 7 \times 13 \times 13 \times 71$ **6** 1   **7** 36, 49, 64
8 a $x = 1$ **b** $x = 15$ **9** 50   **10** (e) 106

page 41 **Exercise 30** Ⓜ

1 £7055   **2** 10
3 $11^2 = 121$, $111^2 = 12\,321$, $1111^2 = 1\,234\,321$
  Predict $11\,111^2 = 123\,454\,321$
  and $111\,111^2 = 12\,345\,654\,321$

4 ①——③ is one answer.   **5 a** 66 666   **b** 82   **c** 455 551   **6**

  ⑱——㉝

7 500   **8** 5 and 28; 32   **9** $(1 + 2) \times 3 + (4 \times 5) + 6 + (7 \times 8) + 9$ is a possible answer.

page 42 **Exercise 31** Ⓜ

1 a 4 **b** 5

2 a

3	−4	1
−2	0	2
−1	4	−3

3 $x = 57°$ or $54°$ **4** 216 cm^3 **5 a** $\dfrac{1}{994}, \dfrac{1}{949}, \dfrac{1}{499}$ **b** $\dfrac{91}{94}$

6 a £6576 **b i** 8 **ii** 99 **7** 8

page 43 **Exercise 32** Ⓜ

1

0·5	−	0·01	→	0·49
+		×		
3·5	×	10	→	35
↓		↓		
4	÷	0·1	→	40

2

5·2	−	1·8	→	3·4
−		÷		
4·2	×	5	→	21
↓		↓		
1	+	0·36	→	1·36

3

0·7	×	30	→	21
×		−		
10	−	−19	→	29
↓		↓		
7	−	49	→	−42

4

−12	×	−6	→	72
÷		+		
4	+	7	→	11
↓		↓		
−3	+	1	→	−2

5

−8	÷	4	→	−2
×		÷		
−2	+	8	→	6
↓		↓		
16	−	$\frac{1}{2}$	→	$15\frac{1}{2}$

6

27	+	−7	→	20
×		÷		
$\frac{1}{3}$	÷	$\frac{1}{3}$	→	1
↓		↓		
9	+	−21	→	−12

page 44 **Test yourself**

1 a 7 hrs 10 mins **b** £378 **c** 80 sandwiches **2 a** 6 **b** $2^4 \times 3 \times 5$

3 Blackcurrant: 7 g/100 ml; Fizzy orange: 8.5 g/100 ml. Kelly drinks more sugar.

4 a $6\dfrac{5}{12}$ **b i** $\dfrac{1}{3}$ **ii** It converts to $0.\dot{3}$. **5 a** 1 **b** 7

6 a £60 000 ÷ 20 = £3000 **b** $\dfrac{50 \times 40}{0·5} = 4000$ **7** £3500
   or £60 000 ÷ 15 = £4000
   Both answers are acceptable.

8 a −3 **b** −12 **c** −3 **d** 16 **e** −2 **f** −4 **9** $126°$

10 a 64 **b** 0·06 **c** 144 **c** 5·46

page 46 **Functional Task 1**

Profit of £ 157 711·20

2   Number 2

page 48 **Exercise 1** Ⓜ

1 a 0·25 **b** 0·4 **c** 0·375 **d** $0·41\dot{6}$ **e** $0·1\dot{6}$ **f** $0·\dot{2}8571\dot{4}$

2 a $\dfrac{1}{5}$ **b** $\dfrac{9}{20}$ **c** $\dfrac{9}{25}$ **d** $\dfrac{1}{8}$ **e** $1\dfrac{1}{20}$ **f** $\dfrac{7}{1000}$

3 a 25% **b** 10% **c** 72% **d** 7·5% **e** 2% **f** $33\frac{1}{3}$%

4 $\frac{1}{4}$ **5** 0·58 **6** 1·42 **7** 0·65 **8** 0·32 **9** 0·07 **10** 0·69 **11** 40%

12

	Fraction	Decimal	Percentage
a	$\frac{3}{4}$	0·75	75%
b	$\frac{1}{5}$	0·2	20%
c	$\frac{16}{25}$	0·64	64%
d	$\frac{1}{1000}$	0·001	0·1%
e	$\frac{1}{50}$	0·02	2%
f	$\frac{1}{3}$	0·$\dot{3}$	$33\frac{1}{3}$%

13 45%, $\frac{1}{2}$, 0·6 **14** 4%, $\frac{6}{16}$, 0·38

15 11%, 0·111, $\frac{1}{9}$ **16** 0·3, 32%, $\frac{1}{3}$

17 No **18** 96·8% (1 dp)

19 a 58·9% (1 dp) **b** 55·0% (1 dp)

20 Prices will still rise but at a slower rate, so Pete is not correct.

21 $r = \frac{4}{9}$ **22 a** $\frac{2}{9}$ **b** $\frac{7}{9}$ **c** $\frac{29}{99}$ **d** $\frac{541}{999}$

page 50 **Exercise 2 Ⓜ**

1 £12 **2** £8 **3** £10 **4** £3 **5** £2·40 **6** £24 **7** £45
8 £72 **9** £244 **10** £9·60 **11** $42 **12** $88 **13** 8 kg **14** 12 kg
15 272 g **16** 45 m **17** 40 km **18** $710 **19** 4·94 kg **20** 60 g **21** £204

page 51 **Exercise 3 Ⓜ**

1 £0·28 **2** £1·16 **3** £1·22 **4** £2·90 **5** £3·57 **6** £0·45 **7** £0·93
8 £37·03 **9** £16·97 **10** £0·38 **11** £0·79 **12** £1·60 **13** £13·40 **14** £50
15 £2·94 **16** £11·06 **17** £1·23 **18** £4·40 **19** £11·25 **20** £22·71 **21** £9·19

page 51 **Exercise 4 Ⓜ**

1 £63 **2** £736 **3** £77·55 **4** £104 **5** £1960 **6** £792 **7** £132
8 £45·75 **9** £110·30 **10** £42 **11** £12·03 **12** £9·49 **13** £7·35 **14** £7·01
15 £12·34 **16** £16·92 **17** £31·87 **18** £9·02 **19** £8·88 **20** £14·14

page 52 **Exercise 5 Ⓜ**

1 a £15 **b** 900 kg **c** $2·80 **d** 125 people **2** £32 **3 a** 45 000 **b** 39 600 **c** $\frac{1}{30}$

4 52·8 kg **5 a** 0·53 **b** 0·03 **c** 0·085 **d** 1·22 **6 a** £1·02 **b** £21·58 **c** £2·22 **d** £0·53

7 £248·57 **8** £26182 **9** £8425·60 **10** £73·03 **11** £6825

12 Video £201·40 House £166 500 Boat £2700 Tree £199·50 Phone £55·25 **13** 325

14 £6·30 **15 a** 256 g **b** 72 p **c** 14 cm **d** 30 minutes **e** 14·4 seconds

16 a $1·08 \times P \times 0·9$ **b** 15% increase, 6% increase **c** $1·25 \times Q \times 0·95$

page 54 **Exercise 6 Ⓜ**

1 8% **2** 10% **3** 25% **4** 2% **5** 4% **6** 2·5% **7** 20% **8** 50% **9** 15% **10** 80%
11 25% **12** 20% **13** 12·5% **14** $33\frac{1}{3}$% **15** 80% **16** 5% **17** 6% **18** 20% **19** 5% **20** 2·5%

page 55 **Exercise 7**(M)

1 a 25% profit   **b** 25% profit   **c** 10% loss   **d** 20% profit   **e** 30% profit   **f** 7·5% profit
2 28%   **3** 44·4% (1 dp)   **4** 46·875%   **5** 12%   **6** 5·3%   **7** 44%
8 27·05%   **9** 20%   **10** 14·3% (1 dp)   **11** 21%   **12 a** 25 000   **b** 76·3% (1 dp)

page 56 **Exercise 8**(M)/(H)

1 £6200   **2** £25 500   **3 a** £50   **b** £200   **c** £54   **d** £150   **e** £2400   **4** 210 000
5 85 kg   **6** 8 cm × 10 cm   **7** 200   **8** £56   **9** 400 kg   **10** 350 g   **11** 29 000
12 500 cm   **13** £500   **14** £480 000 000

page 58 **Exercise 9**(M)/(H)

1 a £2200   **b** £2420   **c** £2662   **2 a** £5600 **b** £7024·64   **3** £13 108 (to nearest £)
4 a £36 465·19   **b** £38 288·45   **5 a** £85 500   **b** £81 225
6 a £18 350   **b** £730   **c** £1 079 500   **7** £4913   **8** No   **9** 8 years   **10** 11 years
11 13 years   **12** £30,000 at 8% produces more

13 a

x	1	2	3	4	5	6	7	8	9	10
y	1·08	1·17	1·26	1·36	1·47	1·59	1·71	1·85	2·00	2·16

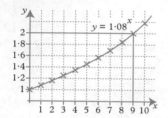

b $x = 9$ **c** 9 years

page 60 **Exercise 10**(M)

1 a £6·10   **b** £6·40   **c** £4·70   **d** £116   **e** £129·30   **f** £0·04
2 a 2·5 h   **b** 4·25 h   **c** 3·75 h   **d** 0·1 h   **e** 0·2 h   **f** 0·25 h   **g** 0·283 h   **h** 1·13 h   **i** 2·56 h
3 a 24·75 h   **b** 22·75 h   **c** 2·9 h   **d** 2·75 h   **e** 2·5 h   **f** 1·75 h

page 61 **Exercise 11**(M)

1 22   **2** 3   **3** 5   **4** 12   **5** 0   **6** 3   **7** 9   **8** 12   **9** 18   **10** 70
11 4   **12** 70   **13** 6   **14** 250   **15** 97   **16** 2   **17** $\frac{1}{2}$ **18** 11   **19** 8   **20** $10\frac{1}{2}$

page 61 **Exercise 12**(M)

1 4·2   **2** 15·9   **3** 0·6   **4** 5·3   **5** 4·0   **6** 12·7   **7** 0·5   **8** 5·6   **9** 14·0   **10** 2·1
11 14·1   **12** 1·2   **13** 9·9   **14** 9·1   **15** 9·5   **16** 0·6   **17** 23·0   **18** 11·4   **19** 7·4   **20** 5·5
21 11·5   **22** 11·7   **23** 10·9   **24** 1·9   **25** 13·0   **26** 4·9   **27** 18·8   **28** 3·4   **29** 2·4   **30** 2·9

page 62 **Exercise 13**(M)

1 3·041   **2** 1460   **3** 0·030 83   **4** 47·98   **5** 130·6   **6** 0·4771   **7** 0·3658
8 37·54   **9** 8·000   **10** 0·6537   **11** 0·037 16   **12** 34·31   **13** 0·7195   **14** 3·598
15 0·2445   **16** 2·043   **17** 0·3798   **18** 0·7683

page 64 **Exercise 14** Ⓜ

1 10·18	**2** −0·061 11	**3** 1·858	**4** 0·8264	**5** 2·717	**6** 4·840
7 10·87	**8** 7·425	**9** 13·49	**10** 0·7392	**11** 1135	**12** 13·33
13 5·836	**14** 86·39	**15** 10·23	**16** 5540	**17** 14·76	**18** 8·502
19 57·19	**20** 19·90	**21** 6·578	**22** 9·097	**23** 0·082 80	**24** 1855
25 2·367	**26** 1·416	**27** 7·261	**28** 3·151	**29** 149·9	**30** 74 020
31 8·482	**32** 75·21	**33** 1·226	**34** 6767	**35** 5·964	**36** 15·45
37 25·42	**38** 2·724	**39** 4·366	**40** 0·2194	**41** 0·000 465 9	**42** 0·3934
43 −0·7526	**44** 2·454	**45** 40 000	**46** 3·003	**47** 0·006 562	**48** 0·1330

page 65 **Exercise 15** Ⓜ

1 $\frac{11}{15}$ **2** $1\frac{1}{6}$ **3** $\frac{16}{21}$ **4** $\frac{23}{30}$ **5** $\frac{5}{6}$ **6** $\frac{13}{16}$ **7** $\frac{5}{28}$ **8** $\frac{19}{30}$ **9** $\frac{19}{20}$ **10** $\frac{1}{6}$

11 $\frac{2}{9}$ **12** $\frac{3}{44}$ **13** $2\frac{5}{6}$ **14** $2\frac{1}{6}$ **15** $5\frac{1}{8}$ **16** $3\frac{11}{12}$ **17** $4\frac{7}{8}$ **18** 6 **19** $8\frac{3}{4}$ **20** 4

21 a $3\frac{9}{20}$ **b** $3\frac{1}{6}$ **c** $\frac{11}{14}$ **d** $\frac{23}{40}$ **e** $\frac{6}{7}$ **f** $\frac{7}{18}$

page 65 **Exercise 16** Ⓜ

1 a 509 cm^2 **b** 507 cm^2

page 66 **Exercise 17 Calculator words** Ⓜ

1 HE LIES	**2** SOS	**3** HEDGEHOG	**4** GOSH	**5** GOBBLE
6 BEG	**7** BIG SLOB	**8** SID	**9** HILL	**10** LESLIE
11 HOBBIES	**12** GIGGLE	**13** LOOSE	**14** BIGGISH	**15** EGGSHELL
16 IGLOO	**17** GLOSS	**18** BILGE	**19** LEGLESS	**20** SHOES
21 SIEGE	**22** HE DID	**23** OBLIGE	**24** LIBEL	**25** BESIEGE
26 HE IS SO BIG	**27** BOOHOO	**28** BOOZE	**29** EEL	**30** GOOSE
31 GOODBIE	**32** HE SELLS	**33** BIBLE	**34** BIGGLES	**35** BOBBLE
36 HEIDI	**37** HIGH	**38** HELLS BELLS	**39** SHE DIES	**40** SOLEIL

page 68 **Exercise 18** Ⓜ/Ⓗ

1 4×10^3 **2** 5×10^2 **3** 7×10^4 **4** 6×10 **5** $2·4 \times 10^3$ **6** $3·8 \times 10^2$

7 $4·6 \times 10^4$ **8** $4·6 \times 10$ **9** 9×10^5 **10** $2·56 \times 10^3$ **11** 7×10^{-3} **12** 4×10^{-4}

13 $3·5 \times 10^{-3}$ **14** $4·21 \times 10^{-1}$ **15** $5·5 \times 10^{-5}$ **16** 1×10^{-2} **17** $5·64 \times 10^5$ **18** $1·9 \times 10^7$

19 $1·67 \times 10^{-24}$ g **20** $2·17 \times 10^8$ **21** $5·1 \times 10^8$ km^2 **22** $2·5 \times 10^{-10}$ cm

23 $6·023 \times 10^{23}$ **24** 3×10^{10} cm/s **25** £$3·6 \times 10^6$ **26** c, a, b **27** 13 **28** 16

29 a 8×10^{11} **b** 5×10^{13} **c** $4·5 \times 10^7$ **d** $6·4 \times 10^2$ **e** $5·5 \times 10^{-4}$ **f** 4×10^9
  g 3×10^4 **h** 4×10^{12}

30 a 23 000 **b** 0·03 **c** 560 **d** 800 000 **e** 0·0022 **f** 900 000 000
  g 0·6 **h** 7000 **i** 3 140 000

31 a $6·5 \times 10^5$ **b** $2·7 \times 10^3$ **c** $3·54 \times 10^6$ **d** $2·8 \times 10^{-2}$ **e** 7×10^{-3} **f** $7·07 \times 10^{22}$

32 a 6×10^3 **b** $1·1 \times 10^{-1}$ **c** $4·5 \times 10^8$ **d** $8·5 \times 10^{-6}$
  e 6×10^{12} (if an English billion) **f** 5×10^{-1}

page 70 **Exercise 19** Ⓜ/Ⓗ

1 $1·5 \times 10^9$ **2** 3×10^8 **3** $2·8 \times 10^{-2}$ **4** 7×10^{-9} **5** 2×10^6

6 4×10^{-6} **7** 9×10^{-2} **8** $6·6 \times 10^{-8}$ **9** $3·5 \times 10^{-7}$ **10** 1×10^{-16}

11 8×10^9 **12** 7.4×10^{-7} **13** 4.9×10^{11} **14** 4.4×10^{12} **15** 1.5×10^3
16 2×10^{17} **17** 1.68×10^{13} **18** 4.25×10^{11} **19** 9.9×10^7 **20** 6.25×10^{-16}
21 7.2×10^7 **22** 6.82×10^{-7} **23** 1.2×10^{-5} **24** 5×10^{-4} **25** 1.1×10^{10} days
26 1.3×10^{-4} m **27 a** Europe 6.50×10^{-2} people/m^2, Asia 6.80×10^{-2} people/m^2 **b** Asia
28 5.57×10^9 **29** 2.4528×10^7 km **30** 3000 secs or 50 mins
31 L $= 6 \times 10^2$ **32** 2.4×10^9 kg **33 a** 9.5×10^{12} km (1 dp) **b** 1.44×10^8 km
34 a 6×10^{101} **b** 20·5 seconds **c** 6.34×10^{91} years.

page 72 Exercise 20 Ⓜ

1 a 12 **b** 1 **c** 19 **d** 12 **e** 4 **f** 6 **g** 16 **h** 24
2 a 8 **b** 12 **c** 30 **d** −1 **e** 1 **f** 4 **g** 25 **h** 5
3 a 3 **b** 7 **c** −2 **d** 27 **e** 2 **f** −2 **g** $\frac{1}{2}$ **h** 2

4

¹2	4		²1	³9
3		⁴1	2	0
	⁵4	1	5	
⁶2	0		⁷2	
2	4		⁸4	5

page 73 Exercise 21 Ⓜ

1 7 **2** 13 **3** 13 **4** 22 **5** 1 **6** −1 **7** 18 **8** −4 **9** −3
10 37 **11** 0 **12** −4 **13** −7 **14** −2 **15** −3 **16** −8 **17** −30 **18** 16
19 −10 **20** 0 **21** 7 **22** −6 **23** −2 **24** −7 **25** −5 **26** 3 **27** 4
28 −8 **29** −2 **30** 2 **31** 0 **32** 4 **33** −4 **34** −3 **35** −9 **36** 4

page 74 Exercise 22 Ⓜ

1 9 **2** 27 **3** 4 **4** 16 **5** 36 **6** 18 **7** 1
8 6 **9** 2 **10** 8 **11** −7 **12** 15 **13** −23 **14** 3
15 32 **16** 36 **17** 144 **18** −8 **19** −7 **20** 13 **21** 5
22 −16 **23** 84 **24** 17 **25** 6 **26** 0 **27** −25 **28** −5

page 74 Exercise 23 Ⓜ

1 −20 **2** 16 **3** −42 **4** −4 **5** −90 **6** −160 **7** −2
8 −81 **9** 4 **10** 22 **11** 14 **12** 5 **13** 1 **14** $\sqrt{5}$
15 4 **16** $-6\frac{1}{2}$ **17** 54 **18** 25 **19** 4 **20** 312 **21** 45
22 22 **23** 14 **24** −36 **25** −7 **26** 1 **27** 901 **28** −30
29 −5 **30** $7\frac{1}{2}$ **31** −7 **32** $-\frac{3}{13}$ **33** $1\frac{1}{3}$ **34** $\frac{-5}{36}$

page 75 Exercise 24 Ⓜ

1 21 **2 a** 54 cm^2 **b** 63 m^2 **3** T $= 1.62$ (2 dp) **4** A $= 395.9$ (1 dp)
5 650 **6** 63·8 **7** E $= 9 \times 10^{12}$ **8** 10·5
9 800 **10 a** 1245 km/h (4 sf) **b** 4·8 °C (1 dp) **c** 1008 km/h
11 a A $= ab + ac - a^2$ **b** 30 **12 a** $S = r - p + q$ **b** $S = 4.1$

page 77 **Exercise 25**Ⓜ

1 0.85 m	**2** 2 400 m	**3** 63 cm	**4** 0.25 m	**5** 0.7 cm	**6** 20 mm
7 1 200 m	**8** 2 m^2	**9** 580 m	**10** 0.815 m	**11** 0·65 km	**12** 125 000 cm^2
13 5 000 g	**14** 4 200 g	**15** 6 400 g	**16** 3 000 g	**17** 800 g	**18** 0.4 kg
19 2 000 kg	**20** 0.25 kg	**21** 500 kg	**22** 620 kg	**23** 0.007 t	**24** 1.5 kg
25 0.8 l	**26** 2 000 ml	**27** 4 500 ml	**28** 6 000 ml	**29** 3 000 cm^3	**30** 2 000 l
31 5 500 l	**32** 500 000 cm^3	**33** 0.001 m^3			

34 a km   **b** ml or l   **c** g   **d** mm   **e** t or kg   **f** m^2

page 78 **Exercise 26**Ⓜ

1 180 cm	**2** 45 l	**3** 10 kg	**4** 8·8 pounds	**5** 5 miles
6 0·5 kg	**7** 20 miles	**8** 135 cm	**9** 30 kg	**10** 17·5 pints
11 25 miles	**12** 1750 pints	**13** 6·6 pounds	**14** $\frac{5}{16}$ miles (or 0·3125 miles)	**15** 200 gallons
16 5 000 miles	**17** 50 mph	**18** The market		

19 A one pound coin has a mass of about 10 g.

20 The width of the class room is about 7m.

21 A can of pepsi contains about 500 ml.

22 The distance from London to Birmington is about 100 miles.

23 The thickness of a one pound coin is about 3 mm.

24 a 3 kg   **b** 15 l   **c** 3 pounds

page 79 **Exercise 27**Ⓜ

1 a $2\frac{1}{2}$ hours   **b** 3 hours $7\frac{1}{2}$ mins   **c** 75 seconds   **d** 4 hours

2 45 m/s   **3 a** 75 km/h   **b** 14 mph   **c** 50 m/s

4 a 120 km   **b** 30 miles   **c** 4500 m   **d** 50 400 m

5 a 3 hours $7\frac{1}{2}$ minutes   **b** 76·8 km/h   **6 a** 4 hours 27 mins   **b** 23·6 km/h (1 dp)

7 a 6·67 m/s (2 dp)   **b** 6·33 m/s   **c** 123·19 secs (2 dp)   **8** 1230·8 km/h (1 dp)

9 3 hours   **10** 100 seconds   **11** 1·5 minutes   **12** 600 m

13 53·3 secs (1 dp)   **14** 5 cm/sec   **15** 1 minute   **16** 120 mph

page 82 **Exercise 28**Ⓜ

1 10 g/cm^3 **2** 240 g   **3** 35 m^3 **4** 0·6 kg   **5** 250 people/km^2

6 a 6000 people/km^2 **b** 5·58 × 10^{10} **7** 1000 kg **8** 3000 kg/m^3 **9** 0·8 g/cm^3

10 a 20 m/s   **b** 108 km/h   **c** 1·2 cm/s   **d** 90 m/s

page 84 **Exercise 29**Ⓗ

1 195·5 cm   **2** 36·5 kg   **3** 3·25 kg   **4** 95·55 m   **5** 28·65 seconds

6

	Measurement	Half unit	Lower bound	Upper bound
a	Temperature in a fridge = 2 °C to the nearest degree	0·5 °C	1·5 °C	2·5 °C
b	Mass of an acorn = 2·3 to 1 dp	0·05 g	2·25 g	2·35 g
c	Length of telephone cable = 64 m to nearest m	0·5 m	63·5 m	64·5 m
d	Time taken to run 100 m = 13·6 s to nearest 0·1 s	0·05 s	13·55 s	13·65 s

7 B   **8** 175 g and 185 g   **9** 2650 m   **10 a** No   **b** 1 cm

11 a $16.5 \leqslant m < 17.5$ **b** $255.5 \leqslant d < 256.5$ **c** $2.35 \leqslant e < 2.45$ **d** $0.335 \leqslant m < 0.345$
e $2.035 \leqslant v < 2.045$ **f** $11.95 \leqslant x < 12.05$ **g** $81.35 \leqslant T < 81.45$ **h** $0.25 \leqslant M < 0.35$
i $0.65 \leqslant m < 0.75$ **j** $51\,500 \leqslant n < 52\,500$

12 No, as the card could be $11.54\dot{9}$ cm long and the envelope only 11·5 cm long.

page 86 Exercise 30 (H)

1 a 7·5 cm, 8·5 cm, 10·5 cm **b** 26·5 cm **2** 46·75 cm^2 **3 a** 7   **b** 5   **c** 10   **d** 4 **e** 2
f 5   **g** 2   **h** 24   **4** 250 cm   **5** 47·5 m^2 **6** 18·0375 m^2
7 a 13 cm   **b** 11 cm   **c** 3 cm   **d** 12·5 cm   **8 a** 10·5 **b** 4·3   **9 a** 11 **b** 1 **c** 0·6
10 56 cm^2 **11** 55·71 (2 dp)   **12** 17·20 m/s (2 dp)   **13** 3·30 (2 dp)   2·87 (2 dp)

page 88 Exercise 31 (M)

1 a €30·80   **b** $114·80   **c** 2200 Rand   **d** ¥315   **e** $4·10   **f** €1·39
2 a £324·68   **b** £1524·39   **c** £240   **d** £584·42 **e** £42   **f** £0·40
3 £3·84   **4** Britain £15 000   France £13 500   USA £15 172
5 3·25 Swiss francs to the £   **6** 20   **7 a** 256   **b** 65 536   **8** 4·68 × 10^5
9 a $\frac{3}{19}$ **b** $\frac{1}{6}$ **10** £10 485·76

page 89 Exercise 32 (M)

1 a 1·25 m/min   **b** 2·08 cm/sec (2 dp)   **2** 77·5% (1 dp)   **3** 225 mm
4 a 35 pints   **b** 5·14 litres (2 dp)   **5** 43·5 mm

6 a

3	−4	1
−2	0	2
−1	4	−3

b 12   **7** 210 kg   **8 a** 10·45   **b** 51 mins   **9** 221
10 a £884 (nearest £)   **b** 8400 ÷ 40 = 210 gallons
200 × 5 = 1000 litres
1000 × £1 = £1000

page 90 Exercise 33 (M)/(H)

1 a 0·0018 km/h   **b** 1·8 × 10^{-3} km/h   **2** $n = 12$ **3** £6·36 × 10^{10}
4 a £9·79 **b** £32   **c** £44·20   **d** £45   **5 a** 323 g   **b** 23   **c** 121p   **d** 29   **6** 6 m
7 190 ml   **8** 400   **9** 18:52   **10 a** 8·9 m/s (1 dp)   **b** 769·5 mph   **11** $\frac{1}{8}$ **12** 37   **13** 7^{77}

page 92 Test yourself

1 32%   **2** 53·6% (3 sf)   **3** £8450   **4** 1.7   **5** £8 400
6 $\frac{34}{99}$ **7 a** 4 × 10^7 **b** 0·000014   **c** 3 × 10^{14} **8 a** 2·122 226 697   **b** 2·12
9 a 9.476841579 **b** 9·48   **10 a** 1.2   **b** 500   **11** 12.7 g
12 28·3 cm/s   **13 i** 100·5 mm   **ii** 101·5 mm
14 19   **15** 8·1 secs (2 sf)   **16** 51 bags

page 95 Functional Task 2

Task 1 €206 430
Task 2 Profit of €64 320
Task 3 19 yrs 5·5 months

3 Algebra 1

page 95 **Exercise 1** Ⓜ

1 a $2x - 6$ **b** $(x + 4)^2$ **c** $\dfrac{2(x - 5)}{3}$ **d** $7x + n$ **e** $4h - t$ **f** $2(x + y)$

2 a $2a + b$ **b** $n^2 - n$ **c** $(x + 2)^2$ **d** $(w - x)^2$ **e** $\dfrac{(n + p)^3}{a}$ **f** $[3(t - 1)]^2$

3 a $8w$ kg   **b** nw kg   **4** £$\dfrac{n}{3}$ **5** $\dfrac{x}{n}$ pence   **6** $\dfrac{y}{n}$ kg   **7** £$\dfrac{p}{5}$

8 $y + n + 6$ years old   **9** £$(n + r)w$ **10 a** $\dfrac{x}{t}$ metres   **b** $\dfrac{xn}{t}$ metres   **11** $\dfrac{100n}{x}$ **12** $\dfrac{100n}{x + 1}$ pence

page 98 **Exercise 2** Ⓜ

1 $5x + 8$ **2** $9x + 5$ **3** $7x + 4$ **4** $7x + 4$ **5** $7x + 7$

6 $8x + 12$ **7** $12x - 6$ **8** $2x + 5$ **9** $2x - 5$ **10** $2x - 5$

11 $13a + 3b - 1$ **12** $10m + 3n + 8$ **13** $3p - 2q - 8$ **14** $2s - 7t + 14$ **15** $2a + 1$

16 $x + y + 7z$ **17** $5x - 4y + 4z$ **18** $5k - 4m$ **19** $4a + 5b - 9$ **20** $a - 4x - 5e$

21 a $3x^2 + 1$ **b** $2a + 5ab$ **c** $x^3 - 7x + 4x^2$ **d** $4a^2 - 7a$ **22** B and D

23 $x^2 + 3x + 3$ **24** $3x^2 + 6x + 5$ **25** $2x^2 + 6x - 7$ **26** $3x^2 + x + 12$ **27** $2x^2 + x + 3$

28 $2x^2 - x$ **29** $x^2 - 4x - 2$ **30** $3x^2 - 2x - 2$ **31** $4y^2 + 5x^2 + x$ **32** $12 + x$

33 $2 - 6y + 5y^2$ **34** $3ab - 3b$ **35** $2cd - 2d^2$ **36** $4ab - 2a^2 + 2a$ **37** $2x^3 + 5x^2$

38 $11 + x^2 + x^3$ **39** $3xy$ **40** $p^2 - q^2$

page 99 **Exercise 3** Ⓜ

1 True   **2** False unless $n = 1\frac{1}{2}$ **3** True   **4** True   **5** False unless $n = 0$ or $\pm\sqrt{3}$ **6** True

7 False unless $m = n$ **8** True   **9** False unless $n = 0$ or $\pm\sqrt{3}$ **10** False unless $m = 0$ or $\dfrac{1 \pm \sqrt{5}}{2}$

11 False unless $c = 1\frac{1}{2}$ **12** True   **13** True   **14** False unless $n = \pm 2$

15 True   **16** False unless $n = 0$ **17** True   **18** False unless $n = 0$ or 1

19 a $n + n, 4n - 2n$ **b** $n \times n^2, n \times n \times n$ **c** $3n \div 3, n^2 \div n$ **d** $4 \div n$

     e Multiple answers e.g. $2 \times n \times n$

20 a $n \to \boxed{\times 6} \to \boxed{-1} \to 6n - 1$ **b** $n \to \boxed{\times 8} \to \boxed{+10} \to 8n + 10$

     c $n \to \boxed{\div 2} \to \boxed{+3} \to \dfrac{n}{2} + 3$ **d** $n \to \boxed{\times 2} \to \boxed{+5} \to \boxed{\times 3} \to 3(2n + 5)$

     e $n \to \boxed{\times 2} \to \boxed{-4} \to \boxed{\div 5} \to \dfrac{(2n - 4)}{5}$ **f** $n \to \boxed{\text{square}} \to \boxed{+4} \to \boxed{\div 7} \to \dfrac{(n^2 + 4)}{7}$

21 3 **22** 1 **23** n **24** n **25** $2a + b + c$ **26** $2n^2$ **27** $2mn$ **28** n^2 **29** 3

30 a^3 **31** n **32** $3t - 3p + 3$ **33** 1 **34** $4n + 2$ **35** $2n + 8$ **36** n **37** $\dfrac{1}{a}$ **38** $3ab$

page 100 **Exercise 4** Ⓜ

1 a $6x$ **b** $15a$ **c** $-10t$ **d** $1000a$ **e** $2x^2$ **f** $4x^2$ **g** $2x^3$ **h** $20t^2$

2 a $4x + 2$ **b** $10x + 15$ **c** $a^2 - 3a$ **d** $2n^2 + 2n$ **e** $-4x - 6$ **f** $2x^2 + 2xy$

     g $-3a + 6$ **h** $pq - p^2$ **i** $ab + a^2$ **j** $2b^2 - 2b$ **k** $2n^2 + 4n$ **l** $6x^2 + 9x$

3 $A = G, B = D, C = E, F = H$

4 a $6ab$ **b** $3x^4$ **c** $2xy$ **d** $10pq$ **e** $15xy$ **f** $18x^3$ **g** $24a^4$ **h** $2a^2b$

     i $3xy^2$ **j** $5c^2d$ **k** a^2b^2 **l** $2x^2y^2$ **m** $3d^3$ **n** $10x^2y$ **o** $-6a^2$ **p** $2a^2b$

5 $6x^2 + 12x$ cm^2

6 a $7x + 10$ **b** $8x + 2$ **c** $5a - 3$ **d** $11a + 17$ **e** $8a - 10$ **f** $8t + 4$ **g** $x + 4$ **h** $x + 6$

7 a $l = 2x - 2$ **b** $l = x - 1$ **c** $l = x + 2$ **d** $l = x - 3$ **e** $l = x + 3$ **f** $l = x + 3$

8 a $2x^2 + 4x + 6$ **b** $2x^2 + 2x + 5$ **c** $3a^2 + 6a - 4$ **d** $5y^2 + 4y - 3$ **e** $5x^2 + 2x$ **f** $a^2 + 2a$

page 102 Exercise 5 Ⓜ/Ⓗ

1 $x^2 + 4x + 3$ **2** $x^2 + 5x + 6$ **3** $y^2 + 9y + 20$ **4** $x^2 + x - 12$ **5** $x^2 + 3x - 10$

6 $x^2 - 5x + 6$ **7** $a^2 - 2a - 35$ **8** $z^2 + 7z - 18$ **9** $x^2 - 9$ **10** $k^2 - 121$

11 $2x^2 - 5x - 3$ **12** $3x^2 - 2x - 8$ **13** $2y^2 - y - 3$ **14** $49y^2 - 1$ **15** $x^2 + 8x + 16$

16 $x^2 + 4x + 4$ **17** $x^2 - 4x + 4$ **18** $4x^2 + 4x + 1$ **19** $2x^2 + 6x + 5$ **20** $2x^2 + 2x + 13$

21 $5x^2 + 8x + 5$ **22** $2y^2 - 14y + 25$ **23** $10x - 5$ **24** $-8x + 8$

25 a $4x(x + 5) = 4x^2 + 20x$ **b** $(2x + 5)^2 = 4x^2 + 20x + 25$ **c** 25

26 $x^3 + 5x^2 + 6x + x^2 + 5x + 6 = x^3 + 6x^2 + 11x + 6$

27 a $x^3 + 8x^2 + 19x + 12$ **b** $x^3 + 3x^2 + 3x + 1$ **c** $x^3 + 3x^2 - 4$

28 d Yes. $n = 10, 11, 21, 22$ etc.

page 103 Exercise 6 Ⓜ/Ⓗ

1 $x(x + 5)$ **2** $x(x - 6)$ **3** $x(7 - x)$ **4** $y(y + 8)$

5 $y(2y + 3)$ **6** $2y(3y - 2)$ **7** $3x(x - 7)$ **8** $2a(8 - a)$

9 $3c(2c - 7)$ **10** $3x(5 - 3x)$ **11** $7y(8 - 3y)$ **12** $x(a + b + 2c)$

13 $x(x + y + 3z)$ **14** $y(x^2 + y^2 + z^2)$ **15** $ab(3a + 2b)$ **16** $xy(x + y)$

17 $2a(3a + 2b + c)$ **18** $m(a + 2b + m)$ **19** $2k(x + 3y + 2z)$ **20** $a(x^2 + y + 2b)$

21 $x(7x + 1)$ **22** $4y(y - 1)$ **23** $p(p - 2)$ **24** $2a(3a + 1)$

25 $4(1 - 2x^2)$ **26** $5x(1 - 2x^2)$ **27** $\pi(4r + h)$ **28** $\pi r(r + 2)$

29 $\pi r(3r + h)$ **30** $x(3y + 2)$ **31** $xk(x + k)$ **32** $ab(a^2 + 2b)$

33 $bc(a - 3b)$ **34** $ae(2a - 5e)$ **35** $ab(a^2 + b^2)$ **36** $x^2y(x + y)$

37 $2xy(3y - 2x)$ **38** $3ab(b^2 - a^2) = 3ab(b - a)(b + a)$ **39** $a^2b(2a + 5b)$

40 $ax^2(y - 2z)$ **41** $2ab(x + b + a)$ **42** $yx(a + x^2 - 2yx)$ **43 a** $2x + 8y$ **b** $10a - 2x$

page 104 Exercise 7 Ⓗ

Equation	Expression	Identity	Formula
$7x + 11 = x - 9$	$7y + 10$	$x(x + 1) = x^2 + x$	$V = IR$
$x^2 - 7x = 0$	$x^2 - 3x + 10$	$(x + 1)^2 = x^3 + 2x + 1$	$A = \pi r^2$

page 105 Exercise 8 Ⓜ

1 3 **2** 17 **3** 14 **4** 16 **5** 7 **6** 7 **7** 3 **8** 13 **9** 4

10 0 **11** 31 **12** 8 **13** 10 **14** 8 **15** 8 **16** 3 **17** 5 **18** 3

19 2 **20** 4 **21** 1 **22** 0 **23** 2 **24** 1 **25** $\frac{1}{5}$ **26** 7 **27** 200

28 7 **29** 0 **30** $\frac{1}{8}$

page 106 **Exercise 9** Ⓜ

1 8 **2** 9 **3** 7 **4** 10 **5** $\frac{1}{3}$ **6** 10 **7** $1\frac{1}{2}$ **8** -1
9 $-1\frac{1}{2}$ **10** $\frac{1}{3}$ **11** $\frac{99}{100}$ **12** 0 **13** 1000 **14** $-\frac{1}{1000}$ **15** 1 **16** -7
17 -5 **18** $1\frac{1}{6}$ **19** 1 **20** 2 **21** -5 **22** -3 **23** $-1\frac{1}{2}$ **24** 2
25 1 **26** $3\frac{1}{2}$ **27** 2 **28** -1 **29** $10\frac{2}{3}$ **30** 1·1 **31** -1 **32** 2

page 107 **Exercise 10** Ⓜ/Ⓗ

1 35 **2** 130 **3** 14 **4** $\frac{2}{3}$ **5** $3\frac{1}{3}$ **6** $-2\frac{1}{2}$ **7** 3 **8** $1\frac{1}{8}$ **9** $\frac{3}{10}$
10 $-1\frac{1}{4}$ **11** 10 **12** 27 **13** 20 **14** 18 **15** 28 **16** -15 **17** $2\frac{1}{2}$ **18** $1\frac{1}{3}$

page 107 **Exercise 11** Ⓜ/Ⓗ

1 $-1\frac{1}{2}$ **2** 2 **3** $-\frac{2}{5}$ **4** $-\frac{1}{3}$ **5** $1\frac{2}{3}$ **6** 6 **7** $-\frac{2}{5}$ **8** $-3\frac{1}{5}$ **9** $\frac{1}{2}$ **10** -4
11 18 **12** 5 **13** 4 **14** 3 **15** $2\frac{3}{4}$ **16** $-\frac{7}{22}$ **17** $\frac{1}{4}$ **18** 1 **19** 4 **20** -11

page 108 **Exercise 12** Ⓗ

1 $\frac{1}{3}$ **2** $\frac{1}{5}$ **3** $1\frac{2}{3}$ **4** -3 **5** $\frac{5}{11}$ **6** -2 **7** -7 **8** $-7\frac{2}{3}$ **9** 2 **10** 3 **11** 2
12 3 **13** 5 **14** -4 **15** 4 **16** $\frac{3}{5}$ **17** $1\frac{1}{8}$ **18** -1 **19** 1 **20** 1 **21** $1\frac{5}{7}$

page 108 **Exercise 13** Ⓜ

1 3 **2** $\frac{3}{4}$ **3** $4\frac{1}{2}$ **4** $-\frac{3}{10}$ **5** $-\frac{1}{2}$ **6** $17\frac{2}{3}$ **7** $\frac{1}{6}$ **8** 5 **9** 12 **10** $3\frac{1}{3}$
11 $4\frac{2}{3}$ **12** -9

page 110 **Exercise 14** Ⓜ

1 $3\frac{1}{3}$ cm **2** 12 cm **3** 91, 92, 93 **4** 21, 22, 23, 24 **5** 57, 59, 61
6 506, 508, 510 **7** $12\frac{1}{2}$ **8** $x = 20°$ **9** $27\frac{1}{2}$, $18\frac{1}{2}$ **10** 20°, 60°, 100°
11 5, 15, 8 **12** 5 cm **13** Paul $72\frac{2}{3}$ kg, John $64\frac{2}{3}$ kg, David $59\frac{2}{3}$ **14** 7
15 **a** $(4 + 2x)(3 + 2x) - 12 = 4x^2 + 14x$ **b** $14 + 8x$ m **c** $\frac{3}{4}$

page 112 **Exercise 15** Ⓜ

1 $x = \frac{27}{5}$ **2** $53\frac{1}{3}$ or $56\frac{2}{3}$ **3** $2\frac{9}{10}$ **4** 26, 58 **5** 2000 m
6 8 km **7** 21 **8** 23 **9** £3600 **10** 26 cm

page 113 **Exercise 16** Ⓜ/Ⓗ

1 $\frac{1}{4}$ **2** -3 **3** 4 **4** $-7\frac{2}{3}$ **5** -43 **6** 11 **7** $-\frac{1}{2}$ **8** 0 **9** 1 **10** $-1\frac{2}{3}$

11 7 **12** 10 **13** 4 **14** 5

page 114 **Exercise 17** Ⓜ

1 12 cm **2** 5 cm × 15 cm

3 a 26 cm, 13 cm **b** 16 cm, 8 cm **c** 32 cm, 16 cm **d** 9 cm, 4·5 cm **e** 6·5 cm, 3·25 cm

4 a 19 cm, 18 cm **b** 6·5 cm, 5·5 cm **c** 8·7 cm, 7·7 cm

page 116 **Exercise 18** Ⓜ

1 a 9·5 cm **b** 7·6 cm **2 a** 3·4 **b** 4·6 **c** 6·7 **3 a** 1·7 **b** 3·0 **c** 6·1

4 a 5·1 **b** 3·8 **c** 4·0 **d** 3·3 **5** 8·1 cm **6** 3·58 cm **7 a** 3·2 **b** 2·7 **c** 3·6

8 2·15 **9 a** 3 cm^2 **b** $x(x + 3) - 6$ **c** $x = 3·8$ **10 a** $x - 1$ **b** $x^2 - x = 1$ $x = 1·62$

11 6·3 **12** 38·5 **13 a** $2x\sqrt{1 - x^2}$ **b** 0·71 **14 b** 1·5

page 120 **Exercise 19** Ⓜ

1 $w = b + 4$ **2** $w = 2b + 6$ **3** $w = 2b - 12$

4 a

t	m
1	3
2	5
3	7
4	9

b $m = 2t + 1$

5

t	m
1	5
2	8
3	11
4	14

$m = 3t + 2$

6 $s = t + 2$ **7 a** $p = 5n - 2$ **b** $k = 7n + 3$ **c** $w = 2n + 11$

8 $m = 8c + 4$ **9 a** $y = 3n + 1$ **b** $h = 4n - 3$ **c** $k = 3n + 5$

10 a $t = 2n + 4$ **b** $e = 3n + 11$ **c** $e = 1·5t + 5$

11 a £7800 **b** $R = 400N + 5000$ **12** $n^2 - n$

page 123 **Exercise 20** Ⓜ

1 a $10n$ **b** $5n$ **c** $n + 2$ **d** $2n + 1$ **e** $30n$ **f** $6n - 1$ **g** $3n - 2$

2 a $10 \to 20 \to 23$ **b** $20 \to 60 \to 61$

$n \to 2n \to 2n + 3$ $n \to 3n \to 3n + 1$

3 a

n	5n	Term
1	5	6
2	10	11
3	15	16
4	20	21
5	25	26
⋮	⋮	⋮
n	5n	5n + 1

4 b A $3n - 2$ B $4n + 2$ C $7n - 2$

5 a 80 **b** 76 **c** $8n$ **d** $8n - 4$

6 a $6n + 2$ **b** $6n - 1$

7 a $10 \to 10 \times 11 \to 110$ **b** $10 \to 10^2 + 1 \to 101$

$n \to n(n + 1) \to n^2 + n$ $n \to n^2 + 1 \to n^2 + 1$

8 $n^2 + 4$ **9 a** 10^2 **b** n^2 **10 a** 10×11 **b** $n(n + 1)$

11 a $\frac{10}{11}$ **b** $\frac{n}{n + 1}$ **12 a** $\frac{5}{10^2}$ **b** $\frac{5}{n^2}$

b i 51 **ii** $5n + 1$ **13 a** 10×12 **b** $n(n + 2)$

page 126 **Exercise 21** Ⓜ

1

x	0	1	2	3	4	5	6
y	3	5	7	9	11	13	15

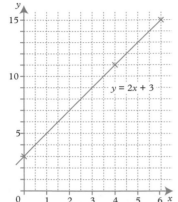

$y = 2x + 3$

2

x	0	1	2	3	4	5
y	−1	1	3	5	7	9

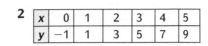
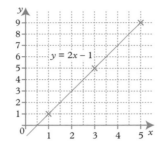

$y = 2x - 1$

3

x	−3	0	3
y	−5	1	7

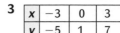

$y = 2x + 1$

4

x	−3	0	3
y	−13	−4	5

$y = 3x - 4$

5

x	−2	0	4
y	10	8	4

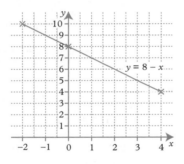

$y = 8 - x$

6

x	−2	0	4
y	14	10	2

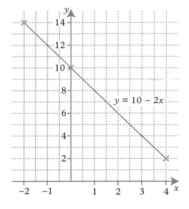

$y = 10 - 2x$

7

x	−3	0	3
y	1	$2\frac{1}{2}$	4

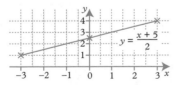

$y = \dfrac{x + 5}{2}$

8

x	−3	0	3
y	−15	−6	3

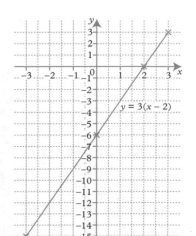

$y = 3(x - 2)$

9

x	−3	0	3
y	$2\frac{1}{2}$	4	$5\frac{1}{2}$

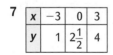

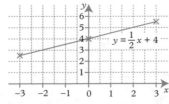

$y = \dfrac{1}{2}x + 4$

10

x	−2	0	4
y	−7	−3	5

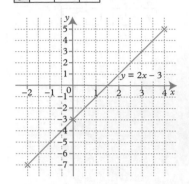

$y = 2x - 3$

11

x	−2	0	4
y	18	12	0

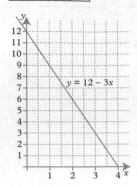

$y = 12 - 3x$

12

x	−1	0	4
y	10	8	0

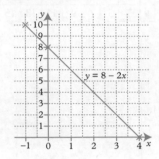

$y = 8 - 2x$

13

x	0	50	100	150	200	250	300
C	35	45	55	65	75	85	95

a 180 miles

b $C = \dfrac{x}{5} + 35$

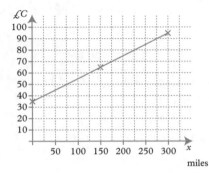

14

h	0	1	2	3
C	18	33	48	63

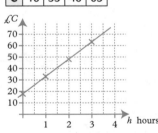

15

s	0	80	160
C	50	530	1010

a £190

b 158 km/h

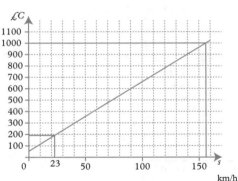

page 128 **Exercise 22** Ⓜ/Ⓗ

1 a $(4, 10)$ **b** $(6, 7)$ **c** $(1, 3)$ **d** $(1, 3)$ **2 a** 5 **b** 13 **c** $\sqrt{74}$

3 $A(0, 12)$ $B(6, 0)$ **a** $(3, 6)$ **b** $\sqrt{180}$ or $6\sqrt{5}$ **4** $(4a, a)$

page 129 **Exercise 23** Ⓜ

1 $\dfrac{1}{5}$, $\dfrac{5}{2}$, $-\dfrac{4}{3}$ **2** $\dfrac{4}{5}$, $-\dfrac{1}{6}$, -5 **3 a** 3 **b** $\dfrac{3}{2}$ **c** 4 **d** 5   **4** $a = 3\dfrac{1}{2}$

5 a $\dfrac{n+4}{2m-3}$ **b** -4 **c** $1\dfrac{1}{2}$ **6 a** 1 and -1, $\dfrac{1}{2}$ and -2, 3 and $-\dfrac{1}{3}$ **b** -1 **c** -1 **d** the same

7 Missing numbers are 5, $-\dfrac{1}{2}$, $-\dfrac{2}{3}$, $\dfrac{1}{5}$ and $-\dfrac{4}{3}$ **8** -1

page 131 **Exercise 24** (M)/(H)

1 a 1　**b** 3

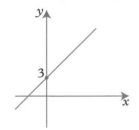

2 a 1　**b** −2

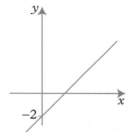

3 a 2　**b** 1

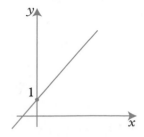

4 a 2　**b** −5

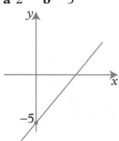

5 a 3　**b** 4

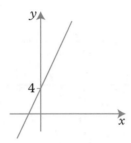

6 a $\frac{1}{2}$　**b** 6

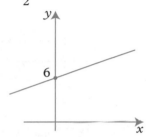

7 a 3　**b** −2

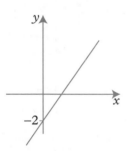

8 a 2　**b** 0

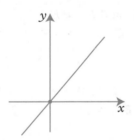

9 a $\frac{1}{4}$　**b** −4

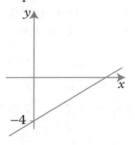

10 a −1　**b** 3

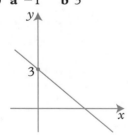

11 a −2　**b** 6

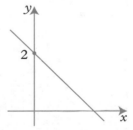

12 a −1　**b** 2

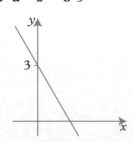

13 a −2　**b** 3

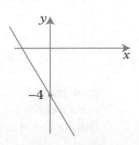

14 a −3　**b** −4

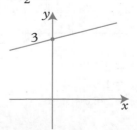

15 a $\frac{1}{2}$　**b** 3

16 a $-\dfrac{1}{3}$ **c** 3

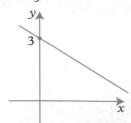

17 a 4   **c** -5

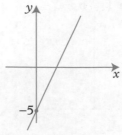

18 a $\dfrac{3}{2}$ **b** -4

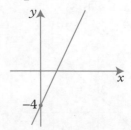

19 a 10   **b** 0

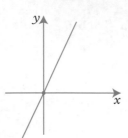

20 a 0   **b** 4

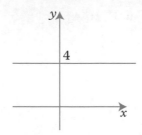

21 $A: y = 3x - 4$
  $B: y = x + 2$

22 $C: y = \dfrac{2}{3}x - 2$
  $D: y = -2x + 4$

23 a F and G; C and D **b** B and H; A and E **c** I
24 a $A(0, -8)$; $B(4, 0)$ **b** 2   **c** $y = 2x - 8$

page 133 **Exercise 25** Ⓜ/Ⓗ

1 $y = 3x + 7$ **2** $y = 2x - 9$ **3** $y = -x + 5$ **4** $y = 2x - 1$ **5** $y = 3x + 5$
6 $y = -x + 7$ **7** $y = \dfrac{1}{2}x - 3$ **8** $y = 2x - 3$ **9** $y = 3x - 11$ **10** $y = -x + 5$

11

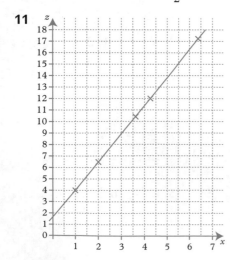

$a = 2.5$
$c = 1.5$
$z = 2.5x + 1.5$

12

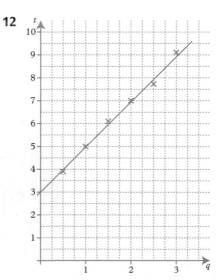

gradient $= 2$
t-intercept $= 3$
Estimate $m = 2$, $c = 3$

page 134 **Exercise 26** Ⓜ

1 a 40 km **b** 60 km **c** York and Scarborough **d** 15 minutes **e i** 11·00 **ii** 13·45
  f i 40 km/h **ii** 60 km/h **iii** 100 km/h
2 a 45 mins **b** 09·15 **c** 60 km/h **d** 100 km/h **e** 57·1 km/h (1 dp)
3 a 09:00 **b** 64 km/h **c** 40 km/h **d** 70 km **e** 80 km/h

4

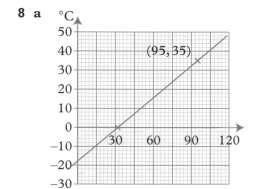

5

6

7 a £15 **b** £1.50 per mile

8 a

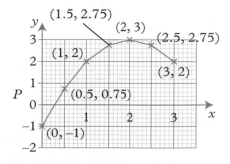

9 a 740 pence **b** £280 **c** £14 000
10 a 5 km per litre **b** 28 m.p.g **c** 12 gallons
11 a 2000 **b** 286 **c** $1600 \leqslant x \leqslant 2400$

b i 104 °F
  ii 14 °F
  iii 10 °C

page 139 **Exercise 27** Ⓜ

1 B   **2 a** C   **b** A   **c** D   **d** B   **3** D

4

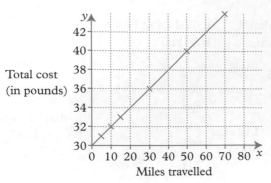
Total cost (in pounds) / Miles travelled

5 X → B   Y → C   Z → A

6 P   Q   R

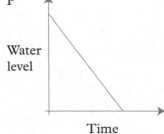   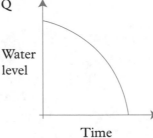   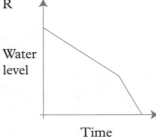

Water level / Time   Water level / Time   Water level / Time

7

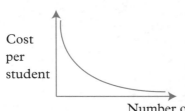
Cost per student / Number of students

No, it is not linear.

8 a 6 gallons   **b i** 40 mpg   **ii** 30 mpg   **c** $33\frac{1}{3}$ mpg

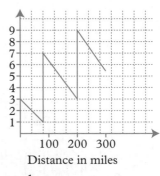
Petrol in tank (gallons) / Distance in miles

$5\frac{1}{2}$ gallons

page 143 **Exercise 28** Ⓜ/Ⓗ

1 a $x = 3, y = 7$ **b** $x = 1, y = 3$ **c** $x = 11, y = -1$ **2** $x = 2, y = 4$
3 $x = 2, y = 3$ **4** $x = 3, y = 1$ **5** $x = 1, y = 5$ **6** $a = 5, b = 3$
7 a $x = 4, y = 0$ **b** $x = 1, y = 6$ **c** $x = -2, y = -3$ **d** $x = 8, y = -1$ **e** $x = -0.6, y = 1.2$

page 145 **Exercise 29** Ⓜ/Ⓗ

1 $x = 2, y = 1$ **2** $x = 4, y = 2$ **3** $x = 3, y = 1$ **4** $x = -\frac{1}{3}, y = -2\frac{1}{3}$ **5** $x = 3, y = 2$

6 $x = 5, y = -2$ **7** $x = 2, y = 1$ **8** $x = 5, y = 3$ **9** $x = 3, y = -1$ **10** $a = 2, b = -3$

11 $a = 5, b = \frac{1}{4}$ **12** $a = 1, b = 3$ **13** $m = \frac{1}{2}, n = 4$ **14** $w = 2, x = 3$ **15** $x = 6, y = 3$

16 $x = \frac{1}{2}, z = -3$ **17** $m = 2, n = 1$ **18** $c = \frac{39}{23}, d = \frac{-58}{23}$

page 146 **Exercise 30** Ⓜ/Ⓗ

1 $x = 2, y = 4$ **2** $x = 1, y = 4$ **3** $x = 2, y = 5$ **4** $x = 3, y = 7$ **5** $x = 5, y = 2$

6 $a = 3, b = 1$ **7** $x = -2, y = 3$ **8** $x = 4, y = 1$ **9** $x = \frac{5}{7}, y = 4\frac{3}{7}$ **10** $x = 1, y = 2$

11 $x = 2, y = 3$ **12** $x = 4, y = -1$ **13** $x = 1, y = 2$ **14** $a = 4, b = 3$ **15** $x = 4, y = 3$

16 $x = 5, y = -2$ **17** $x = 3, y = -1$ **18** $x = 5, y = 0\cdot2$

page 147 **Exercise 31** Ⓜ/Ⓗ

1 $9\frac{1}{2}$ and $5\frac{1}{2}$ **2** 6 and 3   **3** 5 and 2   **4** 8 and 5   **5** $? = 10\frac{1}{2}, \star = 7\frac{1}{2}$

6 $54°, 63°, 63°$ **7** $c = £3, p = £4$ **8** TV costs £200, DVD player costs £450

9 $b = 3\frac{1}{2}$ oz, $w = 2$ oz   **10** $(100b + 140j)$ **11** $m = 4, c = -3$

12 15 two pence coins; 25 five pence coins   **13** 14 ten pence coins; 7 fifty pence coins

14 20   **15** Current 4 m/s, Herring 10 m/s   **16** $\frac{5}{7}$ **17** mouse is 3, owner is 10

18 $a = 1, b = 2, c = 5$ **19** $a = 1, b = 3, c = 4$

page 149 **Exercise 32** Ⓗ

1 a $x = -1, y = 3$ or $x = 4, y = 8$ **b** $x = 3, y = 13$ or $x = 5, y = 27$

2 a $x = 3, y = 7$ or $x = 5, y = 17$ **b** $x = \frac{1}{2}, y = \frac{1}{2}$ or $x = 4, y = 32$ **3** $(-3, -2)$ and $(2, 3)$

4 $x = -2, y = -4$ or $x = 4, y = 2$ **5** $(1, -4)$ and $(4, 8)$

6 Solving simultaneously gives $2x^2 + 20x + 99 = 0$ which has no solutions. However, it is easy to see that there are no solutions from a sketch graph:

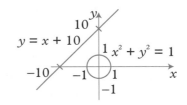

7 b $(3, 4), (-4, -3)$ **8** $(2, 8)$

page 150 **Test yourself**

1 a $\frac{2}{3}$ **b** $n(n - 3)$ **c** $2x^2 + 9x - 5$ **d** $m^4 - 2m^2 + m$ **2 a** $3(x - 2y)(x + 2y)$ **b** $\frac{x}{x + 3}$

4 a $3bc$ **b** $2x + 5y$ **c** m^3 **d** $6np$ **5** $2\cdot26$ **6** $4\cdot2(4\cdot15)$

7 a $4n + 1$ **b** Diagram 50   **8 a** n^2 **b** $7n - 9$ **9** $y = -2x + 5$

10 $x = 1, y = 2$ or $x = -\frac{1}{2}, y = \frac{17}{4}$.   **11 a** 3   **b** $4n + 6$ **12 a** $-5, -1, 3, 7$ **b** 4   **c** 100

13 a $(90 - \frac{1}{2}x)°$ **b** $p = 4, q = -1$ **14** $x = -3, y = 5$ or $x = -5, y = 3$

15 a $5x + 4y$ **b** $x = 1\frac{2}{5}, y = -1\frac{1}{2}$ **c i** $nx + (n - 1)y$ **ii** 15

16 a $x = \frac{9}{2}$ **b** $5y - 27$ **17 a** $3xy$ **b** $5a + 9b$

Page 154 **Functional Task 3**

Task 1 Phil: healthy weight, Annie: over weight, Kate: under weight,
Jack: healthy weight, Stuart: very over weight, Hans: obese.

Task 2 **1** 21 weeks **2** 41 weeks

4 Shape, space and measures 1

page 157 **Exercise 1** Ⓜ

1 70° **2** 70° **3** 48° **4** $a = 40°$, $b = 140°$ **5** 60° **6** $x = 122°$, $y = 116°$ **7** 135°
8 $a = 28°$ **9** 20° **10** $e = 70°$, $f = 75°$ **11** $a = 36°$, $x = 36°$ **12** $a = 60°$, $b = 40°$
13 72° **14** 98° **15** 80° **16** $x = 95°$, $y = 50°$ **17** $a = 87°$, $b = 74°$
18 $a = 65°$, $c = 103°$ **19** $a = 70°$, $b = 60°$ **20** 65° **21** 46°
22 a $x = a$, $y = b$ **b** $a + b + c = x + y + c = 180°$ (angles on a straight line)
23 a $p + q + r = 180°$ **b** $q + x = 180°$ **c** $x = 180° - q = p + r$
24 136° **25** 80° **26** 66° **27** $27\frac{1}{2}°$

page 161 **Exercise 2** Ⓜ

2 a 115° **b** 90° **c** 80° **4** Rhombus **5** True **6** False
7 a Rectangle **b** Square **c** Trapezium **d** Parallelogram
8 a Trapezium **b** Rectangle **c** Square **d** Parallelogram **e** Isosceles triangle
9 Any trapezium with 2 right angles

page 162 **Exercise 3** Ⓜ

1 C(5, 1), D(0, 0) **2 a** 180° **b** 180° **c** Sum of angles $= a + c + d + f + e + b$
$$= a + b + c + d + e + f$$
$$= 180° + 180° = 360°$$

3 a 72° **b** 108° **c** 80° **4 a** 40° **b** 30° **c** 110°
5 a 116° **b** 32° **c** 58° **6** 53° **7 a** 26° **b** 26° **c** 77°
8 150° **9** 110° **10 a** 54° **b** 72° **c** 36° **11 a** 60° **b** 15° **c** 75°

page 165 **Exercise 4** Ⓜ

1 a $a = 80°$, $b = 70°$, $c = 65°$, $d = 86°$, $e = 59°$
  b $a + b + c + d + e = 80° + 70° + 65° + 86° + 59° = 360°$ as required.
2 Interior angle of a hexagon is 120°. $360° \div 120° = 3$ (an integer)
  Interior angle of a pentagon is 108°. $360° \div 108° = 3.3$ (not an integer)
3 a $a = 36°$ **b** 144° **c** No **4 a i** 40° **ii** 20° **iii** 8° **iv** 6° **b i** 140° **ii** 160° **iii** 172° **iv** 174°
5 $p = 101°$, $q = 79°$, $m = 70°$, $n = 130°$, $x = 70°$ **6** 24 sides **7** 9 sides **8** 20 sides
9 b i 1080° **ii** 1440° **10 a** 150° **b** 30° **c** $\frac{360°}{12} = 30°$ **11** 20 sides

page 167 **Exercise 5** Ⓜ

1 A and G, B and O, C and F, H and P, I and N, J and K

page 169 **Exercise 6** Ⓗ

1 Yes (SSS) **2** Yes (SAS) **3** No **4** Yes (AAS) **5** No **6** Yes (AAS) **7** CDA, DEB, EAC, ABD

8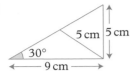

9 AB = BC
AD = DC
BD is common
Therefore, △ABD ≡ △CBD (SSS)
∠A corresponds to ∠C & therefore
under the congruency, A = C.

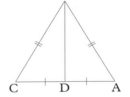

10 ∠OAT = 90° = ∠OBT; OA = OB (radii); OT is common.
Therefore, △OAT ≡ △OBT (RHS). Therefore AT = BT as required.

11 Since LY = LX, LM = LN and angle L is common,
the two triangles are congruent (SAS).

12 ∠AXD = ∠BCD   (corresponding angles)
∴ ∠AXD = ∠ADX
∴△ADX is isosceles (base angles equal)

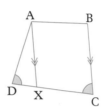

13 ∠NYM = ∠MYZ and ∠MZN = ∠NZY.
But triangle XZY is isosceles, so ∠Z = ∠Y.
Thus ∠NYM = ∠MYZ = ∠MZN = ∠NZY.
Therefore, △MYZ ≡ △NZY (AAS i.e. ∠NZY = ∠MYZ,
∠NYZ = ∠MZY and ZY is common)
Therefore YM = ZN.

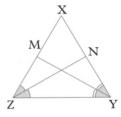

14 a DX = XC, so △DXC is isosceles, i.e. angle C = angle D. But angle
C = ∠XAB and angle D = ∠XBA. Therefore, ∠XAB = ∠XBA so △ABX
is isosceles. Therefore AX = BX.
b AX = BX and XC = XD, so AX + XC = BX + XD, i.e. AC = BD
c DZ = DV + VZ = ZC + VZ = VC
AC = BD (proven in **b**) and angle D = angle C as DXC is isosceles.
Therefore, △DBZ ≡ △CAV (SAS).

15 ∠ABX = ∠XDC (alternate angles)
∠BAX = ∠XCD (alternate angles)
AB = CD   (parallelogram)
∴ △ABX ≡ △CDX (AAS)
∴ BX = XD and AX = XC

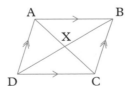

16 PQ = QA
QR = QB
Angle PQR = 90°
Angle AQR = 90° − 60° = 30°
Therefore, ∠AQB = 30° + 60° = 90° = ∠PQR
∴ △PQR ≡ △AQB (SAS)
∴ PR = AB as required.

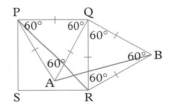

page 172 **Exercise 7** Ⓜ

Students' own constructions

page 173 **Exercise 8** Ⓜ

1 P moves on a circle center X radius 3 cm.

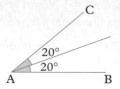

2

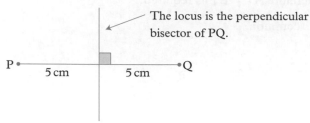

The locus is the perpendicular bisector of PQ.

3 The locus is the angle bisector of A.

4 a Locus is SQ. **b** Locus is shaded area.

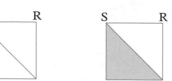

c Locus is shaded area inside a circle centre S, radius 3 cm.

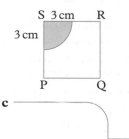

5 a ——————— **b** ——————⌐ **c** ———————⌐

6 A moves on a circle centre B radius 3 and C on a circle centre B radius 4.

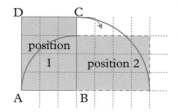

7 Inside of 2 circles, one centre L radius 4, the other centre C radius 2.

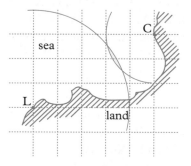

8

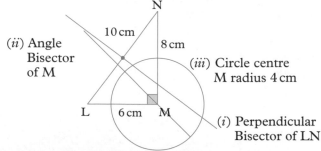

(ii) Angle Bisector of M

(iii) Circle centre M radius 4 cm

(i) Perpendicular Bisector of LN

9 Locus is a circle, diameter AB.

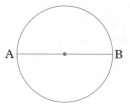

10 X is the intersection of the perpendicular
bisector of the trees with the angle bisector
of the lower left corner.

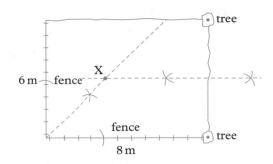

page 175 **Exercise 9** Ⓜ

1 Locus is shaded area.

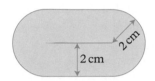

2 D   10 cm   C

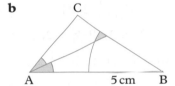

3 a Locus is angle bisector of A.

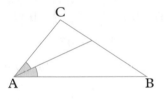

b

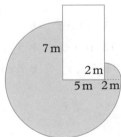

4 a

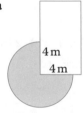

b

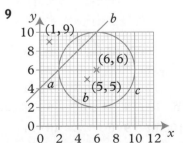

5

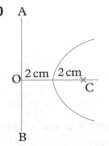

6 A plane containing the perpendicular bisector of the line joining the 2 fixed points.

7 A circle diameter AB

8 Students' own construction

9

10 A

11 A

page 177 **Exercise 10** (M)

1 10 cm **2** 4·1 cm **3** 10·6 cm **4** 5·7 cm **5** 4·2 cm
6 9·9 cm **7** 4·6 cm **8** 5·2 cm **9** 9·8 cm (2 sf) **10** 9·8 cm (2 sf)
11 3·5 m (2 sf) **12** 40·3 km (3 sf) **13** 12·7 cm (3 sf) **14** 5·7 cm (2 sf) **15 a** 5 **b** 40
16 9·5 cm (2 sf) **17** 32·6 cm (3 sf) **18 a** 14·1 cm (3 sf) **b** 57·1 cm^2 (3 sf) **19** 374 cm^2 (3 sf)

page 180 **Exercise 11** (H)

1 5·4 (2 sf) **2** PQ = 5·83, QR = 6, PR = 5·83; Yes. **3** 6·6 (2 sf) **4** 5·6 (2 sf)
5 8·7 (2 sf) **6** 5·7 (2 sf) **7** 6·6 (2 sf) **8** $x = \sqrt{5} = 2\cdot2$ (2 sf)
9 a 5 cm **b** 7·8 cm (2 sf) **10 a** 6·4 cm (1 dp) **b** 13·6 cm (1 dp) **11** 6·34 m (3 sf)
12 4·6 cm (2 sf) **13 a** 7·5 (1 dp) **b** 12·5 (1 dp) **c** 14·9 (1 dp) **14** 24 cm
15 10 feet **16 b** 8·6 (1 dp) **c** 16·4 (1 dp) **17 a** $x = 4$ **b** 20·6 (1 dp)
18 a i 13 **ii** 25 **iii** 9 **b** Enter results from (a) into table. **c** a is an odd number, c is always $b + 1$.
d 11, 60, 61; 13, 84, 85; 15, 112, 113; Yes they do. **19** 29·5 cm (1 dp)

page 183 **Exercise 12** (M)/(H)

1 A (2, 4, 0) B (0, 4, 3) C (2, 4, 3) D (2, 0, 3)
2 a B (3, 0, 0) C (3, 4, 0) Q (3, 0, 2) R (3, 4, 2) **b i** (0, 2, 0) **ii** (0, 4, 1) **iii** $\left(1\frac{1}{2}, 4, 0\right)$
    c i $\left(1\frac{1}{2}, 2, 0\right)$ **ii** $\left(1\frac{1}{2}, 2, 2\right)$ **iii** $\left(1\frac{1}{2}, 4, 1\right)$ **d** $\left(1\frac{1}{2}, 2, 1\right)$
3 a C (2, 2, 0) R (2, 2, 3) B (2, −2, 0) P (0, −2, 3) Q (2, −2, 3) **b i** $\left(2, -2, 1\frac{1}{2}\right)$ **ii** (1, −2, 3)
4 i 5 **ii** 5·83 **iii** 6·40 **5 i** $\left(2, 3\frac{1}{2}, 5\right)$ **ii** $\left(2, 7, 2\frac{1}{2}\right)$ **iii** $\left(2, 3\frac{1}{2}, 0\right)$ **6** (20, 45, 5)
7 a 4 **b** 5 **c** $5\sqrt{2}$ **8** Square-based pyramid **9 a** 10 **b** 45° **10** 29·9 m

page 185 **Exercise 13** (M)

1 34 cm^2 **2** 33 cm^2 **3** 54 cm^2 **4** 18 cm^2 **5** 20 cm^2 **6** 25 cm^2 **7** 23 cm^2
8 31 cm^2 **9** 35 cm^2 **10** 54 cm^2 **11** 51 cm^2 **12 a** 4 cm **b** $5\frac{1}{2}$ cm **c** 4·3 cm **13** 248 cm^2
14 33 cm^2 **15** 39 cm^2 **16** 39 cm^2 **17 a** 6 cm **b** 5 cm **c** 7 cm **19** Mark
20 a A 16 cm B 14 cm C 18 cm

page 189 **Exercise 14** (M)

1 42 cm^2 **2** 22 cm^2 **3** 103 cm^2 **4** 60·5 cm^2 **5** 143 cm^2 **6** 9 cm^2 **7** 13 cm **8** 15 cm **9** 2500
10 a 25 cm^2 **b** 21 cm^2 **c** 11·8 cm^2 (1 dp) **d** 19·7 cm^2 (1 dp) **e** 21 cm^2 **f** 16·4 cm^2 (1 dp)
11 24 cm^2 **12** 40 cm^2 **13** 32 cm^2 **14** 46 cm^2 **15** 47 cm^2 **16** 81·75 cm^2

17 $\sin C = \dfrac{h}{b}$
    $\therefore h = b \times \sin C$
    $\therefore$ Area of $\triangle ABC = \dfrac{1}{2} ab \sin C$

page 190 **Exercise 15** (M)

1 80 m^2 **2 a** 18 cm **b** 8 cm^2 **3 a** 8·5 m **b** 10 m **c** 4 m **4** 45 cm^2 **5** 2·4 cm
6 a $\dfrac{1}{3}$ **b** $\dfrac{4}{9}$ **c** 25 cm^2 **7** Check student's work. **8** 1100 m **9** 6 **10** 14

11 1849   **12** 10 cm   **13 a** 60°   **b** Triangle = 3·90 cm², Hexagon = 23·38 cm²
14 123·6 m²   **15** 57·1 cm² **16** 10·7 cm   **17** 4·1 cm   **18** 4·85 cm **19** 7·23 cm
20 a $\frac{360°}{n}$ **b i** $\frac{1}{2}\sin\left(\frac{360°}{n}\right)$ **ii** $\frac{n}{2}\sin\left(\frac{360°}{n}\right)$ **c** 3·1395; 3·14157 It is approaching π. **21** 18·7 cm

page 194 **Exercise 16** Ⓜ

1 a 34·6 cm **b** 25·1 cm **c** 37·7 cm   **d** 15·7 cm   **e** 10π cm   **f** 13π cm   **g** 2π cm   **h** 4·4π cm
2 13 mm or 1·3 cm **3** 8·5 m (2 sf) **4** 400 m **5** 22·6 cm (3 sf) **6** 212   **7** 360 + 90π cm
8 a 95·0 cm² **b** 50·3 cm² **c** 113 cm² **d** 19·6 cm² **e** 78·5 cm² **f** 133 cm² **g** 3·14 cm² **h** 15·2 cm²
9 29·5 cm² (3 sf)   **10** 124·7 cm² (1 dp)   **11** 100 − 25π cm²   **12** 4·6 kg (2 sf)

page 196 **Exercise 17** Ⓜ Answers are given to an appropriate degree of accuracy.

1 a 31·4 cm   **b** 78·5 cm²   **2 a** 21·4 cm   **b** 28·3 cm²   **3 a** 128·5 cm   **b** 981·7 cm²
4 a 56·6 cm   **b** 190·1 cm²   **5 a** 53·7 m   **b** 198·5 m²   **6 a** 28·1 m   **b** 54·7 m²
7 a 20·6 cm   **b** 24·6 cm²   **8 a** 25·1 cm   **b** 31·4 cm²   **9 a** 20·3 cm   **b** 24·6 cm²
10 a 37·2 cm   **b** 92·9 cm²   **11 a** 25·1 cm   **b** 13·7 cm²   **12 a** 25·1 cm   **b** 25·1 cm²
13 a 18·8 cm   **b** 12·6 cm²

page 198 **Exercise 18** Ⓜ Answers are given to an appropriate degree of accuracy.

1 2·2 cm   **2** 0·32 m   **3 a** 11·5 m   **b** 41·4 km   **c** 3·7 m   **d** 2·4 m   **e** 1·6 × 10⁷ km
4 9·3 cm   **5** 17·8 cm   **6** 14·2 mm   **7** 497 000 km²   **8** 42·9
9 30 discs can be cut.   **a** 1508 cm² **b** 508 cm²   **10** 5305   **11** 29
12 40·8 m²   **13 a** 80   **b** 7   **14** 5·4 cm   **15** 117·7 m² **16 a** 32·99 cm   **b** 70·88 cm²
17 a 98 cm² **b** 14 cm² **18** 796 m² **19 a** 4 × 4 **b** radius 2 **c** side is 4√3 or 6·93 **20** 57·5°
21 He actually needed to take off 2π metres, so 6 m would give a good estimate.   **22** 1·72 cm

page 201 **Exercise 19** Ⓜ

1 2 : 1   **2** 112 cm²   **3** $\frac{7}{16}$ **4** 0·586 m   **5** 20%   **6** 7·172 cm   **7** 112·5°
8 60°   **9** 55·4 cm²   **10** 8 cm   **11** 4 : 1   **12 a** 2·41 (3 sf)   **b** 1·85 (3 sf)   **c** 1 + √2

page 204 **Exercise 20** Ⓗ Answers are given to an appropriate degree of accuracy.

1 a 2·09 cm **b** 7·85 cm   **c** 4·36 cm   **2** 15·36 cm   **3 a** 8·7 cm²   **b** 40·6 cm²   **c** 9·4 cm²
4 a 6·28 cm **b** 61·1 cm² **c** 11·64 cm **d** 150·5 cm² **e** 3945 cm²   **5 a** 7·1 cm² **b** 19·5 cm²
6 17·8 cm   **7** 74·2 cm³ **8** 5·94 cm² **9 a** 3·98 cm **b** 74·9°   **10 a** 12 cm   **b** 30°
11 a 30°   **b** 10·5 cm

page 206 **Exercise 21** Ⓗ

1 a 85·9°   **b** 57·3°   **c** 6·25 cm   **2** 30·6 cm²   **3** 57·3°
4 a 36°   **b** Arc DC = $\frac{36}{360}$ × 2πr
$$= \frac{36}{360} \times 10\pi$$
$$= \pi \quad \therefore \text{ Perimeter} = 5 \times \text{Arc DC} = 5\pi$$
5 a 6·1 cm **b** 27·6 cm **c** 28·6 cm² **6** 1850 metres **7 a** 18 cm **b** 38·2° **8 a** 10 cm **b** 43·0°

9 a Area $\triangle ABC = \frac{1}{2} AC \times AB$

$$= \frac{1}{2} \times 1 \times 1 \tan x$$

$$= \frac{1}{2} \tan x$$

∴ Shaded area in this $\triangle$

$$= \frac{1}{2} \tan x - \frac{x}{360} \times \pi \times 1^2$$

$$= \frac{1}{2} \tan - \frac{\pi x}{360}$$

This equals area of sector ECD

$$\therefore \frac{1}{2} \tan x - \frac{\pi x}{360} = \frac{\pi x}{360}$$

$$\therefore \frac{1}{2} \tan x = \frac{\pi x}{180}$$

$$\therefore 90 \tan x = \pi x$$

$$\therefore \left(\frac{90}{\pi}\right) \tan x = x$$

b $x = 66.8$

10 Let A = unshaded area and r = radius of circle.
Then A + area ② = area of semi-circle

$$= \frac{1}{2} \pi r^2$$

A + area ① = area of $\triangle ABC$

$$= \frac{1}{2} BC \times AC$$

$$= \frac{1}{2} \times 2r \times 2r \tan x$$

$$= 2r^2 \tan x$$

Since area ① = area ②

$$\therefore 2r^2 \tan x = \frac{1}{2} \pi r^2$$

$$\therefore \tan x = \frac{1}{4} \pi$$

$$\therefore \tan x = \frac{\pi}{4}$$

$$x = 38.1°$$

page 209 Exercise 22 (H)

1 a 72.6 cm^2 **b** 24.5 cm^2 **c** 48.1 cm^2 **2 a** 5.08 cm^2 **b** 82.8 cm^2 **c** 5.14 cm^2
3 a $60°, 9.06 \text{ cm}^2$ **b** $106.3°, 11.2 \text{ cm}^2$ **4** 3 cm **5** 3.97 cm **6 a** 13.49 cm^2 **b** 404.7 cm^3
7 a 129.9 cm^2 **b** 184 cm^2 **8** 19.6 cm^2 **9** $0.31 r^2$ (or $r^2 (\pi - 2\sqrt{2})$)
10 a 8.4 cm **b** 54.5 cm **c** 10.4 cm **11** 81.2 cm^2

page 212 Exercise 23 (M)

1 $48 \text{ cm}^3, 88 \text{ cm}^2$ **2** $30 \text{ cm}^3, 62 \text{ cm}^2$ **3** $60 \text{ cm}^3, 104 \text{ cm}^2$ **4** 125 cm^3 **5** 50 minutes
6 a $x = 3 \text{ cm}$ **b** $x = 5.5 \text{ cm}$ **c** $x = 2.5 \text{ cm}$ **7 b** $1\,000\,000$ or 10^6 **c** 6 m^2

page 214 Exercise 24 (M)

1 a 150 cm^3 **b** 60 cm^3 **c** 110 cm^3 **d** 94.5 cm^3 **e** 57 cm^3 **f** 32 cm^3
2 c 192 cm^2 **d** 177.4 cm^2 (1 d.p) **e** 116 cm^2 **f** 64.8 cm^2
3 2400 cm^3 **4 a** 200 m^2 **c** 2400 m^3 **5 a** 62.8 cm^3 **b** 113.1 cm^3 **c** 502.7 cm^3
6 a 502.7 cm^3 **b** 760.3 cm^3 **7** 10 litres **8** 7.1 kg (2 sf) **9 a** 141.4 cm^3 **b** 25.1 cm^3
10 769.7 cm^3 **11** $x = 7\frac{7}{15} \text{ g/cm}^3$ (or $7.46°$)

page 216 Exercise 25 (M)

1 3.98 cm **2** 6.37 cm **3** 1.89 cm **4** 9.77 cm **5** 7.38 cm
6 1273 cm **7** 4.24 litres **8** 106 cm/s **9** Volume $= 1570 \text{ cm}^3$ Weight $= 12.6 \text{ kg}$
10 No, approximately 2300 bricks are needed. **11** 1.2 cm **12** 53 times **13** 191 cm

page 218 Exercise 26 (H)

1 40 cm^3 **2** 33.5 cm^3 **3** 66.0 cm^3 **4** 89.8 cm^3 **5** 339 cm^3 **6** 144 cm^3
7 4.71 kg **8** 262 cm^3 **9** 359 cm^3 **10** 235 cm^3 **11** 5 m **12** $3.6 \times 10^{-3} \text{ mm}$

13 10 balls of radius 2 cm   **14** 415 cm^3 **15** 1·8 mins   **16** 488 cm^3 **17** 37 500 000
18 1·05 cm^3 **19** $10\frac{2}{3}$ cm^3 **20** 1·93 kg

page 220 Exercise 27 (H)

1 a 3·91 cm   **b** 2·43 cm   **c** 7·16 cm   **2** 6·4 cm   **3** 23·9 cm   **4 a** 125   **b** $2·7 \times 10^7$
5 a 0·36 cm   **b** 0·43 cm   **6 a** 6·7 cm   **b** 39·1 cm   **7** NC = 4 cm   Volume = 4·2 cm^3
8 53·6 cm^3 **9** 74·5 cm^3 **10** 122·5 cm^3 **11** 54·5 litres   **12 a** 2·9 m^3 **b** 1·7 m
13 a 16π cm   **b** 8 cm   **c** 6 cm   **14** 1481 cm^3

page 223 Exercise 28 (H)

1 a 36π cm^2 **b** 40π cm^2 **c** 60π cm^2 **d** $1·4\pi$ cm^2 **2 a** 40π **b** 28π **c** 32π **d** 24π
3 £3 870   **4** £233   **5** 303 cm^2 **6** 675 cm^2 **7** $1·62 \times 10^8$ years   **8 a** 1·59 cm   **b** 4·77 cm
9 1·64 cm   **10** 2·12 cm   **11** 3·46 cm   **12** 94 cm^3 **13** 44·6 cm^2 **14** 377 cm^2
15 slant ht. = 20 cm   base radius = 10 cm   **16 a** 3·72 cm   **b** 41·9 cm^2 **17** 122·6 cm^2 **18** 297·6 cm^3

page 227 Exercise 29 (M)/(H)

1 A and G, C and J, K and M, L and F   **3** a, c, e, and f   **4** C   **5** 12 cm   **6** 9 cm
7 $a = 2·5$ cm, $e = 3$ cm   **8** $6\frac{3}{4}$ cm   **9** $3\frac{1}{5}$ cm   **10** $t = 5·25$ cm, $y = 5·6$ cm
11 7·7 cm   **12** No   **13 a** Yes   **b** No   **c** No   **d** Yes   **e** Yes   **f** No   **g** No   **h** Yes
14 a Angle B is common to both, and they each have a right angle ∴ the 3rd angle is also equal.
b 11·2 cm   **c** 4·2 cm
15 6 cm   **16** 6 cm   **17** 4·5 cm

page 230 Exercise 30 (H)

1 16 m   **2** 3·75 m   **3** 10·8 m
4 a ∠AOB = ∠DOC (vertically opposite)
∠OAB = ∠OCD (alternate angles)
∠OBA = ∠ODC (alternate angles)
∴ triangles are similar
b AO = 2 cm, DO = 6 cm
5 $m = 6, n = 6$ **6** $w = 6\frac{2}{3}, v = 5\frac{1}{3}$
7 $x = 1\frac{1}{11}$ or $3\frac{2}{3}$ **8** $x = 0·618$; ratio of sides = 1·618   **9** $x = 5$
10 a △ABD is similar to △ACB   **b** $y^2 = qz$ **c** $x^2 + y^2 = pz + qz = z(p + q) = zz = z^2$ as required
$$\therefore \frac{AB}{AC} = \frac{AD}{AB}$$
$$\therefore \frac{x}{z} = \frac{p}{x}$$
$$\therefore x^2 = pz$$

11 a ∠A is common. ∠E = ∠C and ∠D = ∠B (corresponding angles)
∴ △ADE is similar to △ABC

b ∠EXD = ∠BXC (vertically opposite)
∠EDX = ∠XCB (alternate)
∠DEX = ∠CBX (alternate)
∴ △DXE is similar to △BXC

c As the triangles in (a) are similar,
$$\frac{AD}{AB} = \frac{ED}{CB}$$
As the triangles in (b) are similar,
$$\frac{ED}{CB} = \frac{EX}{XB}$$
Therefore, $\frac{AD}{AB} = \frac{EX}{XB}$ as required.

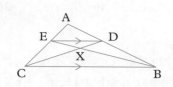

page 233 **Exercise 31 Ⓜ/Ⓗ**

1 16 cm^2 **2** 27 cm^2 **3** $11 \cdot 25 \text{ cm}^2$ **4** $14 \cdot 5 \text{ cm}^2$ **5** 180 cm^2 **6** 12 cm^2

7 8 cm **8** 18 cm **9** $4\frac{1}{2} \text{ cm}$ **10** $7\frac{1}{2} \text{ cm}$ **11** $9, 12, 15 \text{ cm}$

12 a $500\,000 \text{ cm}^2$ **b** 50 m^2 **13** $4\frac{1}{2}$ hours   **14** 150 tiles   **15** 360 tiles   **16** $5r$

17 a $16\frac{2}{3} \text{ cm}^2$ **b** $10\frac{2}{3} \text{ cm}^2$ **18 a** 25 cm^2 **b** 21 cm^2 **19** 24 cm^2 **20** Less

21 $6 \cdot 3 \text{ cm}$ (2 sf)   **22** $\sqrt{2}$ or $1 \cdot 414$ (3 dp)   **23** $\frac{4}{9}$

page 238 **Exercise 32 Ⓜ/Ⓗ**

1 480 cm^3 **2** 540 cm^3 **3** 160 cm^3 **4** 4500 cm^3 **5** 81 cm^3 **6** 11 cm^3

7 16 cm^3 **8** $85\frac{1}{3} \text{ cm}^3$ **9** 4 cm **10** 21 cm **11** $4 \cdot 6 \text{ cm}$ **12** 9 cm

13 $6 \cdot 6 \text{ cm}$ **14** $4 \cdot 5 \text{ cm}$ **15** $168 \cdot 75 \text{ cm}^3$ **16** $106 \cdot 3125 \text{ cm}^3 \left(106\frac{5}{16}\right)$ **17** 12 cm

18 a $2:3$ **b** $8:27$ **19** $8:125$ **20** 60 cm **21** £12·80

22

	On model	On actual ship
Length	42 cm	21 m
Capacity of hold	500 cm³	62·5 m³
Area of sails	700 cm²	175 m²
Number of cannon	12	12
Deck area	370 cm²	92·5 m²

23 54 kg
24 240 cm^2
25 $9 \cdot 375$ litres $\left(9\frac{3}{8}\right)$
26 $2812 \cdot 5 \text{ cm}^2$
27 100 g

page 241 **Test yourself**

1 $7x - 60 = 360$
$x = 60°$

2 $144°$

3 70 cm^2

4 ∠ABC = ∠BCD (alternate angles)
∠CAB = 90° = ∠CDB
(angles in a semi-circle)
Side BC is common.
Therefore, △ABC ≡ △DCB (AAS)

5 First construct an
equilateral triangle,
then bisect one
of the angles.

6

7 $\sqrt{x^2 + (y - 2)^2} = y$

Therefore, $x^2 + (y - 2)^2 = y^2$

Therefore, $x^2 + y^2 - 4y + 4 = y^2$

Therefore, $x^2 = 4y - 4$

Therefore, $4y = x^2 + 4$

Therefore, $y = \frac{1}{4}x^2 + 1$

8 $250\,\text{cm}^2$ **9** $10 \cdot 7\,\text{cm}$ (1 dp) **10** $53 \cdot 8\,\text{cm}^2$ **11** $15 \cdot 9\,\text{cm}$ (1 dp) **12** $26 \cdot 9\,\text{cm}$ (1 dp)

13 $18 + 4\pi\,\text{cm}$ **14** Length $= 22 \cdot 99\,\text{cm}$ (2 dp) Height $= 10\,\text{cm}$ **15** Students' own construction

16 a $160\pi\,\text{cm}^3$ **b** $28\pi\,\text{cm}^2$ **17** $a = 18\,\text{cm}$, $b = 3 \cdot 5\,\text{cm}$ **18** $2129 \cdot 6\,\text{cm}^3$ **19** $7 \cdot 2\,\text{cm}$

20 a $628\,\text{cm}^3$ **b** $0 \cdot 628\,\text{litres}$

5 Algebra 2

page 247 Exercise 1 Ⓜ

1 $x = e - b$ **2** $x = m + t$ **3** $x = a + b + f$ **4** $x = A + B - h$ **5** $x = y$

6 $x = b - a$ **7** $x = m - k$ **8** $x = w + y - v$ **9** $x = \dfrac{b}{a}$ **10** $x = \dfrac{m}{h}$

11 $x = \dfrac{(a + b)}{m}$ **12** $x = \dfrac{(c - d)}{k}$ **13** $x = \dfrac{(e + n)}{v}$ **14** $x = \dfrac{(y + z)}{3}$ **15** $x = \dfrac{r}{p}$

16 $x = \dfrac{(h - m)}{m}$ **17** $x = \dfrac{(a - t)}{a}$ **18** $x = \dfrac{(k + e)}{m}$ **19** $x = \dfrac{(m + h)}{u}$ **20** $x = \dfrac{(t - q)}{e}$

21 $x = \dfrac{(v^2 + u^2)}{k}$ **22** $x = \dfrac{(s^2 - t^2)}{g}$ **23** $x = \dfrac{(m^2 - k)}{a}$ **24** $x = \dfrac{(m + v)}{m}$ **25** $x = \dfrac{(c - a)}{b}$

26 $x = \dfrac{(y - t)}{s}$ **27** $x = \dfrac{(z - y)}{c}$ **28** $x = \dfrac{a}{h}$ **29** $x = \dfrac{2b}{m}$ **30** $x = \dfrac{(cd - ab)}{k}$

31 $x = \dfrac{c}{a} + b$ **32** $x = \dfrac{e}{c} + d$ **33** $x = \dfrac{n^2}{m} - m$ **34** $x = \dfrac{t}{k} + a$ **35** $x = \dfrac{k}{h} + h$

36 $x = \dfrac{n}{m} - b$ **37** $x = 2a$ **38** $x = \dfrac{d}{c} - a$ **39** $x = \dfrac{e}{m} - b$

page 248 Exercise 2 Ⓜ

1 $x = tm$ **2** $x = en$ **3** $x = pa$ **4** $x = tam$ **5** $x = abc$

6 $x = ey^2$ **7** $x = a(b + c)$ **8** $x = t(c - d)$ **9** $x = m(s + t)$ **10** $x = k(h + i)$

11 $x = \dfrac{ab}{c}$ **12** $x = \dfrac{mz}{y}$ **13** $x = \dfrac{hc}{d}$ **14** $x = \dfrac{em}{n}$ **15** $x = \dfrac{hb}{e}$

16 $x = c(a + b)$ **17** $x = m(h + k)$ **18** $x = \dfrac{mu}{y}$ **19** $x = t(h - k)$ **20** $x = (a + b)(z + t)$

21 $x = \dfrac{e}{t}$ **22** $x = \dfrac{e}{a}$ **23** $x = \dfrac{h}{m}$ **24** $x = \dfrac{cb}{a}$ **25** $x = \dfrac{ud}{c}$

26 $x = \dfrac{m}{t^2}$ **27** $x = \dfrac{h}{\sin 20°}$ **28** $x = \dfrac{e}{\cos 40°}$ **29** $x = \dfrac{m}{\tan 46°}$ **30** $x = \dfrac{b^2 c^2}{a^2}$

page 249 **Exercise 3** Ⓜ

1 $x = a - y$
2 $x = h - m$
3 $x = z - q$
4 $x = b - v$
5 $x = k - m$

6 $x = \dfrac{h - d}{c}$
7 $x = \dfrac{y - c}{m}$
8 $x = \dfrac{k - h}{e}$
9 $x = \dfrac{a^2 - d}{b}$
10 $x = \dfrac{m^2 - n^2}{t}$

11 $x = \dfrac{v^2 - w}{a}$
12 $x = y - y^2$
13 $x = \dfrac{k - m}{t^2}$
14 $x = \dfrac{b - e}{c}$
15 $x = \dfrac{h - z}{g}$

16 $x = \dfrac{c - a - b}{d}$
17 $x = \dfrac{v^2 - y^2}{k}$
18 $x = \dfrac{d - h}{f}$
19 $x = b - \dfrac{c}{a}$
20 $x = m - \dfrac{n}{h}$

21 $c - \dfrac{t}{m} = a$
22 $p - \dfrac{w}{v} = a$
23 $q - \dfrac{e}{d} = a$
24 $b^2 - r^2 = a$
25 $a = x - 2f^2$

26 $a = \dfrac{B - ED}{A}$
27 $a = \dfrac{D - BN}{E}$
28 $a = \dfrac{h - bx}{f}$
29 $a = \dfrac{v^2 - dC}{h}$
30 $a = \dfrac{TN}{M} - B$

31 $a = \dfrac{BM + ef}{fN}$
32 $a = M - \dfrac{EF}{T}$
33 $\dfrac{2}{(3y - 1)} = x$
34 $\dfrac{5}{(4z + 2)} = x$
35 $\dfrac{A}{C - B} = x$

36 $\dfrac{V}{H - G} = x$
37 $\dfrac{r}{n + t} = x$
38 $\dfrac{b}{q - d} = x$
39 $\dfrac{m}{t + n} = x$
40 $\dfrac{b}{d - h} = x$

41 $\dfrac{d}{C - e} = x$
42 $\dfrac{m}{r - e^2} = x$
43 $\dfrac{n}{b - t^2} = x$
44 $\dfrac{d}{mn - b} = x$
45 $\dfrac{N}{2M} - P = x$

46 $\dfrac{B}{6A} - c = x$
47 $\dfrac{m^2}{n - p} = x$
48 $\dfrac{q}{w - t} = x$

page 250 **Exercise 4** Ⓜ/Ⓗ

1 $x = \pm\sqrt{\dfrac{h}{c}}$
2 $x = \pm\sqrt{\dfrac{f}{b}}$
3 $x = \pm\sqrt{\dfrac{m}{t}}$
4 $x = \pm\sqrt{\dfrac{a + b}{y}}$

5 $x = \pm\sqrt{\dfrac{t + a}{m}}$
6 $x = \pm\sqrt{a + b}$
7 $x = \pm\sqrt{t - c}$
8 $x = \pm\sqrt{z - y}$

9 $x = \pm\sqrt{a^2 + b^2}$
10 $x = \pm\sqrt{m^2 - t^2}$
11 $x = \pm\sqrt{a^2 - n^2}$
12 $x = \pm\sqrt{\dfrac{c}{a}}$

13 $x = \pm\sqrt{\dfrac{n}{h}}$
14 $x = \pm\sqrt{\dfrac{z + k}{c}}$
15 $x = \pm\sqrt{\dfrac{c - b}{a}}$
16 $x = \pm\sqrt{\dfrac{e + h}{d}}$

17 $x = \pm\sqrt{\dfrac{m + n}{g}}$
18 $x = \pm\sqrt{\dfrac{z - y}{m}}$
19 $x = \pm\sqrt{\dfrac{f - a}{m}}$
20 $x = \pm\sqrt{b^2 - a^2}$

21 $4z^2 = x$
22 $9y^2 + 2 = x$
23 $D^2 - C = x$
24 $\dfrac{c^2 - b}{a} = x$

25 $\dfrac{b^2 + t}{g} = x$
26 $d - t^2 = x$
27 $n - c^2 = x$
28 $c - g^2 = x$

29 $\dfrac{D - B}{A} = x$
30 $\pm\sqrt{g} = x$
31 $\pm\sqrt{B} = x$
32 $\pm\sqrt{A + M} = x$

33 $\pm\sqrt{C - m} = k$
34 $\pm\sqrt{\dfrac{n}{m}} = k$
35 $\dfrac{ta}{z} = k$
36 $\pm\sqrt{a - n} = k$

37 $\pm\sqrt{A + B^2} = k$
38 $\pm\sqrt{t^2 - m} = k$
39 $\dfrac{M^2}{A^2} - B = k$
40 $\dfrac{N}{B^2} = k$

41 $a^2 - t^2 = k$
42 $\dfrac{4}{\pi^2} - t = k$
43 $\pm\sqrt{\dfrac{b + c^2}{a}} = k$
44 $\pm\sqrt{x^2 - b} = k$

page 251 **Exercise 5** Ⓜ/Ⓗ

1 $y = \dfrac{2x + 5p}{3}$
2 $y = 3$
3 $\dfrac{D - B}{2N}$
4 $\dfrac{E + D}{3M}$
5 $\dfrac{2b}{(a - b)}$
6 $\dfrac{e + c}{m + n}$

7 $\dfrac{3}{x + k}$
8 $\dfrac{C - D}{R - T}$
9 $\dfrac{z + x}{a - b}$
10 $\dfrac{nb - ma}{m - n}$
11 $\dfrac{d + bx}{x - 1}$
12 $\dfrac{a - ab}{b + 1}$

13 $\dfrac{d-c}{c+d}$ **14** $\dfrac{M(b-a)}{a+b}$ **15** $\dfrac{n^2-mn}{m+n}$ **16** $\dfrac{m^2+5}{z-m}$ **17** $\dfrac{n^2+2}{n-1}$ **18** $\dfrac{e-b^2}{b-a}$

19 $\dfrac{3x}{a+x}$ **20** $\dfrac{e-c}{a-d}$ **21** $\dfrac{d}{(a-c-b)}$ **22** $\dfrac{ab}{(m+n-a)}$ **23** $\dfrac{t-5}{a-b}$ **24** $2x$

25 $\dfrac{1}{3}V$ **26** $\dfrac{a(b+c)}{b-2a}$ **27** $\dfrac{5x}{3}$ **28** $-\dfrac{4z}{5}$ **29** $\dfrac{mn}{p^2-m}$ **30** $\dfrac{n(m+1)}{4+m}$

page 251 **Exercise 6 M/H**

1 a $a=\dfrac{v-u}{t}$ **b** 2 **2** $x=\dfrac{360A}{\pi r^2}$ **3 a** $k=\dfrac{Py}{m}$ **b** $y=\dfrac{mk}{P}$ **4 a** $n=pR+d$ **b** 1255

5 $r=\sqrt{\dfrac{A}{\pi}}$ **6** $h=\dfrac{V}{\pi r^2}$ **7** $r=\sqrt{\dfrac{A}{4\pi}}$; $r=3\sqrt{\dfrac{3V}{4\pi}}$

page 252 **Exercise 7 M/H**

1 $x=\dfrac{h+d}{a}$ **2** $y=\dfrac{m-k}{z}$ **3** $y=\dfrac{f}{d}-e$ **4** $k=\dfrac{d}{m}-a$ **5** $m=\dfrac{c-a}{b}$

6 $e=\pm\sqrt{\dfrac{b}{a}}$ **7** $t=\pm\sqrt{\dfrac{z}{y}}$ **8** $x=\pm\sqrt{e+c}$ **9** $y=\dfrac{b+n}{m}$ **10** $z=\dfrac{b}{a}-a$

11 $x=\dfrac{a}{d}$ **12** $k=mt$ **13** $u=mn$ **14** $x=\dfrac{y}{d}$ **15** $m=\dfrac{a}{t}$

16 $g=\dfrac{d}{n}$ **17** $t=k(a+b)$ **18** $e=\dfrac{v}{y}$ **19** $y=\dfrac{m}{c}$ **20** $a=\pm\sqrt{bm}$

21 $m=\dfrac{b}{g}-a$ **22** $g=\dfrac{x^2}{h}-h$ **23** $t=y-z$ **24** $e=\pm\sqrt{\dfrac{c}{m}}$ **25** $x=\dfrac{t}{a}-y$

26 $v=\dfrac{y^2+t^2}{u}$ **27** $k=\pm\sqrt{c-t}$ **28** $w=k-m$ **29** $n=\dfrac{b-c}{a}$ **30** $y=\dfrac{c}{m}-a$

31 $x=pq-ab$ **32** $k=\dfrac{a^2-t}{b}$ **33** $z=\dfrac{w}{v^2}$ **34** $u=t-c$ **35** $c=\dfrac{t}{x}$

36 $w=\dfrac{k}{m}-n$ **37** $m=\dfrac{v-t}{x}$ **38** $y=\dfrac{c}{a}-b$ **39** $c=a-\dfrac{e}{m}$ **40** $a=\pm\sqrt{\dfrac{c}{b}}$

41 $p=\dfrac{a}{q}$ **42** $n=\pm\sqrt{\dfrac{a}{e}}$ **43** $f=\pm\sqrt{\dfrac{h}{m}}$ **44** $x=\pm\sqrt{\dfrac{V}{n}}$ **45** $c=\dfrac{v-t^3}{a}$

46 $y=\dfrac{b^3-a^3}{a}$ **47** $h=\pm\sqrt{\dfrac{b+d}{a}}$ **48** $k=\dfrac{bc-h^2}{h}$ **49** $n=\pm\sqrt{u^2-v^2}$ **50** $z=b-\dfrac{b^3}{m}$

page 253 **Exercise 8 M**

1 a $<$ **b** $>$ **c** $>$ **d** $>$ **e** $>$ **f** $>$

2 a $x>2$ **b** $x\leqslant 5$ **c** $x<100$ **d** $-2\leqslant x\leqslant 2$ **e** $x>-6$ **f** $3<x\leqslant 8$

g $4<x<7$ **h** $-5\leqslant x\leqslant 0$ **i** $-1\leqslant x<3$

3

4 a $A\geqslant 16$ **b** $3<A\leqslant 70$ **c** $150<T<175$ **d** $h\geqslant 1{\cdot}75$ **5 a** True **b** True **c** True

6 $7,-6,8{\cdot}5$ **7** $2<x\leqslant 7$ **8** $3\leqslant x<5$

page 255 **Exercise 9** Ⓜ

1 $x > 13$ **2** $x < -1$ **3** $x < 12$ **4** $x \leqslant 2\frac{1}{2}$ **5** $x > 3$ **6** $x \geqslant 8$ **7** $x < \frac{1}{4}$

8 $x \geqslant -3$ **9** $x < -8$ **10** $x < 4$ **11** $x > -9$ **12** $x < 8$ **13** $x > 3$ **14** $x \geqslant 1$

15 $x < 1$ **16** $x > 2\frac{1}{3}$ **17** $x < -3$ **18** $x > 7\frac{1}{2}$ **19** $x > 0$ **20** $x < 0$ **21** $x > 5$

22 $5 \leqslant x \leqslant 9$ **23** $-1 < x < 4$ **24** $5\frac{1}{2} \leqslant x \leqslant 6$

25 $1\frac{1}{3} < x < 8$ **26** $-8 < x < 2$ **27** $\frac{1}{4} < x < 1\frac{1}{5}$ **28** Students' own answer

page 256 **Exercise 10** Ⓜ

1 $x > 8$ **2** 1, 2, 3, 4, 5, 6   **3** 7, 11, 13, 17, 19   **4** 4, 9, 16, 25, 36, 49   **5** $-4, -3, -2, -1$

6 2, 3, 4, 5, 6, 7, 8, 9, 10, 11, 12   **7** 2, 3, 5, 7, 11   **8** 2, 4, 6, 8, 10, 12, 14, 16, 18

9 1, 2, 3, 4   **10** 5   **11 a** 16   **b** -16 **c** 20   **d** -5 **12** $>$ **13** $\frac{1}{2}$ (other answers are possible)

14 19   **15** 17   **16** $x > 3\frac{2}{3}$ **17 b i** $-10 < x < 10$ **ii** $-9 < x < 9$ **iii** $x > 6$ or $x < -6$

18 $-5 < x < 5$ **19** $-4 \leqslant x \leqslant 4$ **20** $x > 1$ or $x < -1$ **21** $x \geqslant 6$ or $x \leqslant -6$

22 all values of x except $x = 0$ **23** $-2 < x < 2$

24 a $2 \leqslant pq \leqslant 40$ **b** $\frac{1}{2} \leqslant \frac{p}{q} \leqslant 10$ **c** $-2 \leqslant p - q \leqslant 9$ **d** $3 \leqslant p + q \leqslant 14$

25 7   **26** 5   **27** 6   **28** 3, 4, 5   **29** $30° < x < 90°$ **30** $0 \leqslant x < 75 \cdot 5°$

page 259 **Exercise 11** Ⓜ/Ⓗ

1 $x \geqslant 3$ **2** $y \leqslant 2\frac{1}{2}$ **3** $1 \leqslant x \leqslant 6$ **4** $x < 7$ and $y < 5$ **5** $y \geqslant x$

6 $x + y \leqslant 10$ **7** $2x - y \leqslant 3$ **8** $x \leqslant 8, y \geqslant -2, y \leqslant x$

9 a $x \geqslant 0, x + y \leqslant 7, y \geqslant x$ **b** $y \geqslant 0, y \leqslant x + 2, x + y \leqslant 6$

10
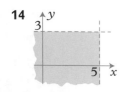

11
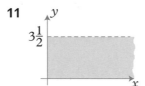

12
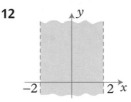

13
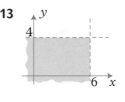

14
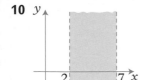

15
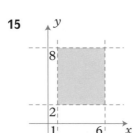

16
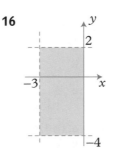

17
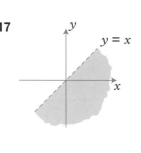

18

19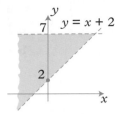

20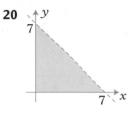

21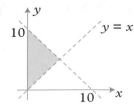

22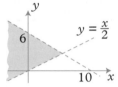

23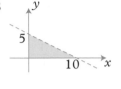

24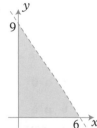

25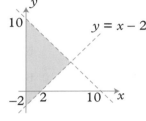

26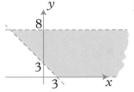

27

28 A $x + y \leqslant 5, y \geqslant x + 1$
B $x + y \leqslant 5, y \leqslant x + 1$
C $x + y \geqslant 5, y \leqslant x + 1$
D $x + y \geqslant 5, y \geqslant x + 1$

29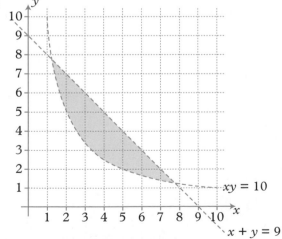

Answers are $(6, 2), (5, 3), (4, 3), (4, 4),$
$(3, 4), (3, 5), (2, 6).$

page 263 **Exercise 12** Ⓗ

1 a $S = ke$ **b** $v = kt$ **c** $x = kz^2$ **d** $y = k\sqrt{x}$ **e** $T = k\sqrt{L}$

2 $k = \frac{3}{2}$ **a** 9   **b** $2\frac{2}{3}$ **3 a** 35   **b** 11   **4 a** 75   **b** ± 4

5

x	1	3	4	$5\frac{1}{2}$
z	4	12	16	22

6

r	1	2	4	$1\frac{1}{2}$
V	4	32	256	$13\frac{1}{2}$

7 $333\frac{1}{3}$ N//cm²   **8** 180 m; 2 secs   **9 a** 675 joules   **b** 1·15 cm

10 a 9000 N   **b** 25 m/s   **11** $p \propto \omega^3$ **12** 50 625 : 1

page 265 **Exercise 13** Ⓗ

1 a $x = k \times \frac{1}{y}$ **b** $s = k \times \frac{1}{t^2}$ **c** $t = k \times \frac{1}{\sqrt{q}}$ **d** $m = k \times \frac{1}{\omega}$ **e** $z = k \times \frac{1}{t^2}$

2 a 6   **b** $\frac{1}{2}$ **3 a** 6   **b** $1\frac{1}{2}$ **4 a** 1   **b** 4

5 a 36   **b** ± 4 **6 a** 6   **b** 16

7

y	2	4	1	$\frac{1}{4}$
z	8	4	16	64

8

t	2	5	20	10
v	25	4	$\frac{1}{4}$	1

9 a 6   **b** 50   **10 a** 2·5 m³   **b** 200 N/m²   **11 a** 3 hours   **b** 48

12 6 cm   **13** 80   **14 a** 2 days   **b** 200 days   **15 a** 20 mins   **b** 2 mins

16

x	1	2	4	10
z	100	$12\frac{1}{2}$	$1\frac{9}{16}$	$\frac{1}{10}$

17

v	1	4	36	10 000
y	12	6	2	$\frac{3}{25}$

page 269 **Exercise 14** Ⓜ/Ⓗ

1 a quadratic, negative x^2 **b** cubic, positive x^3 **c** reciprocal, $y = \frac{k}{x}$ where k is positive

     d cubic, negative x^3 **e** quadratic, positive x^2 **f** exponential, $y = k(a^x)$ where k is positive

2 a 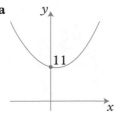   **b** 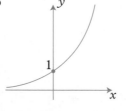   **c**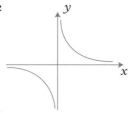

d 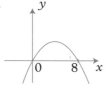   **e**    **f**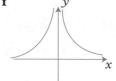

3 i c **ii** b **iii** a **iv** e **v** f **vi** d

4 a 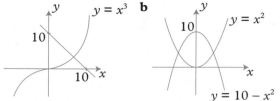 **b** 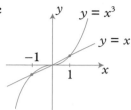 **c** 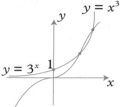 **d**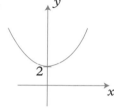

Intersect once   Intersect twice   Intersect 3 times   Intersect twice

page 271 **Exercise 15** Ⓜ

1 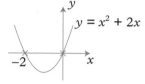 **2**   **3** 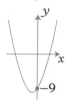 **4**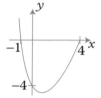

5 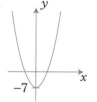 **6** 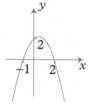 **7**   **8**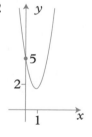

9 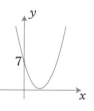 **10** 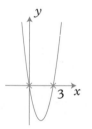 **11** 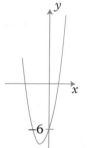 **12**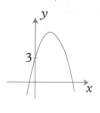

13 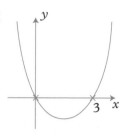 **14** 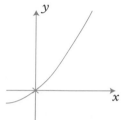 **15** ... **16** ...

17 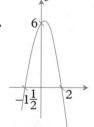 **18** 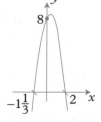 **19** 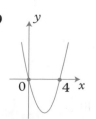 **20**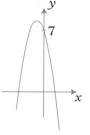

page 272 **Exercise 16** (M)/(H)

1

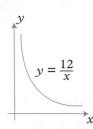

$y = \dfrac{12}{x}$

2

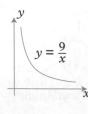

$y = \dfrac{9}{x}$

3

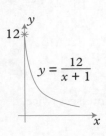

12

$y = \dfrac{12}{x + 1}$

4

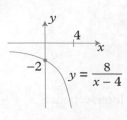

4

−2

$y = \dfrac{8}{x - 4}$

5

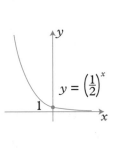

$y = \dfrac{x}{x + 4}$

1

−4

6

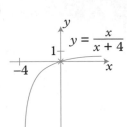

8

$y = \dfrac{x + 8}{x + 1}$

1

7

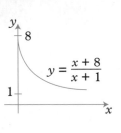

$y = \dfrac{10}{x} + x$

8

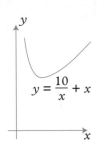

$y = 3^x$

1

9

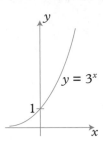

$y = \left(\dfrac{1}{2}\right)^x$

1

10

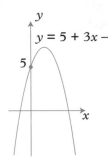

$y = 5 + 3x - x^2$

5

a 7·25
b $x = 3·8$ and $-0·8$

11

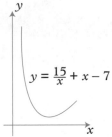

$y = \dfrac{15}{x} + x - 7$

a 0·75
b 1·2

12

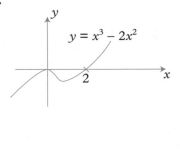

$y = x^3 - 2x^2$

2

a 3·1
b 3·3

13

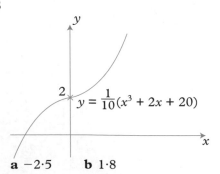

2

$y = \dfrac{1}{10}(x^3 + 2x + 20)$

a −2·5   **b** 1·8

14

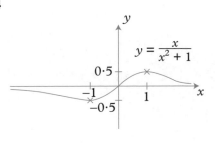

$y = \dfrac{x}{x^2 + 1}$

0·5

−1   1

−0·5

15

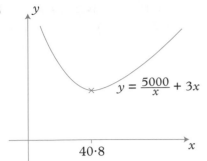

$y = \dfrac{5000}{x} + 3x$

40·8

a 244·9
b 40·8
c $25 < x < 67$

page 274 **Exercise 17** (M)/(H)

1 a $10\cdot7\,\text{cm}^2$ **b** $1\cdot7\,\text{cm} \times 5\cdot3\,\text{cm}$ **c** $12\cdot25\,\text{cm}^2$ **d** $3\cdot5\,\text{cm} \times 3\cdot5\,\text{cm}$ **e** A square
2 a $60 - 2x\,\text{m}$ **b** $x(60 - 2x)\,\text{m}^2$
   c Dome shaped parabola crossing x-axis at $x = 0$ and $x = 30$.   **d** $15\,\text{m} \times 30\,\text{m}$
3 a 2·5 secs   **b** 31·25 m   **c** $2 < t < 3$ seconds

4

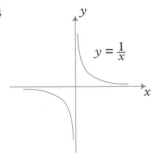

$y = \dfrac{1}{x}$

5

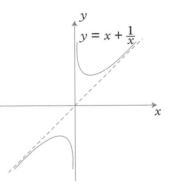

$y = x + \dfrac{1}{x}$

6

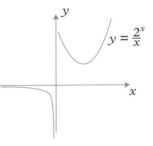

$y = \dfrac{2^x}{x}$

7 a

Number of bacteria

t minutes

b 96 minutes

8

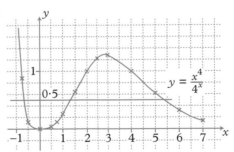

$y = \dfrac{x^4}{4^x}$

a $x = 0$ or $2\cdot9$
b $x = -0\cdot65$, $1\cdot35$ or $5\cdot35$

page 277 **Exercise 18** (H)

1 a $-0\cdot4, 2\cdot4$ **b** $-0\cdot8, 3\cdot8$ **c** $-1, 3$ **d** $-0\cdot4, 2\cdot4$ **2 a** $-2\cdot5, 1\cdot5$ **b** $-0\cdot7, 2\cdot7$ **c** $-2, 2$
3 $-0\cdot3, 3\cdot3$ **4** $0\cdot6, 3\cdot4$ **5** $0\cdot3, 3\cdot7$ **6 c i** $1, 4$ **ii** $0\cdot3, 3\cdot7$
7 a $y = 3$ **b** $y = -2$ **c** $y = x + 4$ **d** $y = x$ **e** $y = 6$
8 a $y = 6$ **b** $y = 0$ **c** $y = 4$ **d** $y = 2x$ **e** $y = 2x + 4$
9 a $y = -4$ **b** $y = 2x$ **c** $y = x - 2$ **d** $y = -3$ **e** $y = 2$
10 b i $-3\cdot3, 0\cdot3$ **ii** $-4\cdot5, 1\cdot5$ **iii** $-3, 1$ **11 a** $-1\cdot6, 3\cdot6$ **b** $-1\cdot3, 2\cdot3$ **c** $-1\cdot4, 3\cdot4$

12 a $1\cdot7, 5\cdot3$ **b** $0\cdot2, 4\cdot8$ **13 a** $-2\cdot4, 0\cdot9$ **b** $-2\cdot8, 1\cdot8$ **14 a** $3\cdot4$ **b** $2\cdot4, 7\cdot7$ **c** $4\cdot2$
15 a 1 **b** 2 **c** 2 **d** 1 **e** 2 **f** 3 **16 a** $-3\cdot8, 3\cdot8$ **c** $-2\cdot8, 2\cdot8$ **17** $0 < k < 4$
18 a $1\cdot7$ **b** $-1\cdot4, 0, 1\cdot4$

Part b is same as $2x - x^3 = 0$
$\therefore x(2 - x^2) = 0$
$\therefore x = 0$ or $x^2 = 2$
i.e $x = 0$ or $\pm\sqrt{2}$

19 a $2\cdot6$ **b** $0\cdot5, 3\cdot3$ **c** $0\cdot6, 5\cdot7$

page 281 **Test yourself**

1 $p = \frac{1}{5}(6 - 3q)$ or $1\frac{1}{5} - \frac{3}{5}q$

2 $x = \frac{1}{8}(6y + 5)$ **3 a** $n > 2$

4 a $x \leqslant 3$ **b** $x > -2$ **c** $-1, 0, 1, 2, 3$

5 $x \leqslant 5, \quad y \geqslant 0, \quad y \leqslant x - 1$

6 a $y = \frac{3\cdot2}{x^2}$ **b** $y = 0\cdot3\dot{5}$

7 a $y = 18x^2$ **b** $y = 4\frac{1}{2}$ **8** $\frac{2}{3}$

9 a

x	-2	-1	0	1	2	3	4
y	11	5	1	-1	-1	1	5

10 a

x	-6	-5	-4	-3	-2	-1	1	2	3	4	5	6
y	-1	-1·2	-1·5	-2	-3	-6	6	3	2	1·5	1·2	1

b

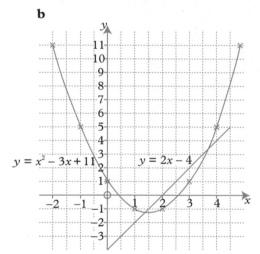

b

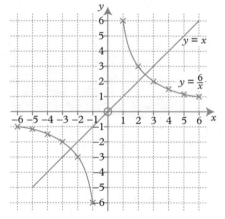

c $-1\cdot25$
d $1\cdot4$ or $3\cdot6$

c If $x^2 = 6$, then dividing each side by x gives $x = \frac{6}{x}$.
Therefore, the solutions to $x^2 = 6$ are where the
graphs of $y = x$ and $y = \frac{6}{x}$ intersect.
d $x = \pm2\cdot45$

11

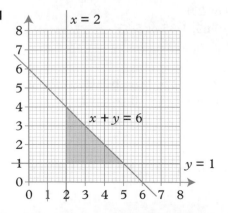

12 a

x	−2	−1	0	1	2	3
y	0·25	0·5	1	2	4	8

b $x = 2·6$

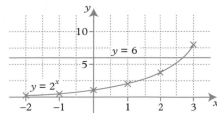

13 a $x > 2$

b

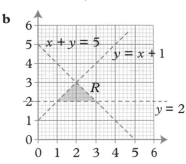

14 a i $y = -x + 13$ **ii** $y = 12·5$ **b** $y = 6·5$

page 284 **Functional Task 4**

Task 1 **1:** 170 cm, **2:** 145 cm, **3:** 150 cm, **4:** 135 cm,
5: 165 cm, **6:** 140 cm, **7:** 150 cm, **8:** 135 cm

Task 2 £8 657·50

6 Shape, space and measures 2

page 287 **Exercise 1**

3

4

5

6

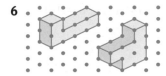

7 Object B

plan view   view from left   view from right

Object C

plan view   view from left   view from right

8

a

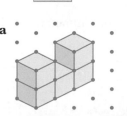

b

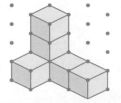

c

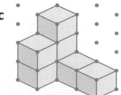

page 289 **Exercise 2**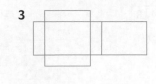

1 **a**, **b** and **d**   **2** For example: **3**

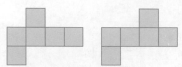

4

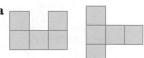

5 **a** Triangular prism, 6 vertices, 5 faces
  b Square-based pyramid, 5 vertices, 5 faces
6 **a** For example: **b** For example:

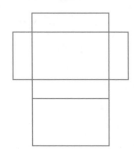

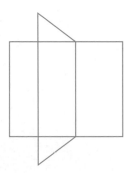

7 **a** 6   **b** 4   **c** $a = 10$, $b = 6$, $c = 10$, $d = 10$ **d** 64 cm^3
8 $a = 1{\cdot}41$, $b = 1{\cdot}41$, $c = 1{\cdot}73$, $x = 1{\cdot}41$, $y = 1{\cdot}73$

page 291 **Exercise 3** Ⓜ

1 **a** 1 **b** 1   **2** **a** 1 **b** 1   **3** **a** 4 **b** 4   **4** **a** 2 **b** 2   **5** **a** 0 **b** 6   **6** **a** 0 **b** 2
7 **a** 0 **b** 2   **8** **a** 4 **b** 4   **9** **a** 0 **b** 4   **10** **a** 4 **b** 4   **11** **a** 6 **b** 6   **12** **a** ∞ **b** ∞
13 **a**   **b**   **15** **a**   **b**

page 292 **Exercise 4** Ⓜ

1 3   **2** **a** 1 **b** 1 **c** 2   **3** **a**   **b** 9   **4** 4 planes of symmetry   **5** 4

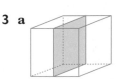

page 293 **Exercise 5** Ⓜ/Ⓗ

2 $\sin x = \dfrac{O}{H} = \dfrac{5}{13}$ **3** $\sin y = \dfrac{O}{H} = \dfrac{6}{10}$ **4** $\sin z = \dfrac{O}{H} = \dfrac{7}{25}$ **5** $\cos w = \dfrac{A}{H} = \dfrac{4}{5}$

6 $\cos x = \dfrac{A}{H} = \dfrac{12}{13}$ **7** $\cos y = \dfrac{A}{H} = \dfrac{8}{10}$ **8** $\tan w = \dfrac{O}{A} = \dfrac{3}{4}$ **9** $\tan x = \dfrac{O}{A} = \dfrac{5}{12}$

10 $\tan y = \dfrac{O}{A} = \dfrac{6}{8}$ **11** $\cos z = \dfrac{A}{H} = \dfrac{24}{25}$ **12** $\tan z = \dfrac{O}{A} = \dfrac{7}{24}$

page 294 **Exercise 6** Ⓜ/Ⓗ

1 3·01 cm **2** 5·35 cm **3** 3·13 cm **4** 7·00 cm **5** 73·1 cm **6** 15·4 cm **7** 5·31 cm
8 7·99 cm **9** 11·6 cm **10** 11·4 cm **11** 961 cm **12** 19·7 cm **13** 46·0 cm **14** 34·9 cm
15 9·39 cm **16** 8·23 cm **17** 35·6 cm **18** 80·2 cm **19** 4·86 cm **20** 6·98 cm

page 296 **Exercise 7** Ⓜ/Ⓗ

1 18·4 cm **2** 9·15 cm **3** 10·7 cm **4** 17·1 cm **5** 13·7 cm **6** 126 cm **7** 6·88 cm
8 11·8 cm **9** 17·6 cm **10** 11·4 cm **11** $x = 5·00$ cm, $y = 5·55$ cm **12** $x = 13·1$ cm, $y = 27·8$ cm
13 4·26 cm **14** 3·50 cm **15** 26·2 cm **16** 8·82 cm

page 297 **Exercise 8** Ⓜ/Ⓗ

1 38·7° **2** 48·6° **3** 31·0° **4** 54·5° **5** 38·7° **6** 17·5° **7** 38·9°
8 59·0° **9** 41·3° **10** 62·7° **11** 54·3° **12** 66·0° **13** 48·2° **14** 12·4°
15 72·9° **16** 56·9° **17** 36·9° **18** 41·8° **19** 78·0° **20** 89·4°

page 299 **Exercise 9** Ⓜ/Ⓗ

1 68·0° **2** 3·65 m **3** 14 m **4** 20·6° **5** 56·7° **6** 15 m **7** 90 cm
8 4·32 cm **9** 7·66 cm **10** 66 km **11** 189 km **12** 25·7 km **13** 180 km **14** 37 m
15 36·4° **16** 10·3 cm **17** 72°, 8·23 cm **18** 71·1°

page 302 **Exercise 10** Ⓜ/Ⓗ

1 91·2° **2** 1·29 m to 2·11 m **3** 2·6 m **4 a** 26 km **b** 23 km
5 a 89 km **b** 179 km **6 a** 484 km **b** 858 km **c** 985 km, 061° **7** 954 km, 133°
8 76·5 m/s **9 a** 11·1 m **b** 11·1 s **c** 222 m **10** 2·9 m **11** 4·4 m
12 a $\sin A = 0·8$, $\cos A = 0·6$, $(\sin A)^2 + (\cos A)^2 = 1$
       $\sin A = 0·385$, $\cos A = 0·923$, $(\sin A)^2 + (\cos A)^2 = 1$
       $\sin A = 0·447$, $\cos A = 0·894$, $(\sin A)^2 + (\cos A)^2 = 1$
   b In each case $(\sin A)^2 + (\cos A)^2 = 1$.   **c** Yes
13 a $\dfrac{360°}{n}$ **b** $\dfrac{180°}{n}$ **c** $MB = \sin\left(\dfrac{180°}{n}\right)$, $AB = 2\sin\left(\dfrac{180°}{n}\right)$ **d** $2n \sin\left(\dfrac{180°}{n}\right)$
   e 6·282, 6·283; perimeter $\approx 2\pi$ as the polygon is almost a circle.
14 3·13 m **15** 25·6 cm **16** 27 cm

page 306 **Exercise 11** Ⓗ

1 a 13 cm **b** 13·6 cm **c** 17·1° **2 a** 4 m **b** 38·9° **c** 11·2 m **d** 19·9°
3 a 8·49 cm **b** 8·49 cm **c** 10·4 cm **d** 35·3° **4 a** 14·1 cm **b** 18·7 cm **c** 69·3°
5 a AC = 11·3 cm, AY = 12·4 cm **b** 23·8° **6 a** ZX = 10 cm, KX = 9·4 cm **b** 26·6° **c** 32·5°
7 a 4·47 cm **b** 7·48 cm **c** 15·5° **8** 58·0° **9 a** 57·5° **b** 61·0°
10 a $\dfrac{h}{\tan 25°}$ **b** $\dfrac{h}{\tan 33°}$ **c** 22·7 m **11** 22·6 m **12** 55 m

page 308 **Exercise 12** Ⓜ

1

2

3 a–b

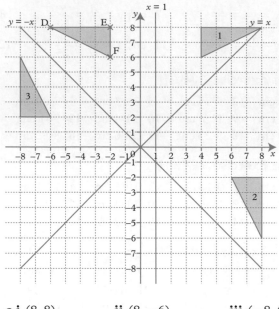

c i $(8, 8)$ **ii** $(8, -6)$ **iii** $(-8, 6)$

4 a–e

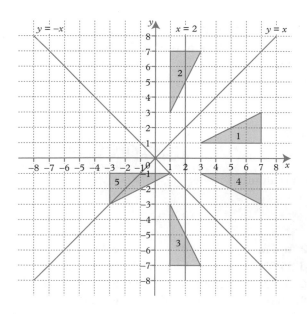

f $(-3, -1), (-3, -3), (1, -1)$

5 a–e

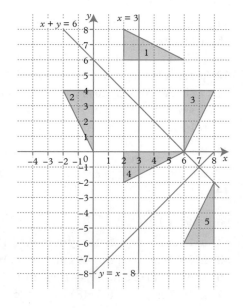

f $(8, -2), (8, -6), (6, -6)$

6 a–b

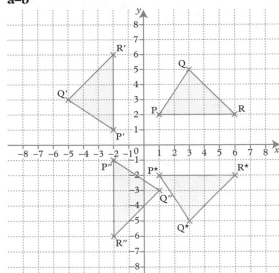

c P′(−2, 1), P″(−2, −1), P⋆(1, −2)

7 a–d

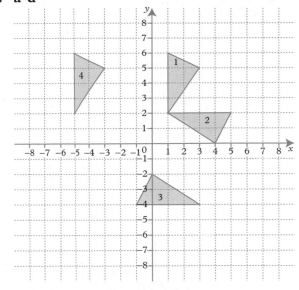

e (−5, 2), (−5, 6), (−3, 5)

8 a

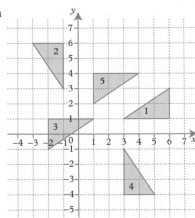

b i   Rotation 90° anticlockwise, centre (0, 0)
  ii   Rotation 180°, centre (2, 1)
  iii Rotation 90° clockwise, centre (2, 0)
  iv   Rotation 180°, centre $\left(3\frac{1}{2}, 2\frac{1}{2}\right)$
  v   Rotation 90° anticlockwise, centre (6, 1)
  vi   Rotation 90° clockwise, centre (1, 3)

page 310 **Exercise 13 Ⓜ/Ⓗ**

1 a $\begin{pmatrix} 7 \\ 3 \end{pmatrix}$ **b** $\begin{pmatrix} 0 \\ -9 \end{pmatrix}$ **c** $\begin{pmatrix} 9 \\ 10 \end{pmatrix}$ **d** $\begin{pmatrix} -10 \\ 3 \end{pmatrix}$ **e** $\begin{pmatrix} -1 \\ 13 \end{pmatrix}$ **f** $\begin{pmatrix} 10 \\ 0 \end{pmatrix}$ **g** $\begin{pmatrix} -9 \\ -4 \end{pmatrix}$ **h** $\begin{pmatrix} -10 \\ 0 \end{pmatrix}$

2

3

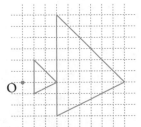

4

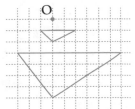

5

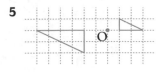

6
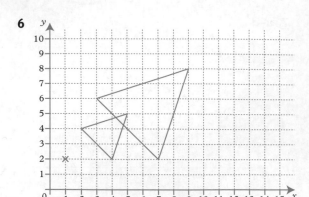

7
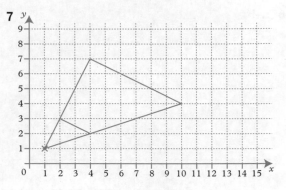

8
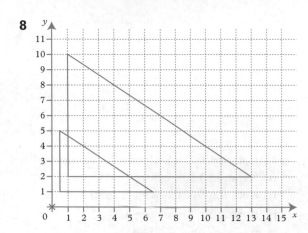

9
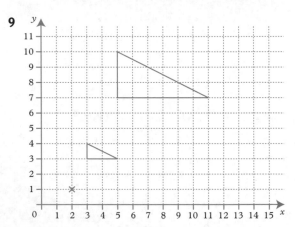

10
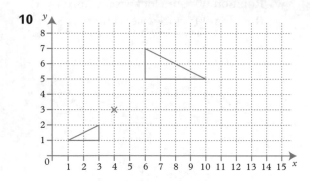

11
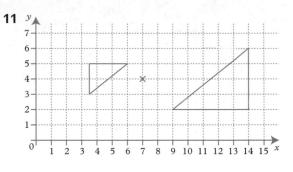

12 a **iii** **b** $1\frac{1}{2}$

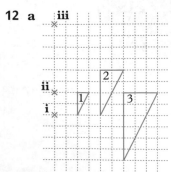

13 a Enlargement scale factor 2, centre $(6, 1)$
   b Enlargement scale factor 3, centre $(7, 1)$
   c Enlargement scale factor 4, centre $(4, 1)$
   d Enlargement scale factor 2, centre $(0, 1)$
   e Enlargement scale factor $\frac{2}{3}$, centre $(10, 1)$
   f Enlargement scale factor $\frac{1}{4}$, centre $(4, 1)$

g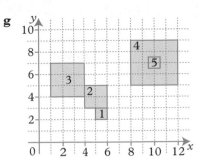

14 a Rotation 90° clockwise, centre $(0, -2)$ **b** Reflection in $y = x$
   c Translation $\begin{pmatrix} 3 \\ 7 \end{pmatrix}$ **d** Enlargement scale factor 2, centre $(-5, 5)$
   e Translation $\begin{pmatrix} -7 \\ -3 \end{pmatrix}$ **f** Reflection in $y = x$

15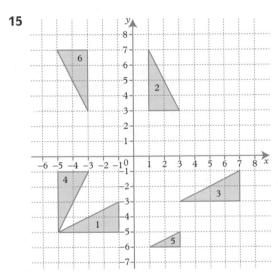

   a Rotation 90° clockwise, centre $(0, 0)$
   b Translation $\begin{pmatrix} 8 \\ 2 \end{pmatrix}$
   c Reflection in $y = x$
   d Enlargement scale factor $\frac{1}{2}$, centre $(7, -7)$
   e Rotation 90° anticlockwise, centre $(-8, 0)$
   f Enlargement scale factor 2, centre $(-1, -9)$
   g Rotation 90° anticlockwise, centre $(7, 3)$

16

   a Enlargement scale factor $1\frac{1}{2}$, centre $(1, -4)$
   b Rotation 90° clockwise, centre $(0, -4)$
   c Reflection in $y = -x$
   d Translation $\begin{pmatrix} 11 \\ 10 \end{pmatrix}$
   e Enlargement scale factor $\frac{1}{2}$, centre $(-3, 8)$
   f Rotation 90° anticlockwise, centre $\left(\frac{1}{2}, 6\frac{1}{2}\right)$
   g Enlargement scale factor 3, centre $(-2, 5)$

page 313 **Exercise 14** Ⓜ/Ⓗ

1 a–b

c Reflection in $x = 4$

2 a–b

c Rotation 90° clockwise, centre $(0, 0)$

3

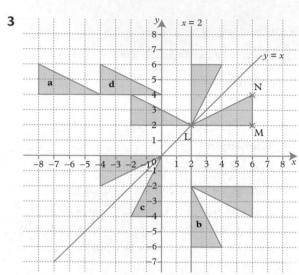

a $(-4, 4)$ **b** $(2, -2)$ **c** $(0, 0)$ **d** $(0, 4)$

4

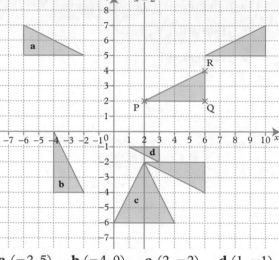

a $(-2, 5)$ **b** $(-4, 0)$ **c** $(2, -2)$ **d** $(1, -1)$

5

a Reflection in $x = 0$
b Rotation 180°, centre $(-2, 2)$
c Rotation 90° clockwise, centre $(2, 2)$

6

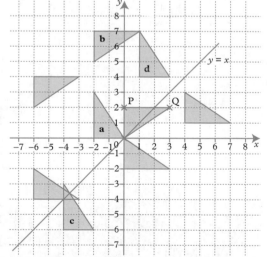

a Rotation 90° anticlockwise, centre $(0, 0)$
b Translation $\begin{pmatrix} -2 \\ 5 \end{pmatrix}$
c Rotation 90° anticlockwise, centre $(2, -4)$
d Rotation 90° anticlockwise, centre $\left(-\frac{1}{2}, 3\frac{1}{2}\right)$

7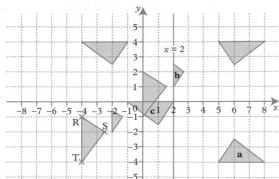

a Rotation 90° anticlockwise, centre $(2, 2)$

b Enlargement scale factor $\frac{1}{2}$, centre $(8, 6)$

c Rotation 90° clockwise, centre $\left(-\frac{1}{2}, -3\frac{1}{2}\right)$

8 A Reflection in $x = 2$ **B** Rotation 180°, centre $(1, 1)$ **C** Translation $\begin{pmatrix} 6 \\ -2 \end{pmatrix}$

D Reflection in $y = x$ **E** Reflection in $y = 0$ **F** Translation $\begin{pmatrix} -4 \\ -3 \end{pmatrix}$

G Rotation 90° anticlockwise, centre $(0, 0)$ **H** Enlargement scale factor 2, centre $(0, 0)$

page 317 **Exercise 15 Ⓗ**

1 d   **2** 2c   **3** 3c   **4** 3d   **5** 5d   **6** 3c   **7** −2d
8 −2c   **9** −3c   **10** −c   **11** c + d   **12** c + 2d   **13** 2c + d   **14** 3c + d
15 2c + 2d   **16** $\overrightarrow{QI}$ **17** $\overrightarrow{QU}$ **18** $\overrightarrow{QH}$ **19** $\overrightarrow{QB}$ **20** $\overrightarrow{QF}$ **21** $\overrightarrow{QJ}$
22 a 2a + b **b** 2a + 2b **c** −a − b **d** 4a + 2b **e** 2a − 2b   **f** 2a + b
23 a $\overrightarrow{CO}$ **b** $\overrightarrow{TN}$ **c** $\overrightarrow{FT}$ **d** $\overrightarrow{KC}$ **24 a** −a   **b** a + b   **c** 2a − b **d** b − a
25 a a + b **b** a − 2b **c** b − a **d** −a − b **26 a** −a − b **b** 3a − b **c** 2a − b **d** b − 2a
27 a a − 2b **b** a − b   **c** 2a   **d** 3b − 2a **28 a** or d   **29** A and D, B and E, C and F

page 319 **Exercise 16 Ⓗ**

1 a a   **b** b − a   **c** 2b   **d** −2a   **e** 2b − 2a   **f** b − a
  g a + b   **h** b   **i** 2a − b   **j** a − 2b
2 a a   **b** b − a   **c** 3b   **d** −2a   **e** 3b − 2a   **f** $1\frac{1}{2}$b − a
  g a + $1\frac{1}{2}$b   **h** $1\frac{1}{2}$b   **i** 2a − b   **j** a − 3b
3 a 2a   **b** b − a   **c** 2b   **d** −3a   **e** 2b − 3a   **f** b − $1\frac{1}{2}$a
  g $1\frac{1}{2}$a + b   **h** $\frac{1}{2}$a + b   **i** 3a − b   **j** a − 2b
4 a $\frac{1}{2}$a   **b** b − a   **c** 4b   **d** −$1\frac{1}{2}$a   **e** 4b − $1\frac{1}{2}$a   **f** $2\frac{2}{3}$b − a
  g $\frac{1}{2}$a + $2\frac{2}{3}$b   **h** $2\frac{2}{3}$b − $\frac{1}{2}$a   **i** $1\frac{1}{2}$a − b   **j** a − 4b
5 $\frac{1}{2}$s − $\frac{1}{2}$t   **6** $\frac{1}{3}$a + $\frac{2}{3}$b **7** a − b + c **8** 2m + 2n
9 a b − a   **b** b − a   **c** 2b − 2a   **d** b − 2a   **e** b − 2a   **f** 2b − 3a
10 a y − z   **b** $\frac{1}{2}$y − $\frac{1}{2}$z   **c** $\frac{1}{2}$y + $\frac{1}{2}$z   **d** −x + $\frac{1}{2}$y + $\frac{1}{2}$z   **e** −$\frac{2}{3}$x + $\frac{1}{3}$y + $\frac{1}{3}$z **f** $\frac{1}{3}$x + $\frac{1}{3}$y + $\frac{1}{3}$z

11 a i $2\mathbf{b} - 2\mathbf{a}$ **ii** $2\mathbf{c} - 2\mathbf{b}$ **iii b** **iv** $\mathbf{c} - \mathbf{a}$ **v** $\mathbf{c} - \mathbf{a}$
   b Parallel lines of equal length **c** Parallelogram
12 a i $\mathbf{a} - \mathbf{b}$ **ii** $\frac{1}{3}\mathbf{a} - \frac{1}{3}\mathbf{b}$ **iii** $\frac{2}{3}\mathbf{b} - \frac{1}{6}\mathbf{a}$ **iv** $2\mathbf{b} - \frac{1}{2}\mathbf{a}$
   b $\overrightarrow{CE} = 3\overrightarrow{CD}$, so CE and CD are parallel. They have a common point, C, so
   C, D, E lie on the same straight line.

page 323 **Exercise 17** Ⓗ

1 a 1, 0·866, 0·5, 0, −0·5, −0·866, −1, −0·866, −0·5, 0, 0·5, 0·866, 1
   b Accurate graph of $y = \cos x$
2 Accurate graph of $y = \sin x$
3 0, 0·364, 0·839, 1·732, 5·671, −5·671, −1·732, −0·839, −0·364, 0,
   0·364, 0·839, 1·732, 5·671, −5·671, −1·732, −0·839, −0·364, 0
   a The tangent is very large and positive to the left of these points, then
   very large and negative to the right. Tangent is undefined
   at 90° and 270°.
   b–c Accurate graph of $y = \tan x$ with asymptotes.
4 cos 60° = cos 300°, sin 50° = sin 130°, sin 70° = sin 110°
5 tan 45° = tan 225°, sin 180° = cos 270°, cos 30° = cos 330°, tan 60° = tan 240°
6 162° **7** 153° **8 a** 140° **b** 110° **c** 50° **9** 290° **10** 315°
11 a 350° **b** 304° **c** 60°
12 220° **13** 160° **14** 82° **15** 315° **16** 240° **17** 250°
18 sin 270° = cos 180°, sin 90° = cos 360°, tan 20° = tan 200°, tan 150° = tan 330°

page 325 **Exercise 18** Ⓗ

1 58·0°, 122·0° **2** 20·5°, 159·5° **3** 53·1°, 306·9° **4** 45°, 225°
5 a 19·8°, 160·2° **b** 72·0°, 108·0° **c** 30°, 150° **d** 60°, 120°
6 a 46·1°, 133·9° **b** 72·5°, 287·5° **c** 78·7°, 258·7° **d** 220·5°, 319·5°
7 30°, 150°, 210°, 330° **8** 45°, 135°, 225°, 315°

9

	30°	60°	120°	150°
sin	$\frac{1}{2}$	$\frac{\sqrt{3}}{2}$	$\frac{\sqrt{3}}{2}$	$\frac{1}{2}$
cos	$\frac{\sqrt{3}}{2}$	$\frac{1}{2}$	$-\frac{1}{2}$	$-\frac{\sqrt{3}}{2}$
tan	$\frac{1}{\sqrt{3}}$	$\sqrt{3}$	$-\sqrt{3}$	$-\frac{1}{\sqrt{3}}$

10 a $0° < x < 180°$ **b** $90° < x < 270°$
11 Accurate graphs of $\sin\theta$, $\cos\theta$ and $\tan\theta$.
12 a For example, −270°, 90°, 450° **b** For example, −270°, −90°, 90°, 270°
13 a 0 **b** 1
14 a $\cos\theta = \frac{b}{c}$, $\tan\theta = \frac{a}{b}$ $\sin\theta \div \cos\theta = \frac{a}{c} \div \frac{b}{c} = \frac{a}{b}$
   b i −1, −1 **ii** 1·732, 1·732 **iii** −0·839, −0·839
15 sin 130° **16 a** 45°, 225° **b** 30°, 60°, 210°, 240° **c** 15°, 75°, 135°, 195°, 255°, 315°

17

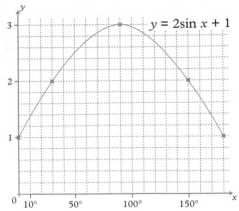

$y = 2\sin x + 1$

a 41°, 139°
b 30°, 150°
c 63°

18

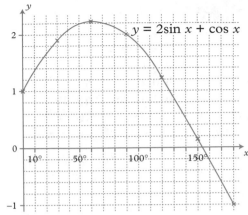

$y = 2\sin x + \cos x$

a i 15°, 111°
   ii 153°
b 2·24

19

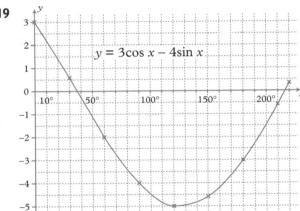

$y = 3\cos x - 4\sin x$

a 48°, 205°
b 37°, 217°

page 328 **Exercise 19** (H)

 1 6·38 m   **2** 12·5 m   **3** 5·17 cm   **4** 40·4 cm   **5** 7·81 m   **6** 6·68 m   **7** 8·61 cm
 8 9·97 cm   **9** 8·52 cm   **10** 15·2 cm

page 329 **Exercise 20** (H)

 1 35·8°   **2** 42·9°   **3** 32·3°   **4** 37·8°   **5** ∠R = 35·5°, ∠T = 48·5°   **6** 68·8° **7** 64·6°
 8 34·2°   **9** 50·6°   **10** 39·1°   **11** 39·5°   **12** 21·6°   **13** 72·5°, 107·5°

page 331 **Exercise 21** (H)

 1 6·24 cm   **2** 6·05 cm   **3** 5·47 cm   **4** 9·27 cm   **5** 10·1 cm   **6** 8·99 cm   **7** 5·87 cm
 8 4·24 cm   **9** 11·9 cm   **10** 154 cm   **11** 25·2°   **12** 71·4°   **13** 115·0°   **14** 111·1°
 15 24·0°   **16** 92·5°   **17** 99·9°   **18** 38·2°   **19** 137·8°   **20** 34·0°

page 333 **Exercise 22** Ⓗ

1 a 50·2 km   **b** 055°   **2** 35·6 km   **3** 25·2 km

4 $a = 10·3$ cm, $b = 11·6$ cm, $c = 4·86$ cm, $d = 37·2°$, $e = 94·1°$, $f = 42·6°$ **5** 92·9°   **6** 40·4 m

7 54·7 m   **8** 14·5 cm   **9 a** 9·85 km   **b** 086°

10 a 29·6 km   **b** 050°   **11** 141 km   **12 a** 10·8 m   **b** 72·6°   **c** 32·6°   **13** 048°, 378 km

14 a 62·2°   **b** 2·33 km   **15** 101°   **16** 8·85 m

17 a $c^2 + 36 - 6c = 28$ **b** $c = 2$ or 4 **c** The sine rule gives the same answers.

18 a $\cos C = \frac{x}{b} \Rightarrow x = b \cos C$ **b** $h^2 = b^2 - x^2$ **c** $c^2 = h^2 + (a - x)^2$

   d $h^2 = c^2 - (a - x)^2$ (part **c**)
         $= c^2 - a^2 - x^2 + 2ax$
   and $h^2 = b^2 - x^2$ (part **b**)
   So $c^2 - a^2 - x^2 + 2ax = b^2 - x^2$
   $\Rightarrow c^2 = a^2 + b^2 - 2ax$
   $\Rightarrow c^2 = a^2 + b^2 - 2ab \cos C$ (using $x = b \cos C$ from part **a**)

19 9·64 m   **20** 48·2°   **21 a** 5·66 cm   **b** 4·47 cm   **c** 3·74 cm   **22** 70·2°

page 338 **Exercise 23** Ⓜ/Ⓗ

1 $a = 27°, b = 30°$ **2** $c = 20°, d = 45°$ **3** $c = 58°, d = 41°, e = 30°$ **4** $f = 40°, g = h = 55°$

5 $a = 32°, b = 80°, c = 43°$ **6** $y = 34°$ **7** $t = 43°$ **8** $a = 92°$

9 $b = 42°$ **10** $c = 46°, d = 44°$ **11** $e = 49°, f = 41°$ **12** $g = 76°, h = 52°$

13 $x = 48°$ **14** $y = 32°$ **15** $x = 22°$ **16** $a = 36°, x = 36°$

page 340 **Exercise 24** Ⓜ/Ⓗ

1 $a = 94°, b = 75°$ **2** $c = 101°, d = 84°$ **3** $x = 92°, y = 116°$ **4** $c = 60°, d = 45°$

5 $h = 37°$ **6** $m = 118°$ **7** $e = 36°, f = 72°$ **8** $y = 35°$

9 $x = 18°$ **10** $m = 90°$ **11** $a = 30°$ **12** $x = 22·5°$

13 $n = 58°, t = 64°, w = 45°$ **14** $a = 32°, b = 40°, c = 40°$

15 $a = 18°, c = 72°$ **16** $x = 55°$ **17** $e = f = g = 41°$ **18** $z = 8°$

19 $x = 30°, y = 115°$ **20** $x = 80°, z = 10°$

page 342 **Exercise 25** Ⓜ/Ⓗ

1 $a = 18°$ **2** $b = 53°$ **3** $c = 77°$ **4** $x = 40°, y = 65°, z = 25°$

5 $c = 30°, e = 15°$ **6** $f = 50°, g = 40°$ **7** $h = 70°, i = 40°, k = 40°$ **8** $m = 108°, n = 36°$

9 $x = 50°, y = 68°$ **10** $a = 74°, b = 32°$ **11** $e = 36°$ **12** $k = 63°, m = 54°$

13 $k = 50°, m = 50°, n = 80°, p = 80°$ **14 a** p **b** $2p$ **c** $90 - 2p$

15 $x = 70°, y = 20°, z = 55°$ **16** $\angle$RQP = 90° $\angle$PRQ = a

page 344 **Exercise 26** Ⓜ/Ⓗ

1 $a = 137°$ **2** $b = 120°$ **3** $c = d = 30°, e = 27°$ **4** $f = 49°$

5 $g = 18°$ **6** $h = 110°$ **7** $i = 52°, j = 128°$ **8** $k = 45°, l = 135°$

9 $m = 35°, n = 50°$ **10** $p = 69°$ **11** $q = 78°$ **12** $r = 200°, s = 100°$

13 $a = 45°$ **14** $b = 30°, c = 60°$ **15** $d = 72°$ **16** $e = 19°$ **17** $f = 50°$

18 $g = 58°, h = 90°$ **19** $i = 45°$ **20** $j = 96°, k = 68°$ **21** $m = 28°$

page 345 **Test yourself**

1

2 a 24·1 cm (l.d.p.)   **b** 14·7 cm (l.d.p.)

3 a $\sqrt{9^2 + 4^2 + 5^2} = \sqrt{122} = 11\cdot05$ cm (2 dp)   **b** 26·9°

4 a 30·7°   **b** 120·7°   **5** Rotation 180° about (1,0)

6 a 12·9 m (l.d.p.)   **b** 21·0° (l.d.p.)

7 a Enlargement with scale factor $\frac{1}{3}$, centre (2, 1)   **b** 4 cm^2

8 Rotation 90° clockwise about point (2, 1)

9 a, b, c

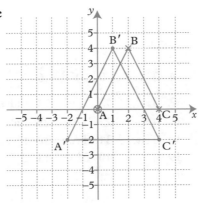

10 a 3q   **b** −p + 2q

11 a i a + b
b i

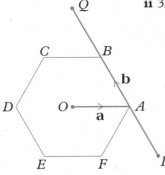

ii b − a   **iii** a + 2b
ii 3b

12 a a + $1\frac{1}{2}$b   **b** a − $\frac{1}{2}$b

c $\frac{1}{2}$a − $\frac{1}{4}$b   **d** $\frac{1}{2}$a + $\frac{3}{4}$b

13 a BD = 8 sin54° = 6·47 cm^2 (2 dp)
b 20·9 cm^2 **c** 6·71 cm

14 a i 114°, 246°   **ii** 41°, 319°

15 a 72 cm   **b** 129.8 cm
16 114·8° (l.d.p)

b

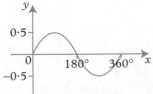

17 a $\angle ABC = \frac{1}{2}$ angle at centre $= 54°$

$\angle ADC = 180° - \angle ABC$ (opposite angles of a cyclic quadilateral)

$\therefore \angle ADC = 180° - 54° = 126°$ as required

b $\angle CAO = \frac{1}{2}(180° - 108°)$ as $\triangle OAC$ is isosceles

$= 36°$

$\angle OAE = 90°$ (Tangent and radius)

$\therefore \angle CAE = 90° - 36° = 54°$

18 a $47°$ **b** $43°$ **19 a** $34°$ **b** $34°$ **c** $17°$ **d** $73°$ **e** $56°$ **20** $13{\cdot}95$ cm

7 Algebra 3

page 354 **Exercise 1** (H)

1 3^4 **2** $4^2 \times 5^3$ **3** 3×7^3 **4** $2^3 \times 7$ **5** 10^{-3} **6** 5^{-4} **7** $15^{\frac{1}{2}}$ **8** $1000^{\frac{1}{2}}$

9 a^8 **10** $17^{\frac{1}{3}}$ **11** 11^{-1} **12** $5^{\frac{3}{2}}$ **13** True **14** False **15** True **16** True **17** False

18 False **19** False **20** True **21** True **22** True **23** False **24** True **25** False **26** True

27 True **28** True **29** True **30** True **31** True **32** False **33** 5^6 **34** 6^5 **35** 2^{13}

36 3^4 **37** 6^5 **38** 8 **39** 6^4 **40** 2^5 **41** 2^{19} **42** 10^{100} **43** 5^{-5} **44** 3^{-10}

45 3^7 **46** 2^7 **47** 7^2 **48** 5^{-1}

page 355 **Exercise 2** (H)

1 27 **2** 1 **3** $\frac{1}{9}$ **4** 25 **5** 2 **6** 4 **7** 9 **8** 2 **9** 27

10 3 **11** $\frac{1}{3}$ **12** $\frac{1}{2}$ **13** 1 **14** $\frac{1}{5}$ **15** 10 **16** 8 **17** 32 **18** 4

19

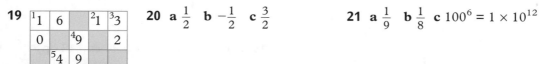

¹1	6		²1	³3
0		⁴9		2
	⁵4	9		
⁶8	0		⁷9	⁸6
⁹1	2	5		4

20 a $\frac{1}{2}$ **b** $-\frac{1}{2}$ **c** $\frac{3}{2}$

21 a $\frac{1}{9}$ **b** $\frac{1}{8}$ **c** $100^6 = 1 \times 10^{12}$

22 10 **23** 1000 **24** $\frac{1}{1000}$ **25** $\frac{1}{9}$ **26** 1 **27** $1\frac{1}{2}$

28 $\frac{1}{25}$ **29** $\frac{1}{10}$ **30** $\frac{1}{4}$ **31** $\frac{1}{4}$ **32** $100\,000$ **33** 1

34 $\frac{1}{32}$ **35** $\frac{1}{10}$ **36** $\frac{1}{5}$ **37** $1\frac{1}{2}$ **38** 1 **39** 9

40 $1\frac{1}{2}$ **41** $\frac{3}{10}$ **42** 64 **43** $\frac{1}{100}$ **44** $1\frac{2}{3}$ **45** $\frac{1}{100}$

46 a $\frac{1}{n}$

page 357 **Exercise 3** (H)

1 $A = F$; $B = D$; $C = H$; $E = G$

2 a True **b** True **c** False **d** True **e** False **f** True

3 **a** $3 + 2\sqrt{2}$ **b** $7 - 4\sqrt{3}$ **c** 18 **d** $14 - 6\sqrt{5}$ **e** $6 + 4\sqrt{2}$ **f** 8

4 **a** True **b** True **c** False **d** True **e** False **f** True

5 **a** 4 **b** $4\sqrt{2}$ **c** $10\sqrt{3}$ **d** $9\sqrt{2}$ **e** $5\sqrt{5}$ **f** $5\sqrt{3}$

 g $5\sqrt{5}$ **h** $\sqrt{3}$ **i** 2 **j** $\frac{3}{2}$ **k** $\frac{5}{2}$ **l** $\frac{4}{3}$

6 **a** $2\sqrt{3}$ **b i** $4\sqrt{2}$ **ii** $4\sqrt{3}$ **iii** $3\sqrt{2}$ **iv** $\frac{2\sqrt{3}}{3}$

7 **a** $4(\sqrt{2}-1)$ **b i** $2(\sqrt{3}-1)$ **ii** $\sqrt{5}-2$ **iii** $\frac{10(\sqrt{7}-2)}{3}$

page 358 Exercise 4 Ⓗ

1 3^6 **2** 5^{12} **3** 7^{10} **4** x^6 **5** 2^{-2} **6** 1 **7** 7^2 **8** y^2 **9** 2^{21}

10 10^{99} **11** x^7 **12** y^{13} **13** z^4 **14** z^{100} **15** m **16** e^{-5} **17** y^2 **18** w^6

19 y **20** x^{10} **21** 1 **22** w^{-5} **23** w^{-5} **24** x^7 **25** a^8 **26** k^3 **27** 1

28 x^{29} **29** y^2 **30** x^6 **31** z^4 **32** t^{-4} **33** $4x^6$ **34** $16y^{10}$ **35** $6x^4$ **36** $10y^5$

37 $15a^4$ **38** $8a^3$ **39** 3 **40** $4y^2$ **41** $2{\cdot}5y$ **42** $32a^4$ **43** $108x^5$ **44** $4z^{-3}$ **45** $2x^{-4}$

46 $\frac{5}{2}y^5$ **47** 1 **48** $21w^{-3}$ **49** $2n^4$ **50** $2x$

page 359 Exercise 5 Ⓗ

1 $x = 3$ **2** $x = 4$ **3** $x = -1$ **4** $x = -2$ **5** $x = 3$ **6** $x = 3$

7 $x = 1$ **8** $x = \frac{1}{5}$ **9** $x = 0$ **10** $x = -4$ **11** $x = 2$ **12** $x = -5$

13 $x = 1$ **14** $x = \frac{1}{18}$ **15** **a** $(0, 1)$ **b** $x = 0$ **16** **a** $a = -3$ **b** $b = 1$ **c** $c = 5$

17 **a** $p = 2\frac{1}{2}$ **b** $q = 1000$ **c** $r = 1\frac{3}{4}$ **18** **a** $n = 4$ **b** $n = -1$ **c** $n = 0$

19 $x = 2$ or -4 **20** **a** $x = 3{\cdot}60$ **b** $x = 5{\cdot}44$

21 **a**

n	1	2	3	4	5	6	7	8	9	10
Last digit of 7^n	7	9	3	1	7	9	3	1	7	9

b i 1 **ii** 7 **iii** 3

22 21 **23** **a** 512 **b** 6 hours **c** 2^{21} **24** **a** 976 **b** 20 hours

25 **a** $2519{\cdot}42$ **b** 9 years **26** $x^6 = (x^3)^2 = (-1)^2 = 1$; $x^7 = xx^6 = x$; $x^{60} = (x^6)^{10} = 1$; $x^{61} = xx^{60} = x$

page 362 Exercise 6 Ⓜ/Ⓗ

1 $(x + 2)(x + 5)$ **2** $(x + 3)(x + 4)$ **3** $(x + 3)(x + 5)$ **4** $(x + 3)(x + 7)$

5 $(x + 2)(x + 6)$ **6** $(y + 5)(y + 7)$ **7** $(y + 3)(y + 8)$ **8** $(y + 5)^2$

9 $(y + 3)(y + 12)$ **10** $(a - 5)(a + 2)$ **11** $(a - 4)(a + 3)$ **12** $(z + 3)(z - 2)$

13 $(x - 7)(x + 5)$ **14** $(x - 8)(x + 3)$ **15** $(x - 4)(x - 2)$ **16** $(y - 3)(y - 2)$

17 $(x - 3)(x - 5)$ **18** $(a - 3)(a + 2)$ **19** $(a + 9)(a + 5)$ **20** $(b - 7)(b + 3)$

21 $(x - 4)^2$ **22** $(y + 1)^2$ **23** $(y - 7)(y + 4)$ **24** $(x - 5)(x + 4)$

25 $(x - 20)(x + 12)$ **26** $(x - 11)(x - 15)$ **27** $(y - 9)(y + 12)$ **28** $(x + 7)(x - 7)$

29 $(x + 3)(x - 3)$ **30** $(x + 4)(x - 4)$ **31** $2(x + 2)(x + 4)$

32 **a** $2(x + 5)(x - 3)$ **b** $3(x + 2)(x + 5)$ **c** $3(x + 3)(x + 5)$ **d** $2(n + 2)(n - 5)$

 e $5(a - 2)(a + 3)$ **f** $4(x + 4)(x - 4)$

page 363 Exercise 7 Ⓜ/Ⓗ

1 $(2x + 3)(x + 1)$ **2** $(2x + 1)(x + 3)$ **3** $(3x + 1)(x + 2)$ **4** $(2x + 3)(x + 4)$

5 $(3x + 2)(x + 2)$ **6** $(2x + 5)(x + 1)$ **7** $(3x + 1)(x - 2)$ **8** $(2x + 5)(x - 3)$

9 $(2x + 7)(x - 3)$ **10** $(3x + 4)(x - 7)$ **11** $(3x + 2)(2x + 1)$ **12** $(x - 3)(3x - 2)$

13 $(3y - 5)(y - 2)$ **14** $(3y - 1)(2y + 3)$ **15** $(2x + 1)(5x + 2)$ **16** $(6x - 1)(x - 3)$

17 $(4x + 1)(2x - 3)$ **18** $(3x + 2)(4x + 5)$ **19** $(4y - 3)(y - 5)$ **20** $(6x - 15)(x - 2)$

page 363 Exercise 8 Ⓜ/Ⓗ

1 $(y + a)(y - a)$ **2** $(m + n)(m - n)$ **3** $(x + t)(x - t)$ **4** $(y - 1)(y + 1)$

5 $(x - 3)(x + 3)$ **6** $(a - 5)(a + 5)$ **7** $\left(x + \frac{1}{2}\right)\left(x - \frac{1}{2}\right)$ **8** $\left(x + \frac{1}{3}\right)\left(x - \frac{1}{3}\right)$

9 $(2x - y)(2x + y)$ **10** $(a - 2b)(a + 2b)$ **11** $(5x - 2y)(5x + 2y)$ **12** $(3x - 4y)(3x + 4y)$

13 $\left(2x - \frac{z}{10}\right)\left(2x + \frac{z}{10}\right)$ **14** $x(x + 1)(x - 1)$ **15** $a(a - b)(a + b)$ **16** $x(2x + 1)(2x - 1)$

17 $2x(2x + y)(2x - y)$ **18** $y(y + 3)(y - 3)$ **19** $1\,200\,000$ **20** $12\,000\,000$

21 $10\,000 - 9 = (100 + 3)(100 - 3) = 103 \times 97$

page 364 Exercise 9 Ⓜ/Ⓗ

1 $x = -3$ or $x = -4$ **2** $x = -2$ or $x = -5$ **3** $x = 3$ or $x = -5$ **4** $x = 2$ or $x = -3$

5 $x = 2$ or $x = 6$ **6** $x = -3$ or $x = -7$ **7** $x = 2$ or $x = 3$ **8** $x = 5$ or $x = -1$

9 $x = 2$ or $x = -7$ **10** $x = 2$ or $x = -\frac{1}{2}$ **11** $x = -4$ or $x = \frac{2}{3}$ **12** $x = 1\frac{1}{2}$ or $x = -5$

13 $x = 1\frac{1}{2}$ or $x = \frac{2}{3}$ **14** $x = 7$ or $x = \frac{1}{4}$ **15** $x = \frac{3}{5}$ or $x = -\frac{1}{2}$ **16** $x = 8$ or $x = 7$

17 $x = \frac{1}{2}$ or $x = \frac{5}{6}$ **18** $x = 7$ or $x = -9$ **19** $x = -1$ **20** $x = 3$

21 $x = -5$ **22** $x = 7$ **23** $x = -\frac{1}{3}$ or $x = \frac{1}{2}$ **24** $x = 2$ or $x = -1\frac{1}{4}$

25 $x = 13$ or $x = -5$ **26** $x = -3$ or $x = \frac{1}{6}$ **27** $x = -2$ or $x = \frac{1}{10}$ **28** $x = 1$

29 $x = \frac{2}{9}$ or $x = -\frac{1}{4}$ **30** $x = -\frac{1}{4}$ or $x = \frac{3}{5}$ **31** $x = 2, x = -2, x = 1$ or $x = -1$

32 $x = 2, x = -2, x = 3$ or $x = -3$ **33** $x = 2, x = -2, x = \frac{1}{2}$ or $x = -\frac{1}{2}$ **34** $x = 2$ or $x = 1$

page 365 Exercise 10 Ⓜ/Ⓗ

1 $x = 0$ or $x = 3$ **2** $x = 0$ or $x = -7$ **3** $x = 0$ or $x = 1$ **4** $x = \frac{1}{3}$ or $x = 0$

5 $x = 4$ or $x = -4$ **6** $x = 7$ or $x = -7$ **7** $x = \frac{1}{2}$ or $x = -\frac{1}{2}$ **8** $x = \frac{2}{3}$ or $x = -\frac{2}{3}$

9 $y = 0$ or $y = -1\frac{1}{2}$ **10** $a = 0$ or $a = 1\frac{1}{2}$ **11** $x = 0$ or $x = 5\frac{1}{2}$ **12** $x = \frac{1}{4}$ or $x = -\frac{1}{4}$

13 $x = \frac{1}{2}$ or $x = -\frac{1}{2}$ **14** $x = 0$ or $x = \frac{5}{8}$ **15** $x = 0$ or $x = \frac{1}{12}$ **16** $x = 6$ or $x = 0$

17 $x = 11$ or $x = 0$ **18** $x = 0$ or $x = 1\frac{1}{2}$ **19** $x = 0$ or $x = 1$ **20** $x = 0$ or $x = 4$

page 366 Exercise 11 Ⓗ

1 $x = -\frac{1}{2}$ or $x = -5$ **2** $x = -\frac{2}{3}$ or $x = -3$ **3** $x = -\frac{1}{2}$ or $x = -\frac{2}{3}$

4 $x = 3$ or $x = \frac{1}{3}$ **5** $x = 1$ or $x = \frac{2}{5}$ **6** $x = 1\frac{1}{2}$ or $x = \frac{1}{3}$

7 $x = -0·63$ or $x = -2·37$ **8** $x = -0·27$ or $x = -3·73$ **9** $x = 0·72$ or $x = 0·28$

10 $x = 6·70$ or $x = 0·30$ **11** $x = 0·19$ or $x = -2·69$ **12** $x = 0·85$ or $x = -1·18$

13 $x = 0{\cdot}61$ or $x = -3{\cdot}28$ **14** $x = 4$ or $x = -1\frac{2}{3}$ **15** $x = 5$ or $x = -1\frac{1}{2}$

16 $x = 3{\cdot}56$ or $x = -0{\cdot}56$ **17** $x = 0{\cdot}16$ or $x = -3{\cdot}16$ **18** $x = 2\frac{1}{3}$ or $x = -\frac{1}{2}$

19 $x = -\frac{1}{3}$ or $x = -8$ **20** $x = 1\frac{2}{3}$ or $x = -1$ **21** $x = 2{\cdot}28$ or $x = 0{\cdot}22$

22 $x = -0{\cdot}35$ or $x = -5{\cdot}65$ **23** $x = -\frac{2}{3}$ or $x = \frac{1}{2}$ **24** $x = 2{\cdot}58$ or $x = -0{\cdot}58$

25 a $A(0{\cdot}21, 0)$, $B(4{\cdot}79, 0)$ **b** The graph doesn't go through $y = 0$.

26 not less than zero

page 367 **Exercise 12** (H)

1 $x = 2$ or $x = -3$ **2** $x = -3$ or $x = -7$ **3** $x = 2$ or $x = -0{\cdot}5$

4 $x = 1$ or $x = 4$ **5** $x = \frac{1}{2}$ or $x = -1\frac{2}{3}$ **6** $x = -0{\cdot}39$ or $x = -4{\cdot}28$

7 $x = 6{\cdot}16$ or $x = -0{\cdot}16$ **8** $x = 3$ **9** $x = 2$ or $x = -1\frac{1}{3}$

10 $x = -3$ or $x = -1$ **11** $x = -22{\cdot}66$ or $x = 0{\cdot}66$ **12** $x = 2$ or $x = -7$

13 $x = 7$ or $x = \frac{1}{4}$ **14** $x = \frac{3}{5}$ or $x = -\frac{1}{2}$ **15** $x = 0$ or $x = 3\frac{1}{2}$

16 $x = \frac{1}{4}$ or $x = -\frac{1}{4}$ **17** $x = 1{\cdot}27$ or $x = -2{\cdot}77$ **18** $x = 1$ or $x = -\frac{2}{3}$

19 $x = 2$ or $x = -\frac{1}{2}$ **20** $x = 0$ or $x = 3$ **21** Yes

22 a $x = 3$ or $x = -\frac{1}{2}$ **b** $x = 6$ or $x = -4$

23 a $x = -1$ **b** $x = 0{\cdot}6258$ **c** $x = 0{\cdot}5961$ **d** $x = 0{\cdot}2210$

page 369 **Exercise 13** (H)

1 $(x + 4)^2 - 16$ **2** $(x - 6)^2 - 36$ **3** $\left(x + \frac{1}{2}\right)^2 - \frac{1}{4}$ **4** $(x + 2)^2 - 3$

5 $(x - 3)^2$ **6** $(x + 1)^2 - 16$ **7** $(x + 8)^2 - 59$ **8** $(x - 5)^2 - 25$

9 $\left(x + \frac{3}{2}\right)^2 - \frac{9}{4}$ **10 a** $x = 6$ or $x = 2$ **b** $x = -3$ or $x = -7$ **c** $x = -1$ or $x = 5$

11 a $x = 0{\cdot}65$ or $x = -4{\cdot}65$ **b** $x = 3{\cdot}56$ or $x = -0{\cdot}56$ **c** $x = 0{\cdot}08$ or $x = -12{\cdot}08$

12 Involves the square root of a negative number.

13 $f(x) = (x + 3)^2 + 3$ **14** $g(x) = x^2 - 7x + \frac{1}{4}$ **15 a** $2(x + 1)^2 - 1$
as $(x + 3)^2 \geqslant 0$ $= \left(x - \frac{7}{2}\right)^2 - 12$ **b** $3(x - 1)^2 - 1$
then $f(x) \geqslant 3$ as $\left(x - \frac{7}{2}\right)^2 \geqslant 0$ **c** $2\left(x + \frac{1}{4}\right)^2 + \frac{15}{8}$

                              then $g(x) \geqslant -12$

16 a 3   **b** $x = -2$ **c** $\frac{1}{3}$ **17 a** $\frac{3}{4}$ **b** $x = \frac{1}{2}$ **c** $\frac{4}{3}$

18 Simplifies to $(x + 2)^2$ which is never negative.

page 370 **Exercise 14** (H)

1 8, 11   **2** 11, 13   **3** 12 cm   **4** 6 cm   **5** $x = 11$

6 8, 9, 10   **7 a** $x(x - 1)(x + 1) = 15x$ **c** $x(x + 4)(x - 4)$ **d** 3, 4, 5

8 156·5 km   **9** $x = 4$ **10** 10 cm, 24 cm   **11** 4·3 seconds   **12** 2·5 cm

13 $x = 1$ m   **14** 9 cm or 13 cm   **15 a** $n(n + 2)$ **b** $n(n + 2) = 255$, $n = 15$ **16** $x = 2$ cm

17 $x = 3\,\text{cm}$ or $x = 9.5\,\text{cm}$ **18** $x = 6\,\text{cm}$ **19** $\dfrac{40}{x}$ hours, $\dfrac{40}{x-2}$ hours, $10\,\text{km/h}$

20 $4\,\text{km/h}$ **21** $20\,\text{mph}$ **22** $4, -1$ **23** $2, 5$ **24** $\dfrac{3}{4}$

25 a y and $2x$ **b** $2x^2 = y^2$ **c** $x = 59.5\,\text{cm},\ y = 84.1\,\text{cm}$ **d** $29.7\,\text{cm}$

page 374 **Exercise 15** Ⓗ

1 a $\dfrac{5}{7}$ **b** $5y$ **c** $\dfrac{1}{2}$ **d** 4 **e** $\dfrac{x}{2y}$ **f** 2 **g** $\dfrac{a}{2}$ **h** $\dfrac{2b}{3}$

2 A and G, B and E, C and H, D and F

3 a a **b** $\dfrac{10m}{3}$ **c** $\dfrac{2xy}{9}$ **d** $\dfrac{4}{aq}$ **e** $\dfrac{8}{3}$ **f** $\dfrac{y^2}{5x}$ **g** $2a$ **h** 1

4 a $\dfrac{a}{5b}$ **b** a **c** $\dfrac{7}{8}$ **d** $\dfrac{3}{4-x}$ **e** $\dfrac{5+2x}{3}$ **f** $\dfrac{3x+1}{x}$ **g** $\dfrac{4+5a}{5}$ **h** $\dfrac{b}{3+2a}$

5 a 6 **b** 4 **c** $2a$ **6** A and F, B and H, C and E, D and G

7 a $\dfrac{x+2y}{3xy}$ **b** $\dfrac{6-b}{2a}$ **c** $\dfrac{4a+2b}{b}$ **d** $x-2$

8 a $\dfrac{x+2}{x-3}$ **b** $\dfrac{x}{x+1}$ **c** $\dfrac{x+4}{2(x-5)}$ **d** $\dfrac{x+5}{x-2}$ **e** $\dfrac{x+3}{x+2}$ **f** $\dfrac{x+5}{x-2}$

page 378 **Exercise 16** Ⓗ

1
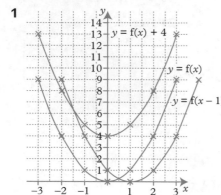

2
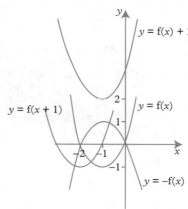

3 a $A(0, -2)$ **b** $A(7, 0)$

4 a $A(-2, -1), B(0, -3), C(2, 0)$ **b** $A(0, 1), B(2, 3), C(4, 0)$ **c** $A(-1, 1), B(0, 3), C(1, 0)$

5
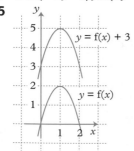

6
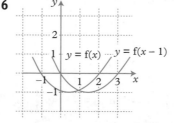

7
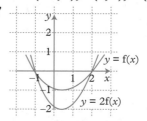
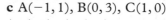

8
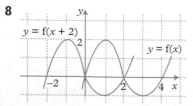

9
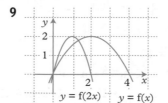

10
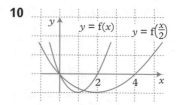

11

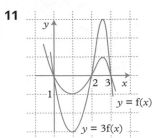

$y = f(x)$

$y = 3f(x)$

12

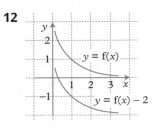

$y = f(x)$

$y = f(x) - 2$

13

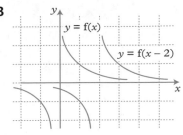

$y = f(x)$

$y = f(x - 2)$

14

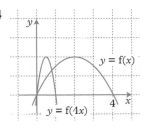

$y = f(x)$

$y = f(4x)$

15

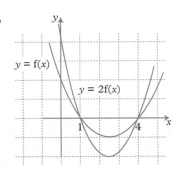

$y = f(x)$

$y = 2f(x)$

16

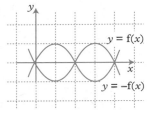

$y = f(x)$

$y = -f(x)$

page 379 **Exercise 17 (H)**

1

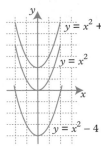

$y = x^2 + 2$

$y = x^2$

$y = x^2 - 4$

2

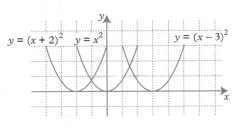

$y = (x + 2)^2$ $y = x^2$ $y = (x - 3)^2$

3

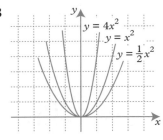

$y = 4x^2$

$y = x^2$

$y = \frac{1}{2}x^2$

4

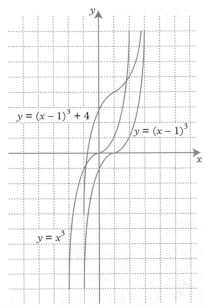

$y = (x - 1)^3 + 4$

$y = (x - 1)^3$

$y = x^3$

5

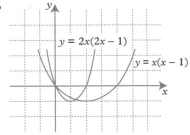

$y = 2x(2x - 1)$

$y = x(x - 1)$

6

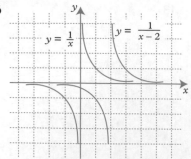

$y = \frac{1}{x}$ $y = \frac{1}{x - 2}$

7

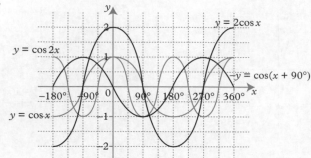

$y = 2\cos x$

$y = \cos 2x$

$y = \cos(x + 90°)$

$y = \cos x$

8 a, c

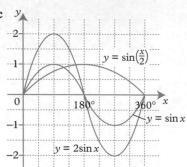

$y = \sin\left(\frac{x}{2}\right)$

$y = \sin x$

$y = 2\sin x$

b, d

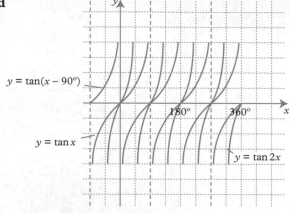

$y = \tan(x - 90°)$

$y = \tan x$

$y = \tan 2x$

9

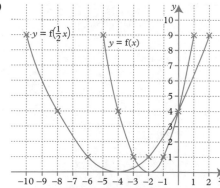

$y = f\left(\frac{1}{2}x\right)$

$y = f(x)$

Stretch parallel to the x-axis, scale factor 2.

10 a $A(1, 5)$ **b** $A(5, -4)$

11 a $y = x^2 + 3x + 5$ **b** $y = (x - 2)^2 + 3(x - 2)$
  c $y = -x^2 - 3x$

12 a $a = 2, b = 3$
  b $y = g(x)$ is $y = x^2$ shifted 2 units right and 3 units up

13 a $A(2, 4)$ **b** $B\left(0, \frac{1}{5}\right)$, $C(3, 1)$ $D(-3, 1)$

page 382 **Test yourself**

1 a i 10 **ii** $4\sqrt{5}$ **b** $\frac{403}{999}$ **2 a** $x = 20$ **b** $y = \frac{1}{3}$ **c** $2ab(3b-1)$ **d** $(3x-4)(x + 3)$

3 $\dfrac{2\sqrt{3} + 3}{3}$ or $\dfrac{2}{3}\sqrt{3} + 1$ **4** $p = \dfrac{5}{4}; q = 4$ **5** $b = -5, c = -14$

6 a $x(2x + 1) + 5x = 95$ **b** $5\cdot55$ **7 a** $(2n - 3)(n + 1)$ **b** 17×11
  $2x^2 + 6x - 95 = 0$

8 a Area $= (x + 5)(x - 1) - \frac{1}{2} \times 4 \times 3$ **b i** $x^2 + 4x - 11 = 59$
$= x^2 + 4x - 5 - 6$ $\therefore x^2 + 4x - 70 = 0$
$= x^2 + 4x - 11$ **ii** 6·6 cm (1 dp)

9 a $(0, 12)$ **b** $(2, 16)$ **c** $(-2, 0)$

10 a i $\dfrac{7}{x - 3}$ **ii** $\dfrac{81x^8}{y^{12}}$ **iii** $\dfrac{x}{4 + x}$

11 a i $(0, 1)$ **ii** $(90°, 0)$ **b** 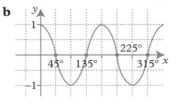   **12 a** $(-2, 3)$ **b** $(1, 3)$ **c** $(1, 3)$ **d** $(2, 6)$

13 a, b 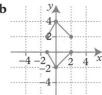   **14 a** $(\sqrt{32} + \sqrt{2})^2 = 32 + 2\sqrt{2}\sqrt{32} + 2$ **b** $5\sqrt{3}$ cm
$= 34 + 2\sqrt{64}$
$= 34 + 16$
$= 50$

8   Handling data

page 388 **Exercise 1** Ⓜ

1 a Median = 5, Mean = 6, Mode = 4, Range = 9
  b Median = 7, Mean = 9, Mode = 7, Range = 19
  c Median = 8, Mean = 6·5, Mode = 9, Range = 11
  d Median = 3·5, Mean = 3·5, Mode = 4, Range = 8
2 a 10   **b** 13   **3** 6   **4 a** 5   **b** 35   **5** 76 kg   **6** 7°C
7 a 16 years **b** 170 cm **c** 50 kg **8** 2, 45   **9** 4 people   **10 a** False   **b** True   **c** False   **d** Possible
11 a Mean = £33 920, Median = £22 500, Mode = £22 500   **b** Mean
12 a Mean = 157·125 kg, Median = 91 kg   **b** The mean. No, it's too high.
13 a Mean = 74·5 cm, Median = 91 cm
  b Yes. It is more appropriate to use the median, as the mean has been distorted by
    a few short plants.
14 78 kg   **15** 35·2 cm   **16 a** 2   **b** 9   **17 a** 20·4 m   **b** 12·8 m   **c** 1·66 m
18 55 kg   **19** For example: 4, 4, 5, 8, 9   **20** For example: 2, 4, 4, 4, 6
21 12   **22** $3\frac{2}{3}$ **23 a** N **b** Mean = $N^2 + 2$, Median = N^2 **c** 2

page 391 **Exercise 2** Ⓜ

1 96·25 g   **2** 51·9 p   **3** 4·82 cm
4 a Mean = 3·025, Median = 3, Mode = 3
  b Mean = 17·75, Median = 17, Mode = 17
5 a Mean = 6·62, Median = 8, Mode = 3   **b** Mode
6 a 9   **b** 9   **c** 15   **7 a** 5   **b** 10   **c** 10   **8** $\dfrac{ax + by + cz}{a + b + c}$

page 394 **Exercise 3** Ⓜ

1 a

Number of words	Frequency f	Midpoint x	fx
1–5	6	3	18
6–10	5	8	40
11–15	4	13	52
16–20	2	18	36
21–25	3	23	69
Totals	20	—	215

b 10·75   **2** 68·25   **3** 3·77

4 a 181 cm
  b The raw data is unavailable and an assumption has been made with the midpoint of each interval.

page 397 **Exercise 4** Ⓜ

1 For example, Isola, because the snowfall is greater on average.
2 a 20%   **b** 20–24 and 30–34
3 a–b

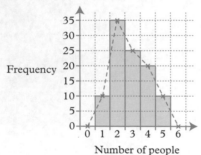

4 a 62
  b The mean weight is lower for A. The range is higher for B.
  c For example, B could be rugby, as weight of player varies with position.
5 a The fertiliser made the plants taller, on average.
  b The fertilise had no effect on the weights.
6 Height has been used to represent production, but the width of the picture also varies, making the difference 100 k much greater.
7 The profit axis does not start at zero, so the increase in profit looks much greater than it is.

8 a i $\frac{1}{8}$ **ii** $\frac{1}{6}$ **iii** $\frac{5}{12}$ **iv** $\frac{1}{12}$ **v** $\frac{1}{8}$ **vi** $\frac{1}{12}$ **b i** £15   **ii** £20   **iii** £50   **iv** £10
9 a £425   **b** £150   **c** £250   **d** £75   **10 a i** 8 min   **ii** 34 min   **iii** 10 min   **b** 18°
11 270°, 12°, 23°, 54°, 1°   **12** 18°, 54°, 54°, 234°   **13 a** 22·5% **b** 45°, 114°   **14** 8

page 402 **Exercise 5** Ⓜ

1 a

Stem	Leaf
1	5
2	3 9 7 8
3	5 9 6 2 8
4	1 0 7 5 8 2 6
5	2 4 1 9 3
6	5 6

b 51

Key: 2|3 means 23

2 a   Stem | Leaf
2 | 0 4 5 8
3 | 1 7 9
4 | 0 4 6
5 | 2 5 8 9
6 | 1 5 7 8
7 | 3 5

Key: 3|1 means 31

b   Stem | Leaf
2 | 2 8 9
3 | 0 5 8
4 | 1 4 6 7 7
5 | 3 4 9
6 | 7
7 | 2

Key: 2|2 means 22

3 a 50 kg   **b** 15   **c** 50 kg

4 a 4·5   **b** 5·3   Stem | Leaf
1 | 4 8
2 | 4 4 8
3 | 1 3 3 7 8
4 | 0 5 5 6 6 7 9
5 | 1 2 5 8
6 | 2 3 7

Key: 3|7 means 3·7

5 a 13   **b** 78   **c** The women's pulse rates were higher on average (median 78 compared with 62). The spread was greater for men (range 53 compared with 45).

page 404 **Exercise 6** Ⓜ

1 a Pearce family: Median weight = 54 kg; Range = 20 kg
Taylor family: Median weight = 63 kg; Range = 40 kg

b The median weight for the Pearce family is lower than that for the Taylor family and the range for the Pearce family is much less than that for the Taylor family. The weights of the Taylor family are much more spread out.

2 a Class 10 M: Median mark = 73; Range = 31
Class 10 S: Median mark = 58; Range = 47

b The median mark for Class 10 S is lower than that for Class 10 M and the range for Class 10 S is greater than that for Class 10 M. The marks of Class 10 S are more spread out.

page 406 **Exercise 7** Ⓜ

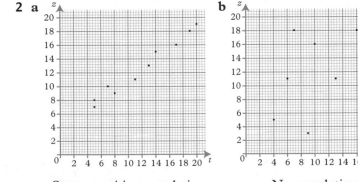

Strong positive correlation   No correlation   Weak negative correlation

3 a No correlation   **b** Strong negative correlation
c No correlation   **d** Weak positive correlation

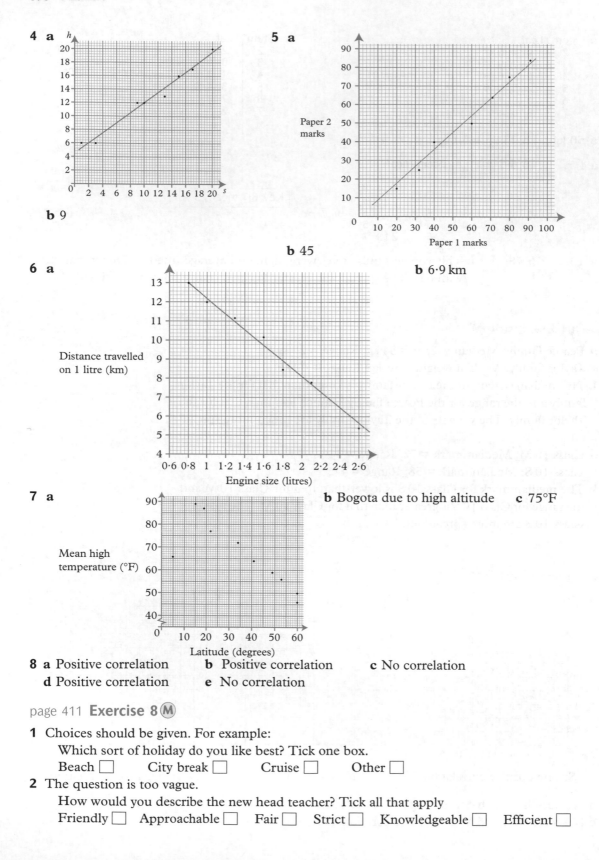

4 a

b 9

5 a

Paper 2 marks

Paper 1 marks

b 45

b 6·9 km

6 a

Distance travelled on 1 litre (km)

Engine size (litres)

b Bogota due to high altitude   **c** 75°F

7 a

Mean high temperature (°F)

Latitude (degrees)

8 a Positive correlation   **b** Positive correlation   **c** No correlation
d Positive correlation   **e** No correlation

page 411 Exercise 8 Ⓜ

1 Choices should be given. For example:
   Which sort of holiday do you like best? Tick one box.
   Beach ☐   City break ☐   Cruise ☐   Other ☐
2 The question is too vague.
   How would you describe the new head teacher? Tick all that apply
   Friendly ☐   Approachable ☐   Fair ☐   Strict ☐   Knowledgeable ☐   Efficient ☐

3 The options overlap, and there is no option for those who don't watch TV.

How long do you watch television each day?

Less than 1 hour ☐

At least 1 hour but less than 2 hours ☐

At least 2 hours but less than 3 hours ☐

At least 3 hours ☐

4 There is no middle option.

How much would you pay to use the new car park?

☐ Less than £1 per day

☐ £1–£2·49 per day

☐ More than £2·50 per day

5 This is a leading question.

Which is the most important subject at school?

Tick one box.

English ☐   Maths ☐   History ☐   Science ☐

PE ☐   Languages ☐   Art ☐   Other ☐

6 'Often' is too vague.

How often do you or your parents hire DVDs
from a shop?

At least once a week ☐

Not every week, but at least once a month ☐

Less than once a month ☐

7 Leading question.

Do we get too much homework?

Yes ☐   No ☐

8 They should not ask for personal information like name and age.
The first question's options are too vague. Options should be given
for the second question. The third question is a leading question
and should be rephrased. The last question only needs yes/no options.

For example:

> • How much television do you watch each day?
>   Less than 1 hour ☐
>   At least 1 hour but less than 2 hours ☐
>   At least 2 hours but less than 3 hours ☐
>   At least 3 hours ☐
> • What is your favourite type of programme on TV?
>   Soaps ☐
>   Films ☐
>   Documentaries ☐
>   Dramas ☐
>   Other ☐
> • Do you like watching MTV?
>   Yes ☐   No ☐   Haven't seen it ☐
> • Do you like nature programmes?
>   Yes ☐   No ☐

9 What is your favourite type of TV programme?
Tick one box.

Comedy ☐ Romance ☐ Sport ☐ Documentary ☐ Other ☐

page 413 **Exercise 9** Ⓜ

1 squares: shaded 33·3%, unshaded 66·6%
triangles: shaded 50%, unshaded 50%

2 a

	Football	Hockey	Swimming	Total
Boys	5	2	1	8
Girls	2	3	2	7
Total	7	5	3	15

b 8   **c** 25%

3 a

	Girls	Boys	Total
Can cycle	95	120	215
Cannot cycle	179	82	261
Total	274	202	476

b 35%   **c** 59%

4 a

	Men	Women	Total
Had accident	75	88	163
Had no accident	507	820	1327
Total	582	908	1490

b 13%   **c** 10%
d Men are more likely to have accidents.

page 418 **Exercise 10** Ⓗ

1 16 year 7, 16 year 8, 16 year 9, 6 year 10, 6 year 11   **2** A 10, B 25, C 10, D 5, E 50

page 420 **Exercise 11** Ⓜ

1 She is only asking students who have school meals. It would be better to include students who choose not to have school meals as well.
2 People who are going to Europe are more likely to be positive about it. It would be better to choose an unbiased location.
3 17-year-old drivers will not be included in the sample (as they are not on the electoral roll), so it is biased.
4 Working people are unlikely to be at the supermarket at this time, so the sample is biased.

page 422 **Exercise 12** Ⓜ

1 a Females under 30 = 8 = 57·6°
Males under 30 = 9 = 64·8°
Females 30 or over = 10 = 72°
Males 30 and over = 23 = 165·6°
Most employees are males 30 and over.

b 240, 270, 300, 690

2 a

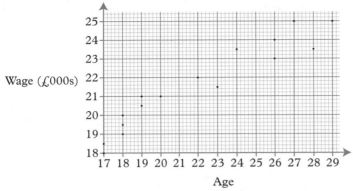

b Positive correlation between employees under 30 and wage. No correlation between employees 30 and over and wage.

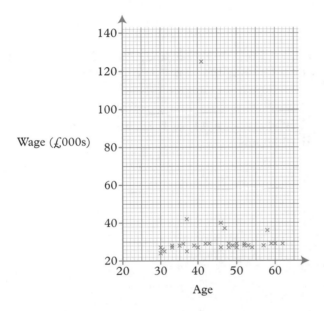

3 a

Wage, £	Mid-value, x	Frequency, f	fx
15 000–20 000	17 500	4	70 000
20 000–25 000	22 500	12	270 000
25 000–30 000	27 500	29	797 500
30 000–35 000	32 500	0	0
35 000–40 000	37 500	2	75 000
40 000–45 000	42 500	2	85 000
125 000–130 000	127 500	1	127 500
Total		50	1 425 000

b £28 500   **c** £28 430

d The mean from the raw data is more reliable. The value from the table is an estimate.

4 Frequency

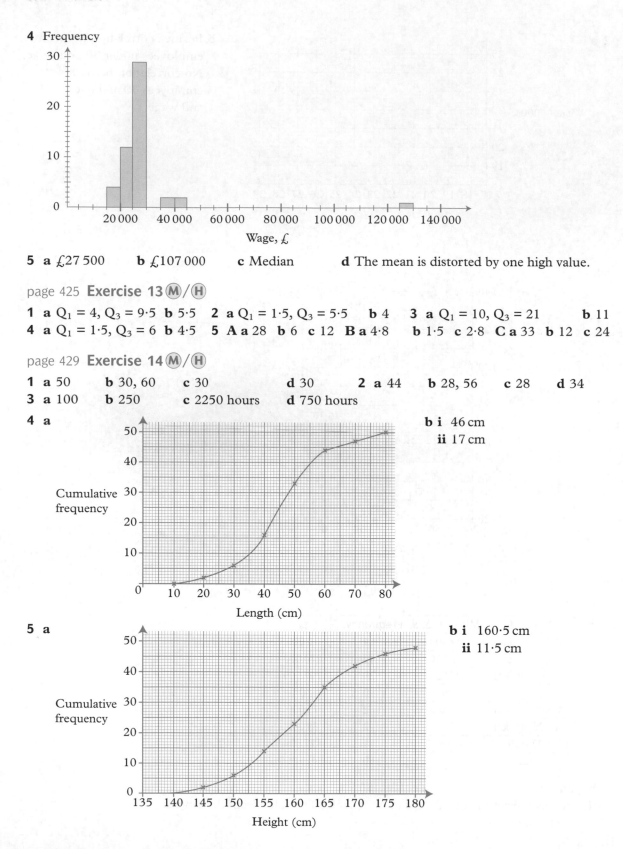

5 a £27 500   **b** £107 000   **c** Median   **d** The mean is distorted by one high value.

page 425 **Exercise 13** Ⓜ/Ⓗ

1 a $Q_1 = 4$, $Q_3 = 9.5$ **b** 5.5   **2 a** $Q_1 = 1.5$, $Q_3 = 5.5$ **b** 4   **3 a** $Q_1 = 10$, $Q_3 = 21$ **b** 11
4 a $Q_1 = 1.5$, $Q_3 = 6$ **b** 4.5   **5 A a** 28 **b** 6 **c** 12 **B a** 4.8   **b** 1.5 **c** 2.8 **C a** 33 **b** 12 **c** 24

page 429 **Exercise 14** Ⓜ/Ⓗ

1 a 50   **b** 30, 60   **c** 30   **d** 30   **2 a** 44   **b** 28, 56   **c** 28   **d** 34
3 a 100   **b** 250   **c** 2250 hours   **d** 750 hours
4 a

b i 46 cm
ii 17 cm

5 a

b i 160·5 cm
ii 11·5 cm

6 a

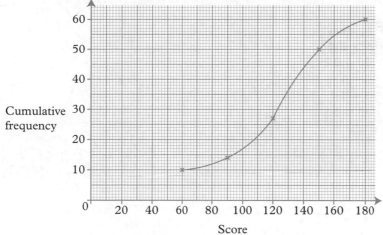

b i 125
  ii 52
c Boris, his interquartile range is smaller.

7 a

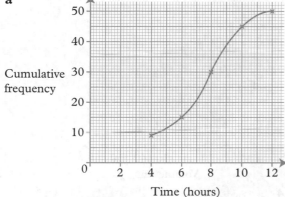

b i 7·4 hours   **ii** 3·5 hours   **c** 90%

8 a

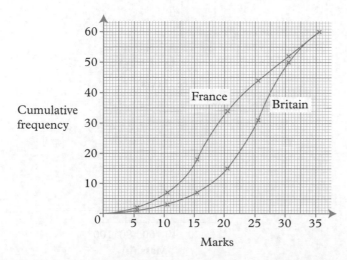

b Britain 25, France $19\frac{1}{2}$

c Britain $8\frac{1}{2}$

d The British results are higher on average, and less spread out.

page 433 **Exercise 15** (H)

1

Length l (in mm)	Frequency	Frequency density (f.d.)
$0 \leqslant l < 20$	5	$5 \div 20 = 0\cdot25$
$20 \leqslant l\ 25$	5	$5 \div 5 = 1$
$25 \leqslant l\ 30$	7	$7 \div 5 = 1\cdot4$
$30 \leqslant l < 40$	3	$3 \div 10 = 0\cdot3$

2

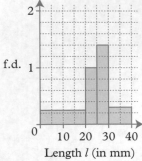

Length l (in mm)

Volume (mm³)	Frequency	Frequency density
$0 \leqslant V \leqslant 5$	5	$5 \div 5 = 1$
$5 \leqslant V \leqslant 10$	3	$3 \div 5 = 0\cdot6$
$10 \leqslant V \leqslant 20$	12	$12 \div 10 = 1\cdot2$
$20 \leqslant V \leqslant 30$	17	$17 \div 10 = 1\cdot7$
$30 \leqslant V \leqslant 40$	13	$13 \div 10 = 1\cdot3$
$40 \leqslant V \leqslant 60$	5	$5 \div 20 = 0\cdot25$

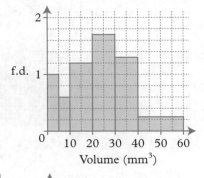

Volume (mm³)

3

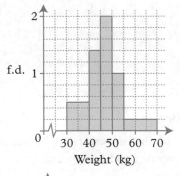

Weight (kg)

4

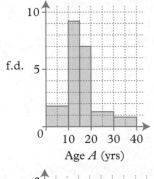

Age A (yrs)

5

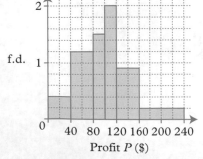

Profit P ($)

6

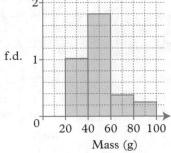

Mass (g)

page 436 **Exercise 16 (H)**

1 a i 10 **ii** 20 **b** 65 **2** 55 **3** 135 **4** 28 eggs **5 a** 6 **b** 8 **c** 54 **6 a** 6 cm **b** 9 cm **c** 3 cm

page 438 **Test yourself**

1 $30 < C \leqslant 40$ **2 a** 7·6 mins   **b** $6 < t \leqslant 8$

3 a 3   **b** 10 marks

4 30 mm

5 a Question not specific about time period; Answer boxes not quantified
or covering all options.

  b Students' own answer.

6 a How many hours a week, on average, do you read for pleasure?

    ☐ $\frac{1}{2}$ hour or less

    ☐   More than $\frac{1}{2}$ hour but no more than 2 hours

    ☐   More than 2 hours but no more than 5 hours

    ☐   More than 5 hours but no more than 10 hours

    ☐   More than 10 hours

  b 320 pupils

7 10 pupils

8 a

Geology	Physics	Chemistry	Biology
43	57	57	80

  b Check that these students have not
already opted for a teaching career.

9 a It is a running total of the frequency values.

  b i 415 people   **ii** 54 times

10 a 6·08 hours **b**

No. of hours (h)	Frequency
$0 < h \leqslant 2$	10
$0 < h \leqslant 4$	25
$0 < h \leqslant 6$	55
$0 < h \leqslant 8$	90
$0 < h \leqslant 10$	115
$0 < h \leqslant 12$	120

c 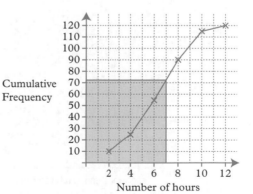 **d** 72

11 The median of the boys marks is higher than the girls (by 4).
The girls marks are more widespread than the boys. The inter-quartile
range for the girls is 25 (i.e. 55–30) whereas for the boys, it is
17 (55–38).

12 32

13

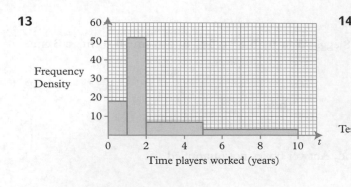

14 a c

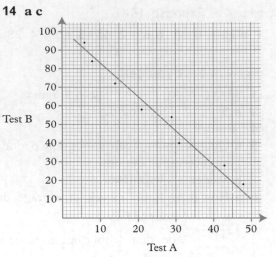

b Strong negative correlation   **d** 68

9 Probability

page 446 **Exercise 1** Ⓜ

1 B   **2** C   **3** A   **4** B or C   **5** C   **6** A   **7** B   **8** B   **9** C   **10** D

11

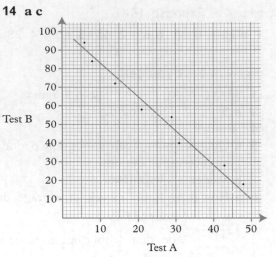

13 Nick, as the relative frequency should be close to 0·1 with his number of spins.

page 448 **Exercise 2** Ⓜ

1 a $\frac{1}{13}$ **b** $\frac{1}{52}$ **c** $\frac{1}{4}$ **2 a** $\frac{1}{9}$ **b** $\frac{1}{3}$ **c** $\frac{4}{9}$ **d** $\frac{2}{9}$ **3 a** $\frac{5}{11}$ **b** $\frac{2}{11}$ **c** $\frac{4}{11}$

4 a $\frac{4}{7}$ **b** $\frac{3}{17}$ **c** $\frac{11}{17}$ **5 a** $\frac{2}{9}$ **b** $\frac{2}{9}$ **c** $\frac{1}{9}$ **d** 0 **e** $\frac{5}{9}$ **6 a** $\frac{1}{13}$ **b** $\frac{2}{13}$ **c** $\frac{1}{52}$ **d** $\frac{5}{52}$

7 a $\frac{3}{13}$ **b** $\frac{5}{13}$ **c** $\frac{8}{13}$ **8** Megan   **9 a** $\frac{1}{10}$ **b** $\frac{3}{10}$ **c** $\frac{3}{10}$ **10** 6 blue, 3 white

11 a 1   **b** There may be a small number of blue balls, and Asif just didn't pick them.

12 a $\frac{1}{7}$ **b** $\frac{4}{7}$ **c** $\frac{6}{7}$ **13 a** $\frac{1}{6}$ **b** $\frac{1}{3}$ **c** $\frac{1}{6}$ **14 a i** $\frac{5}{13}$ **ii** $\frac{6}{13}$ **b i** $\frac{5}{12}$ **ii** $\frac{1}{12}$

15 a $\frac{1}{12}$ **b** $\frac{1}{40}$ **c** $\frac{1}{4}$ **16 a** $\frac{x}{12}$ **b** 3   **17** 0·14   **18 a i** $\frac{1}{4}$ **ii** $\frac{1}{4}$ **iii** $\frac{1}{4}$ **b** $\frac{1}{4}$ **c** $\frac{2}{9}$

page 453 Exercise 3 Ⓜ

1 a 150   **b** 50   **2 a** 25   **b** No   **3** 50   **4** 40

5 a $\frac{3}{8}$ **b** 25   **c** Possibly not, but you need more trials to be certain.

6 a $\frac{1}{2}$ **b** $\frac{1}{4}$

7 a 15   **b** 105   **8 a** 20   **b** 5   **c** 50   **d** 40   **9 a** 0·2146   **b** 1073

page 455 Exercise 4 Ⓜ

1 a HHH, HHT, HTH, THH, HTT, THT, TTH, TTT   **b** $\frac{1}{8}$

2 HHHH, HHHT, HHTH, HTHH, THHH, HHTT, HTHT, THHT, HTTH, THTH, TTHH, HTTT, THTT, TTHT, TTTH, TTTT (16)

3 a

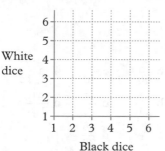

    b 4   **c** $\frac{1}{9}$ **4 a** $\frac{1}{4}$ **b** $\frac{1}{4}$

5 WX, WY, WZ, XW, XY, XZ, YW, YX, YZ, ZW, ZX, ZY; $\frac{1}{6}$

6 a

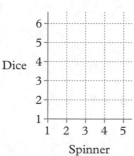

    b $\frac{1}{2}$ **7 a** $\frac{1}{12}$ **b** $\frac{1}{36}$ **c** $\frac{5}{18}$ **d** $\frac{1}{6}$

                  8 a $\frac{1}{12}$ **b** $\frac{5}{36}$ **c** $\frac{2}{3}$ **d** $\frac{1}{12}$ **e** $\frac{1}{36}$

                  9 $\frac{2}{3}$ **10 a** 64   **b** $\frac{1}{64}$ **11** Yes

                  12 No, it is slightly biased in favour of Y.

                  13 Lose 20p   **14 a** $\frac{1}{144}$ **b** $\frac{1}{18}$

page 458 Exercise 5 Ⓜ/Ⓗ

Students' own answers

page 462 Exercise 6 Ⓗ

1 $\frac{4}{5}$ **2 a** $\frac{1}{13}$ **b** $\frac{12}{13}$ **c** $\frac{3}{13}$ **d** $\frac{10}{13}$ **3** $\frac{35}{36}$ **4** 0·76

5 0·494   **6 a** $\frac{1}{4}$ **b** $\frac{3}{4}$ **c** $\frac{3}{8}$ **d** $\frac{5}{8}$ **e** 0   **f** 1

7 a $\frac{5}{11}$ **b** $\frac{7}{22}$ **c** $\frac{17}{22}$ **d** $\frac{15}{22}$ **8** $\frac{1}{18}$ **9 a** 0·3   **b** 0·9

10 $\frac{1}{3}$ **11** $\frac{1}{24}$ **12 a i** 0·24   **ii** 0·89   **b** 575

13 a i Exclusive   **ii** Not Exclusive   **b** $\frac{11}{15}$ **14 a i** Exclusive   **ii** Exclusive   **iii** Not Exclusive   **b** $\frac{3}{4}$

15 a 0·8   **b** 0·7   **c** The events are not mutually exclusive.

16 The events are not mutually exclusive.

page 466 **Exercise 7** Ⓗ

1 a p(A) = $\frac{1}{13}$, p(B) = $\frac{1}{6}$ **b** $\frac{1}{78}$ **2 a** $\frac{1}{2}$ **b** $\frac{1}{2}$ **c** $\frac{1}{4}$ **3** $\frac{1}{10}$

4 a $\frac{1}{78}$ **b** $\frac{1}{104}$ **c** $\frac{1}{24}$ **5 a** $\frac{1}{16}$ **b** $\frac{1}{169}$ **c** $\frac{9}{169}$

6 a $\frac{1}{16}$ **b** $\frac{25}{144}$ **7 a** $\frac{1}{121}$ **b** $\frac{9}{121}$ **8** $\frac{8}{1125}$ **9 a** $\frac{1}{288}$ **b** $\frac{1}{72}$

10 a $\frac{1}{9}$ **b** $\frac{4}{27}$ **11** $\frac{1}{24}$ **12** $\frac{1}{128}$ **13** $\frac{1}{144}$

14 a $\left(\frac{1}{6}\right)^{20} = 2\cdot7 \times 10^{-16}$ **b** $\left(\frac{5}{6}\right)^{n}$ **c** $1 - \left(\frac{5}{6}\right)^{n}$

page 469 **Exercise 8** Ⓜ/Ⓗ

1 a $\frac{49}{100}$ **b** $\frac{9}{100}$ **2 a** $\frac{9}{64}$ **b** $\frac{15}{64}$ **3 a** $\frac{1}{4}$ **b** $\frac{3}{50}$ **c** $\frac{2}{50}$ **4 a** $\frac{7}{15}$ **b** $\frac{1}{15}$

5 a $\frac{1}{12}$ **b** $\frac{1}{6}$ **c** $\frac{1}{3}$ **d** $\frac{2}{9}$ **6 a** $\frac{1}{216}$ **b** $\frac{125}{216}$ **c** $\frac{25}{72}$ **d** $\frac{91}{216}$

7 a $\frac{1}{64}$ **b** $\frac{5}{32}$ **c** $\frac{27}{64}$ **8 a** $\frac{1}{6}$ **b** $\frac{1}{30}$ **c** $\frac{1}{30}$ **d** $\frac{29}{30}$

9 a $\frac{9}{16}$ **b** $\frac{1}{16}$ **10 a** 6   **b** $\frac{1}{3}$ **11 a** $\frac{4}{9}$ **b** $\frac{1}{24}$

12 a $\frac{3}{20}$ **b** $\frac{9}{20}$ **13 a** $\frac{6}{6840}$ **b** $\frac{60}{6840}$ **c** $\frac{120}{116\,280}$

page 473 **Exercise 9** Ⓗ

1 a $\frac{90}{999\,000}$ **b** $\frac{979\,110}{999\,000}$ **c** $\frac{19\,800}{999\,000}$ **2 a** $\frac{3}{20}$ **b** $\frac{7}{20}$ **c** $\frac{1}{2}$ **3 a** 5   **b** $\frac{1}{64}$

4 a $\frac{1}{220}$ **b** $\frac{1}{22}$ **c** $\frac{3}{11}$ **d** 5   **5 a** $\frac{3}{5}$ **b** $\frac{1}{3}$ **c** $\frac{2}{15}$ **d** $\frac{2}{21}$ **e** $\frac{1}{7}$

6 a 0·00781   **b** 0·511   **7 a** $\frac{21}{506}$ **b** $\frac{455}{2024}$ **c** $\frac{945}{2024}$

8 a $\frac{x}{x+y}$ **b** $\frac{x(x-1)}{(x+y)(x+y-1)}$ **c** $\frac{2xy}{(x+y)(x+y-1)}$ **d** $\frac{y(y-1)}{(x+y)(x+y-1)}$

9 a $\frac{x}{z}$ **b** $\frac{x(x-1)}{z(z-1)}$ **c** $\frac{2x(z-x)}{z(z-1)}$

10 a $\frac{1}{125}$ **b** $\frac{1}{125}$ **c** $\frac{1}{10\,000}$ **d** $\frac{3}{500}$ **11 a** $\frac{1}{49}$ **b** $\frac{1}{7}$ **12 a** $\frac{1}{52}$ **b** 1

page 475 **Exercise 10** Ⓗ

1 $\frac{3}{10}$ **2** $\frac{9}{140}$ **3 a** $\frac{1}{5}$ **b** $\frac{18}{25}$ **c** $\frac{1}{20}$ **d** $\frac{2}{25}$ **e** $\frac{77}{100}$ **4 a** 0·07 **b** 0·64 **c** 0·29

5 $\frac{4}{35}$ **6 a** $\frac{19}{25}$ **b** 38 **7** 27

page 477 **Test yourself**

1 0·35 **2 a** 0·3 **b** 175

3 a 252 students

  b 0·252

4 a i $\frac{7}{20}$ **ii** A different experiment usually produces different results **b** 250

5 a 36 **b** 0·47 **6 a** 0·25 **b** 60

7 a 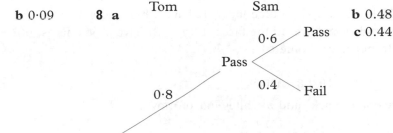 **b** 0·09 **8 a** 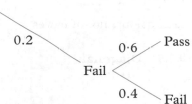 **b** 0.48 **c** 0.44

9 a i $\frac{2}{9}$ **ii** $\frac{1}{3}$ **iii** 0 **iv** $\frac{2}{3}$ **b** 120 **10** $\frac{3}{8}$ or 0·375

10 Using and applying mathematics

The results are given here as a check for teachers or students working on their own.

page 482 **10.1.1 Opposite corners**

The difference in an $n \times n$ square with 3 columns is $(3n - 3)^2$.

page 483 **10.1.2 Hiring a car**

Hav-a-car is cheapest up to 774 miles. Between 774 and 1538 miles,
Snowdon rent-a-car is cheapest. Gibson is cheapest over 1538 miles.

page 483 **10.1.3 Half-time score**

12 different scores were possible at half-time.
The general rule is $(a + 1)(b + 1)$ for a final score $a - b$.

page 484 **10.1.4 Squares inside squares**

Area is always $a^2 + b^2$.

page 485 **10.1.5 Maximum box**

The corner squares should be $\frac{1}{6}$th the size of the square sheet. For a rectangle $a \times 2a$, cut out $\approx \frac{a}{4\cdot732}$.

page 485 **10.1.6 What shape tin?**

Smallest surface area is $393\cdot8\,\text{cm}^2$ when $r = 4\cdot57\,\text{cm}$. In general, height = $2 \times$ radius for smallest surface area.

page 486 **10.1.7 Painting cubes**

27 cubes: 8 have 3 blue faces, 12 have 2 blue faces, 6 have 1 blue face, 1 does not have a blue face. n^3 cubes: 8 have 3 blue faces, $12(n-2)$ have 2 blue faces, $6(n-2)^2$ have 1 blue face, $(n-2)^3$ do not have a blue face.

page 486 **10.1.8 Discs**

a For n blacks and n whites, no. of moves $= \dfrac{n(n+1)}{2}$

b no. of moves $= \dfrac{n(n+2)}{2}$

c For n of each colour, no. of moves $= \frac{3}{2}n(n+1)$

page 487 **10.1.9 Diagonals**

880 squares

page 487 **10.1.10 Chess board**

30 squares on a 4×4 board
204 squares on an 8×8 board
$n^2 + (n-1)^2 + \ldots + 2^2 + 1^2 = \frac{1}{6}n(n+1)(2n+1)$ squares on an $n \times n$ board

page 487 **10.1.11 Find the connection**

They all tend to a constant number. Cube root of x, seventh root of x, $\frac{1}{\sqrt[3]{x}}$, $(\sqrt[3]{x})^2$.

page 487 **10.1.12 Spotted shapes**

$A = i + \frac{1}{2}p - 1$ (Pick's theorem)

page 490 **Exercise 1 Ⓗ**

1 False, $n = 2$

2 False, $x = \frac{1}{2}$

3 False, $n = 0$

4 False, $n = -3$

5 False, 1, 2, 2

6 False, $x = \frac{1}{2}$

7 False, a square

8 False, $\sqrt{2}$

9 Not proven

10 False, $333\,333\,331 \div 17 = 19\,607\,843$

11 False, $n = -\frac{1}{2}$

12 Not proven

13 $r = \frac{1}{24}(n^4 - 6n^3 + 23n^2 - 18n + 24)$

page 492 **Exercise 2** Ⓗ

1 a If today is Monday, then tomorrow is Tuesday.
   b If it is raining, then there are clouds in the sky.
   c If Abraham Lincoln was born in 1809, then Abraham Lincoln is dead.
2 a True **b** True **c** True **d** False **e** True **f** False **g** True **h** True **i** False **j** True

page 494 **Exercise 3** Ⓗ

1 Let $b = 2m$ for an integer m
   so $ab = a \times 2m = 2 \times am$
   so ab is even

2 Let $a = 2n + 1$ and $b = 2m + 1$ for integers n, m
   then $ab = (2n + 1)(2m + 1)$
   $= 4nm + 2n + 2m + 1$
   $= 2(2nm + n + m) + 1$
   which is odd

3 Let the numbers be $2n + 1$ and $2m + 1$.
   Then $2n + 1 + 2m + 1 = 2(n + m + 1)$
   which is even.

4 Let the even number be $2m$.
   Then $(2m)^2 = 4m^2$, which is divisible by 4.

5 The nth line is $(n + 1)(n + 3) - n(n + 4)$
   $= n^2 + 4n + 3 - n^2 - 4$
   $= 3$

6 Let the numbers be $x - 1, x, x + 1$ then $(x - 1)(x + 1) = x^2 - 1$.

7 One of the fair consecutive numbers will be a multiple of 4.
   Therefore, the product will be a multiple of 4.

8 Let the numbers be x and $x + 1$.
   Then $x^2 + (x + 1)^2 = x^2 + x^2 + 2x + 1$
   $\qquad\qquad\qquad = 2(x^2 + x) + 1$
   As $2(x^2 + x)$ is an even number, $2(x^2 + x) + 1$ must be odd.

9 Let the numbers be $(n - 2), (n - 1), n, (n + 1)$ and $(n + 2)$.
   Then $(n - 2)^2 + (n - 1)^2 + n^2 + (n + 1)^2 + (n + 2)^2$
   $= 5n^2 + 10 = 5(n^2 + 2)$, which is a multiple of 5.

10 Let the numbers be $x - 1, x$ and $x + 1$.
   Then $(x - 1)^2 + x^2 + (x + 1)^2 - 2$
   $= 3x^2 + 2 - 2$
   $= 3x^2$, which is 3 times a square number.

11 Area ② $= c^2 + 4 \times \frac{1}{2}ab = c^2 + 2ab$
   So area ① $=$ area ②
   $(a + b)^2 = c^2 + 2ab$
   $\therefore a^2 + b^2 + 2ab = c^2 + 2ab$
   $\therefore a^2 + b^2 = c^2$

12 Dividing by $(b - a)$ is not allowed as $b - a = 0$.

page 497 **Exercise 4** Ⓗ

1 Let $\angle CAB = x$

Then $\angle ACB = x$ (isosceles triangle)

$\therefore \angle ABC = 180 - 2x$ (angles in a triangle)

$\therefore \angle CBD = 180 - \angle ABC$ (angles on a straight line)

              $= 180 - (180 - 2x)$

              $= 2x$

So $\angle CBD = 2 \times \angle CAB$

2 $\angle PRS = \angle PQS$ (isosceles triangle)

$\angle PSR = \angle PSQ$ (both 90°)

$\therefore \angle RPS = \angle SPQ$ (two angles in triangles RPS and PSQ are the same so

                 the third angles must also be the same)

$\therefore$ PS bisects $\angle QPR$

3 $\angle ACD = \angle ABD$ (both subtended by the chord AD)

$\angle CDB = \angle BAD$ (both subtended by the chord BC)

$\angle CXA = \angle BXD$ (vertically opposite angles)

So $\triangle ACX$ is similar to $\triangle DBX$.

$\therefore \dfrac{BX}{CX} = \dfrac{DX}{AX}$ (ratio of corresponding sides are equal)

$\therefore AX \times BX = CX \times DX$

4 Let $\angle CAT = x$

Then $\angle ACT = 90 - x$ (angles in a triangle, tangent meets diameter at 90°)

$\therefore \angle ATD = \angle DET = 90 - x$ (alternate angle theorem)

So $\angle ADT = 180 - x - (90 - x) = 90°$ (angles in a triangle)

Similarly $\angle TEB = 90°$

So $\angle DEB = \angle DET + \angle TEB = 180° - x$

Therefore $\angle BAD + \angle DEB = x + 180° - x$

                    $= 180°$

Similarly $\angle ADE + \angle ABE = 180°$

So ADEB is a cyclic quadrilateral.

5 Use the alternate angle theorem to show that two angles in a triangle are x and $90° - x$, so third angle is $180° - (x + 90 - x) = 90°$.

6 $\angle CTA = \angle CTB$ (same angle)

$\angle TCA = \angle ABC$ (alternate angle theorem)

$\therefore \angle CAT = \angle TCB$ (third angle in triangle)

So TCA and TBC are similar.

Therefore $\dfrac{TC}{TB} = \dfrac{TA}{TC}$ (ratio of corresponding sides are equal)

So $TC^2 = TA \times TB$

7 Let $\angle ACD = x$

Then $\angle CAD = 90° - x$ (angles in a triangle)

$\therefore \angle ABC = 90° - x$ (alternate angle theorem)

$\angle CAB = 90°$ (angle in a semicircle)

So $\angle ACB = 180 - 90 - (90 - x) = x$ (angles in a triangle)

So AC bisects $\angle BCD$.

page 503 **Test yourself**

1 $n + (n + 1) + (n + 2) = 3n + 3 = 3(n + 1) =$ a multiple of 3

2 e.g. when $x = 5$, $y = 25 - 5 + 5 = 25$ (not prime)
Other counter-examples are possible.

3 a $1\cdot30$ or $-2\cdot30$
  b When $x = 10$, $y = 100 + 10 + 11 = 121$ which is not a prime as it is
  divisible by 11. Other counter-examples are possible.

4 a $5n$
  b i $5n + 5(n + 1) = 5n + 5n + 5 = 10n + 5 = 5(2n + 1)$
      which is $5 \times$ an odd number, and therefore odd.
    ii $5n \times 5(n + 1) = 25n(n + 1)$.
    If n is odd, then this equals $25 \times$ odd $\times$ even $= 25 \times$ even $=$ an even.
    If n is even, then this equals $25 \times$ even $\times$ odd $= 25 \times$ even $=$ an even.
    $\therefore$ the product is always even.

5 a i If n is odd, then $n(n + 1) =$ an odd $\times$ an even $=$ an even number.
      If n is even, then $n(n + 1) =$ an even $\times$ an odd $=$ an even number.
    ii $2n$ is a multiple of 2 and therefore even.
      $\therefore 2n + 1 =$ an even number $+ 1 =$ an odd number
  b $(2n + 1)^2 = 4n^2 + 4n + 1$
  c The square of any odd number can be written as $(2n + 1)^2$.
    $(2n + 1)^2 = 4(n^2 + n) + 1 = 4n(n + 1) + 1$
    $$= 4 \times \text{an even number} + 1$$
    $$= \text{an even number} + 1$$
    $$= \text{an odd number}$$

6 $(n + 1)^2 - (n - 1)^2 = n^2 + 2n + 1 - (n^2 - 2n + 1)$ **7** AB = BC; AM = MC
  $$= n^2 + 2n + 1 - n^2 + 2n - 1$$ $\angle$BAM $= \angle$BCM
  $$= 4n$$ Using the SAS rule, the two triangles
  $$= \text{a multiple of 4}$$ are congruent.

11   Revision

page 505 **Revision exercise 1**

1 £25·60, £6·70, 4, £55·30				
2 a 30, 37	**b** 12, 10	**c** 7, 10	**d** 8, 4	**e** 26, 33
3 £8	**4** £92	**5 a** 1810 seconds	**b** 72·4 seconds	**6** 0·8 cm
7 a £13	**b** £148	**c** £170	**8 a** 5·89	**b** 6	**c** 7
9 a -11	**b** 23	**c** -10	**d** -20	**e** 6	**f** -14
10 a 3	**b** 5	**c** -6	**d** -7		
11 a $x = 9$	**b** $x = 11$	**c** $x = 3$	**d** $x = 7$	**12 a** and **c**	

page 506 **Revision exercise 2**

1 a

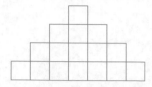

b 1, 4, 9, 16   **c** Square numbers   **d** 49

2 a $x = 7$ **b** $x = \frac{1}{4}$ **c** $x = \frac{4}{5}$ **3 a** $\frac{3}{8}$ **b** $\frac{5}{8}$

4 a $\frac{2}{11}$ **b** $\frac{5}{11}$ **c** $\frac{9}{11}$

5 a

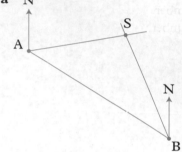

b 86 km

6 a Reflection in $y = 0$
b Reflection in $x = -1$
c Reflection in $y = x$
d Rotation 90° clockwise, centre $(0, 0)$
e Reflection in $y = -1$
f Rotation 180°, centre $(0, -1)$

7

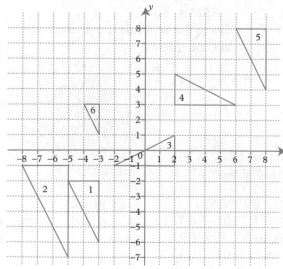

a Enlargement scale factor $1\frac{1}{2}$, centre $(1, -4)$
b Rotation 90° clockwise, centre $(0, -4)$
c Reflection in $y = -x$
d Translation $\begin{pmatrix} 11 \\ 10 \end{pmatrix}$
e Enlargement scale factor $\frac{1}{2}$, centre $(-3, 8)$
f Rotation 90° anticlockwise, centre $\left(\frac{1}{2}, 6\frac{1}{2}\right)$
g Enlargement scale factor 3, centre $(-2, 5)$

8 26 cm²

page 507 **Revision exercise 3**

1 a $t + t + t = 3t$ **b** $a^2 \times a^2 = a^4$ **c** $2n \times n = 2n^2$

2 a 9   **b** 50   **c** $(7 \times 11) - 6 = 72 - 1$

3 a–b

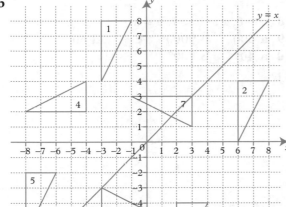

4 17·7 cm

5 a £5·20   **b** 29 minutes

6 $a = 45°, b = 67·5°$

7 a £44 000   **b** 198   **c** £220   **d** £43 560

8 a 4   **b** 19

9 a i Consett   **ii** Durham   **iii** Consett

    b i 55 km   **ii** 40 km

    c i 80 km/h   **ii** 55 km/h   **iii** 70 km/h

    iv 80 km/h   **d** $1\frac{3}{4}$ hours

10 a $1\frac{2}{3}$ **b** 20 cm

  c $(6, 0), (2, -8), (-8, 2), (1, -5), (-1, 3)$

11 A Cross-channel swimmer   B Car ferry from Calais   C Hovercraft from Dover

    D Train from Dover   E Marker buoy outside harbour   F Car ferry from Dover

12 a 560 kg   **b** 57 kg

page 509 **Revision exercise 4**

1 A $y = 6$ B $y = \frac{1}{2}x - 3$ C $y = 10 - x$ D $y = 3x$ **2 a** 5   **b** $\frac{4}{7}$ **3** 6

4 a 220°   **b** 295°   **5 a** 14   **b** 18   **c** 28   **6** 253   **7** £5250   **8** 0·375

9 a $\frac{9}{20}$ **b** $\frac{11}{24}$ **10** iii   **11 a** 1 : 50 000   **b** 1 : 4 000 000

12 a $s = rt + 3t$ **b** $r = \frac{s - 3t}{t}$ **13** 10   **14** $y \leqslant 3x, y \geqslant 2, x + y \leqslant 6$

page 510 **Revision exercise 5**

1 5%   **2** 21%   **3** 6   **4** 25   **5 a** 8   **b** 140   **c** 33   **d** 42   **e** 6   **f** −6

6 $\frac{1}{6}$ **7 a** 9π cm²   **b** 9   **8** $33\frac{1}{3}$ mph   **9** 9   **10 a** 5·45   **b** 5   **c** 5   **11** 6 cm

12 a $6x + 15 < 200$ **b** 29

page 511 **Revision exercise 6**

1 a 20·8%   **b** £240   **2** A $4y = 3x - 16$ B $2y = x - 8$ C $2y + x = 8$ D $4y + 3x = 16$

3 a 0·005 m/s   **b** 1·6 s   **c** 172 800 m or 172·8 km   **4** $c = 5, d = -2$

5 a Reflection in $x = \frac{1}{2}$ **b** Reflection in $y = -x$ **c** Rotation 180°, centre $(1, 1)$

6 152°   **7** 45°, 225°   **8 a** $\frac{3}{5}$ **b** $w = \frac{k - ky}{y}$ **9 a** 50   **b** 50

page 512 **Revision exercise 7**

1 $x < \dfrac{3}{4}$ **2** $1{\cdot}5 \times 10^{11}$ **3 a** $(x + 3)(x + 5)$ **b** $(x + 3)(x - 2)$ **c** $5x(x - 6)$

4 $20\,\text{cm}^2$ **5 a** $\dfrac{1}{28}$ **b** $\dfrac{15}{28}$ **c** $\dfrac{3}{7}$ **6 a** 30 **b** 75

7

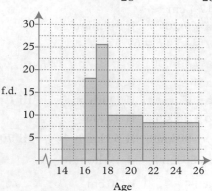

f.d.

Age

8 $x \geqslant 0,\ x + y \leqslant 7,\ y \geqslant x - 2$

9 a 600 **b** 9000 **c** 3 **d** 60

10 a $40°$ **b** $100°$

11

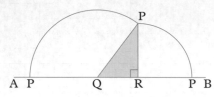

page 513 **Revision exercise 8**

1 a $x = 5$ or -3 **b** $x = 6$ or 2 **2** -4 **3** They arrive at the same time. **4** $250\,\text{cm}^3$

5 a $3\dfrac{1}{3}\,\text{cm}$ **b** $1620\,\text{cm}^3$ **6 a** $a = 50°$ **b** $b = 128°$ **c** $c = 50°,\ d = 40°$ **d** $y = 40°,\ x = 10°$

7 $\dfrac{1}{12}$ **8** $\dfrac{35}{48}$ **9 a** $\dfrac{1}{9}$ **b** $\dfrac{1}{12}$ **c** 0 **10** $1\dfrac{11}{20}\,b$

page 514 **Revision exercise 9**

1 a

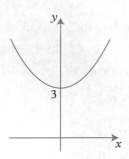

b

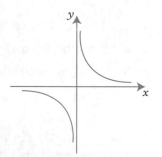

c

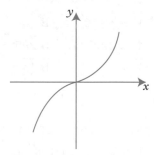

2 a $1\,\text{m}^2$ **b** $1000\,\text{cm}^3$ **3 a** $y = \dfrac{1}{2},\ -\dfrac{1}{2}$ **b** $y = \dfrac{7}{11}$ **4** $\dfrac{5}{16}$

5 $\angle\text{DAE} = \angle\text{BAE}$ (as AE bisects angle A)
$\angle\text{AED} = \angle\text{BAE}$ (alternate angles)
So triangle ADE is isosceles.
Thus DE = AD and AD = BC as ABCD is a parallelogram.
So DE = BC

6 a $1\dfrac{5}{6}$ **b** $0{\cdot}09$ **7 a** $z = x - 5y$ **b** $m = \dfrac{11}{k + 3}$ **c** $z = \dfrac{T^2}{C^2}$

8 a

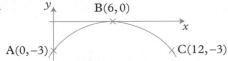

b

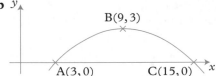

c

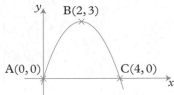

9 $x = 13$ **10 a** $a - c$ **b** $\frac{1}{2}a + c$ **c** $\frac{1}{2}a - \frac{1}{2}c$ NM is parallel to CA, and half as long.

11 a $x = 3$ **b** $x = 0$ or 5 **c** $x = 0$ **12** $x = 0.65$ or 3.85

page 515 Revision exercise 10

1 $201 \, \text{cm}^2$, $32.2 \, \text{cm}^2$ **2 a** 99p per litre   **b** 760 miles   **3 a** $7.21 \, \text{cm}$ **b** $9.22 \, \text{cm}$ **c** $7.33 \, \text{cm}$
4 $x = \pm 1.9$ **5 a** 2.088 **b** 3.043 **6** $7.62 \, \text{cm}$ **7 a** $34.56 \, \text{m}^3$ **b** $14.1 \, \text{m}^3$
8 a -14.88 **b** 58.44 **c** $\frac{22}{35}$ **d** $3\frac{3}{4}$ **9** 41.7%
10 a $198 \, \text{cm}^3$ **b** $1.36 \, \text{cm}^3$ **c** 145

page 516 Revision exercise 11

1 20.75 litres   **2 a** $12.2 \, \text{cm}$ **b** $61.1 \, \text{cm}^2$ **3** 2.3×10^9 **4** 33.1%
5 $9.95 \, \text{cm}$ **6** Maximum $= 123.25 \, \text{cm}^2$, Minimum $= 101.25$ **7** 0.72%
8 a 0.5601 **b** 3.215 **c** 0.6161 **d** 0.4743 **9 a** 84 **b** 19.2 **10** $0.335 \, \text{m}$
11 $x = 1.24$ **12 a** 1.755 **b** $60.3°$

page 517 Revision exercise 12

1 The 1 lb jar   **2 a** $290°$ **b** $60°, 300°$ **3 a** 0.340 **b** 4.08×10^{-6} **c** 64.9 **d** 0.119
4 a $45.6°$ **b** $58.0°$ **c** $3.89 \, \text{cm}$ **d** $33.8 \, \text{m}$ **5** $3.63 \, \text{m}$
6 Fig 1 $3.43 \, \text{cm}^2$ Fig 2 $4.57 \, \text{cm}^2$ **7** $273 \, \text{cm}^3$ **8 a** $14.1 \, \text{cm}$ **b** $35.3°$
9 a $\dfrac{x}{x + 5}$ **b** $\dfrac{x^2}{(x + 5)^2}$ **10** 68p   **11** $1.552 \, \text{m}$ **12** $4.12 \, \text{cm}$
13 a 0 **b** 16 **c** -8

page 518 Revision exercise 13

1 2.1×10^{24} tonnes   **2** $17 \, \text{kg}$ **3 a** $x = 14.1 \, \text{cm}$, $48.3 \, \text{cm}$ square   **b** $1930 \, \text{cm}^2$
4 a $x = -1.25$ or 2.92 **b** $x = 0.76$ or 5.24 **5** 4.74 **6** $5.39 \, \text{cm}$
7 Mean $= 13.6$, Median $= 11.5$ **8** $45.2 \, \text{km}$, $33.6 \, \text{km}$ **9 a** $6.63 \, \text{cm}$ **b** $41.8°$
10 $5.14 \, \text{cm}^2$ **11** AB $= 41.2 \, \text{cm}$, AD $= 9.93 \, \text{cm}$ **12** $8.06 \, \text{cm}$ **13** $89°$

page 519 Revision exercise 14

$8, 9, 16, 20, 23, 29, 36, 41, 42$

Index